DNA sequencing

LABORATORY TECHNIQUES IN BIOCHEMISTRY AND MOLECULAR BIOLOGY

Edited by

T.S. WORK – *Cowes, Isle of Wight (formerly N.I.M.R., Mill Hill, London)*

R.H. BURDON – *Department of Biochemistry, University of Glasgow*

ELSEVIER BIOMEDICAL PRESS
AMSTERDAM · NEW YORK · OXFORD

DNA SEQUENCING

J. Hindley

Department of Biochemistry,
Medical School,
University of Bristol,
University Walk,
Bristol BS8 1TD, England

with a contribution by

R. Staden
MRC Laboratory of Molecular Biology,
Cambridge, England

ELSEVIER BIOMEDICAL PRESS
AMSTERDAM · NEW YORK · OXFORD

ISBN – series: 0 7204 4200 I
– vol. 10: 0 444 80497 8
– vol. 10, paperback: 0444 80385 8

1st edition	1983
2nd printing	1985
3rd printing	1986
4th printing	1988
5th printing	1988

Published by:

ELSEVIER BIOMEDICAL PRESS
P.O. BOX 211
1000 AE AMSTERDAM, THE NETHERLANDS

Sole distributors for the U.S.A. and Canada:

ELSEVIER SCIENCE PUBLISHING CO., INC.
52 VANDERBILT AVENUE
NEW YORK, N.Y. 10017

Printed in The Netherlands

Acknowledgements

The writing of this book was only made possible in its present form by the generous help afforded by Dr. F. Sanger and his colleagues, both past and present, at the M.R.C. Laboratory of Molecular Biology, Cambridge and Dr. Allan Maxam of Harvard University, both in supplying me with unpublished data and in granting permission to make extensive use of their procedures and observations. I am also indebed to Dr. Andrew J.H. Smith for permission to make use of material in his Ph.D. thesis (University of Cambridge, 1980) in connection with the description of the exonuclease III procedure for making single-stranded template DNA, to Drs. Jan Maat, Isabelle Seif and B. Barrell for autoradiographs of sequencing gels and to Dr. N.L. Brown of this Department for the detailed description of the depurination and ribosubstitution reactions and the accompanying autoradiographs.

Finally I wish to thank Miss G.A. Phear for checking the manuscript and several colleagues for discussions concerning particular points of technique and experimental observations.

To Frances, Judith, and Joanna

Contents

CHAPTER 1

Introduction

1.1. *Preliminary remarks*

The present art of DNA sequencing has its origins in a variety of different fields of nucleic acid enzymology and chemistry. Indeed as early as 1970 our knowledge and understanding of these fields was, in theory, sufficiently far advanced to anticipate the development of the modern rapid methods but two obstacles had first to be overcome to convert these ideas into reality. The first was the problem of separating the oligonucleotides, generated in the sequencing reactions, in a rapid convenient and reproducible manner and displaying them as an ordered set of fragments according to their chain length. While the technique of homochromatography, in which a random mixture of polynucleotides of all possible chain lengths is used to develop a chromatogram (Brownlee and Sanger, 1969), was an important step in this direction, it was through the development of gel electrophoretic techniques that this problem was finally solved. All the methods to be described rely on the extraordinary resolving power of polyacrylamide gels run under denaturing conditions to achieve the final separations; much effort has gone into perfecting such systems so as to optimise their resolving properties. There is no doubt that, despite the power of this system, it remains the limiting step as far as sequence determination is concerned and further technical advances can be expected.

The second obstacle in thinking about new methods was conceptual rather than practical. Most biochemists with an eye on DNA sequencing had a background in either RNA or protein sequencing and it is clear that the earlier methods developed for DNA sequencing were essentially an extrapolation of the methods used for RNA sequencing (cf. Brownlee, 1972). A radical rethinking of the entrenched 'divide, fractionate and analyse' approach can, in retrospect, be seen as the other prerequisite for development of modern methods. This was, in a sense, anticipated in some of the later developments in RNA sequencing such as the end-labelling of polynucleotides with polynucleotide kinase (Szekeley and Sanger, 1969) and the use of sequential degradation methods.

The real breakthrough was the appreciation of the limitations inherent in trying to modify RNA sequencing methods to suit the DNA problem. In this respect it is perhaps fortunate that enzymes with the type of base specificity exhibited by the RNA endonucleases are not available for degrading DNA and this also played its part in driving the development of the modern chemical and chain termination procedures for DNA sequence analysis.

As with any other type of chemical or biochemical analysis of biological macromolecules the first hurdle is the isolation of the molecular species in sufficient quantity and purity. DNAs isolated from bacterial plasmids, bacteriophages and small animal viruses e.g. polyoma and SV40, are usually homogeneous single species and the preparation of a few hundred micrograms (equivalent to approx. 100 pmol of these DNAs) is a routine procedure. Since DNA sequencing procedures depend on radiochemical labelling, achieved by one means or another, to exhibit and relate the fractionated products, the sensitivity of the methods can be remarkably high; thus about 300 μg of DNA from a typical small plasmid such as pBR322 is sufficient for the chemical sequence analysis of one of the antibiotic resistance genes. However, as one moves up the evolutionary scale to progressively more complex organisms the isolation of particular genomic regions becomes a problem in its own right and it is usually necessary to employ molecular cloning methods to both isolate, and identify, and subsequently amplify the DNA fragment in question. *E. coli* has a haploid DNA content corresponding to a DNA of molecular weight 2.4×10^9.

This is equivalent to about 4×10^6 base pairs, about 1000 times larger than that of a small plasmid or virus. The DNA content of a typical mammalian cell nucleus is in turn about 1000 times greater than that in the *E. coli* cell. Moreover the DNA segments of such macromolecules are for all practical purposes indistinguishable from each other. Thus to isolate physically 100 pmol of a gene (10^3–10^4 base-pairs) from the total human genome (10^9 base pairs) would require in excess of 100 g of DNA from 100 kg of tissue—even supposing physical and chemical methods were somehow capable of resolving a single gene-sized piece from several hundred thousand others. In theory, and now with increasing frequency in practice, this problem has been solved by recombinant DNA techniques and nowadays DNA sequencing is inextricably linked with gene cloning. However, it is not the purpose of this book to discuss methods for gene cloning per se as a technique for isolating the DNA segment in question and we shall assume that as the starting material we already have the DNA either as a pure sample or cloned in phage or plasmid vector. The size of the DNA segment is the most important factor in determining the strategy of our sequencing approach. A fragment of a few hundred nucleotides may be directly amenable to sequencing whereas a more complex molecule of several thousand base pairs will usually require some sub-cloning of its restriction fragments to obtain defined sequenceable pieces.

The revolution in DNA sequencing technology has also dramatically affected our thinking and approach to RNA sequencing. The traditional RNA technology perhaps reached its culmination in the sequence determination of the genome of phage MS-2, a small coliphage containing a single stranded RNA genome of 3500 nucleotides. However, the effort needed to carry through this undertaking made it unlikely that similar approaches would form the basis of routine methods for studying the sequence of large mRNA species, or the genomes of influenza or the type ‘C’ tumour viruses. Application of the principles used for DNA sequencing have now made such problems much less daunting and a brief description of the methods currently used for RNA sequencing as based on DNA technology, have been included in this monograph.

One final point, which needs emphasis, is that at the time of writing the methods documented are the most up to date in current use. However, judging by past experience, it is likely that these will be further improved and amended as sequencing objectives become more ambitious and complex.

1.2. Organisation of the book

The modern rapid methods for DNA sequencing fall into two broad groups depending on the procedures used to generate and relate sets of labelled oligonucleotides which, after resolution by gel electrophoresis, permit the DNA sequence to be deduced.

The first group of methods employs a primed synthesis approach in which a single stranded template, containing or comprising the sequence of interest, is copied to produce the radioactively labelled complementary strand. By using chain-terminating inhibitors, and also by other methods, sets of partially elongated molecules are produced which can be fractionated on denaturing gels and the patterns of labelled bands obtained used to deduce the sequence. This procedure is very adaptable and though originally devised for sequencing naturally occurring single stranded DNAs, highly efficient procedures have now been developed for generating single stranded templates from any duplex DNA.

The forerunner of the primed synthesis methods was the ‘plus and minus’ method developed by Sanger and Coulson in 1975 and this marked the first real breakthrough in the search for new and efficient ways of sequencing DNA. This method is the subject of Chapter 2. By itself this procedure does have distinct limitations and additional back-up procedures had to be employed to confirm regions of the sequences deduced. One of these, the depurination method originally introduced by Burton and Petersen, proved a particularly useful adjunct and a description of this method is included as applied to the analysis of pyrimidine clusters in the products of a primed synthesis reaction. Another development which made the method more flexible was the single site ribo-substitution reaction and examples of the use of this technique are discussed. Chapter 2 is completed by a description of the classical 'wandering-spot' method for analysing short DNA sequences (Sanger et al., 1973). This method depends on the identification, by mobility changes in a two-dimensional fractionation procedure, of sets of oligonucleotides produced by the stepwise degradation of a labelled DNA with spleen phosphodiesterase (a $5' \rightarrow 3'$ exonuclease) and venom phosphodiesterase (a $3' \rightarrow 5'$ exonuclease). Though this method is more usefully applied to end-labelled, or uniformly labelled DNA, it is included here since, with the depurination method, it was the mainstay of the earlier DNA sequencing procedures and is still widely used for particular purposes.

Chapter 3 is concerned with the further developments in primed synthesis methods made possible by the introduction of the chain terminating dideoxy-nucleotides as specific inhibitors. A description of the background of this method is followed by detailed experimental methods and an account of the application of this method to DNA sequencing using the different primer–template combinations that can be obtained, starting from duplex DNA by taking advantage of the remarkable versatility of exonuclease III. This enzyme can be used to prepare either single stranded templates or primers, and in conjunction with the use of restriction enzymes to cleave out and select particular fragments the procedure has been developed into an extremely powerful and generally applicable sequencing procedure. DNA polymerase in the presence of dideoxynucleoside triphosphates can also be used to generate specific sets of fragments from 5′-end-labelled duplex DNA which has been nicked with DNAaseI, and this has been developed into a set of useful and versatile sequencing methods. These are also considered in Chapter 3.

Chapter 4 describes the application of primed synthesis methods to sequencing single stranded DNAs prepared by cloning restriction fragments into derivatives of the single stranded DNA phage

M13. The development of these phage vectors solved the general problem of preparing single stranded template from any double stranded DNA. The further development of new vectors with a variety of cloning sites together with the use of universal flanking primers which anneal adjacent to the cloned sequence have made this into the currently first choice primed-sequencing method. At the time of writing, this is under active development and new procedures in which the random selection of recombinant clones is replaced by a more structured approach can soon be expected. Sequencing via cloning into M13 is probably the most rapid and versatile procedure at the present. One way in which this can be structured by using exonuclease III treated restriction fragments as primers is discussed in this chapter. Current approaches to RNA sequencing, by analysis of cDNA transcripts synthesized by the use of reverse transcriptase are also reviewed in Chapter 4 with emphasis on the most recent developments.

Chapter 5 is devoted to the Maxam–Gilbert chemical method for sequencing 5′ or 3′ end-labelled DNA. The description and discussion is based on their most recent recommendations and a detailed experimental protocol is included. The final section is a discussion of the strategies which may be used for sequencing double stranded DNA by combining the Maxam–Gilbert approach with methods which permit the ordered sequencing and restriction mapping of DNA and ways for comparing and pinpointing regions of sequence divergence in related DNAs.

With this wealth of techniques available the main problem is likely to be one of selection. As far as possible the advantages of the different methods are discussed to help in choosing the most appropriate method for a particular problem.

At the end of this volume a series of appendices give useful information regarding sources of enzymes etc.

Chapter 6 describes the computer methods developed by Dr. Roger Staden at the MRC Laboratory of Molecular Biology, Cambridge, for handling the sequence data produced by the rapid 'shotgun' sequencing techniques described in Chapter 4. Since sequence data can be accumulated at a rate of up to 1,000 nucleotides a day the major problem soon becomes one of handling and keeping track of the sequences obtained. A computer can store the data and with the aid of suitable programmes can edit, analyse and print out the data in different forms. Sequences can be compared, matched up, modified and searched for overlaps or other common subsequences as each new piece of data is entered from the gel reading. Searches for repeated sequences, palindromes and restriction enzyme recognition sites can be carried out and additional programmes can search for regions of secondary structure such as hairpin loops, and translate DNA sequences into amino acid sequences. As the complexity of sequencing objectives increase so does the value of programmes capable of handling the immense amount of data and analysing it for particular features. Chapter 6 describes these programmes and the appendix contains a step by step description of how to run the programmes.

1.3. General background to DNA sequencing

The procedures described in this book make extensive use of enzymes which can synthesize, modify or degrade DNA. All the procedures rely on the incorporation of one or more [^{32}P]-labelled nucleotide residues into the DNA molecule to be sequenced and the sequence is finally deduced by the examination of autoradiographs of electrophoretic separations on polyacrylamide gels in which the different labelled fragments are ordered in a size-dependent manner. The ability to dissect DNA sequences into sequenceable sized fragments in a defined and specific fashion, the separation and identification of these fragments and their cloning into a vector DNA are all techniques which are inextricably involved in the modern sequencing methods. For the reader not familiar with this field the remainder of this chapter describes in a general way, the scientific background to the various enzymatic reactions used and their application to sequencing problems. In addition a description is given of the restriction endonucleases (restriction enzymes) for cleaving DNA sequences to produce fragments with defined termini for sequencing or cloning, or as a means of obtaining a 'restriction map' of the DNA sequence in question.

1.3.1. DNA synthesis with DNA polymerase I

DNA Polymerase I (DNA Pol I) from *E. coli* is the best characterized of all the DNA polymerases and, as far as DNA sequencing is concerned, is by far the most widely used. The purified enzyme is a single polypeptide chain of molecular weight 109,000 daltons containing approximately 1000 amino acid residues. The turnover number is about 667 nucleotides polymerized per molecule of enzyme per min at 37°C. One atom of zinc per molecule of enzyme appears to be required for the activity of the enzyme.

DNA polymerases (like RNA polymerases) are unique among enzymes in that the choice of substrate is determined by the template. *E. coli* polymerase can copy eukaryotic DNA and animal polymerases can copy bacterial DNA sequences given the appropriate DNA template. This means that DNA sequences, irrespective of their origin, can be accurately copied using *E. coli* DNA Pol I and this is the basis of the 'primed-synthesis' DNA sequencing methods.

The synthesis of DNA has two basic features. First, a phosphodiester bond is formed between the 3′-hydroxyl group (3′-OH) at the growing end of a DNA chain (the *primer* DNA strand) and the 5′-phosphate group of the incoming deoxynucleotide. The reaction involves a nucleophilic attack by the 3′-OH group at the innermost (α)-phosphate of the incoming deoxynucleoside triphosphate with the release of pyrophosphate. The direction of chain growth is $5' \rightarrow 3'$ (Fig. 1.1.). Secondly, each deoxynucleotide residue added to the growing end of the primer is selected by its ability to base-pair with the complementary nucleotide on the DNA *template* strand. The sequence of nucleotides added to the primer is therefore exactly complementary to the sequence of the template strand (Fig. 1.1.). DNA polymerase is unable to initiate

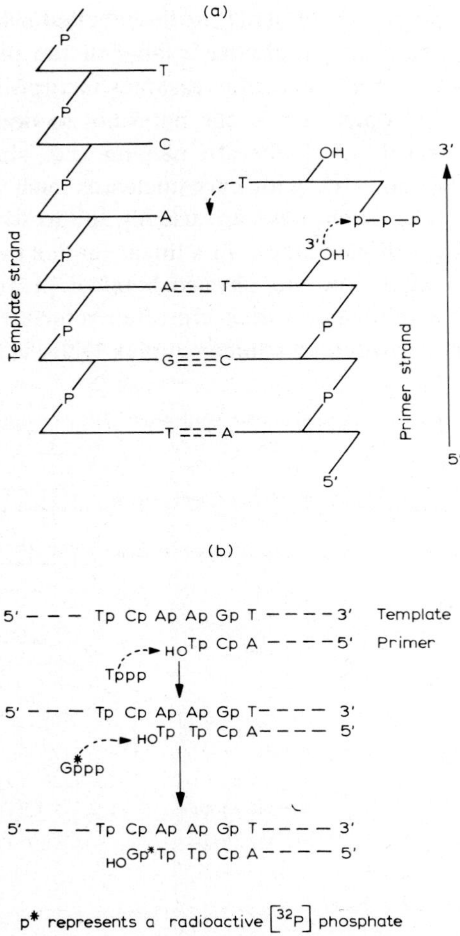

Fig. 1.1. Mechanism of chain extension catalysed by DNA polymerase on a primed template. (For discussion see text.)

the synthesis of new DNA strands in the absence of a free 3′-OH group. If one (or more) of the deoxynucleotide triphosphates are labelled with [^{32}P] in the innermost phosphate then the resulting synthesized DNA strand will be radioactively labelled. Thus, in

Fig. 1.1b., the use of α-[^{32}P] dGTP as the labelled substrate results in the incorporation of a radioactive label in the phosphodiester band to the 5′-side of all guanosine residues incorporated.

Free 3′-hydroxyl ends may occur naturally in double stranded DNA or as a result of deliberate nicking (i.e. single stranded breakage) of the molecule with endonucleases such as DNAase I (Figs. 1.2. and 1.3.). Such nicks are usually found to be randomly distributed along a duplex DNA. In a linear duplex the presence of nicks has little effect on the physicochemical properties of the molecule and for that reason they are often referred to as hidden breaks. To discover whether a linear duplex DNA is nicked or not

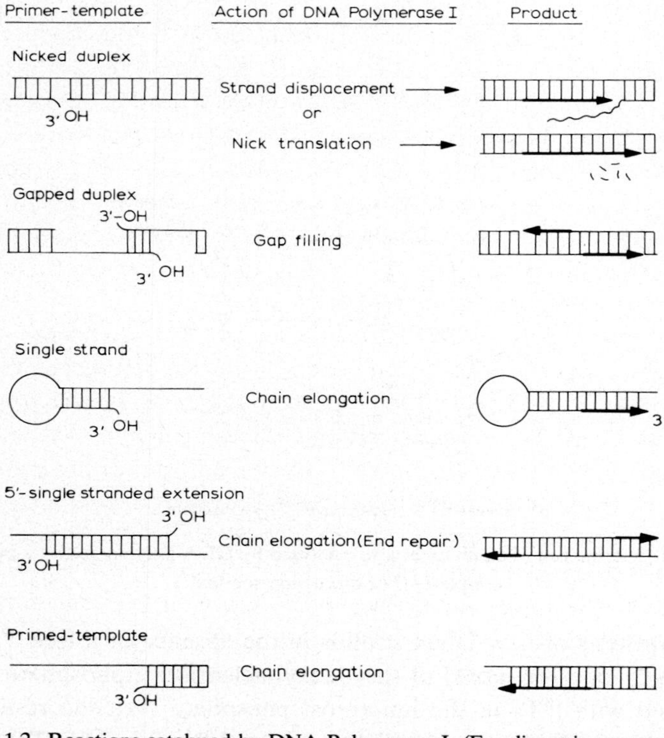

Fig. 1.2. Reactions catalysed by DNA Polymerase I. (For discussion see text.)

usually requires denaturation at high pHs followed by a size fractionation by alkaline sucrose gradient centrifugation or gel electrophoresis. In contrast, a single nick introduced into a covalently closed circular DNA molecule which is normally supercoiled (e.g. most plasmid DNAs), results in the relaxation of the molecule to yield the open-circular configuration which differs sufficiently in its hydrodynamic properties to permit its identification and separation by electrophoresis on agarose gels. Another important difference is the fact that the dye, ethidium bromide, can bind in much greater amount to open circular DNA molecules than to supercoiled DNA. The density of the open circular (and linear) duplex DNA is thereby reduced relative to that of the supercoiled DNA and centrifugation on caesium chloride density gradients readily resolves the supercoiled DNA (of high density) from the lower density configurations. This is the basis of the most important preparative method we have for isolating recombinant DNA plasmids from crude cell lysates.

Since the initiation of DNA synthesis has an obligatory requirement for a free 3′-hydroxyl end it is important to consider the types of structures which can yield potential primer-template complexes. Figure 1.2. shows a number of examples which serve as substrates for *E. coli* DNA Pol I.

Nicked duplexes. Endonucleases, such as DNAase I cause single stranded nicks more or less at random in DNA strands. A nick is defined as a break in a DNA strand that can be repaired by the enzyme DNA ligase which requires adjacent 3′-hydroxyl and 5′-phosphate groups. Such nicked molecules are efficient templates for *nick translation* in which the free 3′-hydroxyl ends are elongated in the $5' \rightarrow 3'$ direction with the concomitant displacement and hydrolysis of the original strand in the path of the extending chain. If one or more of the deoxynucleotide triphosphates are [^{32}P]-labelled, the nick translated product will be radioactively labelled. This is an important method for labelling DNA to a high specific activity and is the basis of one of the sequencing procedures described. It also provides a method for preparing

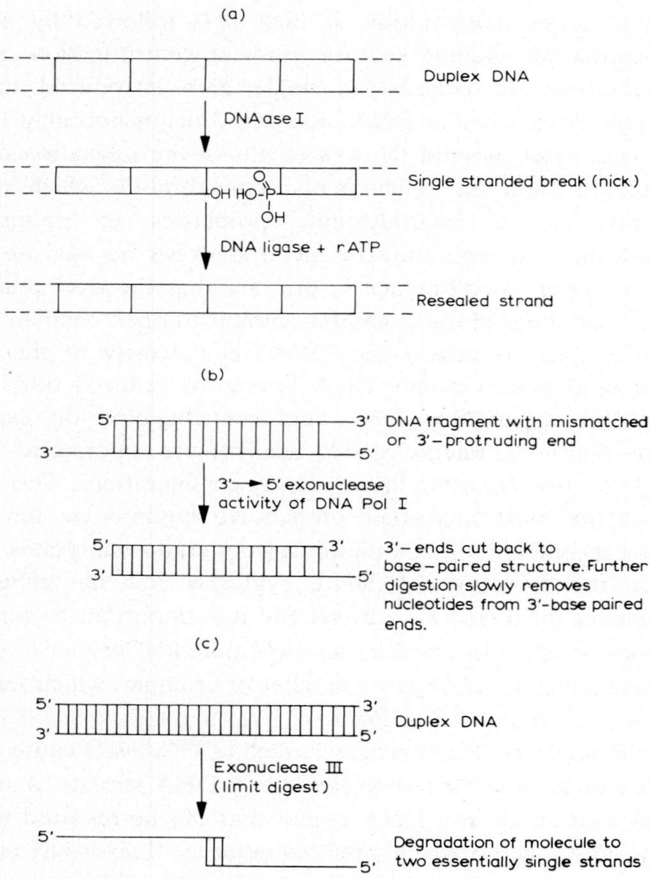

Fig. 1.3. (a) Action of DNAase I and DNA ligase; (b) The 3′→5′ exonuclease activity of DNA Pol I; (c) Action of exonuclease III. (For discussion see text.)

radioactive 'probes' used in a wide variety of hybridization techniques or labelling sub-microgram quantities of DNA for restriction mapping (see Section 1.3.10.).

Gapped duplexes. The nicks introduced into single strands can be broadened into gaps by the action of exonuclease III (Fig. 1.3c.)

an enzyme specific for double stranded DNA which degrades each strand in a $3' \rightarrow 5'$ direction. Gapped duplexes are even more efficient substrates for nick translation but the method is not widely used since over-digestion of a nicked duplex with exonuclease III can result in almost complete breakdown of the DNA. In another context, however, the use of exonuclease III to degrade intact double stranded DNAs into two essentially single stranded half molecules has been developed into a valuable sequencing procedure and this is described in Section 3.4.

Single strands. Single stranded DNA molecules are not active as primer-templates unless they have some partial duplex regions. These may occasionally arise by the fold back of the strand upon itself to bring short regions of complementary sequence into apposition (Fig. 1.2.). When this occurs chain extension from the 3′-hydroxyl end is possible leading to the formation of a hairpin shaped single stranded DNA. This self-priming may cause some difficulty when long single strands are used as primers in the primed synthesis sequencing procedure (see below and Chapters 3 and 4). With short primers (of approx. 15–30 nucleotides) this difficulty is rarely encountered.

Duplex DNA with 5′-extensions. A number of restriction endonucleases cleave duplex DNAs yielding fragments with 5′-single stranded extensions (Fig. 1.2.). Many of the commonly used hexanucleotide recognition enzymes, such as *Eco*RI, *Bam*H1, *Hind*III and *Bgl*II all yield fragments terminating in 5′-single stranded extensions of four nucleotides. DNA Pol I will efficiently catalyse chain extension from the recessed 3′-hydroxyl group to fill in the gap and yield a *flush* or *blunt-ended* duplex DNA. If one or more of the deoxynucleotide triphosphates are [^{32}P]-labelled, the repaired fragments will be labelled. This is a very useful and widely used method for labelling restriction fragments. This procedure for filling-in the ends is also used as a way of modifying restriction fragments for blunt-ended ligation into cloning vectors.

Primed single strands. A single stranded DNA can be converted into an efficient template for DNA synthesis by annealing a short

(>15 nucleotides) oligonucleotide to a region of complementary sequence on the template strand. The oligonucleotide primers may be synthesized *de novo* or obtained following denaturation of restriction fragments derived from a region of duplex DNA corresponding to the single stranded template. This approach is the basis of the primed-synthesis method for DNA sequencing described in Chapter 3.

1.3.2. Other properties of E. coli DNA polymerase I relevant to DNA sequencing

In addition to the polymerase function of DNA Pol I the purified enzyme also possesses discrete $3' \rightarrow 5'$ and $5' \rightarrow 3'$ exonucleolytic activities. Under particular conditions these activities degrade DNA strands with the release of 5′-mononucleotides (Fig. 1.3b.).

The $3' \rightarrow 5'$ exonuclease activity plays an important role in polymerization in proof reading the base pair formed at each polymerization step. The enzyme checks the nature of each base-paired primer terminus before the polymerase proceeds to add the next nucleotide to the primer. It thus supplements the capacity of the polymerase to match the incoming nucleotide substrate to the template. A mismatched terminal nucleotide on the primer activates a site on the enzyme which results in the hydrolysis of the phosphodiester bond and the removal of the mismatched residue. The function of this $3' \rightarrow 5'$ exonuclease activity is therefore to recognize and cleave incorrectly or non-base paired residues at the 3′-end of DNA chains. It will therefore degrade single stranded DNA and frayed or non-base paired residues at the ends of duplex DNA molecules provided they terminate in a 3′-hydroxyl group.

The $5' \rightarrow 3'$ exonuclease activity is much less specific in its requirements. A base-paired terminus is all that is required for its activity, irrespective of whether the 5′-terminal end is phosphorylated or not. The $5' \rightarrow 3'$ nuclease can remove oligonucleotides up to ten residues long from the 5′-end and the activity of this degradative function is markedly enhanced by concomitant DNA synthesis. It is this property of the enzyme which allows the highly efficient nick translation of DNA (Fig. 1.2.); polymerization at a nick is coordinated with the $5' \rightarrow 3'$ nuclease action so as to move (translate) the nick linearly along the helix. As mentioned earlier this is a useful way for preparing highly radioactive DNA. A double stranded DNA (either linear or circular) nicked with DNAase I and incubated with DNA Pol I in the presence of $[\alpha\text{-}^{32}P]$ deoxynucleoside triphosphates can yield a labelled product containing more than 10^8 cpm per μg DNA. For other purposes, in particular the primed synthesis DNA sequencing procedures, the presence of this $5' \rightarrow 3'$ nuclease activity is a disadvantage. This is because the analysis of the various chain terminated products of the extension reaction depend upon them all having an invariant 5′-end. Clearly, random cleavage of nucleotides from this end will give fragments of variable length and so invalidate the basis for their separation.

1.3.3. The Klenow sub-fragment of DNA polymerase I

As described above, *E. coli* DNA Pol I is a single polypeptide chain of about 1000 amino acid residues. Treatment of the intact protein with the proteolytic enzyme subtilisin results in the cleavage of the polymerase into two fragments, a longer fragment of 76 kdal which contains the polymerase and $3' \rightarrow 5'$ exonuclease activities, and a smaller one (36 kdal) with only the $5' \rightarrow 3'$ exonuclease activity. These two fragments are readily separated and for most sequencing purposes (other than nick translation) the longer sub-fragment is used. Preparations of this fragment, usually referred to as ‘Klenow polymerase’ or ‘DNA polymerase (Klenow sub-fragment)’ are available from many suppliers. The preparation supplied by Boehringer Ltd. has been consistently reliable.

1.3.4. Phage T_4-polymerase

Another enzyme, which shares many of the properties of the DNA polymerase (Klenow sub-fragment) is the phage T_4 encoded polymerase. The enzyme is a single polypeptide chain similar in size to Pol I of *E. coli* but which totally lacks the $5' \rightarrow 3'$

exonuclease activity of DNA Pol I. For that reason T_4-polymerase cannot use a nicked duplex as a primer-template but requires a primed single stranded template. The $3' \rightarrow 5'$ exonuclease activity of T_4-polymerase is much more active than that of Pol I and the enzyme can readily degrade single stranded DNA to the terminal dinucleotide. Though not generally used for sequencing by primed synthesis it does provide a superior alternative to the Klenow polymerase for labelling and blunt ending restriction fragments with 3'-overhanging single stranded extensions or frayed ends. The nuclease action, coupled with the polymerase action leads to turnover of the terminus and may thus be used to label it.

1.3.5. Chain terminating inhibitors

In the presence of Klenow polymerase, a primed template and the four deoxynucleoside triphosphates, the primer strand is extended in the $5' \rightarrow 3'$ direction until either the end of the template is reached or the triphosphate pool is exhausted. For sequencing it is necessary to start with a spectrum of primer extended fragments of all possible lengths and this is generally achieved by including in the reaction mixture an analogue of one of the deoxynucleotides which, when incorporated at the growing end of a chain, effectively stops further extension of that chain. It is also important that the terminated chain is resistant to breakdown by the $3' \rightarrow 5'$ exonuclease activity of the polymerase and the most useful analogues that fit these requirements are the dideoxynucleoside triphosphates. Since these lack the 3'-hydroxyl group on the deoxyribose, the incorporation of a dideoxynucleotide stops chain growth at that point since a 3'-H is not a substrate for the polymerase (see Section 3.1. for structural formulae). Dideoxynucleotides are not substrates for the $3' \rightarrow 5'$ nuclease so once terminated, a chain is resistant to both further extension or degradation. By adjusting the relative concentrations of a dideoxynucleotide (e.g. dideoxyadenosine triphosphate, ddA) to its cognate deoxynucleotide (deoxyadenosine triphosphate, dA) the relative frequency of chain termination events can be varied over a wide range. Increasing the ddA/dA ratio gives an increased frequency of termination and correspondingly shorter products, and *vice versa*. Since ddA only competes with dA for incorporation, all the prematurely terminated chains will end in a ddA residue, corresponding to a T (deoxythymidine residue) in the template strand. Whether a ddA- or a dA-residue is incorporated is a random event and the product of such a reaction is a pool of fragments, all of which possess the same 5′-end (corresponding to the 5′-end of the primer) and terminate with ddA on the 3′-end. Such a pool of fragments when separated according to chain length by electrophoresis on an acrylamide gel give a series of bands (or a ladder) in which each band corresponds to one size class of fragment terminated at the identical position. The distance between successive bands is a measure of the number of nucleotide residues between each A-residue. If four separate reactions are set up, each containing one of the four dideoxynucleotides, and the products fractionated in parallel by polyacrylamide gel electrophoresis then every termination event will give rise to a band in its corresponding channel. Simply reading the gel from the bottom to the top, noting in which of the four sequencing channels each successive band occurs gives the sequence of nucleotides polymerized onto the primer. In order to label the bands so that the pattern can be visualised, α-[^{32}P] dATP is included in the reaction mixture. This is the background to the primed synthesis DNA sequencing procedures described in Chapters 3 and 4.

1.3.6. Sequencing using end-labelled DNA

In these procedures the DNA fragment to be sequenced is initially labelled, at either the 3′- or 5′-ends with [^{32}P] and the DNA chain is cleaved by a set of base-specific reactions to yield families of end-labelled fragments which are separated from each other by electrophoresis on sequencing gels, in exactly the same way as the primed-synthesis fragments are identified.

The most widely used procedure for cleaving the end-labelled DNA fragments are the base-specific chemical degradation reactions introduced by Maxam and Gilbert. These procedures are applicable to both double stranded or single stranded DNAs, whether labelled at either the 3′- or 5′-ends. These methods are fully described in Chapter 5. In essence the principle of the procedure is similar to that of the primed synthesis methods. A DNA fragment, uniquely labelled at one end is subjected to a base-specific attack which randomly cuts the molecule at, for example, guanosine (G) residues. By controlling the severity of the attack one in every, say fifty, G residues are removed in a random manner. Sets of fragments are thus produced, but only those which retain the original labelled end will be detected on the gel. The pattern of bands obtained corresponds to a set of DNA fragments with a common labelled end and terminating at a position corresponding to the nucleotide preceding the excised G residue (Figs. 5.1a and b.). Repeating the reactions using base-specific cleavages directed against the other three nucleotides bases and electrophoresing all four samples in parallel on a gel yields a set of staggered bands in which fragments of all possible nucleotide lengths are represented in the appropriate channels. Reading the sequence merely involves noting in which channel each band occurs, as in the primed-synthesis method. In practice, as discussed in Chapter 5, the procedure is not quite as simple as this, but the principle is the same.

End-labelled duplex DNA fragments can also be sequenced by an enzymatic procedure which yields sets of fragments following random nicking of the DNA with DNAase I followed by a specific trimming of the resultant 3′-hydroxyl ends with DNA polymerase (Klenow sub-fragment) in the presence of one of the dideoxynucleoside triphosphates. This procedure is a particularly revealing example of the use made of the twin polymerase and $3' \rightarrow 5'$ exonuclease activities of this enzyme. In the absence of any deoxynucleotides the $3' \rightarrow 5'$ nuclease removes residues from the 3′-end of each nick until a nucleotide is released corresponding to the single dideoxynucleoside triphosphate added to the mixture. The polymerase activity of the enzyme then inserts the dideoxynucleotide at this point. Since this residue is resistant to both further nuclease digestion and extension, sets of radioactive fragments are obtained, all of which possess the same labelled end and terminate at the 3′-end with a particular dideoxynucleotide residue. This enzymatic approach thus yields sets of specifically terminated end-labelled fragments formally equivalent to the chemical degradation procedure. One drawback of this method is that it is only applicable to 5′-end labelled duplex DNA since the enzymic processing necessarily requires the presence of the template strand. This procedure, and variations on it, are described in Chapter 3.

1.3.7. T4-polynucleotide kinase

The specific 5′-end labelling of DNA restriction fragments is achieved by a phosphorylation reaction catalysed by T4-kinase which transfers ^{32}P from [γ-^{32}P] rATP to the 5′-hydroxyl end of deoxyribonucleotide (and ribonucleotide) chains. Restriction endonuclease fragments terminate in 5′-phosphoryl groups so it is usually necessary to remove initially the 5′-phosphate with alkaline phosphatase (a phosphomonoesterase) before proceeding with the labelling step. However, this is not always necessary since polynucleotide kinase can catalyse an exchange reaction between a 5′-phosphorylated terminus and [γ-^{32}P] rATP (Chapter 5)

$$\text{—ApGpCpTp}^{5'} \xrightarrow{\text{alkaline phosphatase}} \text{—ApGpCpT}_{OH}$$

$$\xrightarrow[{[\gamma\text{-}^{32}P]\text{ rATP}}]{\text{T4 Kinase}} \text{—ApGpCpTp}^{*}$$

[γ-^{32}P] rATP can be prepared or purchased with specific activities in excess of 1500 Ci/mmol and 1–2 pmol of DNA fragments, end-labelled at this specific activity, are sufficient for sequencing the fragment.

1.3.8. Restriction endonucleases

Restriction endonucleases are a diverse class of enzymes found in a wide range of bacteria which are specific for cleaving DNA. Two classes of restriction endonucleases are known. Type I endonucleases cleave DNA in a random manner yielding heterogenous sets of fragments and require ATP and S-adenosylmethionine for their activity. Typical examples of Type I restriction enzymes are those isolated from *E. coli* K12 (EcoK) and *E. coli* B (EcoB). Of much more interest as far as dissecting DNA molecules is concerned are the Type II restriction enzymes. These cleave DNAs at specific sites, determined by the nucleotide sequence, and yield specific fragments whose ends are defined by the specificity of the enzyme. Neither ATP nor S-adenosylmethionine are required for their activity, usually magnesium ions being the only cofactor required. The central role of these enzymes in genetic analysis has prompted a massive search for varieties of different specificity and over 200 restriction endonucleases have now been characterized. Many of the best characterized enzymes are available commercially but, in general, they are not difficult, and much cheaper to purify in the laboratory from the appropriate bacterial strains.

1.3.9. Properties of type II restriction endonucleases

These enzymes all share the ability to recognize specific, symmetrical sequences of four or more nucleotides and they generally, though not always, cleave within the recognition sequence. The requirement that the recognized sequences are symmetrical, or more precisely, possess a 2-fold axis of rotational symmetry, is another way of stating that the sequence of base pairs in the recognition site is palindromic and the cleavage sites are symmetrically positioned with respect to the two fold axis. Figure 1.4. shows the specificities of some restriction endonucleases; note that the cuts introduced may be even, giving blunt (flush) ended products, or staggered giving products with either 5′- or 3′-single stranded overhangs. A particular tetranucleotide recognition sequence will occur, on average, every $4^4 = 256$ nucleotides giving a mean fragment length of 256 base pairs. A given hexanucleotide

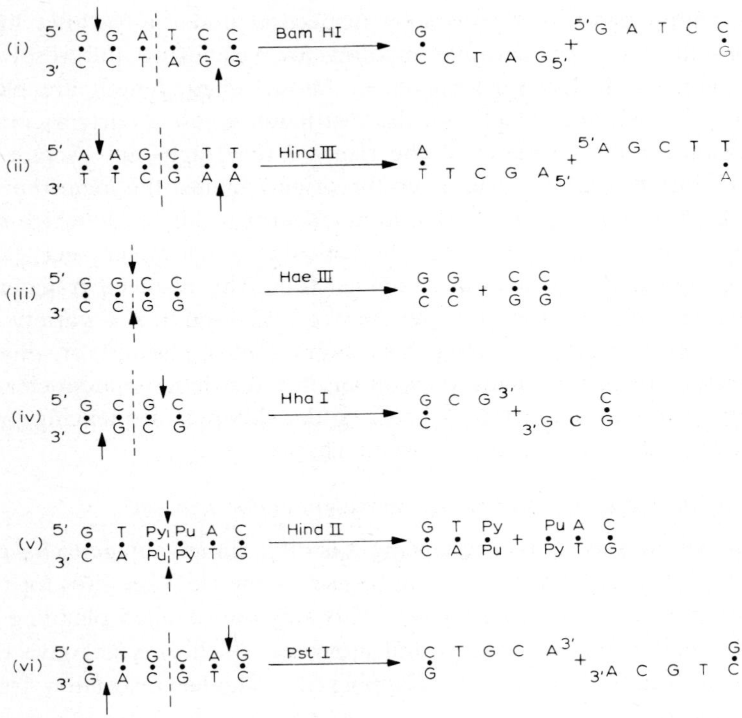

Fig. 1.4. Specificities of some restriction endonucleases. The two fold axis of symmetry is indicated by the dotted line and the arrows indicate the cleavage sites. Note that in (i) and (ii) the products possess 5′-single stranded extensions of four nucleotides; in (iii) and (v) the products are flush (blunt) ended and in (iv) and (vi) the products have 3′-single stranded extensions.

will occur, on average, every $4^6 = 4096$ nucleotides thus giving much longer fragments. The important point is that restriction endonucleases provide very precise tools for cleaving DNAs into sets of defined fragments. The products of digestion are usually displayed by electrophoresis on agarose or polyacrylamide gels where the different fragments are resolved essentially according to their molecular weight. Staining the gel with the fluorescent dye ethidium bromide followed by inspection under ultraviolet light provides a very sensitive method for visualizing the resolved fragments. If DNA fragments of known chain length are electrophoresed in parallel with the restriction digest a fairly accurate estimate may be made of the sizes of the fragments. A plot of mobility (distance moved from the origin) against the logarithm of the chain length gives, within limits, a straight line from which the sizes of the fragments can be calculated. However, gel electrophoresis is not just an analytical tool. The resolved fragments may be eluted from the gel matrix and used for a variety of purposes including sequencing, cloning into plasmid or phage vectors, or further digestion with another restriction endonuclease. These are all important aspects of the different sequencing procedures described in the following chapters.

1.3.10. Mapping DNA using restriction endonucleases

Before starting DNA sequencing it is often an advantage to have a 'map' of the DNA sequence in respect of the cleavage sites for the commoner restriction enzymes. This aids the detailed planning of the sequencing procedures and indicates which enzymes give the most suitable cuts for overlapping the smaller restriction fragments.

Several different approaches can be used for this purpose but initially the first set of experiments will be to digest the unknown DNA with a series of different enzymes to determine the number of cleavage sites for each enzyme. For large DNAs it may not be possible to obtain a precise figure for this since some of the products may be of similar size and poorly resolved on the gel or the smallest fragments may be overlooked unless a heavily loaded gel is used, since the intensity of ethidium bromide staining is proportional to chain length. This latter problem however is readily overcome by end-labelling the DNA fragments with [^{32}P] using the methods described above. The intensity of bands is then proportional to their molar ratio rather than chain length. The inclusion of DNA markers of known size on the gel also permits estimates to be made of the chain lengths of the different fragments.

Having established which enzymes cut the DNA in question one way to proceed is to cleave the DNA with a mixture of two enzymes of different specificity. The products of such a digest, run in parallel with the products of the individual digests will show which fragments from the first enzyme contain cleavage sites for the second enzyme and vice versa. If the original DNA fragment is end-labelled with [^{32}P], restriction fragments derived from the two ends of the molecule will be labelled and thus identified. Using different pairs of enzymes a map can often be deduced from the data showing the relative positions of the different cleavage sites. Defined DNA fragments may also be eluted after gel electrophoresis, digested with the second enzyme and then re-electrophoresed to resolve the fragments.

Another approach uses partial restriction endonuclease digests. Under conditions where the DNA is deliberately underdigested with the enzyme sets of fragments are obtained corresponding to partial digestion products. As the digestion proceeds these are gradually broken down to yield the final products and this process may be followed by removing samples at various stages of the digestion and running the samples in parallel on a gel. The formation and subsequent breakdown of the partial products may therefore be followed and deductions made about the sequence organisation of the fragments. This procedure may be simplified by analysing the partial digests from a DNA fragment labelled at one end only. This can be done by first labelling both 5′-ends of the duplex DNA to be mapped and then cleaving with a restriction endonuclease which ideally gives a single cut near one end of the molecule. The larger singly labelled fragment is then isolated by gel electrophoresis and subjected to partial digestion with an enzyme known to introduce several cuts into the DNA. The partial products are finally resolved by electrophoresis and the gel autoradiographed. Only those products containing the singly labelled end are detected and the pattern of bands gives directly a restriction site map. Full descriptions of these techniques are given in Danna (1980) and Smith and Birnsteil (1976).

1.3.11. Cloning DNA sequences into plasmid and phage vectors

The ability to clone DNA sequences, from whatever origin, into simple plasmids or phages for propagation in a suitable bacterial host has become the mainstay of most sequencing techniques. In the first place cloning permits the selection, identification and amplification of particular genomic sequences as a means of providing the raw material for sequencing and secondly cloning in single stranded phages such as phage M13 or fd yields the single stranded templates vital to the primed synthesis sequencing procedures. Recombinant DNA techniques are therefore inextricably involved with sequencing methods and this section gives a brief discussion of the ideas and principles involved in so far as sequencing is concerned.

Plasmids and DNA phages both share the essential attribute of replicating their DNA in a suitable bacterial host independently of the host chromosome. Though varying widely in size and genetic complexity they provide self replicating DNA molecules which can be introduced at will into suitable hosts and multiply to produce from a few up to more than a thousand usually identical copies. Within the DNA of these *vectors* are sequences, either between genes or in non-essential genes, into which foreign DNA sequences can be introduced. Provided no essential function is interfered with, the inserted DNA is replicated along with the vector. The isolation of plasmids or phages from a bacterial culture is a relatively straightforward procedure and milligram quantities of the recombinant DNA can often be obtained from a 1-l culture. DNA segments intended for sequencing are often initially available in only infinitesimally small quantities and this procedure provides a general method for preparing sufficient material to embark on a sequencing project. The essential requirements for a DNA cloning experiment are:

1. A suitable DNA vehicle (vector, replicon) which can replicate in

living cells after the insertion of a foreign DNA sequence.
2. The DNA molecule to be replicated, or a collection of them.
3. A method of inserting the DNA into the vehicle.
4. A means of introducing the recombinant DNA into the host cell (transformation or transfection).
5. A method for screening or genetic selection of cells which contain and replicate the desired recombinant DNA.

1.3.12. Cloning vehicles (vectors)

Plasmids

The most widely used cloning vehicles are bacterial plasmids, particularly those which can be propagated in *E. coli* cells. Many of the naturally ocurring plasmids have been modified in various ways to make them more suitable as cloning vehicles, the desired properties being firstly, a relaxed mode of replication such that the plasmid replicates autonomously and accumulates in high numbers in cells in which protein synthesis has been inhibited by chloramphenicol, and secondly, the inclusion of specific restriction endonuclease cleavage sites at precise locations within the plasmid into which the foreign DNA can be inserted. For example plasmid pBR322 (Chapter 5) contains genes coding for tetracycline resistance (Tc^R) and ampicillin resistance (Ap^R). The plasmid therefore confers a Tc^R, Ap^R phenotype to the host bacteria. Within the Tc^R gene are unique cleavage sites for the restriction endonucleases *Bam*HI and *Sal*I. Insertion of foreign DNA at either of these sites inactivates the Tc^R gene so that the cells now assume the ampicillin resistant tetracycline sensitive phenotype (Ap^RTc^S). Replica plating Ap^R cells on tetracycline-agar plates therefore provides a selection for Ap^RTc^S cells which contain the recombinant plasmid. Similarly the Ap^R gene contains a unique *Pst*I cleavage site thus permitting the selection of recombinants obtained by cloning into that site. Many plasmid vehicles are now available which take advantage of this insertional inactivation of an antibiotic resistance gene as a means of selecting for recombinant plasmids, e.g. pBR325 and pACYC184 contain a unique *Eco*RI restriction site within the chloramphenicol resistance gene.

Bacteriophages
Bacteriophage λ and its derivatives have been widely used for cloning DNA and have a number of advantages over plasmid vehicles for particular purposes. These stem mainly from the ability to screen large numbers of recombinants for a particular inserted DNA fragment by nucleic acid hybridization techniques and from the much greater efficiency with which recombinant phages can infect the host bacteria and produce large yields of the desired recombinant. In preparing 'libraries' of cloned DNA fragments, representing the total genome of a species (from *E. coli* to man), millions of individually packaged and replicated phage λ recombinants can be stored in a single solution. The derivative λgtWES.λB has been extensively used for the replication of cloned DNA sequences from warm-blooded animals.

Detailed reviews describing the varieties of plasmids and phages which are used for cloning and amplifying DNA sequences are found in Methods in Enzymology, (1979), Vol. 68. This whole volume is devoted to techniques used in recombinant DNA.

Phage M13 and its derivatives
This is described briefly since it is now the vehicle of choice for obtaining single-stranded recombinant DNA molecules for sequencing by the primed-synthesis method described in Chapter 4.

Phage M13mp2 is a single stranded DNA phage which contains the *lac* promotor and operator and the proximal (N-terminal) region of the *lacZ* gene (which codes for the enzyme β-galactosidase). When this phage is used to infect an appropriate *E. coli* host, which contains the distal part of the *lacZ* gene on an episome, the two β-galactosidase fragments produced can complement each other and yield a gal^+ phenotype, i.e. they have the ability to hydrolyse β-galactosides. Such gal^+ cells are readily identified since they can hydrolyse a chromogenic β-galactoside to produce a dark blue colour. In M13mp2 the fifth amino acid codon of the β-galactosidase gene corresponds to an *Eco*RI restriction site. The insertion of a DNA sequence at this point effectively inactivates the gene, by producing a non-functional polypeptide which is unable to complement the C-terminal polypeptide produced from the episome. Such cells are *gal*$^-$ and are unable to hydrolyse the chromogenic substrate thus yielding colourless plaques on suitable agar plates. These plaques therefore all correspond to cells harbouring a recombinant phage and it is a simple matter to pick plaques, grow the cells in a small volume of culture medium and isolate the recombinant phage DNA. Since the DNA is single-stranded it can be used as a template, after annealing to a suitable primer, for sequencing the inserted DNA by the primed-synthesis procedure (Chapter 4).

1.3.13. Insertion of DNA into a cloning vehicle

Though several different methods may be employed to insert a DNA segment into a cloning vehicle the most straightforward procedure uses restriction endonucleases to generate fragments with specific cohesive ends (i.e. with 5′- or 3′- single stranded overhangs) and subsequently DNA-ligase to seal the desired DNA sequence into the vehicle. If the same restriction endonuclease is used both to open the vehicle (e.g. convert a closed circular plasmid into the linear molecule by cleavage at a single site) and to generate the DNA sequences to be cloned the result will be that both species will have naturally self-complementary cohesive ends. The different DNA fragments will randomly associate with each other and in the presence of DNA-ligase these random associations will be converted into stable covalently linked molecules (Fig. 1.5.).

This is a random process but among the products will be recircularized vehicles into which has been ligated the DNA sequence of interest. As discussed earlier an insertion into an antibiotic resistance gene provides a simple selection procedure for isolating the recombinant plasmids. Many variations of this basic

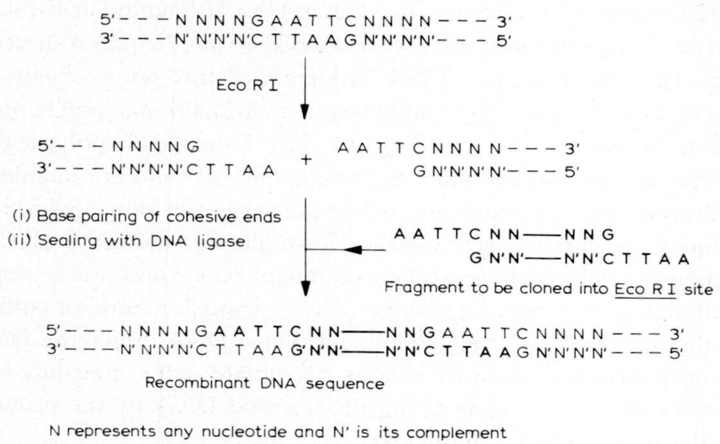

Fig. 1.5. Mechanism of cloning into the EcoRI site. N represents any nucleotide and N′ is its complement. (For discussion see text.)

procedure have been devised for particular purposes, up to date reviews being the volumes on Recombinant DNA (Methods in Enzymology, **68**, 1979) and Genetic Engineering, **1**, ed. R. Williamson, Academic Press (1981).

In some circumstances the aim of the experiment will be to clone a DNA sequence which lacks cohesive ends. Such molecules may, for example, be the cDNA products from the reverse transcription of particular mRNA species and this is currently the most widely used procedure for sequencing eukaryotic mRNAs. Under controlled conditions the reverse transcriptase from Avianmyeloblastosis virus (AMV) will use a mRNA template to synthesize a full-length double-stranded DNA copy (cDNA) of the RNA sequence. The products of this reaction frequently lack precisely defined ends and it is necessary to trim such frayed or uneven termini before proceeding to the cloning step. Incubation with DNA polymerase (Klenow) or T4-polymerase (Section 4.2.2.) will yield molecules with clean blunt-ended termini which can be inserted into the vector either directly, using blunt-ended ligation, or after the addition of artificial linkers to produce cohesive ends (Chapter 4).

1.3.14. Conclusion

All these methods aim at producing a recombinant DNA which can be propagated and amplified at will in order to prepare a sufficient quantity of DNA for sequencing. Amplification in plasmids and phage λ produces double-stranded recombinant DNA from which the segment to be sequenced can usually be excised using restriction endonucleases. Amplification in M13 and its derivatives yields single-stranded recombinant DNA which can be sequenced directly without further excision and purification of fragments. The combination of these techniques together with the sequencing procedures described in the succeeding chapters now permit the sequencing of any desired DNA. Sequencing of molecules in excess of 5000 nucleotides is now almost routine (ϕX174, the yeast 2μ plasmid, human mitochondrial DNA) and a start has been made on much longer molecules (EB virus, 160,000 base pairs). At present, as far as is known to the author, no radically new approaches to DNA sequence analysis are under development and in the forseeable future it seems likely that the main developments will be in the improvement and consolidation of the techniques described.

'Methods in Enzymology', Vol. 65 (Nucleic Acids, Part I) ed. Grossman and Moldave, (1980) and Vol. 68 (Recombinant DNA) ed. Wu, (1979), Academic Press, are valuable sources of information and references on many of the techniques described. 'DNA Replication' by Arthur Kornberg (1980) is a mine of information about the properties of the DNA polymerases. 'Principles of Genetic Manipulation' by Old and Primrose, Second edition (1981), Blackwell Scientific Publications, and 'Genetic Engineering', vols. 1, 2, and 3, ed. R. Williamson (1981, 1982), Academic Press, provide up-to-date reviews of the many and varied cloning procedures.

CHAPTER 2

DNA sequencing

2.1. *DNA Sequencing by primed synthesis methods*

2.1.1. Introduction

In 1973 a new approach to the problem of DNA sequencing using a primed synthesis method was described by Sanger et al. (1973) for the determination of two particular nucleotide sequences in bacteriophage DNA using DNA polymerase primed by synthetic oligonucleotides. In this method the oligonucleotide was hybridized to a specific complementary region on the single strand DNA and deoxynucleotides were added sequentially by DNA polymerase to the 3′-OH end of the primer. Using [^{32}P]-labelled deoxyribonucleoside 5′-triphosphates a radioactive complementary copy of a defined region of the template was obtained which was amenable to sequence determination. Two years later an important development of this procedure was introduced (Sanger and Coulson, 1975) in which the analysis of the complementary DNA was achieved purely by the fractionation of all the different size classes of DNA fragments by electrophoresis on high resolution denaturing polyacrylamide gels. This formed the basis of the 'plus and minus' method which was designed primarily for the sequence analysis of the single strand DNA bacteriophage ϕX174. With the advent of the newer dideoxynucleotide chain termination procedures the original 'plus-minus' method has been somewhat eclipsed but is considered here in some detail both on account of the fact that the method was successfully used to determine the entire nucleotide sequence of phage ϕX174 DNA, a total of 5375 nucleotides (Sanger et al., 1977) and also because it still constitutes a relatively rapid and simple method for sequence analysis and illustrates the principle on which the modern chain-termination procedure is based. In the sequence analysis of ϕX174 DNA restriction fragments generated by ten different restriction endonucleases were required as primers, and the non-viral strand, isolated from the double-stranded replicative form of the virion (RF1) was used as template. Further experimental details of the methods are given in Sanger and Coulson (1975), Brown and Smith (1977) and Barrell (1978).

2.1.2. Principle of the 'plus and minus' method

(*i*) *Primed synthesis*

A DNA primer (commonly a restriction fragment or synthetic oligonucleotide) is hybridized to the single-stranded DNA template and the primer is extended to a limited degree with DNA polymerase I in the presence of all four deoxynucleoside 5′-triphosphates one of which is α-[^{32}P]-labelled. (Fig. 2.1.) Samples of the reaction mixture are removed at varying times corresponding to different degrees of extension of the primer, and the reactions terminated by the addition of EDTA. The various samples are then combined, the DNA polymerase removed by phenol extraction, and the extended polynucleotide chain, still hybridized to the template, is separated from the excess deoxynucleoside triphosphates by gel filtration on Sephadex or agarose. The product is therefore a mixture of partially elongated fragments which should be as random as possible in their chain length. The natural asynchrony of the extension reaction together with the removal of samples at different times usually serves to give a reasonably random set of partially elongated fragments. Ideally the product is a mixture of polynucleotides in which all possible chain lengths of the complementary strand are present corresponding to an elongation of the primer of 0 up to 200 nucleotides.

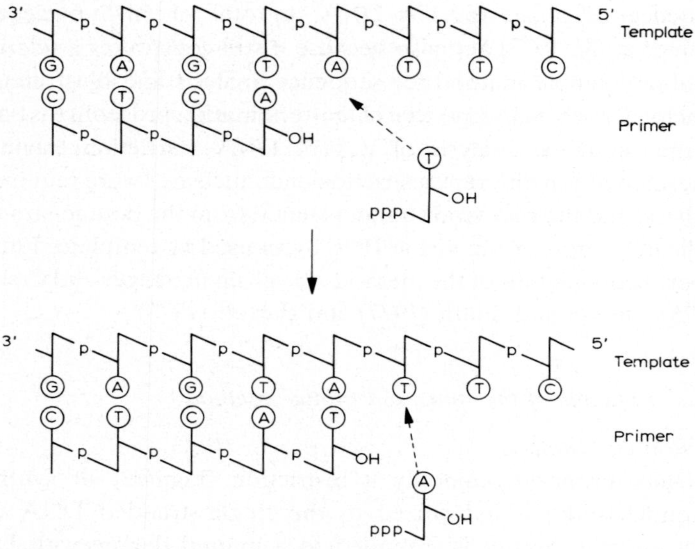

Fig. 2.1. Chain extension reaction catalysed by DNA polymerase I. Chain growth occurs in the $5' \rightarrow 3'$ direction by the stepwise addition of nucleotides to the free 3'-OH end of the primer strand.

(*ii*) *The 'minus' reaction*

In their original work on the 'sticky ends' of phage λ DNA, Wu and Kaiser (1968) showed that, in the absence of one of the four deoxynucleoside triphosphates, DNA polymerase would accurately catalyse chain extension up to the point where the missing nucleotide should have been incorporated. The 'minus' reaction utilizes the same principle. The random mixture of oligonucleotides, still annealed to the DNA template is reincubated with DNA polymerase in the presence of only three deoxynucleoside triphosphates. Synthesis then proceeds as far as the missing triphosphate on each chain. For example if dTTP is the missing triphosphate (the –T system), each chain will terminate at its growing 3'-end at a position immediately preceding the requirement for a T residue (corresponding to an A residue in the template strand). Four separate reactions are set up, each with one of the triphosphates missing. After incubation the reaction mixtures are denatured to separate the nascent strands from the template and the four samples simultaneously analysed by electrophoresis on a high resolution polyacrylamide gel in the presence of 8M urea and the separated oligonucleotides visualized by autoradiography. Figure 2.2. illustrates the principle of the method.

On the acrylamide gel mobility is essentially proportional to size

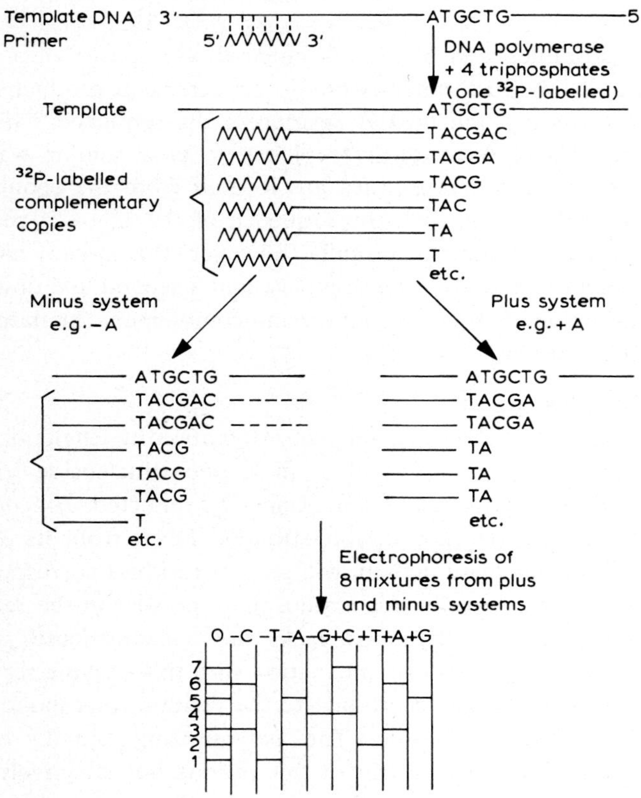

Fig. 2.2. Principle of the 'plus and minus' method for sequencing DNA (from Sanger and Coulson, 1975). For description see text.

so that the various oligonucleotide products, all of which should have a common 5′-end (see later discussion about this point) will be arranged in a ladder according to size. Each oligonucleotide should be resolved from its neighbour, which contains one more residue, by a distinct space. As would be expected the resolution falls off with increasing chain length and it is this factor which usually determines how far the autoradiograph may be read. The separation between two fragments of say 150 and 151 nucleotides will be much less than that between two fragments of 20 and 21 nucleotides.

The autoradiograph of the –A channel will consist of a set of bands, each of which corresponds to an extension product up to, but not including, the next A residue in the sequence. Thus the positions of the A residues are located. In a similar way the positions of the other residues are located from the sequencing channels and, in principle, the sequence of the DNA is read off from the autoradiograph. Usually, however, this system alone is not sufficient to establish the sequence and a second line of attack, the 'plus' system is used to confirm and complement the data from the 'minus' system.

(iii) The 'plus' system

This system takes advantage of the observation by Englund (1971, 1972) that, in the presence of a single deoxynucleoside triphosphate, DNA polymerase from phage T4 infected *E. coli* (T4 polymerase) will degrade double-stranded DNA from its 3′-ends but this 3′-exonuclease activity will stop at residues corresponding to the single deoxynucleoside triphosphate present in the reaction mixture. Since T4 polymerase lacks the 5′-exonuclease activity found in *E. coli* polymerase incubation with this enzyme serves to trim back each elongated product to the residue corresponding to the added deoxynucleotide. The polymerizing activity of the enzyme catalyses the turnover of this residue but effectively halts the progress of the 3′-exonuclease activity. In the 'plus' reaction this method is applied to four further samples of the primer-template complex isolated under (i) above. Samples are incubated with T4 polymerase and a single triphosphate and, after denaturation, the products are analysed directly by gel electrophoresis. Thus, in the +A system (Fig. 2.2.) only dATP is added and all the chains will consequently terminate with a 3′-terminal A residue. The bands observed on the gel will therefore correspond to a set of fragments representing all the extension products which terminate with A. As shown in Fig. 2.2., the products from the 'plus' reaction will be one residue longer than the corresponding band in the 'minus' A reaction.

(iv) The polyacrylamide gel: interpretation of results
The analysis of the products from the 'plus' and 'minus' reactions demands an acrylamide gel capable of resolving oligonucleotides which differ in length by only one residue. A 12% polyacrylamide slab gel in a Tris-borate buffer containing 8M–urea is usually employed for this purpose. Since the adequate resolution of products is the main limiting factor in this type of analysis and since this in turn depends on obtaining a sharp autoradiograph it is important to use extremely thin gels to avoid the blurring which would unavoidably result from the high energy β-emission of [^{32}P] embedded in a thicker gel. This point is discussed in more detail in section 3.3.

In the example shown in Fig. 2.2. we define the smallest oligonucleotide in the −T channel as band 1. This means that the *next* residue after the 3′-terminus of the oligonucleotide corresponding to band 1 will be a T. This is equivalent to, and is confirmed by the presence of a band in the +T channel corresponding to the next longest oligonucleotide. Band 2 occurs in the +T channel and −A channel showing that its 3′-terminus is T and the adjacent nucleotide is an A, thus defining the dinucleotide sequence T–A. The next longest oligonucleotide occurs in the +A and −C channels. This defines the dinucleotide A–C and so estends the sequence to T–A–C. In the example shown the sequence:

5′-T–A–C–G–A–C-3′

can be deduced for the synthesized oligonucleotide.

Typically a sequence of 60–100 nucleotides, starting about 10–20 nucleotides from the 3′-end of the primer sequence, can be obtained from a single gel. In the above example each nucleotide in the sequence is represented by bands in both the + and − channels. However, if a run of two or more identical nucleotides occur, only the first one will be seen in the minus reaction and only the last one of the run will be seen in the equivalent plus reaction. The distance separating the different bands, in either the plus or minus patterns, will define how many residues there are in a run. This can lead to problems in the precise determination of the lengths of longer 'runs' and, partly for that reason, it is advantageous to include a 'zero' channel on the sequencing gel. This is simply an aliquot of the initial reaction (prepared in (i) above) which ideally will contain labelled oligonucleotides of all possible chain lengths and will therefore yield distinct bands corresponding to each residue of a run. Fig. 2.3. shows a diagram of a plus and minus sequencing gel together with its interpretation and Fig. 2.4. is an untouched photograph of a sequencing gel.

In the discussion of the 'minus' method, ((2.1.2.(ii))) the requirement that all the products should have a common 5′-terminus was emphasized. If a synthetic oligonucleotide or a small restriction fragment (<100 nucleotides) is used as a primer for the initial extension, the products of the plus and minus reactions can be analysed directly. Clearly, the smaller the primer that can be used, consistent with its ability to yield a unique primer-template complex in the annealing reaction, the greater the amount of sequence information that can be deduced since the extension reactions can be pushed further and still yield resolvable fragments. When using primers of this sort it is important to maintain the integrity of the 5′-terminus so that the difference in length of the fragments depends only on differences at their 3′ termini. This is achieved by using DNA polymerase lacking the normal 5′-exonuclease activity of DNA polymerase I [such as phage T4 polymerase, or a mutant *E. coli* polymerase I, or *E. coli* poly-

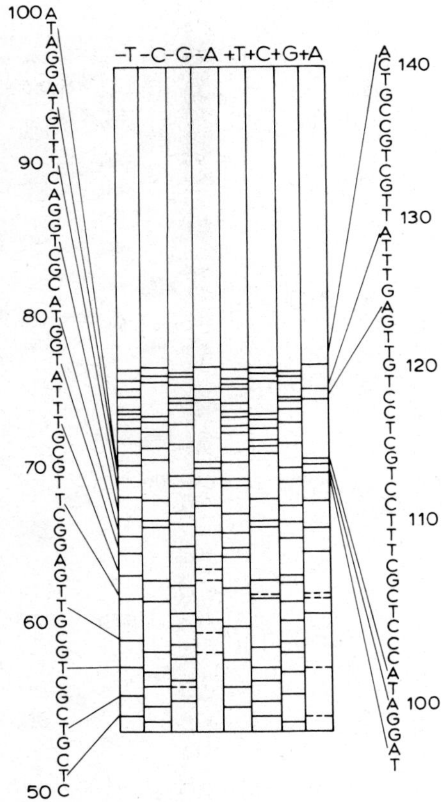

Fig. 2.3. Diagram of a 'plus and minus' sequencing gel with its interpretation. The dotted bands represent unreproducible artifact bands (from Barrell et al., 1976).

merase I 'nach Klenow'* (a subtilisin derived fragment of DNA pol I lacking the 5′-exonuclease activity)] (Klenow and Henningson, 1970; Brutlag et al., 1969). If longer restriction fragments are used as primers it becomes necessary to cleave the primer from the

* See Chapter 1, (Section 1.3.3.) for a description of the properties of this enzyme. In the text this is abbreviated to 'Klenow' polymerase.

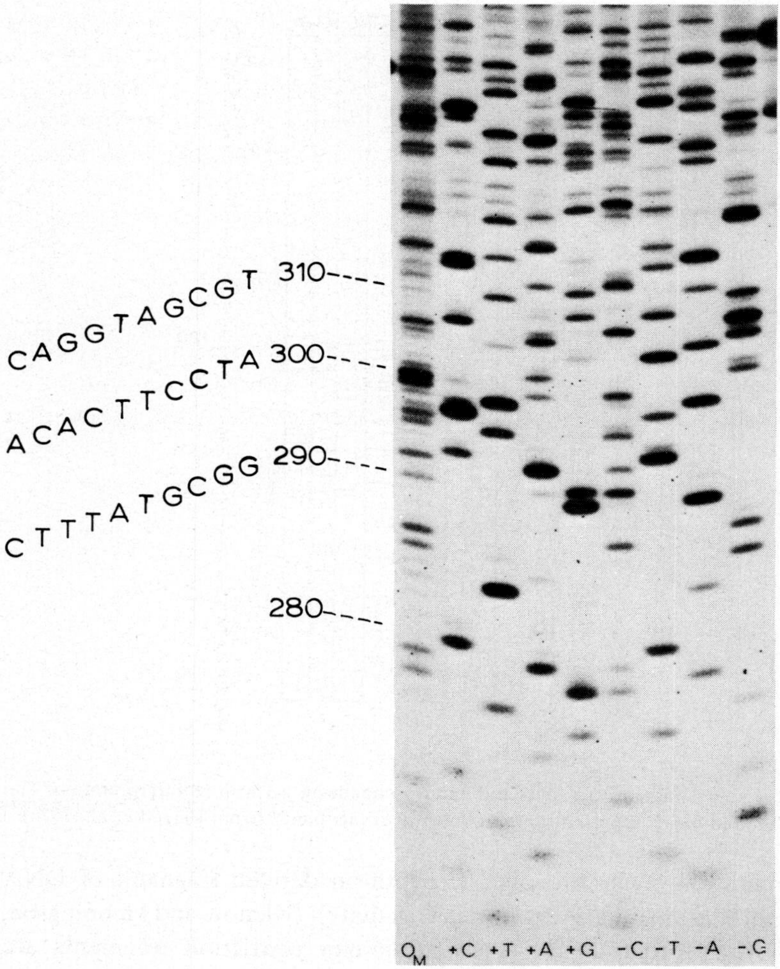

Fig. 2.4. Autoradiograph of a 'plus and minus' sequencing gel. A sequence of 30 nucleotides from positions 280–310 is written alongside (from Brown and Smith, 1977).

elongated cDNA before analysis. This is conveniently done by digestion with the restriction enzyme originally used to prepare the primer. The products for analysis in this case represent only the de novo synthesized sequences and in consequence are theoretically capable of yielding relatively more sequence data than in experiments where the primer remains attached to the analysed products. Fig. 2.5. illustrates this argument in which a *Hae*III produced restriction fragment is used as the primer. Some restriction enzymes, however, are inhibited by the single-stranded DNA present in uncopied regions of the template (e.g. *Alu*I and *Hph*I) and these enzymes cannot therefore be reliably used to cleave the primer from the extended product. One way round this problem is to use the single-site ribosubstitution method (Brown, 1978), cf. Section 2.3.1., in which a single ribonucleotide is incorporated at

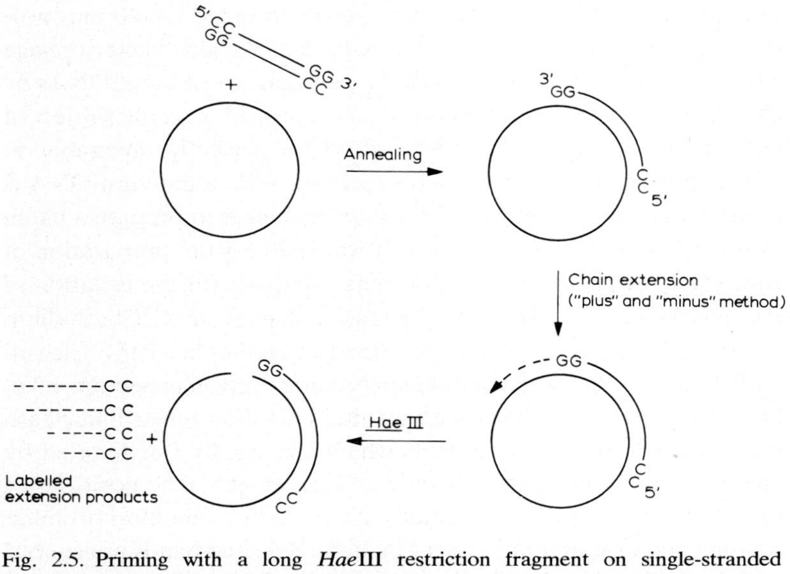

Fig. 2.5. Priming with a long *Hae*III restriction fragment on single-stranded circular DNA. After the chain extension step digestion with *Hae*III removes the unlabelled primer from the heterogeneous set of labelled extension sequences.

the priming site thus allowing the primer to be cleaved from the extended product with ribonuclease or alkali. This method is also useful if the restriction endonuclease used to generate the primer has a second cleavage site within the region to be sequenced. In this case enzymatic cleavage would yield two sets of labelled DNA fragments and a unique sequence would not be obtained. A suitable protocol for applying the single-site ribosubstitution method in this context will be described in Section 2.3.

2.2. *Detailed protocol for the 'plus' and 'minus' primed synthesis method*

(i) Introduction

This method, by its nature, is limited to the sequence analysis of single-stranded DNA molecules such as may be isolated from phages ϕX174, M13, G4 etc. or, more infrequently, by the strand separation of naturally double-stranded DNA. Single-stranded DNAs may now be prepared from any duplex DNA by cloning into bacteriophage M13mp2 or one of its derivatives. The sequencing of these DNAs by the dideoxy-nucleotide chain termination procedure is the subject of chapter 4 but, in theory, they would be perfectly amenable to sequencing by the plus and minus method. Where the viral DNA is naturally single-stranded the only requirement is to prepare a batch of the double-stranded replicative form (RF) for the preparation of restriction fragments to act as primers. Methods for the isolation of RF DNAs based upon their increased density in CsCl-ethidium bromide gradients are widely described (Edgell et al., 1972; Clewell and Helinski, 1969) and a suitable procedure is given in Section 4.3.6. Digestion of the RF DNA with a suitable restriction endonuclease yields a set of defined fragments which are usually fractionated by electrophoresis on polyacrylamide or agarose gels. The positions of the resolved fragments are usually visualized by ethidium bromide staining and examination under UV light. Selected bands are cut out and the DNA eluted from the gel segment. Procedures for this are described in detail in Chapter 5.

(*ii*) *Sanger and Coulson method* (*J. Mol. Biol.* 94: 441–448, 1975)
The composition of mixes, buffers etc. is given at the end of this Section.

Step 1: *Annealing reaction*

- Mix: 5 μl of primer (approx 1–2 pmol in H_2O)
  1 μl DNA template (single strand, approx. 0.4 pmol in H_2O)
  1.25 μl 1 M NaCl
  1.25 μl 10 × Pol mix
  6.5 μl H_2O
  15 μl
- Seal in glass capillary tube (approx. 10 cm long × 1 mm internal diameter). Denature by heating to 100°C for 3 min, anneal by incubating at 67°C for 45 min.

Step 2: *Primed synthesis of* [^{32}P]-*cDNA*

- Dry down 20 μCi α-[^{32}P]-dATP, specific activity about 300 Ci/mMol in a siliconized tube* in vacuo. (This is conveniently done as soon as the annealing reaction is started.)
- To dry dATP add:
  Annealed reaction mixtures from step 1 (15 μl)
  2.5 μl dCTP (0.5 mM)
  2.5 μl dTTP (0.5 mM)
  2.5 μl dGTP (0.5 mM)
  1.5 μl 10 × Pol mix
  24 μl
- Mix contents by sucking up and down in capillary from siliconized tube held at 0°C.
- Start reaction by mixing in 1 μl DNA polymerase I.
- Hold at 0°C.

* Small glass test tubes, 5 cm × 1 cm diameter, are soaked for 15 min in a 2%-solution of dimethyldichlorosilane in trichloroethane (Repelcote) and dried in an oven at 100°C. The tubes are rinsed six times with distilled water, dried, and are ready for use.

- Remove aliquot (approx 15 μl) after 1 min and eject into 25 μl 0.1 M EDTA, pH 7.6 to stop the reaction. After 3 min eject remainder of reaction mixture into the same EDTA. For short primers, say less than 100 nucleotides, DNA polymerase 'nach Klenow' is used for the extension reaction. If longer extension products are required, the incubation time can be increased.

Step 3: *Removal of polymerase and triphosphates*

- To the extension mixture from Step 2 add 25 μl phenol (redistilled, water-saturated).
- Vortex for $\frac{1}{2}$–1 min.
- Extract 5 times with 1 ml-portions of ether to remove phenol.
- Remove last traces of ether with a stream of air or nitrogen. Load the sample onto a column of G-100 Sephadex (3 mm × 200 mm) equilibrated with a degassed buffer containing 10^{-4} M-EDTA, 5×10^{-3} M Tris-HCl, pH 7.5 (The Sephadex column is conveniently prepared in a disposable 1 ml plastic pipette with a siliconized glass wool plug in the tip).
- The polynucleotide is eluted with the break-through volume of the column (flow rate approx 50 μl/min). Fractions of 2–3 drops may be collected and the position of the eluted DNA located by a hand held mini-monitor (Mini-Instruments Ltd., calibrated range of 0–2000 c.p.s.). Appropriate fractions are combined and freeze-dried.
- At this stage the product should register $>$300 c.p.s. on the mini-monitor, equivalent to an incorporation of about 5%–10% of the radioactive nucleotide. Under these conditions the extension of the primer ranges from zero to 150 to 200 nucleotides.

Step 4: *Plus and minus reactions*

- Dissolve the [^{32}P]-labelled extended polynucleotide in 20 μl H_2O.
- Set up 8 capillaries with drawn out tips, resting tip down in siliconized tubes, on ice.

– The eight samples are:

	1	2	3	4	
	+C	+T	+A	+G	(the plus reaction)
and	5	6	7	8	
	−C	−T	−A	−G	(the minus reaction)

– Introduce into the tip of each capillary 2 μl of the polynucleotide solution and 2 μl of the appropriate plus or minus mix (Section 2.2.iv).
– Add 1 μl T4 polymerase to capillaries 1–4, mix as before and incubate at 37°C for 45 min.
– Add 1 μl 'Klenow' polymerase (Section 2.1.) to capillaries 5–8, mix and incubate at 0°C for 45 min.
– The next step depends on whether a short (< 100 nucleotides) or longer polynucleotide was used as a primer. In the former case the reaction is terminated by blowing each reaction mixture into 10 μl formamide-dye mix and the remaining radioactive polynucleotide from step 3 (approx. 4 μl) is added to 10 μl formamide-dye mix. This is the 'zero' sample. The nine samples are now ready to proceed to step 5. Where a longer primer was used, this needs to be cleaved from the product using the appropriate restriction endonuclease. Add 1 μl, (0.5 to 1.0 unit) of the datum restriction endonuclease to each sample, mix, and incubate at 37°C for 30 min. (The datum endonuclease is that which defines the 5′ of the sequence under investigation, which will usually be the endonuclease used to prepare the primer fragment.) Stop the reaction by blowing into 10 μl formamide-dye mix.
– In addition, to provide the reference pattern of oligonucleotides (the zero channel) a sample of the radioactive polynucleotide prepared in *Step 3* is also digested with the restriction endonuclease.
  – Mix: 2 μl radioactive polynucleotide from step 3
  1.5 μl H_2O
  0.5 μl × 10 restriction buffer

1 μl restriction endonuclease (0.5 to 1.0 unit)
Incubate 37°C, 30 min
Stop reaction by blowing into 10 μl formamide-dye mixture.

Step 5: *Gel electrophoresis*

(*i*) *Preparation of the acrylamide gel*

This is conveniently done while the extension product is drying (step 3). The gel is cast in a cell made from two tempered glass plates,* 40 cm × 20 cm separated by two 'perspex' (polymethylmethacrylate sheet) spacers, 1.0–1.5 mm thick, running the length of the gel compartment. The design is essentially that of Studier (1973). The cell is sealed along the bottom and both sides with waterproof tape and immediately after pouring the gel a close-fitting well former giving 12 wells (1.1 cm wide) is inserted and the gel allowed to set in the near horizontal position. Immediately before the gel is required the tape is peeled off the bottom of the cell and the well former carefully removed (care is required not to break the wells). The cell is clamped vertically in the electrophoresis apparatus and the buffer compartments filled with 1 × TBE buffer. The apparatus is illustrated in Section 3.1.2.

- Dissolve 63 g urea (AnalaR) in 15 ml 10 × TBE plus 60 ml 30% acrylamide with gentle warming.
- Add 5 ml 1.6% freshly prepared ammonium persulphate and make up volume to 150 ml with distilled H_2O.
- Degas on water pump.
- Add 75 μl TEMED (N,N,N^1,N^1,-tetramethylethylene diamine). Mix gently.
- Pour gel immediately. The gel usually sets within 20–30 min and can be used after 1 hour. Alternatively the gel can be left overnight, with the well former in place, before use.

---

* Tempered glass plates are made from heat toughened plate glass. They may be purchased (see list of suppliers) or constructed in a glass workshop.

(*ii*) *Running the gel*

- Heat the nine samples from step 4 at 90° for 3 min. Using a pasteur pipette blow fresh 1 × TBE into the sample wells in order to remove urea which has diffused out of the gel.
- Load the samples into the gel wells using a drawn-out capillary tube. A suitable order is:

$$0, +C, +T, +A, +G, -C, -T, -A, -G.$$

- Run the gel at about 600 V until the fast migrating dye (bromophenol blue) is at the bottom of the gel (approx. 4 hr). The gel gets quite hot during electrophoresis ensuring the DNA remains fully denatured.
- Remove one glass plate from the gel and cover the exposed surface with cellophane film. Label with radioactive ink (preferably ^{35}S) and autoradiograph for 1–2 days at −20°C (Appendix IV and Section 3.1.2.). Alternatively the gel may be fixed by immersion in 10% acetic acid for 15–20 min, washed 1–2 min in distilled water, blotted dry with absorbent paper, covered with cellophane and autoradiographed at room temperature.

(*iii*) *Enzymes and chemicals* (*Appendix V*)

DNA polymerase I: (from *E. coli*, Boehringer Grade I, 2–3 units/μl).

'Klenow' polymerase: (*E. coli* DNA polymerase I nach Klenow, Boehringer, 2–3 units/μl)

(*iv*) *Buffers, mixes etc.*

10 × Pol mix: 100 mM–$MgCl_2$, 10 mM-2-mercaptoethanol, 200 mM–Tris HCl, pH 7.4.

G100 buffer: 5 mM–Tris HCl, 0.1 mM EDTA.

10 × Restriction buffer: 500 mM–NaCl, 66 mM–$MgCl_2$, 66 mM-2-mercaptoethanol, 66 mM-Tris HCl, pH 7.4.

10 × TBE buffer: 108 g Tris base, 55 g boric acid, 9.3 g Na_2EDTA. Make up to 1 l with distilled water (pH should be 8.3).

– Minus mixes are made as follows (all volumes in μl):

	−C	−T	−A	−G
dCTP (10 mM)	–	1	1	1
dTTP (10 mM)	1	–	1	1
dATP (10 mM)	1	1	–	1
dGTP (10 mM)	1	1	1	–
10 × restriction buffer:	100	100	100	100
H_2O	300	300	300	300

– Plus mixes:

	+C	+T	+A	+G
dCTP (10 mM)	5			
dTTP (10 mM)		5		
dATP (10 mM)			5	
dGTP (10 mM)				5
10 × restriction buffer:	25	25	25	25
H_2O	70	70	70	70

– Formamide-dye mix: 0.03% xylene cyanol FF, 0.03% bromophenol blue, 25 mM-EDTA in 90% formamide. (Formamide is freshly deionized before making up mix (Section 3.1.2.).
– 30% acrylamide: 29% (w/v) acrylamide, 1% (w/v) bis-acrylamide. Deionized by stirring with Amberlite MB-1 (5 g/100 ml) for 1 hr and filtering.
– Stop mix: 0.03% bromophenol blue, 40% sucrose, 25 mM EDTA in H_2O.

2.3. Additional methods useful in conjunction with plus and minus method

2.3.1. Ribosubstitution method (*Brown, 1978*)

In the protocol described above restriction fragments are used as primers for the synthesis of cDNA, and the same endonuclease is subsequently used to remove the primer and generate a unique 5′-terminus on the cDNA. However, as pointed out earlier, some restriction enzymes are strongly inhibited by the single stranded regions present in the template and cannot be used to cleave at the restriction site. An additional problem arises if a second cleavage site for the same enzyme is present within the sequence copied into the radioactive DNA. Two sets of fragments would be generated and a unique sequence would not be obtained.

In the ribosubstitution method these problems are circumvented by the addition of one or more ribonucleotides between the DNA primer and the radioactive cDNA. This site is susceptible to cleavage with ribonuclease or alkali. This method can also be used in conjunction with other primed synthesis methods for DNA sequencing (Barnes, 1978; Sanger, Nicklen and Coulson, 1977).

The principle of the method is shown in Fig. 2.6. In the presence of Mn^{++} ions, *E. coli* DNA polymerase I will in-

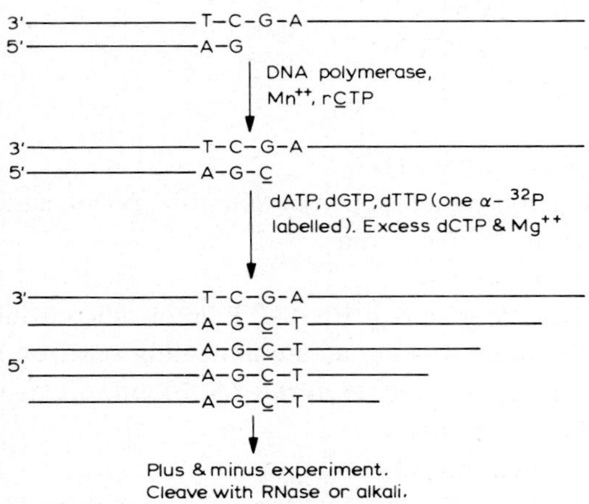

Fig. 2.6. Principle of the single-site ribosubstitution method using an *Alu*I fragment as primer (from Brown, 1978). The inserted ribocytidine residue is underlined C̲.

corporate ribonucleotides into DNA (Berg, Fancher and Chamberlin, 1963). In the single site ribosubstitution reaction a ribonucleotide is incorporated at the 3′-end of a DNA primer in the presence of Mn^{2+}, and with no other triphosphates present. Further ribonucleotide incorporation is effectively suppressed in the subsequent elongation reaction by the addition of deoxyribonucleoside triphosphates and Mg^{2+}.

2.3.2. Experimental procedure

(suitable for the incorporation of riboC at AluI (AG↓CT) or *Hae*III (GG↓CC) priming sites)
(Details of reagents and mixes required follow.)

Step 1: *Annealing reaction.* This is performed exactly as described in Step 1 Section 2.2(ii).

Step 2: *Synthesis of* [^{32}P]*-cDNA*

(a) *ribonucleotide addition*

- Mix: Annealing reaction from Step 1 (15 μl).
  2.5 μl 0.01 M $MnCl_2$
  0.5 μl 10 mM rCTP
  5.0 μl H_2O
  23.0 μl
  in a capillary tube at 0°C.
- Start reaction by adding: 2 μl 'Klenow' polymerase.
- Incubate at 0°C for 5 min.

(b) *extension reaction*

- Dry down 20 μCi α-[^{32}P] dATP in a siliconized tube (this should be done as soon as the annealing reaction is started).
- To the dry dATP add 25 μl rC–dA 'Flood mix' (see below for reagents and mixes).
- Mix and keep at 0°C.
- When the incubation in Step 2(a) is complete the entire contents of the capillary are expelled with mixing into the above 'Flood mix'. Keep at 0°C.

- Take aliquots at 1 min (20 μl)
  3 min (20 μl)
  10 min (remainder)

  and add to the same 5 μl 0.5 M EDTA, pH 7.5.

Step 3: *Removal of polymerase and triphosphates*
Exactly as described in Section 2.2(ii).

Step 4: *Plus and minus reactions*
As described in Section 2.2(ii).

Step 5(i): *Cleavage at ribosubstituted site.*

(i) Cleavage with pancreatic ribonuclease

- The eight plus and minus samples and an aliquot (6 μl) of the original extension product from Step 3 are sealed in their capillary tubes.
- Heat 100°C for 3 min to denature the DNA.
- Cool rapidly in ice-water.
- Open the tubes and add, with mixing, 1 μl of a 10 mg/ml solution of ribonuclease A.
- Incubate at 37°C for 30 min.
- Stop the reaction by adding 10 μl formamide-dye mix at 0°C.
- Heat samples at 90°C for 3 min.

Step 6: *Gel electrophoresis*
As described in Section 1.3(ii).
An alternative to cleavage with ribonuclease in Step 5(i) is to use alkaline hydrolysis.

Step 5(ii): (alternative to 5(i))

- Add 1 μl 2N NaOH to each plus and minus sample and to the 'zero' samples.
- Seal in capillary.
- Heat at 100°C for 20 min.
- Open capillaries and mix with 10 μl formamide-dye mix.
- Apply immediately to gel.

Reagents and mixes

rCdA 'Flood mix': 20 mM-$MgCl_2$, 20 mM-2-mercaptoethanol, 0.1 mM-dTTP, 0.1 mM dGTP, 0.25 mM-dCTP.

Ribonuclease A:	10 mg/ml pancreatic ribonuclease A (Boehringer) in 50 mM Tris HCl, 5 mM-EDTA pH 7.4. Heated at 80°C for 10 min (to inactivate contaminating DNAases). Store frozen.

2.3.3. Discussion

Figure 2.7. shows the result of a 'plus and minus' sequencing experiment using the single-site ribosubstitution method applied to an *Alu*I derived primer annealed to the viral (+) strand of phage ϕX174. No detectable radioactivity is found at the top of the gel indicating virtually quantitative incorporation of ribo-cytidine at the priming site and consequent susceptibility of the product to alkaline cleavage. One drawback of the method is that, in general, the extension reaction does not proceed as far as normally encountered in the 'plus and minus' method. The reason for this is not known but could be due to the inhibitory action of Mn^{++} or rCTP on the DNA polymerase I. Ribonuclease T_1 will cleave at a single rG substitution and ribonuclease A at a single rC or rU substitution. rATP and UTP are less efficiently incorporated at the priming site (Brown, 1978) and most of the sequences reported have used either rC or rG substitution.

2.3.4. Depurination analysis of defined fragments generated by primed synthesis

(i) Introduction

Depurination and depyrimidination are chemical procedures in which the purine or pyrimidine bases, respectively, are removed from the DNA molecule such that the runs of consecutive pyrimidines and purines are left intact. The classical depurination procedure for DNA, devised by Burton and Peterson (1960) involves incubating the DNA with 2% diphenylamine in 60% formic acid. This results in the quantitative elimination of the purine nucleosides leaving pyrimidine tracts as the 3′,5′-diphosphates. These may be separated either by chromatography on DEAE-

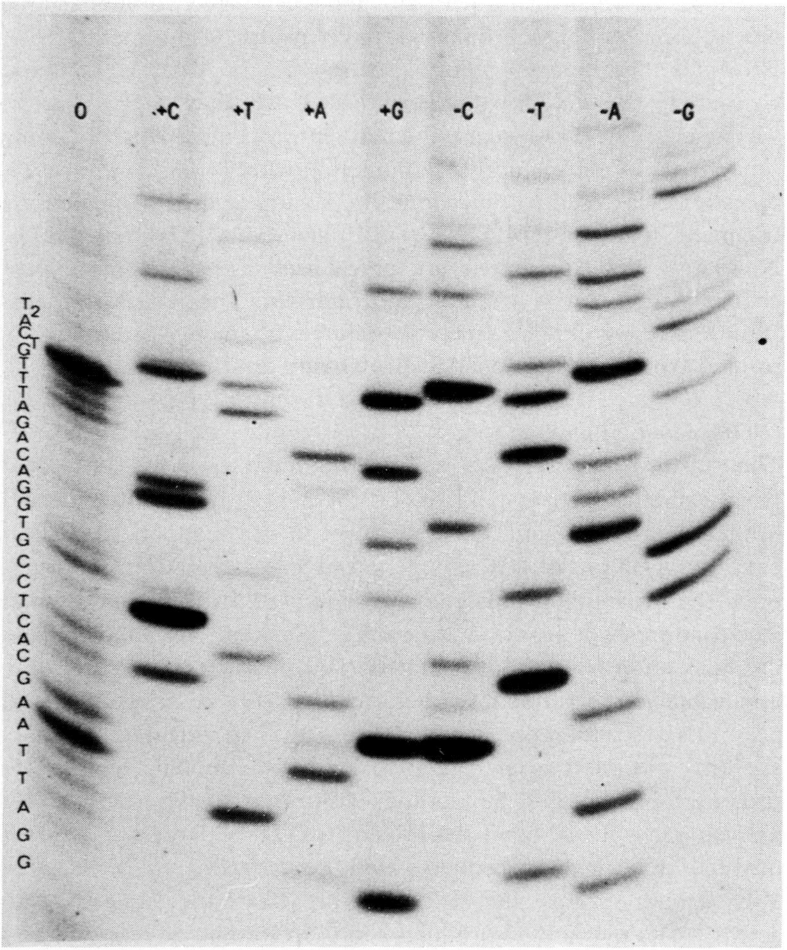

Fig. 2.7. Autoradiograph of a 'plus and minus' DNA sequencing gel after single-site ribosubstitution with rCTP using an *Alu*I primer on ϕX174 viral strand DNA as template (courtesy of Dr. N.L. Brown).

cellulose, or when using [^{32}P]-labelled DNA, by two-dimensional electrophoresis-homochromatography using the method of Brownlee and Sanger (1969), Section 2.3.5. Analysis of these pyrimidine clusters (Burton et al., 1963; Mushynski et al., 1970) was one of the most successful earlier procedures for obtaining sequence information and in conjunction with T4 endonuclease IV digestion, was used to determine the sequence of a 48-nucleotide fragment from ϕX174 DNA (Ziff, Sedat and Galibert, 1973). Nowadays, depurination is not much used as a sequencing procedure *per se* but is valuable in confirming the length and distribution of pyrimidine tracts in defined fragments generated by primed synthesis methods, e.g. Brown and Smith (1977).

(*ii*) *Principle of method*

The starting material is a defined fragment of double-stranded DNA, labelled with α-[^{32}P] dATP (or α-[^{32}P] dGTP) in separate primed syntheses and isolated from a subsequent restriction enzyme digest of the DNA by gel electrophoresis. After depurination, the pyrimidine clusters are separated by two-dimensional electrophoresis-homochromatography (Section 2.3.5.) and the products identified and quantitated. This procedure is clearly only applicable if the primed synthesis extends the primer up to and beyond a restriction enzyme cleavage site since meaningful results are only obtained from a defined starting fragment. Figure 2.8. shows the principle of the method. If two primed syntheses are set up using α-[^{32}P]-labelled dATP and dGTP separately, then all possible depurination products will be labelled singly in their 3′-phosphate groups and will therefore have the same specific activity. The pyrimidine products can therefore be quantitated and their molar ratios obtained. The use of primed synthesis to label the DNA has the advantage over nick translation methods in that the DNA produced is only labelled in one strand. It is therefore known in which strand the depurination products lie and having only one strand labelled means that the quantitation of the products is often trivial as there are only a few. Fractionation of

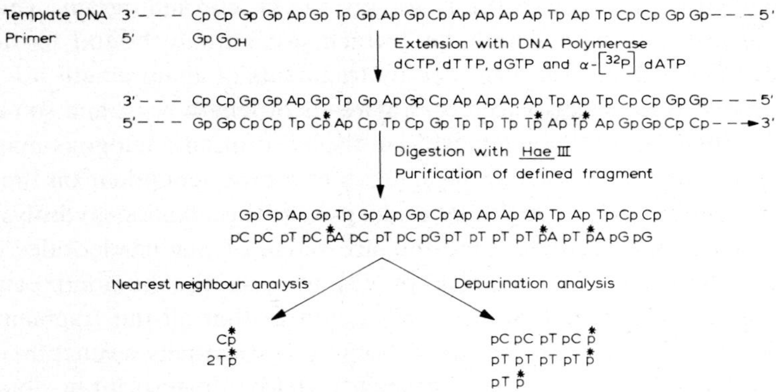

Fig. 2.8. Nearest neighbour analysis and quantitative depurination analysis of a defined product from a primed synthesis reaction. When radioactive dATP (or dGTP) is used in the primed synthesis, depurination analysis will yield pyrimidine tracts each of which terminate in a radioactive 3′-phosphate. Thus only those depurination products which lie 5′-adjacent to the labelled nucleotide will be labelled. Each depurination product will be labelled to the same specific activity thus greatly simplifying the quantitation. Digestion of the labelled product with a mixture of micrococcal nuclease and bovine spleen phosphodiesterase yields the nucleoside 3′-monophosphates. Identification of the labelled products (by paper electrophoresis at pH 3.5) gives the nearest neighbours to the labelled substrate.

the pyrimidine oligonucleotides employs a two-dimensional system in which the fragments are first partially resolved on the basis of their charge differences at pH 3.5 by electrophoresis on an inert support (cellulose acetate) followed by homochromatography in a direction at right angles to the first dimension which further resolves the oligonucleotides largely on the basis of their chain length. In this latter procedure the chromatogram is developed using a complex mixture of oligonucleotides containing a wide range of chain lengths prepared by a partial alkaline hydrolysate of yeast RNA. As the ‘homomix’ moves up the DEAE-thin layer, DNA fragments are progressively displaced by the gradually increasing concentration of unlabelled oligonucleotides such that a gradient of size fragments is set up on the chromatogram. Thus fragments of chain length "n" which are initially bound to the DEAE-thin layer are displaced by fragments of chain length n + 1. These in turn are displaced by fragments of length n + 2 and so on. Absorbed DNA fragments will be displaced in an analogous manner and move up the thin layer plate at a rate dependent on their chain length. The conditions under which the alkaline hydrolysis is carried out will determine the size range of oligonucleotides in the homomix. The inclusion of 7M urea in the homomix and running the chromatogram at 65°C ensures that all the fragments are fully denatured so that their mobility is essentially a function of their chain length only. This procedure yields a fingerprint in which the different pyrimidine tracts are distributed in a (usually) defined pattern with respect to each other and inspection of the autoradiograph is often sufficient to identify the products. This fractionation procedure does not separate isomers of the same base composition but as the longer pyrimidine tracts often only occur once in the sequence under investigation this is not a serious problem. Figure 2.9. shows a diagram of a two-dimensional fractionation of pyrimidine oligonucleotides isolated from bacteriophage fd DNA (Ling, 1972). With small defined fragments the pattern is very much simpler; usually only five to ten products are obtained and their compositions can often be guessed from their position. Where necessary the longer tracts can be sequenced by partial digestion with spleen and venom phosphodiesterases but this is rarely necessary. Molar ratios are estimated from the autoradiograph or by scraping off the DEAE layer containing each spot into liquid scintillant and determining the radioactivity. Figure 2.10. is an autoradiograph showing the results of a depurination analysis on the complementary strands corresponding to a defined segment of ϕX174 (Brown and Smith, 1977).

2.3.5. Detailed procedure

Step 1: *Annealing reaction*

– Mix: 4 μl primer (approx. 1.2 pMol; in H_2O)

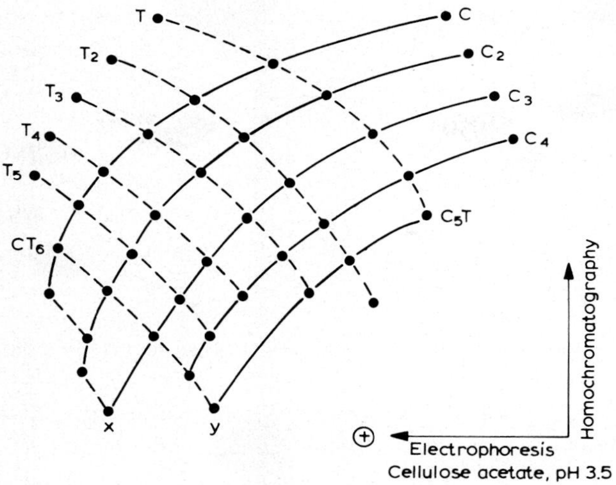

Fig. 2.9. Diagram of a two-dimensional fractionation involving electrophoresis-homochromatography of pyrimidine oligonucleotides isolated from bacteriophage fd DNA. Solid lines join oligonucleotides of the same C content, and broken lines join oligonucleotides of the same T content. Oligonucleotide X thus has the composition C_3T_8 and y is C_5T_6 (after Ling, 1972). The fractionation system gives no information about the sequence of the oligonucleotide.

1 μl template strand (approx. 0.4 pMol; in H_2O)
0.5 μl M NaCl
0.5 μl 10 × Pol mix (Section 2.2, Step 5, iv)

- Seal in capillary
- Heat 100°C for 3 min
- Incubate 67°C for 45 min.

During the incubation it is convenient to dry the dATP* and dGTP*—Step 2 and make gel—Step 3

Step 2: *Synthesis of* ^{32}*P-cDNA*

- Dry down 10 μCi α-^{32}P dATP (dATP*) and 10 μCi α-^{32}P dGTP (dGTP*) in separate siliconised tubes.

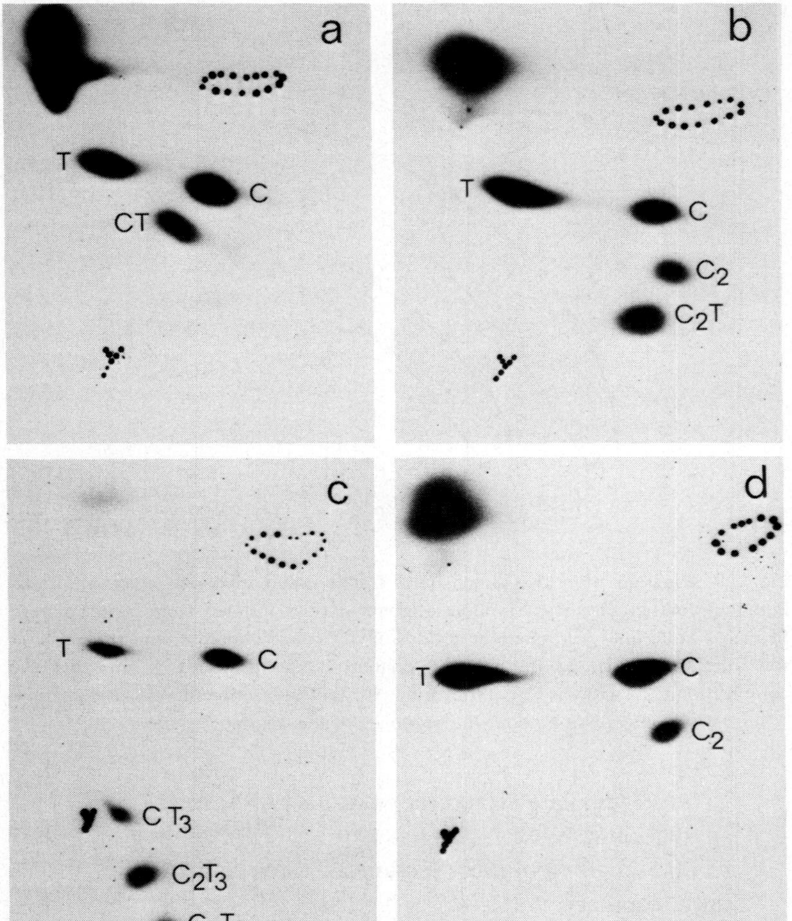

Fig. 2.10. Two-dimensional fingerprints of pyrimidine tracts obtained by specific priming on the ϕX174 viral strand, (a) and (b), and the corresponding region on the complementary strand (c) and (d). For (a) and (c) labelling was with α-[^{32}P]dATP and for (b) and (d) with α-[^{32}P]dGTP. The results were used to confirm and complement the sequence obtained by the 'plus and minus' method (from Brown and Smith, 1977). The composition of each spot e.g. C_2T_3 (an oligonucleotide containing two cytidine and three thymidine residues) is deduced from its relative mobility on the fingerprint. Electrophoresis is from right to left and homochromatography from bottom to top. The dotted outline is the position of the xylene cyanol blue marker and Y the position of the orange G marker.

– To dry dATP* add: Half of annealing reaction from step 1 (~3 μl) 10 μl dA* extension mix

To dry dGTP* add Half of annealing reaction from step 1 (~3 μl) 10 μl dG* extension mix

(Composition of mixes given in Section 2.3.6. below)
– Mix each in a capillary at 0°C.
– Start each reaction by adding 0.5 μl 'Klenow' polymerase (2–3 units/μl)
– Incubate at 0°C for 30 min.
– Stop reaction by adding: 10 μl phenol (water saturated)
20 μl H_2O
(Do not use EDTA).
– Vortex ~30 sec.
– Then vortex with 5 × 1 ml of ether to remove phenol.
– Blow off ether with air or nitrogen.

Step 3: *Cleavage with restriction enzymes and separation of fragments.*

– To each sample add: 2.5 μl 10 × restriction buffer (Section 2.3.6. below)
1 μl appropriate restriction enzyme (approx. 1 unit)
– Incubate at 37°C for 30 min.
– Add 5 μl stop mix
– Load samples onto a 20 cm × 20 cm × 0.15 cm slab gel (8% acrylamide in 1 × TBE; 1–1.5 cm wide wells). Electrophorese at 150 volts until bromophenol blue marker dye is approx. $\frac{3}{4}$ of the way down.
– Remove one of the gel plates; cover the gel with cellophane film; mark the gel with radioactive ink at all four corners; and autoradiograph for approx. 4 hr.
– The gel solution is made up as follows: 20 ml 30% acrylamide : bis-acrylamide (29 : 1); 7.5 ml 10 × TBE; 2.5 ml 1.6% ammonium persulphate; 45 ml H_2O. Degas. Add 50 μl TEMED and pour gel immediately.

Step 4: *Isolation of DNA fragments*

- Identify the fragment bands on the autoradiograph of the gel, and cut out this portion of the autoradiograph.
- Align the autoradiograph on the gel, and using the autoradiograph as a template, cut out the portion of gel containing the radioactive DNA fragment.
- Gently homogenise the gel in a siliconized tube with a glass rod and cover the gel fragments with 0.5 ml of Elution Buffer (0.5 M-NaCl, 0.005M-EDTA, 0.1 M-Tris HCl, pH 8.5). Leave at 37°C for 5 hr or more. (Alternatively the procedure described in Section 5.9., procedure A, can be used.)
- Decant supernatant and gel fragments into a micro-filter tube (see below).
- Wash out siliconized tube with 100 μl H_2O, and add washing to micro-filter.
- Centrifuge briefly to filter.
- Add 2 μl carrier tRNA solution (20 mg/ml; 40 μg) to filtrate. Mix.
- Add 2 vols. cold Ethanol. Mix.
- Freeze sample at −70°C to precipitate nucleic acids.
- Thaw sample at −20°C.
- Centrifuge to pellet nucleic acids.
- Decant off supernatant. Remove excess liquid with a capillary.
- Quickly wash precipitate with 80% ethanol at −20°C. Freeze, thaw, and centrifuge again. This removes most of the salt.
- Decant off supernatant. Remove excess liquid with a capillary.
- Take up precipitate in 20 μl H_2O and transfer to a new siliconized tube.
- Dry down under vacuum.

Step 5: *Depurination of defined fragment*: (*Burton and Petersen, 1960*; *Tate and Petersen, 1975.*)

- To dried sample add: 25 μl H_2O
  1 μl 0.1M EDTA
  50 μl Burton reagent (3% diphenylamine in 98% formic acid)

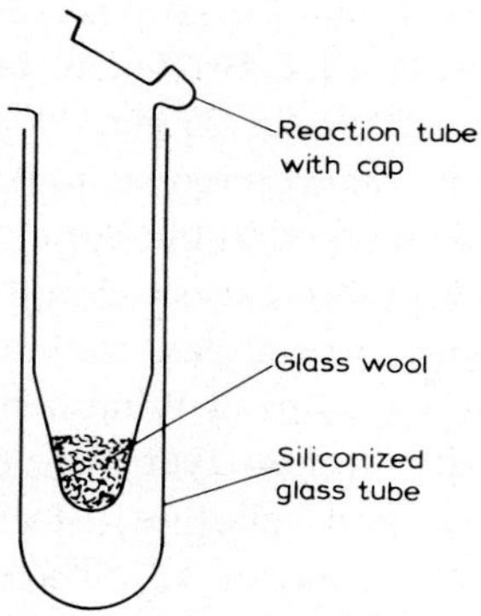

Micro-filter tubes are constructed as follows: A plastic reaction tube is punctured with a hot needle in its base and its cap. Wet glass wool is packed to the bottom of the tube. The reaction tube is then inserted into a siliconized glass tube (50 mm × 13 mm).

- Incubate 30°C, 16 hr.
- Add 100 μl H_2O.
- Extract four times with 2 vol. ice cold ether (saturated with H_2O).
- Transfer into fresh siliconized tube.
- Dry in vacuo.

Step 6: *Separation of depurination products* (*Ling, 1972*)

- Dissolve DNA in 5 μl H_2O.
- Wet a cellulose acetate strip (Schleicher and Schüll, West Germany) 3 cm × 55 cm with pH 3.5 urea buffer (this is made by mixing 47.5 ml 7M urea, 2.5 ml glacial acetic acid, 0.5 ml 0.5M-EDTA).
- Blot excess buffer from ~10 cm of one end.
- Apply sample as a single spot about 5 cm from blotted end and small spots of the marker dye mixture Section 2.3.6. on each side of the sample.
- Allow sample to soak in. Quickly blot the remainder of the strip to remove excess buffer and transfer into an elec-

trophoresis tank such that the strip is completely submerged in the white spirit coolant (cf. Brownlee, 1972).

- Electrophorese at 4500V for 25 min.
- Remove strip from electrophoresis tank, cutting off the ends which had been submerged in the buffer compartments. Allow strip to partially dry (until excess white spirit had dripped off).
- Lay strip lengthways over a glass rod and put water-saturated wicks, made from 3–4 strips of Whatman 3MM paper, on each side of the strip. Lay a thin-layer plate coated with a mixture of DEAE-cellulose and cellulose (Machery, Nagel and Co., West Germany) at a ratio of 1:7.5 face down over the strip and apply gentle weight. In this procedure water from the wicks is drawn up through the cellulose acetate strip into the DEAE thin layer. The oligonucleotides are thereby almost quantitatively transferred onto the thin-layer plate. Allow transfer to proceed for 30–45 min (until the blue marker dye has migrated onto the thin layer).
- Wash the thin layer in 95% ethanol (5 min) and leave to dry. This procedure removes excess urea from the plate.
- Put plate into prewarmed chromatography tank in an oven held at 60–80°C.
- Allow to equilibrate for 30 min.
- Add prewarmed 3% 'homomix*' (as defined below) to submerge bottom 5 mm of thin-layer plate, seal the tank and allow development of chromatogram until the blue marker dye is near the top of the plate (40–60 min).
- Remove plate. Dry at 60–80°C.
- Mark thin layer at each corner with radioactive ink and autoradiograph.

---

* 3% homomix: 30 g yeast RNA is stirred with 300 ml H_2O and 30 ml 10M-KOH for 40 min at room temperature and then neutralized to approx. pH 7.5 with 10M-HCl. The partially hydrolysed RNA is dialysed for 2-4 hr against distilled H_2O. Urea, 420 g, is added to the solution and the volume adjusted to 1000 ml with H_2O.

2.3.6. *Buffers and mixes*

10 × Restriction buffer: 500 mM-NaCl, 66 mM-$MgCl_2$, 66 mM-2-mercaptoethanol, 66 mM-Tris HCl pH 7.4.

dA* and dG* extension mixes are made up as follows:

	*dA**	*dG**
dCTP (10 mM)	1	1
dTTP (10 mM)	1	1
dATP (10 mM)	—	1
dGTP (10 mM)	1	—
10 × pol mix	15	15
H_2O	82	82

Marker dye mixture: Mix equal volumes of 1% Xylene Cyanol F.F. (blue), 2% Orange G (yellow) and 1% Acid Fuchsin (pink) (all from George T. Gurr, Ltd., London.)

2.4. *Discussion and summary*

The plus and minus method represented a major breakthrough in the development of primed synthesis methods for the rapid determination of DNA sequences.

This method made it possible to deduce sequences of about 50–80 residues in a single-stranded DNA relatively quickly where suitable primers were available. Using restriction enzyme fragments (for a list of suitable enzymes see Appendix III), primers can be selected covering any desired region of the DNA and regions of particular interest picked out and sequenced. This approach reached its pinnacle in the determination of the entire sequence of the genome of bacteriophage ϕX174, approximately 5,375 nucleotides (Sanger et al., 1977) and permitted the identification of many of the features responsible for the transcription and translation of the nine known genes of the virus.

Like any method however, there are certain limitations and

drawbacks in its application and before deciding on a sequence strategy it is important to consider these. The major limitation is that it cannot be applied directly to double-stranded DNA, so that a strand separation of either the template or primer must first be carried out. For naturally occurring single-stranded DNAs this problem does not arise and the only requirement is a suitable source of the corresponding replicative (double-stranded) form of the virus for the preparation of the DNA primers. The problem of obtaining single-stranded DNAs from naturally double-stranded molecules has now been solved (Chapters 3 and 4) with the advent of cloning techniques which permit the ready isolation and propagation of recombinant single-stranded DNA phages (e.g. M-13) containing long (up to 4000 nucleotides) inserts. In theory, though this does not appear to have been attempted, such recombinant phages should provide perfectly suitable templates for sequencing by the plus and minus method. A formally very similar approach however utilizes such single stranded templates, in conjunction with a 'universal' primer, and this forms the basis of the most powerful primed-synthesis method currently available. This is described in Chapter 4.

The plus and minus method does have its limitations and, as pointed out by Sanger et al. (1977), cannot be regarded as a completely reliable method in the absence of confirmatory data. The ribosubstitution method and depurination analysis procedures go a long way towards overcoming two of the problems that arise, namely sequencing through additional homologous restriction enzyme sites on the template and confirming the lengths of consecutive runs of pyrimidine nucleotides respectively. A further difficulty arises due to the non-random length distribution of the different cDNA elongation products. Ideally oligonucleotides of all possible lengths should be present in the initial product of synthesis, so that all residues are represented in the plus and minus systems. In practice, this is difficult to achieve and under all conditions studied it was found that certain products are formed in relatively high yield whereas others are absent. This suggests that the DNA polymerase acts at different rates on different sequences and this may be related to the secondary structure of the template (formation of stems and loops). The relative concentrations of the triphosphates used are also important. At a relatively low concentration of the ^{32}P-labelled triphosphate this may constitute the rate-limiting step and "piling-up" frequently occurs before these residues. In general the best results are obtained if synthesis is carried out for short times with a relatively high concentration of polymerase (Sanger et al., 1977). Even so it was frequently found that some of the expected products were missing.

Another problem with the method is the occasional appearance of 'artifact' bands on the gel. These are usually faint bands and can be recognized by the fact that they are not consistent in the plus and minus systems. Also their presence or intensity is not usually reproducible in different experiments. Their occurrence is likely to be due to elements of secondary structure in the template and they are often found clustered together in particular regions on the sequencing gel. Transient intra-strand secondary structure may also affect the mobility of fragments on the gel (e.g. Fiddes, 1976). Whatever their origin, the occurrence of these different artifacts does emphasize the need for caution in interpreting the gels and the importance of obtaining independent confirmatory data. Often this can be obtained by sequencing the complementary DNA strand by the same method. Where this is not possible an alternative DNA sequencing method will need to be used. In some cases advantage can also be taken of amino acid sequence data to ensure that no residues are missed and the correct reading frame is maintained.

Taken all in all the plus and minus method is a powerful technique. However, it does have its limitations and the more recent developments using the dideoxynucleoside triphosphates as specific chain inhibitors have much improved both the speed and accuracy of the method and will usually be the method of choice for primed synthesis sequencing. This is described in the next chapter.

2.5. *Sequence analysis of short DNA fragments*

2.5.1. Introduction

While most DNA sequencing has as its objective the analysis of long sequences the need also arises from time to time of sequencing short fragments of chain length up to about 20 nucleotides. For example the analysis of depurination products (Ling, 1972, and Section 2.3.4.), or the first few nucleotides from the end of a labelled fragment, or the confirmation of the sequence of a chemically or enzymatically synthesized oligonucleotide primer are typical of the problems which can arise.

The usual procedure for sequence analysis of short chains employs a two-dimensional fractionation method in which the DNA is first subjected to partial digestion with an exonuclease under conditions where all the partially degraded products are represented in the mixture, the products are then separated on the basis of their charge and mass by electrophoresis and homochromatography and related to each other on the basis of characteristic shifts in their mobility. This method is essentially a development of the original two-dimensional procedure introduced by Brownlee and Sanger (1969) for the fractionation of oligodeoxynucleotides in which the first dimensional separation was achieved by electrophoresis on cellulose-acetate strips at pH 3.5 followed by homochromatography on DEAE-thin layer plates. This fractionation scheme has been further developed (Ling, 1972; Sanger et al., 1973; Galibert et al., 1974; Jay et al., 1974) to the point where two oligonucleotides, differing in length by one residue can be resolved and related to each other in such a way that the mobility shift is characteristic of that particular residue. This is the basis of the "wandering spot" method for the analysis of short DNA fragments and represents probably the best, and most widely used current technique.

The initial requirement is for a singly end-labelled DNA fragment or a uniformly labelled single-strand. If the fragment molecule is single-stranded e.g. a chemically synthesized primer, all that is required is the introduction of either a 5′-phosphorylated end (T_4-polynucleotide kinase and γ-[^{32}P]ATP; Wu et al., 1976; Maxam and Gilbert, 1977) or labelling the 3′-end with deoxynucleotidyl transferase and α-[^{32}P]CTP (Roychoudhury and Wu, 1980; Maxam and Gilbert, 1980). Where a double-stranded DNA is to be analysed, these labelling schemes will label both strands and it is necessary to separate these before proceeding with the analysis. For long DNAs the strand separation procedure of Maxam and Gilbert (1980) (cf. Chapter 5), is employed whereas for short fragments (up to about 40 nucleotides) homochromatography on DEAE-thin layers in the presence of 7M urea is the method of choice. The separated singly labelled strands are next subjected to partial digestion with either venom phosphodiesterase (for 5′-end labelled DNA) or spleen phosphodiesterase (for 3′-end labelled DNA) or pancreatic DNAase (either end-labelled). Under suitable conditions, determined empirically, all the partial degradation products, uniquely labelled at one end, will be present in the mixture. These are finally resolved by two-dimensional electrophoresis-homochromatography on a thin-layer plate and the pattern of spots interpreted from a set of 'mobility shift' rules. This usually permits an unambiguous derivation of the sequence. The principle of the method is shown in Fig. 2.11.

2.5.2. Procedures (*cf*: *Tu and Wu, 1980*)

1. *End labelling*
   Detailed procedures for 5′- or 3′-end labelling DNA fragments, following the method of Maxam and Gilbert (1980) are given in Chapter 5. Essentially similar protocols are given by Wu et al. (1976) and Roychoudhury and Wu (1980).
2. *Strand separation*
   Long duplex DNA fragments are resolved into their complementary single-strands by denaturation followed by electrophoresis on a 5–8% acrylamide gel (Maxam and Gilbert, 1980; this volume, Chapter 5).
   Smaller fragments (<40 nucleotides) are resolved into single-

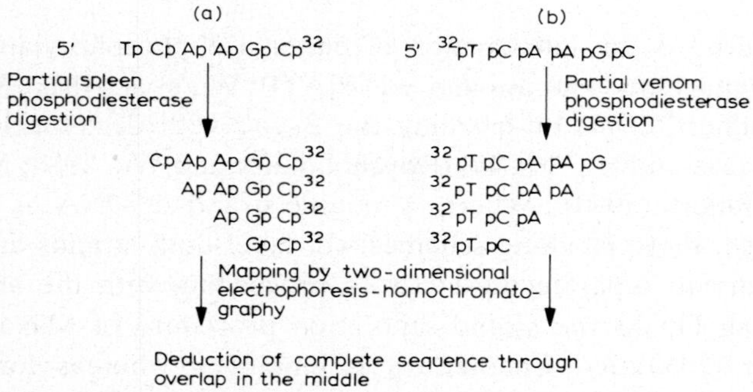

Fig. 2.11. In (a), 3′-end labelling is achieved using deoxynucleotidyl transferase and an α-[^{32}P] ribonucleotide triphosphate followed by elimination of the ribonucleotide residues. In (b), 5′-end labelling is carried out using polynucleotide kinase and γ-[^{32}P]ATP. Partial digestion with spleen phosphodiesterase removes nucleotides sequentially from the 5′-end of the oligonucleotide giving a mixed population of partially degraded molecules each labelled at the 3′-end. Venom phosphodiesterase removes nucleotides from the 3′-end similarly yielding a mixed population of shortened fragments. The products of the reaction are resolved by two-dimensional electrophoresis-homochromatography and the sequence deduced by the characteristic pattern of shifts.

strands by homochromatography on DEAE-thin layers (Tu and Wu, 1980).

Method: Dissolve the DNA fragment, end-labelled with ^{32}P, in 10–20 μl H_2O and apply as a 1–2 cm band to a DEAE-thin layer plate, 250–500 μm thick composed of DEAE-cellulose : cellulose (1 : 7.5). These may be made as required using a thin-layer spreader or purchased (Analtech Inc, or Cel 300 DEAE/HR from Machery-Nagel and Co). The plates are chromatographed in a closed jar at 65°C using a homochromatography solvent composed of 2% partially hydrolysed RNA in 7M urea.* A tracking dye (orange

---

* 2% Homochromatography mix

Suspend 20 g yeast RNA (free acid) in 88 ml H_2O. Chill in ice and add, with stirring, 12 ml 5 N KOH until the RNA has dissolved. Incubate at 37°C for 24 hours with intermittent stirring or shaking. Add 100 ml H_2O and titrate to pH 7.0–7.4 with 1 N HCl. Add 420 g solid urea and make the volume up to 1 l with distilled water. Filter and store frozen.

G) is conveniently run in parallel with the sample and the run discontinued when the dye-marker has reached the top of the plate. The plate is finally dried and autoradiographed.

3. *Elution of DNA*

Regions of the DEAE matrix corresponding to the separated fragments are scraped off the plate using a small glass device, fitted with a sintered glass filter and connected to a water pump. The upper part of the device is then removed, set upright in a rack and the DEAE washed with 1–2 ml 95% ethanol to remove urea. DNA is eluted with 100–200 μl of 30% triethylamine carbonate, pH 10.0 (made by bubbling CO_2 through a mixture of triethylamine (30 ml) and H_2O (70 ml) until a single-phase solution is obtained). The eluate is taken to dryness in vacuo, dissolved in 50 μl H_2O, lyophilized, and the last procedure repeated once more. The mixture, though not entirely salt free is suitable for the exonuclease digests.

4. *Preparation of partial exonuclease digests*

(i) *Venom phosphodiesterase*

A typical reaction mixture (10–20 μl) contains 2 mM Magnesium acetate, 20 mM Tris-acetate, pH 8.0 to 8.3, 25–50 μg carrier RNA (eluted along with the DNA from the DEAE-thin layer), the 5′-end labelled DNA (5000–50,000 cpm) and 1–2 μg

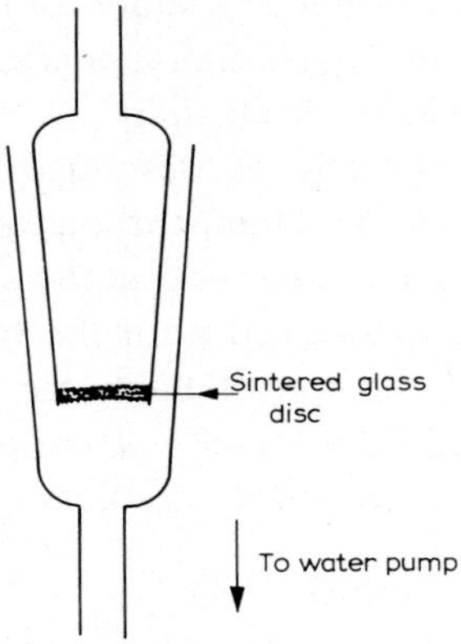

of snake venom phosphodiesterase. The sample is incubated at 37°C and 1 μl aliquots removed at intervals over a 2 h period and pipetted into 10 μl 2 mM EDTA in 0.1 M NH_4OH to stop the reaction. The sample is denatured by heating at 100°C for 3 minutes, [^{14}C]dTMP, 2000–10,000 cpm added as a chromatography marker and the mixture dried in vacuo. The sample is taken up in 5–10 μl H_2O and 2–4 μl applied to a strip of cellulose acetate (cellogel) for electrophoresis. The optimum incubation times and amount of enzyme may vary with the length of DNA fragment. Aliquots removed at intervals of 1, 2, 5, 10, 20, 40, 80 and 120 minutes give a good spread of fragment sizes.

(ii) *Partial spleen phosphodiesterase digestion*

A typical digestion mixture (10–20 μl) contains 3 mM potassium phosphate (pH 6.0), 0.3 mM EDTA, 0.02% Tween 80, 25–50 μg carrier RNA (eluted along with the DNA from the thin layer), the 3′-end labelled DNA and approximately 0.1 unit bovine spleen phosphodiesterase. The mixture is incubated at 37°C and aliquots (1–2 μl) pipetted out into 20 μl 0.1 M NH_4OH (4°C) at intervals of 1, 2, 5, 10, 20, 40, 80 and 120 min. The DNA is denatured at 100°C for 5 min and the mixture taken to dryness in vacuo. The sample is dissolved in 5–10 μl H_2O and 2–4 μl applied to a strip of cellulose acetate for electrophoresis.

5. *Fractionation of partial degradation products by two-dimensional electrophoresis-homochromatography*

   This procedure is exactly as described in Section 2.3.5. for 'Depurination Analysis'. Homochromatography is carried out with the same 2% Homomix used in the strand separation step. Chromatography is continued until the blue marker dye (Section 2.3.6.) is near the top of the plate (approx. 3 h for 20 × 20 cm plates and 10–12 h for 20 × 40 cm plates). The plates are finally thoroughly dried and autoradiographed.

2.5.3. Interpretation of results

The parameters influencing the mobilities of the oligonucleotides on the two-dimensional system are complex. In the first dimension the rate of movement of a molecule is dependent on its charge and also its mass. At pH's between 2 and 5, dTMP has a net charge of −1 whereas the other nucleotides are partially protonated and have a net charge between 0 and −1. Thus the relative mobilities of the different oligonucleotides are a function of the additive charge on the nucleotide components and on the mass of the oligonucleotide. In the second dimension mobility is determined largely by the mass (chain length) of the oligonucleotide. The superimposition of these two effects in the two-dimensional separation results in a map in which the particular nucleotide removed in each successive degradation product can be deduced from the type of mobility shift found. Fig. 2.12. shows the pattern of shifts of an oligonucleotide (n) when shortened by the loss of one residue by exonuclease digestion. In this example the electrophoresis in the first dimension was carried out at pH 2.8 since, as noted by Tu and Wu (1980), this gives an improved resolution between the −G and −A shifts. The inclusion of [^{14}C]pdT in the

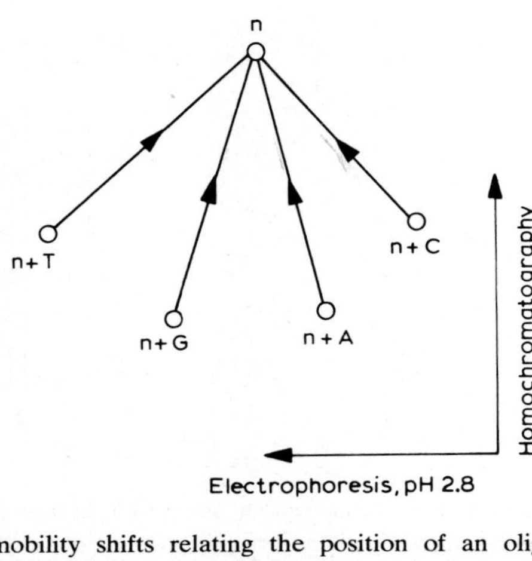

Fig. 2.12. The mobility shifts relating the position of an oligodeoxynucleotide 'n + 1' to the same oligodeoxynucleotide shortened by the terminal removal of a single nucleotide (after Tu and Wu, 1980).

partial digest gives a marker nucleotide to which the relative mobilities of the different fragments can be related. A detailed study of the factors influencing the resolution of the partial products and equations for calculating the mobility shifts and relating these to the observed values are given by Galibert et al. (1974) and more recently by Tu and Wu (1980). Often simple inspection of the pattern of resolved products is sufficient to enable the sequence to be deduced (Fig. 2.13.).

2.5.4. Discussion

This method has found wide application for limited sequencing objectives. Before the advent of the newer rapid techniques it was the main tool for the analysis of small DNA fragments produced by endonuclease digestion e.g. T_4-endonuclease IV, and a combination of these methods with depurination analysis enabled

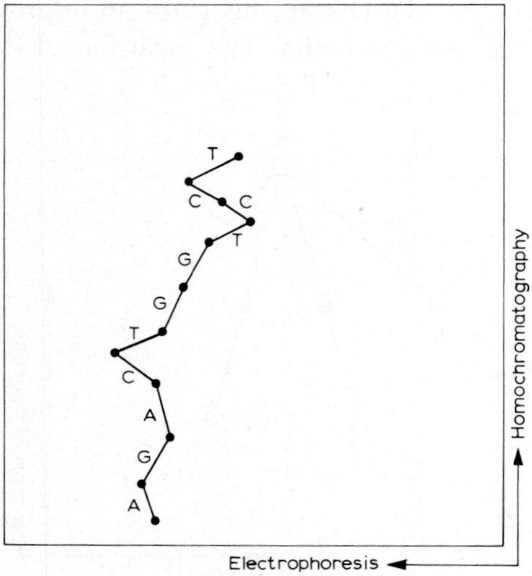

Fig. 2.13. Fractionation of a partial venom phosphodiesterase digest of a 5′-end labelled oligodeoxynucleotide by 2-D electrophoresis-homochromatography
Sequence deduced: $---_pT_pC_pC_pT_pG_pG_pT_pC_pA_pG_pA_p$ using the mobility shifts shown in Fig. 2.12.

Galibert et al. (1974) to deduce the sequence of a 48 nucleotide long fragment from ϕX174. While its use for routine analysis has now been superceded problems can still arise for which the newer procedures are not entirely satisfactory. Confirming the sequence of chemically synthesized oligonucleotides for use as primers and the analysis of short sequences adjacent to restriction endonuclease cleavage sites are conveniently carried out by these methods.

CHAPTER 3

Chain terminator sequencing

3.1. *DNA sequencing with chain terminating inhibitors (Sanger et al., 1977)*

3.1.1. Principle of the method

In the plus and minus method chain termination is brought about by the omission of one or more triphosphates. Atkinson et al. (1969) showed that the inhibitory action of 2′,3′-dideoxythymidine triphosphate (ddTTP) on DNA polymerase I depended on its being incorporated into the growing end of the oligonucleotide chain in place of deoxyribothymidine (dT). Because ddT contains no 3′-hydroxyl group, the chain cannot be extended further so that termination occurs specifically at positions where dT should be incorporated. If a primer-template is incubated with DNA polymerase in the presence of both ddTTP and dTTP together with the other three deoxynucleoside triphosphates (one of which is labelled with ^{32}P as in the plus and minus method) a mixture of partially elongated fragments will be obtained all of which possess the same 5′-end and terminate in ddT at the 3′-end. If the mixture is fractionated by electrophoresis on a denaturing gel a pattern of bands will be obtained corresponding to a series of fragments all terminating in ddT. By using the corresponding inhibitors for the other nucleotides, in separate incubations, and running the samples in parallel on the gel a pattern of bands is obtained from which the sequence can be read off, as in the plus and minus method. In order to obtain a suitable pattern of bands the ratio of the terminating triphosphate to normal triphosphate can be varied. A ratio of dNTP/ddNTP = 1:100 gives a good distribution of bands from which sequences up to 200 or more nucleotides can be read. Decreasing the concentration of ddNTP relative to the cognate dNTP yields progressively longer extension products which, together with longer gel runs at lower acrylamide concentrations, is capable of extending the readable sequence to 300 nucleotides.

An alternative to the ddNTP's (as chain terminators) are the arabinonucleoside triphosphates. Arabinose is a stereoisomer of ribose in which the 3′-hydroxyl group is *trans* with respect to the 2′-hydroxyl group. Such arabinosyl-(ara) nucleotides act as chain-terminating inhibitors of *E. coli* DNA polymerase I in a manner comparable to ddT. The structures of the chain-terminating inhibitors are shown in Figure 3.1. and the principle of the method in Figure 3.2.

3.1.2. Detailed protocol and sequencing with chain-terminating inhibitors

As discussed above the principle of the method is very similar to the plus and minus method. The starting material is a single-stranded DNA containing or comprising the sequence of interest and a suitable primer, or set of primers (which are usually restriction enzyme fragments) capable of hybridizing specifically to particular regions adjacent to or within the sequence to be determined. Sections 3.3. and 4.2.1. describe two methods in current use for the preparation of suitable single-stranded templates starting with naturally double-stranded DNA.

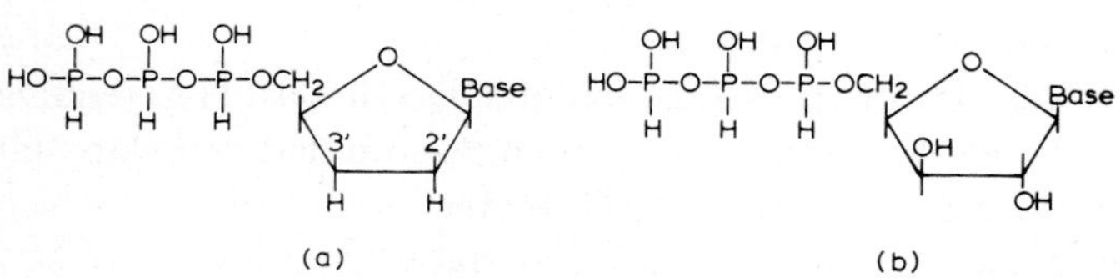

Fig. 3.1. Structures of chain-terminating inhibitors: (a) Dideoxynucleoside-5′-triphosphate; (b) Arabinonucleoside-5′-triphosphate.

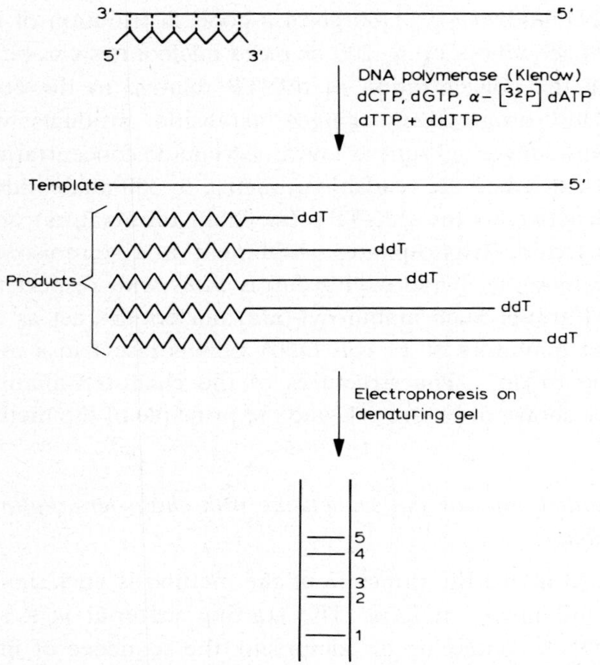

Fig. 3.2. The principle of the dideoxynucleotide chain-termination method for DNA sequencing.

The reagents and buffers used are listed at the end of the section.

Step 1: Hybridization reaction

The single-stranded DNA template and primer are denatured by boiling and then annealed.

- Mix: 2.5 μl restriction fragment dissolved in dist. H_2O (equivalent to about 1 pmol); 0.5 μl single-stranded template (about 0.2 pmol); 0.5 μl 10 × Hin buffer
- Seal in capillary tube. Incubate at 100°C, 3 min
- Anneal 67°C 15 min
- Dilute to 10 μl with 1 × Hin buffer.

Step 2: Synthesis of complementary DNA in presence of chain terminators

Four separate reactions are set up in the tips of drawn out capillary tubes. Each reaction mixture contains the four deoxynucleotide triphosphates one of which in each case is present in limiting amounts corresponding to the dideoxynucleotide added. In the protocol described α-[^{32}P]-dATP is used as the label but, with the appropriate changes in the compositions of the different dNTP mixes, any of the α-[^{32}P] labelled nucleotides could be used. The extension reaction is catalysed using DNA polymerase (Klenow subfragment)

– (i) Dry down 5 μl α-[^{32}P]-dATP (specific activity 300–400 Ci/m mole) under vacuum in a siliconized glass tube (Section 2.2.). Take up residue in 5 μl H_2O and keep on ice.
– (ii) Draw out four capillary tubes, label T, C, G, and A and set horizontally on plasticine support. Into the tip of each capillary add successively the following reaction components:

	T	*C*	*G*	*A*
α-[^{32}P] dATP (1.0 mCi/ml)	1 μl	1 μl	1 μl	1 μl
Annealed sample from Step 1	2 μl	2 μl	2 μl	2 μl
T° mix	1 μl			
C° mix		1 μl		
G° mix			1 μl	
A° mix				1 μl
2.0 mM ddT	1 μl			
0.7 mM ddC		1 μl		
0.7 mM ddG			1 μl	
1.5 mM ddA				1 μl
1 μl Klenow DNA polymerase (0.2 unit)	1 μl	1 μl	1 μl	1 μl

– Mix by gently expelling capillary contents into siliconized glass tubes and drawing up back into capillary tip.
– Incubate 15 min. at room temperature.
– To each sample add 1 μl 0.5 mM dATP, mix as previously. (dATP chase)
– Incubate 15 min., at room temperature. (If this chase step is

omitted some termination at A residues occurs due to the low concentrations of the α-[^{32}P] dATP.)

Step 3: *Analysis of complementary DNA fragments*

(i) If the primer used was a relatively short fragment (<100 nucleosides), the reaction products from Step 2 can be directly analysed on the denaturing gel.

- Expel the contents of each capillary tube into 15 μl of dye-formamide-EDTA mix in four Eppendorf tubes. Heat in water bath 90–100°C for 3 min.
- Load 3–5 μl onto thin 8% sequencing gel (below). Electrophorese at constant 30 mA (~1200 V) until the slow blue marker dye (xylene cyanol FF) reaches the bottom of the gel.
- Autoradiograph as previously.

(ii) If a primer of >100 nucleotides was used this needs to be split off the complementary DNA products before analysis. This is usually achieved by incubation with the same restriction endonuclease originally used to prepare the primer.

Method

Immediately after the dATP chase add 1 μl restriction endonuclease (1 unit) to each sample in the capillaries.

- Mix as previously.
- Incubate 5 min. at 37°C.
- Expel into 15 μl dye formamide EDTA mix etc. (as above).
- Electrophorese samples on 12% sequencing gel in 1 × TBE buffer.

Chemicals, mixes etc.

(*i*) 10 × *Hin buffer* 66 mM Tris-HCl, pH 7.4
66 mM $MgCl_2$
500 mM NaCl
10 mM dithiothreitol

(*ii*) *Dideoxynucleoside triphosphates* (*ddNTP*)

Obtained from P-L Biochemicals or Collaborative Research

Laboratories. Stock solutions are made up in water (4–10 mM) and kept frozen.

(*iii*) *Deoxynucleotide triphosphate* (*dNTP°*) *mixes*
Stocks of 20 mM dNTP's (in 5 mM Tris-HCl, pH 7.4, 0.1 mM EDTA) are stored frozen. The 0.5 mM working solutions are made freshly by dilution into water.

Composition of dNTP° mixes

	T°	C°	G°	A°
0.5 mM dTTP	1 μl	15 μl	15 μl	20 μl
0.5 mM dCTP	15 μl	1 μl	15 μl	20 μl
0.5 mM dGTP	15 μl	15 μl	1 μl	20 μl
10 × Hin buffer	15 μl	15 μl	15 μl	20 μl

Mixes are made up fresh every week.

(*iv*) *Dye-Formamide-EDTA mix*
Gently stir 100 ml formamide with 5 g Amberlite MB1 (mixed bed resin) for 30 min. Remove resin by filtration. Add 0.3 g xylene cyanol FF, 0.3 g bromo-phenol blue and Na_2EDTA to 10 mM. Store at 4°C.

(*v*) 10 × *TBE buffer*
108 g Tris base
9.3 g EDTA
55 g boric acid
make up to 1 litre with distilled H_2O (pH should be 8.3).

(*vi*) 8% *thin* (0.25–0.35 *mm*) *gels*
(*a*) 40% *acrylamide stock solution* (20/1 *acrylamide* : *bis ratio*)
Make up 190 g acrylamide and 10 g bis-acrylamide to 500 ml with distilled H_2O and stir gently with 5 g Amberlite MB1 for 30 min. Remove resin by filtration and keep stock at 4°C.

(*b*) 1.6% *ammonium persulphate*
Make up fresh in distilled H_2O.

The gel is conveniently made during the annealing reaction (Step 1).

(c) Assembly and running of sequencing gel

To 21 g Analar or 'Aristar' urea, add 10 ml deionized 40% acrylamide stock solution, 5.0 ml 10 × TBE buffer and 10 ml distilled H_2O. Stir gently to dissolve urea. Add 1.6 ml 1.6% ammonium persulphate. De-gas for a few minutes (vacuum line or good water pump) and make up volume to 50 ml with H_2O. Add 50 μl TEMED (N,N,N′,N′,-Tetramethylethylenediamine) to commence polymerization. Stir briefly and pour solution into gel cell, insert slot former and leave gel to set raised slightly from the horizontal.

The cell is set up using two tempered glass plates (20 × 40 cm) as described in Chapter 2.2.(iii). The inner surface of one of the plates is siliconized by swabbing with "Repelcote" (dimethyldichlorosilane in trichloroethane) and immediately before assembly both inner surfaces are swabbed with ethanol. The spacers are made from strips of plasticard, 0.25–0.35 mm thick and the plates taped together using waterproof adhesive tape. When set the tape is removed from the bottom of the gel, the slot former removed, and the gel plates clamped onto the electrophoresis apparatus. The reservoirs are filled with 1 × TBE buffer and the gel wells rinsed out with buffer to remove unpolymerized acrylamide. Samples (3–5 μl) are loaded into the wells with a drawn out capillary and electrophoresis commenced immediately at a constant 30 mA. The gels get quite hot, which both speeds up the running time and helps to keep the DNA denatured during the course of the run. Figure 3.3. shows the construction of the gel former (a) and a suitable electrophoresis apparatus (b) (Raven Scientific Ltd.).

(d) Autoradiography

When the run is completed, as judged by the position of the dye markers, the cell is dismantled and the siliconized glass plate removed leaving the gel adhering to the non-siliconized plate. A sheet of used X-ray film is laid on top of the gel, the sandwich inverted and the glass plate removed leaving the gel adhering to the

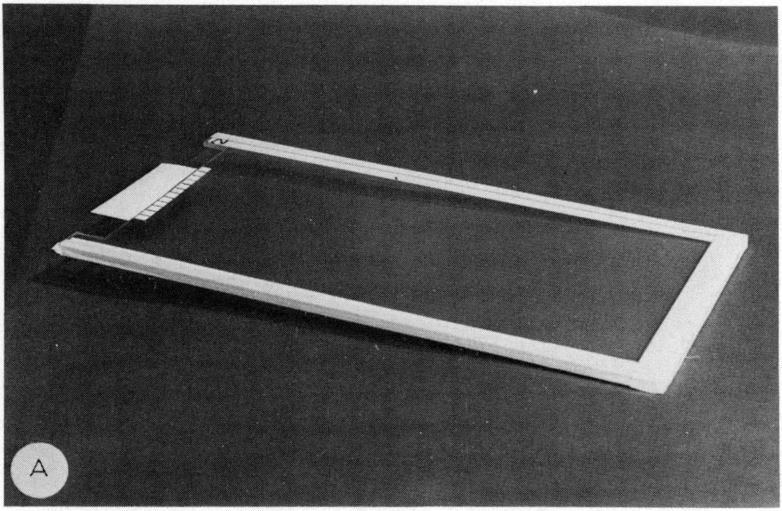

Fig. 3.3. (a) Former for pouring a polyacrylamide slab gel.

film. This is covered with a piece of cellophane (e.g. clingfilm or Saren-wrap). The gel is finally labelled using radioactive ink (^{35}S) and autoradiographed either in a cassette giving an even pressure over the surface of the gel or between weighted steel plates (Appendix 4). Autoradiography is carried out at −20°C to keep the gel frozen. Normally 1–3 days is sufficient. The exposure times may be reduced 5–7 fold by using pre-flashed film backed with an intensifying screen (cf. Appendix 4). This procedure is also useful if low activity samples are to be analysed (e.g. where the specific activity of the α-[^{32}P]-dATP has fallen to <100 Ci/m mol). The only disadvantages are that the use of intensifying screens cause some slight loss of resolution and autoradiographs have to be prepared at −70°C to avoid fading of the latent image. A suitable film for intensification is Fuji RX Medical X-ray film (Fuji Photo Film Co. Ltd., Tokyo 106). Suitable conditions for pre-flashing

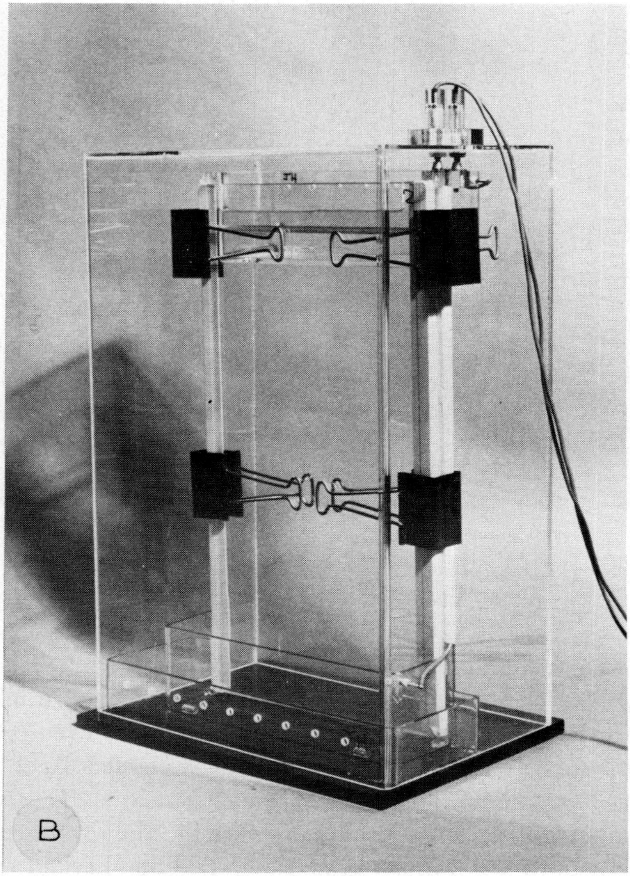

Fig. 3.3. (b) Raven gel electrophoresis apparatus.

need to be determined empirically. Procedures for autoradiography of gels are described in Appendix 4.

(*e*) *Electrophoresis apparatus*

Slab gel apparatus: Raven Scientific Ltd., Sturmer End, Haverhill,

Suffolk CB9 7UV, UK. Other equivalent commercially available equipment may be equally suitable or can be constructed (Appendix 5).

3.1.3. Discussion

In general, sequences of from 15 to about 200 nucleotides or more from the priming site can be determined with reasonable accuracy. Occasionally artifacts are observed but these are usually readily identified and are generally due to contaminants in the fragments. Problems occasionally arise due to "pile-ups" of bands, which are usually caused by the DNA-forming base-paired loops under the conditions of the acrylamide gel electrophoresis. These effects can often be minimized by stringent denaturing conditions and running the gels as hot as possible. A detailed discussion on recognising and avoiding artifacts and hints on reading sequencing gels is given in Chapter 4. Figure 3.4. shows an autoradiograph of a typical sequencing gel. The DNA sequence corresponding to the different bands is written alongside.

It is often found that the intensities of the bands in the different sequencing channels can vary markedly. One important reason for this is the ratio of dNTP/ddNTP in the reaction mixture. The different ddNTP's do not seem to compete with the corresponding dNTP's with equal efficiency for incorporation (and therefore termination) into the polynucleotide. With a given batch of d- and dd-triphosphates it is usually advisable to determine empirically the optimum ratios which give patterns of bands of similar intensity covering the readable portion of the gel. An excess of ddNTP corresponding to any particular lane will result in the pattern fading out in the upper part of the gel whereas excess dNTP will result in heavy blackening in this region of the gel. Another feature often observed is that the frequency of termination is not random. This is particularly apparent in the case of short runs of identical nucleotides where the relative intensities of bands within the run can vary markedly. However, this is not a serious defect of the method provided it is recognised. The spac-

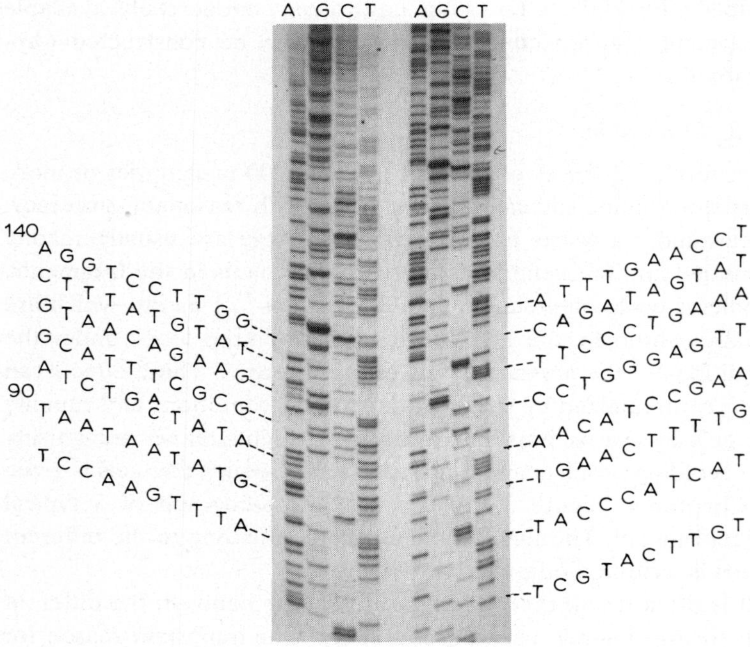

Fig. 3.4. Autoradiograph of a dideoxy-sequencing gel. A second loading (right-hand channels) was made 90 minutes after the initial loading (Section 4.4(iv)). The deduced sequence of 140 nucleotides is written alongside. A third loading (not shown) permitted a sequence of 230 nucleotides to be deduced. One anomalous band is found at position 95 where a faint band in the C channel is found in the same position as a stronger band in the A channel. In a separate priming experiment this C band was not found confirming residue 95 to be an A. Compare the ease of reading this sequence with that obtained using the 'plus and minus' method. The first 40 nucleotides (approximately) from the priming site have been run off the bottom of the gel.

ings between bands will usually draw attention to a faint band which might otherwise be missed (cf. Chapter 4).

In the protocol described an 8% acrylamide sequencing gel was used. This gives a good separation of fragments ranging from about 50 nucleotides to 200 or more. Since the main limitation of the method is the resolving power of the gel it is frequently advantageous to vary the gel concentrations to suit the size distribution of fragments within the samples. Where the primer DNA is cleaved off the sample before analysis a 12% gel will give an improved resolution of fragments less than 100 nucleotides in length. At the other end of the spectrum, a 6% electrophoresis gel frequently enhances the resolution of longer fragments (>250 nucleotides) and permits further reading into the sequence than is possible on an 8% gel. When this is the objective a second set of samples, prepared by halving the relative concentration of ddNTP's in the reaction mixture will increase the intensity of the slower moving bands on the gel.

During the annealing step it is necessary to use up to a five-fold molar excess of primer to template. The annealed mixture will therefore contain both the desired primer-template hybrid and reformed double-stranded primer molecules. These latter components can, under some circumstances, also be elongated and labelled and subsequently interfere with the interpretation of the gel. Primers corresponding to *Eco*RI and *Hin*dIII restriction fragments which possess 5′-sticky ends will be repaired and labelled and produce strong anomalous bands on the sequencing gel which obscure the first few bands of the desired sequence. Another difficulty which may be encountered is also shown in Figure 3.5. Priming of single strands, which can assume some fold back secondary structure, will yield a set of labelled fragments which subsequently confuse the interpretation of the gel and may give rise to a region of mixed sequence adjacent to the priming site. Whatever the reasons it is a consistent observation that the sequence of the first 15–20 nucleotides from the priming site cannot be determined using intact primers of the type described. Both these drawbacks, however, can be overcome by prior treatment of the primer with exonuclease III. This enzyme is a $3' \rightarrow 5'$ exonuclease and is specific for double-stranded DNA. Digestion of a linear duplex proceeds from the 3′-ends of both strands and the

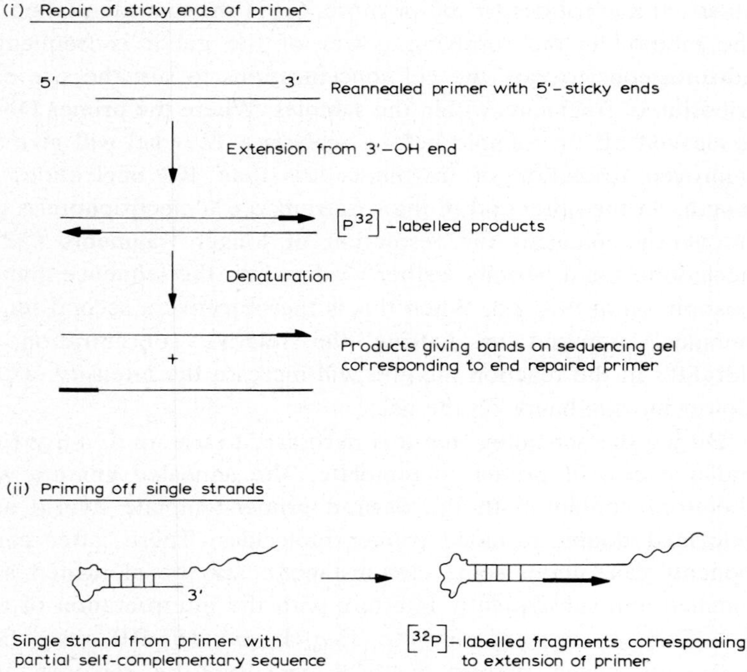

Fig. 3.5. Two mechanisms which can lead to the appearance of extraneous bands on a sequencing gel when using various restriction fragments as primers.

(i) The repair of recessed 3′-ends on the reannealed primer by the filling-in reaction catalysed by the Klenow polymerase gives a set of labelled products which appear as strong extraneous bands masking the first few bands on the sequencing gel.

(ii) Sometimes the primers have sufficient self-complementarity to assume a fold-back structure which can self-prime in the extension reaction. The labelled products may be longer than the primer and yield extraneous bands which obscure the true sequence.

products of a complete digest are essentially two single-stranded molecules, each approximately half the length of the original duplex, with probably only a small amount of complementary sequences remaining at their 3′-ends. An exonuclease III-treated primer therefore consists of two families of single-stranded species which, though not necessarily homogeneous in size do retain intact their original 5′-ends. Figure 3.6. shows the principle of this method. In practice the use of ExoIII-treated primers offers a number of advantages, the main one of which is the ability to sequence right through the restriction site and derive sequences flanking both sides of that site. As this development has found its principal application in sequencing fragments cloned in phage M13mp2 using a "universal" phage derived primer, discussion of this procedure is given in Chapter 4. 'Sequencing of single-stranded DNA fragments cloned in phage M13'.

Currently the "dideoxy method" probably is the method of choice for sequencing DNA by primed synthesis methods and is generally applicable to any DNA that can be obtained in single-stranded form. In the last two years two important advances have been made which have essentially solved the problem of preparing the DNA template in single-stranded form and these are described in the following Section and in Chapter 4. The first method, which

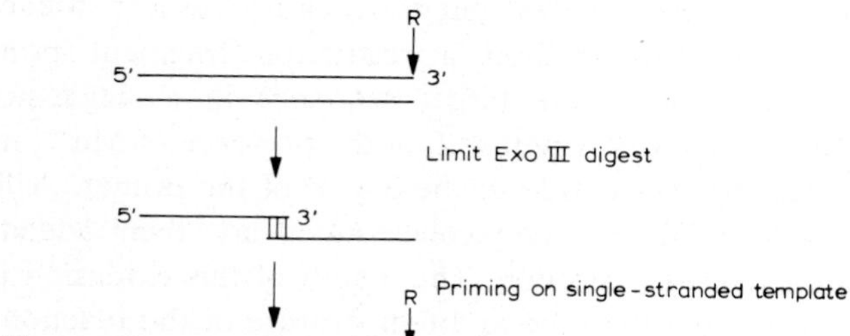

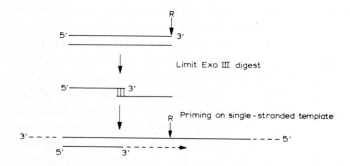

Fig. 3.6. Use of exonuclease III digested restriction fragment for priming of a single-stranded DNA template. Exonuclease III digestion proceeds from the 3′-ends of a double-stranded DNA yielding a heterogeneous population of essentially single-stranded products. Though the 3′-ends are variable the 5′-ends are unchanged so that the extension products from a primed synthesis all possess an invariant 5′-terminus. This procedure permits direct reading through the restriction site R, and additionally allows the deduction of part of the primer sequence.

does not involve any genetic manipulation, uses exonuclease III for preparing single-stranded DNA which is subsequently used as the template in the "dideoxy method" (Smith, 1978). In the second method, double-stranded fragments of the DNA to be sequenced are ligated into the replicative form of phage M13 (a single-stranded filamentous DNA phage) at a unique restriction site within the latter. Clones of infected cells containing the recombinant phage are isolated and subcultured under conditions where the phage is secreted into the culture medium. Finally, the phage particles are recovered from the clarified culture medium, DNA extracted, and used directly as the template for the dideoxy method. A double-stranded DNA fragment, corresponding to a region of the M13 genome adjacent to the cloning site, provides a universal primer for sequencing all the recombinants isolated (Chapter 4).

3.2. Sequencing by partial ribosubstitution

Another method, similar in principle to the above has been developed by Barnes (1978) but this has not, as yet, found wide application. In this method a restriction fragment primer is annealed adjacent to the target sequence in a single-stranded template and DNA polymerase I, in the presence of Mn^{++}, used to add a single ribonucleotide to the 3′-end of the primer. A limited number of ^{32}P-labelled deoxynucleotides are then added in a 'labelling extension' reaction. The length of this extension is controlled by limiting the time and temperature of the reaction or by limiting the concentration of the deoxynucleoside triphosphates added. As in the plus and minus method, the aim is to produce a random set of extension products of different lengths. After removal of the triphosphates by gel filtration, the DNA solution is apportioned into four partial ribosubstitution reactions. Each of these reactions is an unlabelled extension in the presence of Mn^{++}, all four deoxynucleoside triphosphates and one ribonucleoside triphosphate. The ratio of concentrations of the ribonucleotide to the corresponding deoxynucleotide is adjusted empirically to result in about 2% ribonucleotide substitution at each position for that base in the DNA sequence. Chemical or enzymatic cleavage at the substituted ribonucleotides then results in the production of a set

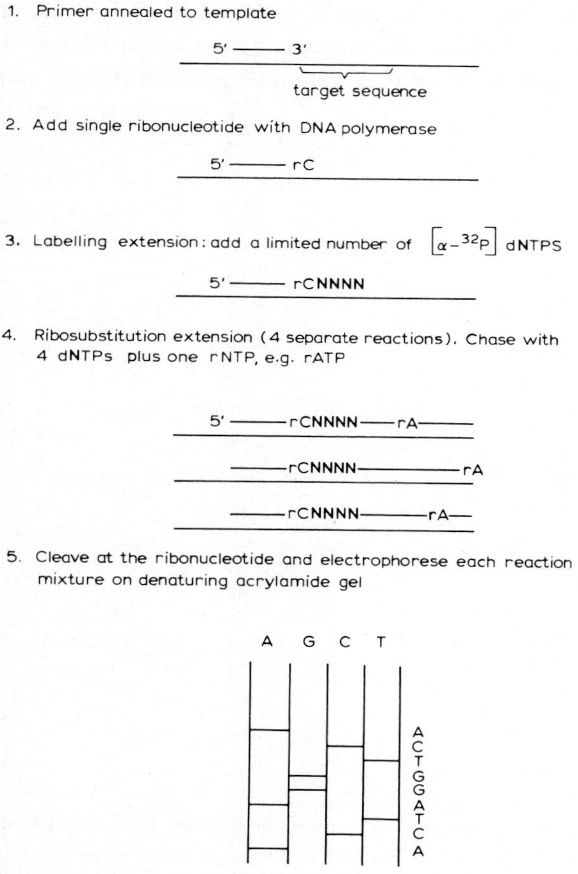

1. Primer annealed to template
5' —— 3'
target sequence
2. Add single ribonucleotide with DNA polymerase
5' —— rC
3. Labelling extension: add a limited number of [α-32P] dNTPS
5' —— rCNNNN
4. Ribosubstitution extension (4 separate reactions). Chase with 4 dNTPs plus one rNTP, e.g. rATP
5' ——rCNNNN——rA——
——rCNNNN————rA
——rCNNNN————rA—
5. Cleave at the ribonucleotide and electrophorese each reaction mixture on denaturing acrylamide gel
A G C T
A
C
T
G
G
A
T
C
A

Fig. 3.7. Principle of the partial ribosubstitution method of Barnes (1978). rC, rA, inserted ribocytidine and riboadenine residues; N, any deoxynucleotide.

of labelled molecules, all with the same 5′-end and terminating in the ribosubstituted nucleotide at the 3′-end. High-resolution electrophoresis on an acrylamide gel is used to analyse each of the four reaction mixtures and produce a ladder of fragments. The sequence can then be read directly from an autoradiograph of the gel.

The principle of this method is shown in Figure 3.7. and Figure 3.8. shows a diagram of a sequencing gel (from Barnes, 1978).

A full description of this method is given in Barnes' paper and it

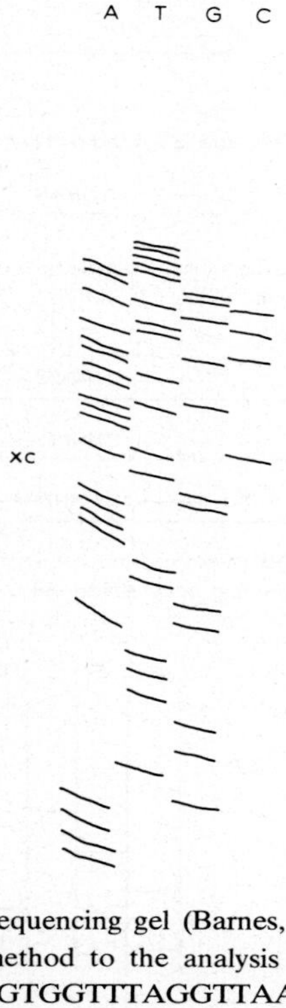

Fig. 3.8. Line diagram of a sequencing gel (Barnes, 1978) showing application of the partial ribosubstitution method to the analysis of a cloned DNA sequence. Sequence deduced: AAAAGTGGTTTAGGTTAAAAGGTATCAAATGAAT-AAGCATTCGATCGGAATTTTT . . . xc, position of the xylene-cyanol FF dye marker.

is unnecessary to discuss it further in any detail. One important feature of the method, which may influence its adoption by the reader, is that it may be adapted to identify and isolate DNA sequences which interact with genetic control proteins such as RNA polymerase or repressors. Since it is possible to cleave the ribosubstituted nucleotides in double-stranded DNA with pancreatic ribonuclease (for U and C residues), bound proteins would protect their recognition site from ribonuclease attack and so identify these particular regions in a DNA sequence. At the time of writing these possibilities do not seem to have been exploited.

3.3. *Direct enzymatic method for sequencing double-stranded restriction fragments using dideoxynucleoside triphosphates* (*Maat and Smith, 1978*)

3.3.1. *Introduction*

In the primed synthesis methods described above the initial requirement in all cases was the formation of a primer-template hybrid prepared by annealing a suitable primer with a single-stranded template. In a great many cases a DNA sequence problem resolves itself into the sequencing of sets of fragments produced by restriction endonucleases and much ingenuity has gone into devising methods by which such fragments can be sequenced directly without resort to the necessity of isolating single-stranded templates by cloning or other methods. The most widely used and best known method for this direct sequencing approach is the Maxam and Gilbert procedure and since its introduction in 1977, it has become the method of choice in many laboratories. The essence of the Maxam-Gilbert approach (see Ch. 5) depends on the base-specific chemical degradation of [^{32}P] end-labelled DNA fragments whereas the method introduced by Maat and Smith uses an enzymatic method to generate a ladder of fragments from 5′-end labelled DNA. Though not as well known as the chemical method it does have the virtue of greater speed and simplicity and deserves to be more widely known.

Principle of the method

The starting material is a 5′ terminally labelled restriction fragment. Such fragments are usually prepared by the action of polynucleotide kinase on a dephosphorylated DNA in the presence of high specific activity γ-[^{32}P]-rATP as the phosphorylating agent. This results in both 5′-ends of the double-stranded DNA being labelled. In order to prepare fragments labelled at only one end, the doubly labelled DNA is further cleaved with another restriction endonuclease and the singly labelled fragments separated by polyacrylamide gel electrophoresis. This procedure is analogous to that used for preparing singly labelled fragments in the Maxam-Gilbert method. However, it is important to note that the Maat and Smith method is only applicable to double-stranded DNAs; 5′-end labelled single-stranded DNA, such as may be obtained by the strand separation of labelled restriction fragments, cannot be sequenced by this method.

The next step involves the limited nicking of both strands of the molecule throughout the sequence by pancreatic DNAaseI. The 3′-hydroxyl groups exposed by each nick are then used to prime chain extension by DNA polymerase I in four separate reactions. Each reaction uses one of the four chain-terminating dideoxynucleoside triphosphates (ddNTPs) in the presence of the four deoxynucleoside triphosphates (dNTPs). Chain extension then proceeds from each nick until terminated, at random, by the incorporation of a dideoxynucleotide. Although the nicking and subsequent chain extension occurs on both strands, since only one is labelled a pattern of radioactive products is produced for only that strand, thus allowing its sequence to be deduced. As in the other sequencing methods the products of the reactions are fractionated on denaturing polyacrylamide gels, thereby resolving the oligonucleotides with the common labelled 5′-end. Though nicking of the fragments with DNAase I is not random (showing a preference for cleaving between purine and pyrimidine nucleotides, dPu-p$\downarrow$dPy) the fact that both deoxy- and dideoxy-NTPs are present in each reaction ensures that chain extension continues through several residues of the same base from the site of the nick. Every residue in the sequence can therefore be expected to give rise to a band.

The principle of the method is shown in Figure 3.9. It might be thought that the 5′→3′ exonuclease activity of the DNA polymerase would result in the removal of 5′-end label from the DNA fragment. However, this does not seem to be a serious problem and experience (Maat and Smith, 1978) has shown that there is no significant loss of labelled material due to hydrolysis of the labelled 5′-end.

This sequencing approach can be applied to DNA fragments of any length and the sequence that can be deduced is limited only by the resolving power of the gel electrophoresis (at present about 300–400 nucleotides long). The conditions to be described are suitable for application to fragments of about 100–150 nucleotides. For longer sequences the DNAaseI concentration may be increased to obtain most of the labelled DNA fragments within a size range below 300–400 nucleotides.

3.3.2. Detailed protocol for nick-translation sequencing method

The starting material is a double-stranded DNA (e.g. a restriction fragment) labelled at a single 5′-end with [^{32}P] of high specific activity (>1000 Ci/m mole). A suitable method for preparing such fragments is described in Chapter 5.

Sequencing procedure

Step 1. DNAase digestion and nick translation

- Mix: 2–5 pmoles [^{32}P] 5′-end labelled DNA in 10 μl H_2O
  1 μl pancreatic DNAase I (at a concentration of 0.1 μg/ml)
- Set up four capillary tubes labelled A, G, C and T (with drawn out tips) and introduce 2.5 μl of the above mixture into each

- Add 2.5 μl of the corresponding base specific mixture (see below)
  0.5 μl DNA polymerase I (2.5 units)
- Mix
- Incubate for 30 min. at room temperature.

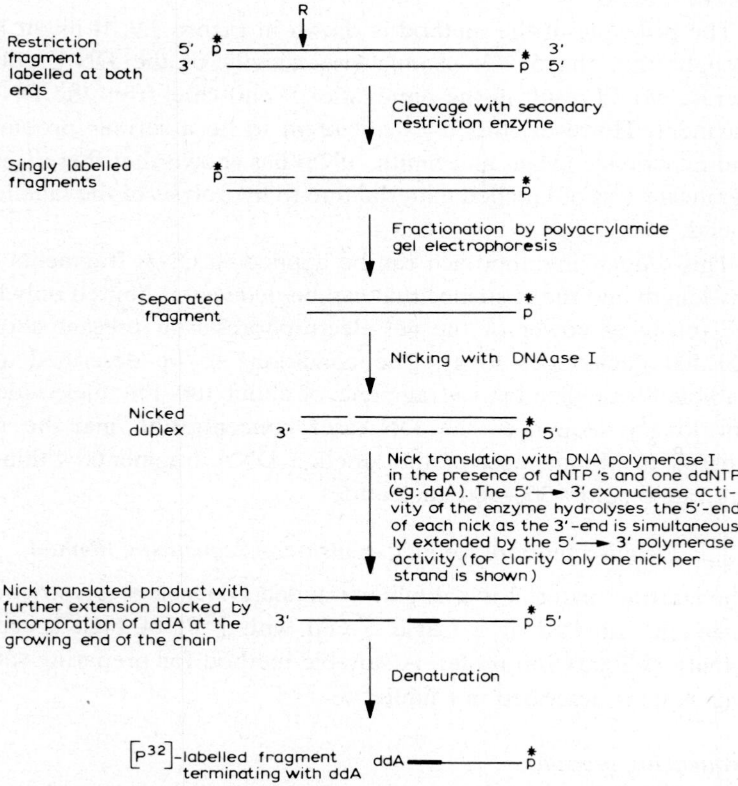

Fig. 3.9. Diagram showing the principle of the nick-translation sequencing procedure. For the purpose of illustration the single unique product generated from one nicked molecule is shown. With a heterogeneous set of nicked duplexes nick-translation will proceed from each gap until terminated by the incorporation of a dideoxynucleotide. Since the corresponding deoxynucleotide is also incorporated in competition with the dideoxynucleotide, the effect of non-random cleavage with DNAase I is minimized.

Step 2. Termination of reaction

– Blow the contents of each capillary into 15 μl formamide containing 10 mM EDTA, 0.05% XCFF and 0.05% BPB
– Mix, remove 5 μl and denature at 100°C for 3 min.

Step 3. Polyacrylamide gel electrophoresis

– Load the samples into 5 mm wide wells of an 8% polyacrylamide gel slab (400 × 200 × 0.35 mm) containing 7 M urea. (The procedure for setting up, running and autoradiographing of the gel is described in Section 3.1.2(iv).)
– Sufficient material is left to allow several loadings on the gel at different times after the electrophoresis has been started. In this way expansion of the band pattern in different regions of the sequence is obtained. On an 8% gel, up to 300 nucleotides from the 5′-end can be resolved.

Composition of the base-specific mixtures

A	G	C	T
ddATP 1 mM	ddGTP 1 mM	ddCTP 1 mM	ddTTP 1 mM
dATP 2 μM	40 μM	40 μM	40 μM
dGTP 40 μM	2 μM	40 μM	40 μM
dCTP 40 μM	40 μM	2 μM	40 μM
dTTP 40 μM	40 μM	40 μM	2 μM

Each mix also contains: 10 mM Tris HCl pH 7.6,
10 mM $MgCl_2$
10 mM dithiothreitol
100 mM NaCl

3.3.3. Discussion

Figure 3.10. shows an example of a sequencing gel prepared using the method described. Inspection of the band pattern shows the remarkable clarity and lack of artifacts produced and also the sharpness and definition of the bands. The method, however, is not

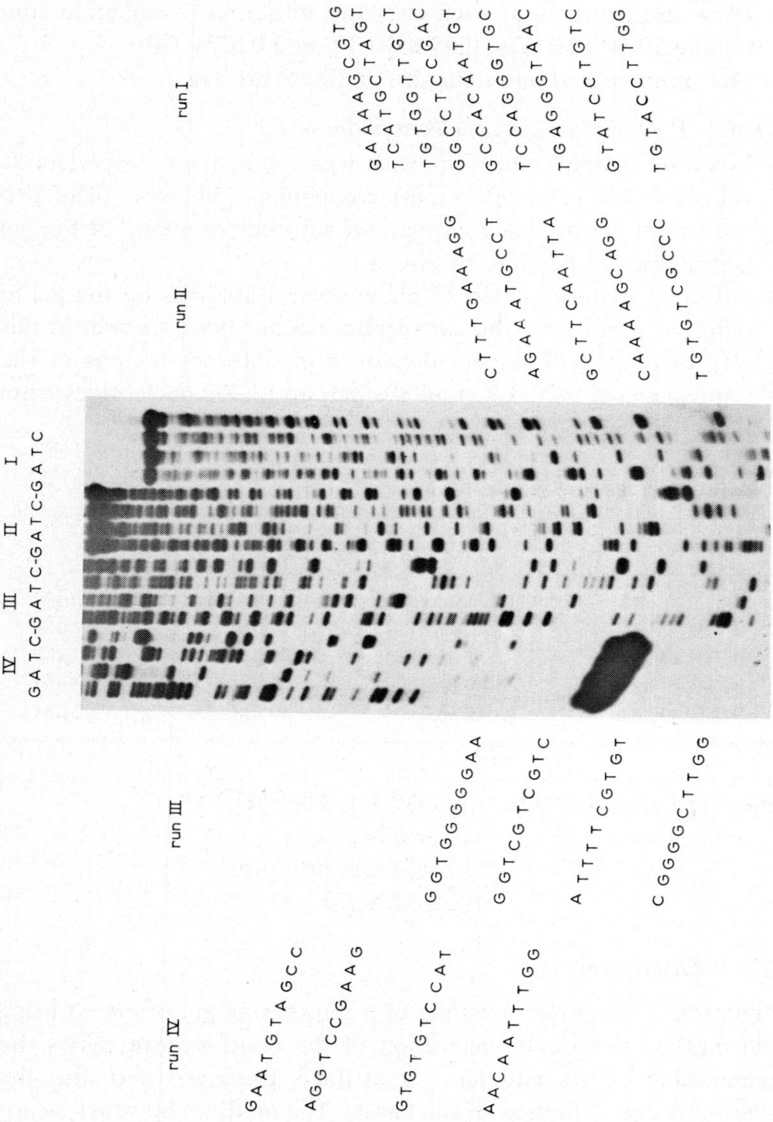
IV III II I
GATC-GATC-GATC-GATC
run I
GAAAAGCGTG
TCATGCTAGT
TGTGGTTGCT
GGCCTCCGAC
GCCACAATGT
TCCAGGGTGC
TGAGGGTAAC
GTATCCTGTC
TGTACCTTGG
run II
CTTTGAAAGG
AGAAATGCCT
GCTTCAATTA
CAAAAGCAGG
TGTGTCGCCC
run III
GGTGGGGAA
GGTCGTCGTC
ATTTTCGTGT
CGGGCTTGG
run IV
GAATGTAGCC
AGGTCCGAAG
GTGTGTCCAT
AACAATTTGG

without its peculiarities and some of these are evident in the autoradiograph.

(a) Every residue gives rise to a band apart from the first 10–20 residues from the 5′-labelled end. In some cases a sequence can be deduced directly after the position of the strong band which runs across all four tracks and appears to have a size corresponding to an oligonucleotide of chain length of 8–10 residues. In other cases not even weak bands are seen directly after the position of this strong artifact band and sometimes sequences cannot be deduced until some 10 nucleotides from this position. The reasons for these phenomena are not completely understood.

(b) Throughout the sequence band intensity is very variable. This particularly happens in a run of the same residue, the intensity decreasing in a 5′ to 3′-direction. This is thought to be due in part to the fact that nicking of the fragment does not occur randomly throughout the sequence and is especially infrequent in runs of the same residue. Chain extension through a run in most cases therefore takes place from the 3′-end of a nick lying adjacent 5′ to the run. Partial incorporation of the chain terminators at each residue in a run results in an exponential fall-off (in a 5′ to 3′-direction) of the numbers of oligo-nucleotides terminated at those residues.

(c) Although not seen in Fig. 3.10., DNA polymerase I sometimes has difficulties in copying certain sequences of the template strand. This leads to pile-ups of bands lying across all tracks of the

---

Fig. 3.10. Autoradiograph of a sequencing gel prepared using the Maat and Smith procedure. The sequence shown is that derived from a 440 nucleotide-long fragment from a *Hin*f1 digest of a 5′-end labelled *Hind*III fragment of adenovirus type 5 DNA. Samples from each base-specific reaction mixture were loaded every 2 hours (runs I, II, III, and IV). Electrophoresis was carried out at a constant current of 30 mA. Nucleotide sequence analysis of the complementary DNA strand revealed one mistake in the sequence as written. At position 2870 (in run III) two C's should be read instead of one. The zone of compression responsible for this error is not very apparent and emphasizes the importance of sequencing both DNA strands. (Courtesy of Dr. J. Maat).

gel. There seems to be no consistent feature of the sequences in these regions.

One feature of this sequencing method which is not completely understood is the precise role played by the 5′→3′ exonuclease activity of the DNA polymerase I. As mentioned earlier, the expected partial loss of the labelled 5′-end does not seem to occur to any significant extent and even more puzzling is the observation that the 'Klenow' sub-fragment derived from DNA polymerase I, which lacks the 5′→3′ exonuclease activity, appears to function equally well for the nick translation. Presumably the enzyme is somehow capable of displacing the 5′-end of the nick as polymerization proceeds from the 3′-end. Experiments have also shown that, with some batches of Klenow polymerase, the DNAase treatment can also be omitted since the polymerase contains sufficient contaminating nucleases to produce random nicks. These observations (Smith, A.J., personal communication) have resulted in a further simplification of the method which, though perhaps not as generally applicable as the procedure described above, have given excellent results in the hands of the author (Hindley and Phear, 1979). This procedure is briefly described in the following section.

3.3.4. A simplified dideoxynucleotide-nick translation sequencing procedure (*Figure 3.11.*)

- Mix: 1–2 p.mol of 5′-end labelled restriction fragment (i.e. labelled at one 5′-terminus only) dissolved in 20 μl H_2O, with 10 μl of a buffer containing 30 mM Tris-HCl, pH 7.6, 30 mM $MgCl_2$ and 10 mM dithiothreitol
  Distribute sample equally into four 0.4 ml plastic Eppendorf centrifuge tubes labelled A, G, C and T
- Add: 2 μl of a 2 mM solution of the appropriate dideoxynucleoside triphosphate to each tube and 3 μl undiluted DNA polymerase (nach Klenow, Boehringer)
- Mix, incubate 90 min. at 40°C
- Chill, add 40 μl 95% ethanol, mix, freeze at −70°C
- Centrifuge (5 min. Eppendorf microfuge)

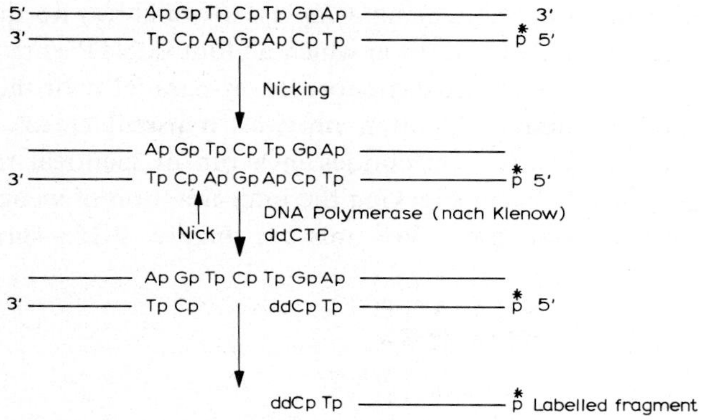

Fig. 3.11. Principle of the simplified dideoxynucleotide-nick translation sequencing procedure. In this method the 5′-end labelled, nicked DNA is incubated with DNA polymerase (Klenow sub-fragment) in the presence of a single dideoxynucleoside triphosphate (e.g. ddC). No deoxynucleotides are added to the reaction mixture and at the temperature of the incubation (37°–42°C) the 3′→5′ exonuclease activity of the enzyme widens the gap by the sequential removal of nucleotides from the 3′-end exposed by the nick. This process continues until a dideoxy-C residue is incorporated corresponding to the G on the complementary strand. This residue is resistant to further exonuclease attack and yields a defined labelled product. Random nicking therefore yields a family of labelled products all terminating in ddC. Repeating the reaction with the other dideoxynucleotides yields similar sets of fragments. Fractionation on a sequencing gel yields ladders of fragments from which the sequence can be deduced. The 'nicking' activity of the DNA polymerase has to be very low relative to the 3′→5′ exonuclease and polymerizing activity to ensure rapid processing of the nicks as they occur. Not all commercial sources of DNA polymerase (Klenow sub-fragment) are suitable for this reaction. In general, the preparation supplied by Boehringer is satisfactory but different batches appear to vary in their endogenous DNAase activity.

- Discard supernatant, dry, dissolve in 10 μl formamide-EDTA-dye mixture (Section 3.1.2.)
- Heat, denature (95–100°C, 3 min.)
- Run 3–5 μl aliquots on sequencing gel (as previously, Section 3.1.2.).

It is usually advantageous to process two additional samples, along with the four sequencing samples, in which (a) no dideoxy-nucleotides are added and (b) in which all four ddNTP's are added. These two samples are electrophoresed in parallel with the other four samples. Sample (a) often provides a useful check on the number of consecutive nucleotides in a run of identical residues and (b), usually helps in checking the interpretation of weak bands and ensuring that none are missed. Figure 3.12. shows an

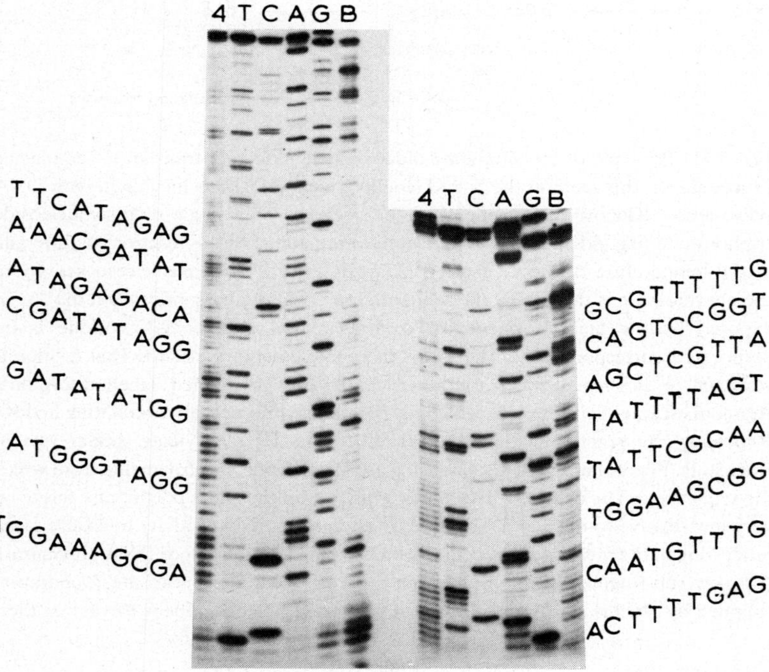

Fig. 3.12. Autoradiograph of sequencing gels prepared using the simplified dideoxy-DNA polymerase method for sequencing 5′-end labelled DNA fragments. The sequences shown were derived from two 5′-end labelled *Taq*I fragments. The first 22 nucleotides were run off the gel. Channel B shows the pattern generated by incubation with Klenow polymerase alone and Channel 4 is the pattern obtained in the presence of all four dideoxynucleoside triphosphates. (From Hindley and Phear, 1979).

autoradiograph of a sequencing gel prepared exactly as described with the interpretation written alongside. One advantage of this procedure is the ethanol precipitation step which effectively frees the sample from salts, excess ddNTP's and smaller oligonucleotides. The samples can consequently be analysed on the thin gel system of Sanger and Coulson (1978) with the attendant advantages of high resolution and sharpness of the resolved bands. Inspection of Fig. 3.12. also shows one disadvantage of the system and that is the wide variation in the intensities of the different bands. This is particularly apparent in the two control channels (4 and B) and presumably results from the failure of the endogenous endonuclease activity in the Klenow polymerase to produce random nicks in the DNA. Another problem encountered is the variation in different batches of Klenow polymerase with respect to their content of non-specific nuclease activity. Some batches seem to catalyse quite efficiently nick translation while others are almost inactive in this respect. Where this is a problem, prior treatment with DNAase I, as described above, should yield a satisfactory nicked duplex (Section 3.3.2.).

As mentioned at the outset, this simplified procedure may not be as consistently reliable as the slightly longer Maat and Smith method. It is capable of giving excellent results, however, and if a nick translation method is chosen as the sequencing tool it may be worth attempting a one-off experiment using the simplified procedure described.

3.3.5. The forward-backward (F-B) procedure (Seif, Khoury and Dhar, 1980)

Introduction

This recently described procedure is essentially a marrying of the Maat and Smith (1978) nick translation method and the 'plus and minus' method of Sanger and Coulson (1975). The reason for this development was to circumvent one recurrent problem of the nick translation methods, namely the uneven distribution of band intensities obtained particularly when a row of repeated nucleotides is encountered (e.g. 5′---A-A-A-A---3′). The intensity of the bands often decreases drastically in the 5′ to 3′-direction, a feature which is accentuated in reactions where the dNTP homologous to the ddNTP is omitted (e.g. as in the 'simplified' version of the nick translation system given above).

In the Forward-Backward procedure the double-stranded DNA fragment, [^{32}P]-labelled at one 5′-end, is subjected to two different reaction schemes. In the Backward reaction, which is analogous to

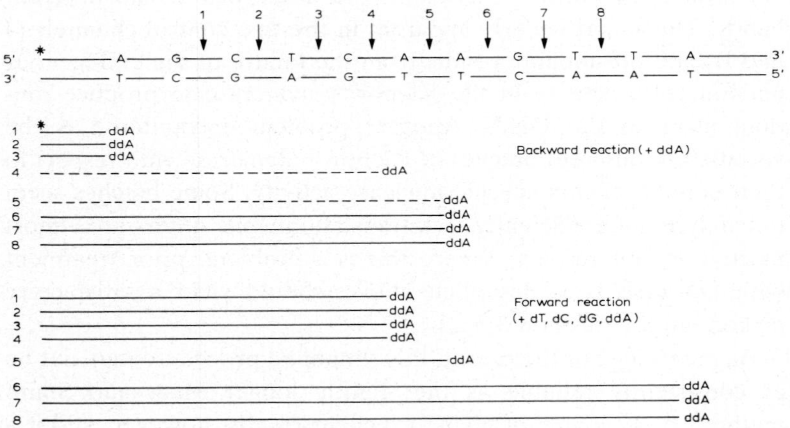

Fig. 3.13. Diagrammatic representation of the Forward-Backward procedure. A double-stranded DNA fragment [^{32}P] labelled (asterisk) at one 5′-end is represented at the top of the figure. DNA polymerase I and a nucleotide chain inhibitor (e.g. ddA) are added. Contaminating DNAases in the PolI preparation produce nicks, indicated by the vertical arrows. From the 3′-end created by each nick, the reaction catalysed by PolI proceeds in the 5′- to 3′-direction (Forward reaction) provided dNTPs (dG, dT, dC) are present; if they are not added the reaction proceeds exonucleolytically in the 3′- to 5′-direction (Backwards). The numbered lines represent the DNA fragments which arise from the similarly numbered DNA nicks. The hypothetical DNA sequence illustrates the complementary results obtained from the Forward and Backward reactions with repeated nucleotides: e.g. the sequence AA. In the Forward reaction the proximal A will be represented by a strong band and the distal A by a weak band. The converse is true for the Backward reaction. The dotted lines 4 and 5 signify those reactions which proceed $5' \rightarrow 3'$ (Forward) in the Backwards procedure.

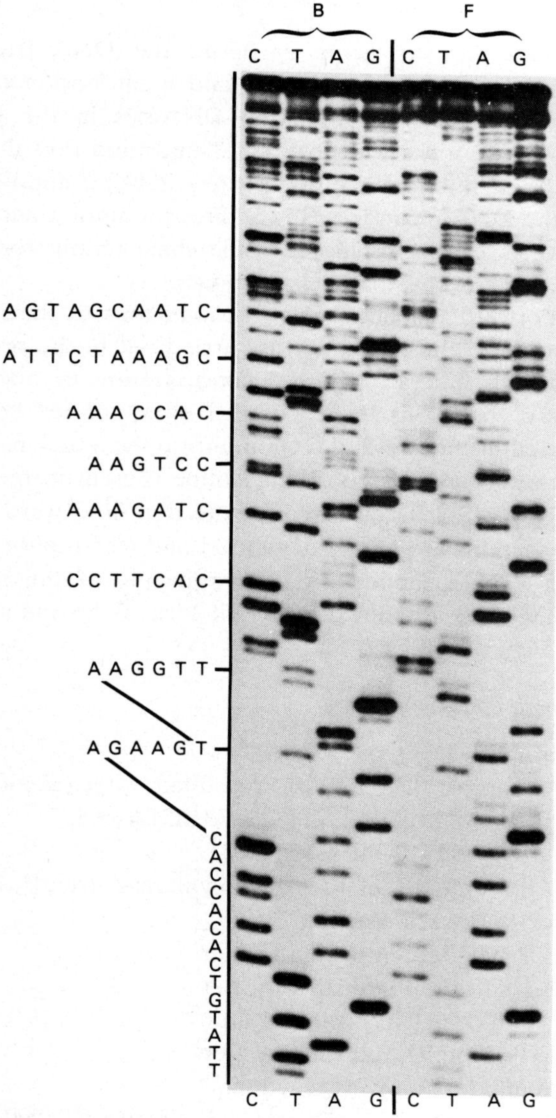

Fig. 3.14. Autoradiograph of a sequencing gel prepared using the procedure of Seif, Khoury and Dhar, (1980). The channels under B and F are the products from the Backwards and Forwards procedures, respectively. (Courtesy of Dr. Isabelle Seif).

the simplified protocol given in 3.3.4., the DNA fragment is incubated with DNA polymerase I and a nucleoside chain terminator (e.g. ddA). Contaminating DNAases in the PolI preparation produce nicks with exposed 3′-ends and PolI then either catalyses the addition of one ddNTP (e.g. ddA) if appropriate, or degrades the DNA chain in a $3' \rightarrow 5'$ direction until a nucleotide is released which can be replaced by the chain terminator (ddA or ddG, ddT, ddC, in four parallel reactions).

In the Forward reaction, which is analogous to the Maat and Smith procedure, the incubation is carried out in the presence of all four dNTPs and an adjusted concentration of one ddNTP. Under these conditions the exposed 3′-ends, formed by nicking, are elongated in the $5' \rightarrow 3'$ direction until polymerization is halted by the incorporation of the dd-nucleotide present in the reaction mixture. The samples generated by the four Backward and four Forward reactions are finally denatured and electrophoresed on a denaturing polyacrylamide gel. The principle of the method is shown in Fig. 3.13. (modified from Seif et al. 1980) and a sequencing gel in Figure 3.14.

3.3.6. Experimental procedure

(a) *Reagents*

[^{32}P]-DNA; approx. 0.2 p.mol/μl in distilled water (Section 3.3.5).

5 × buffer; 33 mM Tris-HCl, pH 7.5, 33 mM $MgCl_2$,
10 mM DTT, 10 mM NaCl.

DNA Pol. I; 5 units/μl, or Klenow polymerase, 1 unit/μl.

ddNTPs for Backward reaction:
- for BA: 1 mM ddA in distilled water
- BG: 1 mM ddG in distilled water
- BT: 1 mM ddT in distilled water
- BC: 1 mM ddC in distilled water

Nucleotide mix for Forward reaction:
- e.g. for FA: 1 mM ddA, 200 μM dG, 200 μM dT, 200 μM dC
- FG: 1 mM ddG, 200 μM dA, 200 μM dT, 200 μM dC
- etc.

(b) *Method*

Each of the four Forward (FG, FA, FT, FC) and Backward (BG, BA, BT, BC) reactions are carried out in a volume of 5 μl containing 2 μl of *DNA* + 1 μl of 5 × *buffer* + 1 μl of *enzyme* + 1 μl of *nucleotide*(s). These volumes are taken from the stock solutions listed. Each reaction mixture (5 μl) is mixed on plastic, drawn into a capillary (unsealed) and incubated at 37°C for 10 min.

To terminate the reaction the capillary contents are ejected into 5 μl formamide dye mix (99% deionized formamide containing 10 mM EDTA, 0.2 mM NaOH, 0.1% xylene cyanol FF dye). After denaturation at 100°C for 30 s the samples are applied directly into gel wells and electrophoresis carried out by the usual method (Section 3.1.2.). Seif et al. (1980) give a recommended order of loading:

FG, FA, FT, FC, BG, BA, BT, BC.

To read sequences of 20 to 200 nucleotides, 12% and 8% acrylamide gels (containing 3.3% bisacrylamide : acrylamide monomer ratio) are used. For sequences of 200–300 nucleotides, a 7% gel (containing 2 to 2.5% bisacrylamide is recommended. Gels (40 × 20 × 0.04 cm) are run at 20 mA (1000–2000 volts) for varying lengths of time (2–7 h) and autoradiographed.

3.3.7. Discussion

The main features of this procedure, as described by Seif et al. (1980) are summarized below:

(i) In the Backward reaction, the distal (3′) nucleotide in a row is the one preferentially replaced by the ddNTP and therefore gives rise to a pattern opposite in band intensity to that of the Forward reaction.

(ii) Authentic bands which are weak in one reaction are usually strong in the complementary reaction and can thus be confirmed.

(iii) Artifact bands do not usually cause problems as they are rarely present in the same position as the Backward and Forward reactions.

(iv) The Backward system usually more clearly indicates the number of nucleotides in a run than does the Forward system.

All these points are discussed in detail in the paper by Seif et al. as are some additional refinements of the procedure which can be used to clarify rare ambiguities. This procedure was successfully used to determine sequence alterations in four early SV-40 temperature sensitive point mutants and five early SV-40 deletion mutants.

These procedures share with the Maxam-Gilbert chemical procedures the requirement for a 5′-end labelled DNA but are clearly only applicable to a duplex molecule. As the procedure is relatively simple and rapid it may well be worth using as an adjunct to the Maxam-Gilbert method (Chapter 5).

3.4. The use of exonuclease III for preparing single-stranded DNA for use as a template in the chain terminator sequencing method (Smith, 1979, 1980)

3.4.1. Introduction

The use of Exonuclease III for preparing single-stranded primers in connection with the chain-termination method has been considered earlier (see 3.1.3.). Exonuclease III can also be used to prepare single-stranded templates from a linear duplex DNA and Smith (1979) pioneered this approach and developed it into a fairly general method for sequencing restriction fragments.

Exonuclease III is a $3' \rightarrow 5'$ exonuclease which is specific for double-stranded DNA (Richardson et al., 1964). Digestion of a linear duplex proceeds from the 3′-ends of both strands and the products of a complete digest are two single-stranded molecules, each about half the length of the original duplex with probably only a small amount of complementarity remaining between them at their 3′-ends. Provided the exonuclease is free from contaminant activities, these single strands should be resistant to further degradation. Smith (1979) showed that these single strands are suitable as templates for sequencing with the chain-terminator method.

The primers used are restriction fragments which originated from within the duplex DNA to be sequenced. In the annealing reaction, to produce the primer-template complex, only one of the primer strands is hybridized to its complementary sequence in the template since the other potential template is destroyed by the Exo III treatment. The result is the formation of a primer-template which can be extended by DNA polymerase in a direction opposite to that of the Exonuclease III attack. This is shown in Fig. 3.15. Two inherent features of the method which can cause problems are also made clear in Fig. 3.15. In the first place priming and chain extension can also occur on the 3′-ends of both template strands when sufficient complementarity remains between the 3′-ends of the template strands to form a base-paired structure. Secondly, a primer which originates from near the centre of the duplex DNA

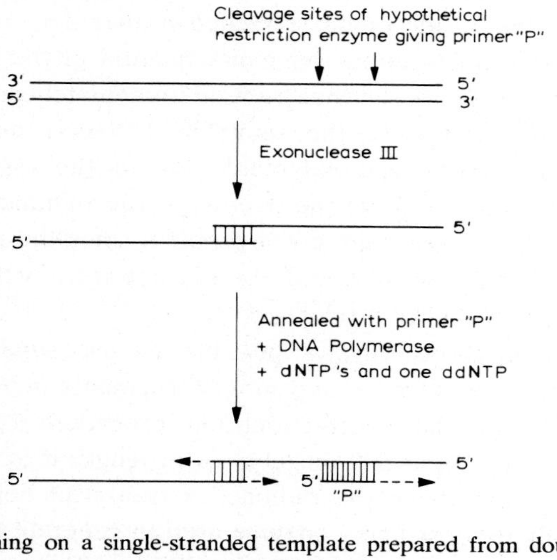

Fig. 3.15. Priming on a single-stranded template prepared from double-stranded DNA by the action of exonuclease III.

could hybridize to both single strands which would then result in chain extension on both templates giving a mixture of two sequences and consequently an unreadable sequencing gel. Little can be done about this latter problem except to select restriction fragments which do not originate from within this region. Priming from the complementary ends of the template, however, is not usually a problem since the products are very much larger than those obtained from the primer and fractionate near the top of the sequencing gel where they do not interfere with the band pattern generated by the primer.

The strategy used for preparing the template DNA depends on whether small or large restriction fragments are to be used as primers for the chain-termination reaction. If small primers are used (generally less than 200 nucleotides in length), the template DNA is prepared by carrying out an Exonuclease III digest on the linear duplex DNA sufficient to expose as single-stranded DNA the sequence to which the primer hybridizes. Since elongation of the primer in the subsequent reaction occurs in the opposite direction to the exonuclease attack, the undegraded region of the template to the 5′-side of the annealed primer does not interfere with the extension reaction. After inactivation of the exonuclease III the degraded DNA can thus be used immediately as a template. Since, in this approach, the template DNA is only partially degraded, a 'repair' type polymerization of the type described above will take place from the 3′-ends of the template DNA. As pointed out this is normally not a problem since the products are of high molecular weight and do not interfere with the fractionation of the smaller products.

However, it is not always possible to use small restriction fragments for this purpose and longer fragments may need to be employed. Since the gel-fractionation procedure only resolves fragments up to about 300 nucleotides in length it is necessary to cleave the primer from the radioactive transcript before analysis and normally the restriction enzyme used to generate the primer is used for this. However, this enzyme will also cleave at other sites which may be present in the double-stranded region of the template, which have become labelled in the self-priming reaction and yield labelled products which interfere with the labelled oligonucleotides resulting from chain extension of the primer. Two methods can be used to get around this problem. The first is to extensively degrade the duplex with exonuclease III so that the amount of self-complementarity remaining at the 3′-ends of the single strands is reduced to only two or three base pairs (Exonuclease III seems unable to completely degrade a duplex DNA into two approximately half-length single strands). This very limited amount of remaining complementarity is not sufficient to allow stable base-pairing at the temperature of the extension reaction and no primed synthesis from these ends is obtained. The other approach is to block the 3′-ends of a DNA which has been only partially degraded with exonuclease III by the addition of a dideoxynucleotide. These ends are inert to chain extension and elongation is subsequently obtained only from the annealed primer (Fig. 3.15.). The removal of these dideoxynucleotide terminators by the hydrolytic $3' \rightarrow 5'$ exonuclease and pyrophosphorolytic activity of the DNA polymerase is fortunately an extremely inefficient process and to all intents and purposes the blocked ends are not elongated.

The procedures for carrying out the chain-termination reaction and fractionation of the products by gel electrophoresis are the standard methods described earlier (Section 3.1.). Smith (1980) makes two recommendations which can be used as an adjunct to these procedures and are designed to circumvent two general problems which can occur in the standard chain-termination procedure. As will be discussed in Chapter 4, there is a tendency for certain bands which occur in the ddC-termination reaction to appear very faint, this being invariably the first C band in a run of C residues. Chain termination with the arabinose-CTP analogue does not give rise to this effect and it may be advantageous to set up a fifth chain-extension reaction using ara-CTP as the chain terminator and subsequently analysing all five samples in parallel on the gel. The second recommendation is to use reverse transcriptase in place of DNA polymerase (Klenow subfragment) where problems occur either due to the 'pile-up' of polymerase molecules at particular regions of the template which the enzyme finds difficult to copy or where one of the non-priming strands of the restriction fragment is capable of forming some internal secondary structure that leaves a non-base-paired 5′-extension. The 3′-end of such a self-priming template is elongated by the Klenow polymerase yielding anomalous, and usually intense, bands on the sequencing gel which obscure the desired sequence. The use of reverse transcriptase (Appendix 1) often overcomes these problems. In the first place where 'pile-ups' do occur these are at different points to those found in the Klenow polymerase-catalysed reaction. Secondly, reverse transcriptase, requires a perfectly base-paired 3′-terminus for elongation to occur, a structure which is unlikely to exist in a partially base-paired single strand.

Using this approach Smith (1979) has successfully sequenced several regions of the phage G4 replicative form DNA as well as other double-stranded cloned sequences. Fig. 3.16. shows one strategy taken from Smith (1979), in which these methods are applied to the sequencing of a cloned *Hind*III fragment, in pMB9, of the repeat unit of *Xenopus laevis* 5s genes. The method is a valuable addition to our armoury of techniques and is described in detail in the following section.

3.4.2. Experimental procedures

1. Exonuclease III digestion

A linear duplex DNA at a concentration of 5 p.mol in 10 μl of exonuclease III buffer (70 mM Tris-HCl, pH 8.0, 1.0 mM $MgCl_2$, 10 mM dithiothreitol) is degraded at an approximate rate of 500 base pairs/hour/duplex end at 20°C using an enzyme concentration of 0.5 units/μl of exonuclease III (Smith, 1980). One unit of enzyme activity is as described by Richardson et al. (1964). Usually the DNA to be sequenced will be a linear duplex of approximately known length derived from a restriction digest. Digestion may be

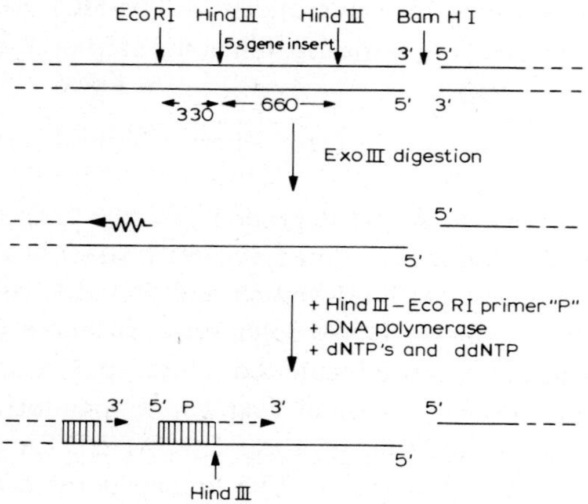

Fig. 3.16. Diagram illustrating the strategy used for sequencing a DNA fragment inserted in the *Hind*III site of plasmid pMB9. Cleavage at the *Bam*HI site exposes two 3′-ends which are degraded with Exonuclease III. The 330 nucleotide *Hind*III-*Eco*RI restriction fragment prepared from pMB9 is the primer, one strand of which hybridizes to the position shown. After chain extension the primer is removed by cutting with *Hind*III. In the extension reaction the relative ratio of ddNTPs/dNTPs were such that all chains terminated within the 5s gene insert, i.e. before the second *Hind*III site is copied. Cleavage of the product with *Hind*III therefore only generates one set of labelled oligonucleotides. Chain extension from the 3′-exo III generated end gives a much larger product which does not fractionate on the gel.

carried out long enough to produce a limit digest, which reduces the product to essentially two non-complementary single strands or, alternatively, just long enough to expose the region of interest in single-stranded form. The exonuclease III reaction may be terminated by heating at 70°C for 15 min. and precipitating the DNA with 3 vol. ethanol at −70°C in the presence of 0.1 M Na acetate, pH 5.5. Alternatively the heating step can be substituted by vortexing with water-saturated phenol, extracting the aqueous phase with ether and precipitating with ethanol as above. The pellet is collected by centrifugation at 10,000 *g* for 10 min, washed once with 95% ethanol, recentrifuged and the pellet finally dried.

It is then redissolved in water or 5 mM Tris-HCl, pH 7.6 to the appropriate concentration for the annealing step or 3′-end blocking reaction.

3′-end blocking (necessary except where a limit Exo III digest is used)

1–5 pmol of exonuclease III degraded DNA is suspended in 10–20 μl of a buffer containing 7 mM Tris-HCl, pH 7.6, 7 mM $MgCl_2$, 50 mM NaCl, 1 mM DTT (H-buffer) and 500 μM with respect to each ddNTP. 2 Units of DNA polymerase (Klenow subfragment) are added and the mixture incubated 1 hour at 37°C. The reaction is conveniently done in a small snap-cap Eppendorf tube and is terminated by the addition of 1 μl of 0.1 M EDTA (neutralized) and heating at 70°C for 15 min. The unincorporated ddNTP's are then removed by chromatography on a small Sephadex G100 column (in a 1 ml plastic pipette) equilibrated in a buffer of 5 mM Tris-HCl, pH 7.6, 0.1 mM-EDTA. The DNA is eluted with the break-through volume of the column and is precipitated with ethanol as previously (Section 2.2.).

An alternative to 3′-end blocking with all four ddNTP's is to employ a single ddNTP in the presence of the other three dNTP's, one of which may be [α-^{32}P]-labelled. This procedure gives a labelled product, at low specific activity, which is useful in following the progress of the subsequent chromatography step. e.g. 5 p.mol of exonuclease III degraded DNA are incubated in 20 μl of H buffer adjusted to 25 μM in dATP, dGTP, dCTP, 500 μM in ddTTP and containing 10 μCi α-[^{32}P] dATP (350 Ci/mmole) in the presence of 2 units Klenow polymerase for 1 h at 20°C. The reaction is terminated and chromatographed on G100 Sephadex as above. The elution of the weakly labelled DNA can be followed by a hand-held monitor (Section 2.2.ii) and collected in the smallest possible volume (100–200 μl). The solution is dried down and redissolved in 10 μl H_2O.

Annealing the primer to the template

The methods used for fractionating restriction fragments by gel electrophoresis and their isolation are described in Chapter 4.

The restriction fragment in question is annealed to the exonuclease III-prepared template as follows:

- 0.5 μl template, at a concentration of about 0.5 pmoles/μl is mixed with 2.5 μl of the double-stranded restriction fragment at a concentration of 0.3 pmol/μl. This gives an approximate ratio of primer to template of 3:1
- 0.5 μl of 10 × concentrated reaction buffer are added (0.5 M NaCl, 66 mM Tris-HCl, pH 7.6, 66 mM $MgCl_2$, 10 mM DTT). The mixture is sealed in a capillary, heated at 100°C in a boiling water bath for 3 min. and the DNA annealed at 67°C for 20 min. The sample is then diluted to 10 μl with 1 × reaction buffer.

Chain extension reaction

This is carried out as described previously (Section 3.1.2.). Two microlitres of the annealed DNA solution are dispensed into five capillary tubes and one microlitre of the different reaction mixes (NTP° mixes) added as before except that the C° mix is added to two of the capillaries. Chain termination in one of the C° capillaries is brought about by the addition of 1 μl of 40 mM arabinosyl-dCTP. The appropriate amounts of ddNTP's are added to the other four capillaries and the extension and chase reactions carried out as normally (Section 3.1.2.). If subsequent cleavage of the primer is not required the reactions are terminated by the addition of 1 μl of 0.2 M EDTA and the sample prepared for loading on the gel.

Removal of the primer

If the restriction fragment used as primer is large (>100–200 nucleotides) it is necessary to cleave this from the radioactive copy DNA before gel electrophoresis. This reaction usually employs the same restriction enzyme that was used to generate the primer.

After completion of the chain extension reaction (above) the addition of EDTA is omitted and 1 unit of the restriction enzyme added to each capillary which is then incubated for a further 5 min. at 37°C. The small amount of DNA in each sample should be completely cut in this time. 1 μl 0.2 M EDTA is added to terminate the reaction.

Alternative chain extension reaction using reverse transcriptase
Reverse transcriptase can be used as an effective alternative to DNA polymerase (Klenow subfragment) in the chain-termination reaction. The following protocol is taken from Smith (1980).

The DNA is annealed as before and 2 μl of the sample introduced into each of four capillary tubes. To each of these is added 1 μl of a reaction mixture composed as follows:

0.5 mM dCTP	20 μl
0.5 mM dGTP	20 μl
0.5 mM dTTP	20 μl
10 × reverse transcriptase buffer*	20 μl

One μl of a different dideoxynucleotide triphosphate solution is then added to each capillary. The concentrations of the stock solutions of each ddNTP which give similar extensions are:

ddATP, 0.1 μM
ddTTP, 0.01 mM
ddCTP, 0.02 mM
ddGTP, 0.01 mM

The 4 μl in each capillary are then used to dissolve 1 μl of dried down α-[^{32}P]dATP (1 μCi/μl, specific activity 350 Ci/mmole). One microlitre of AMV reverse transcriptase (5 units/μl) is diluted 1:5 in 1 × reverse transcriptase buffer, and 1 μl of this diluted preparation added to each capillary and the contents mixed.

The capillaries are incubated at 37°C for 15 min. and at the end of that time 1 μl of 0.5 mM dATP is added to each sample, mixed,

* 0.5 M Tris-HCl, pH 8.5, 0.5 M NaCl, 50 mM $MgCl_2$, 20 mM DTT.

and the incubation continued for a further 15 min. (ATP chase).

The reaction is terminated by the addition of 1 μl of 0.1 M EDTA and the samples prepared for loading onto the sequencing gel.

In the reaction volumes given the concentration of the competing dNTPs in the G, C and T reactions is 25 μM. In the A reaction it is about 0.5 μM. The molar ratios of dNTP to ddNTP are therefore about 10 for the G and T reactions, 5 for the C reaction, and 20 for the A reaction.

3.4.3. Results

Figure 3.17. shows two examples of sequencing gels using templates prepared by Exonuclease III digestion as described above (Smith, 1979). In these examples *Mbo*I restriction fragments from human mitochondrial DNA were cloned into the *Bam*HI site of pBR322 (Fig. 3.19.) the circular duplex molecule opened at either the unique pBR322 *Hind*III or *Sal*I sites and the linear duplex products digested with Exonuclease III to produce single-stranded templates. Small restriction fragment primers were prepared from either the pBR322 sequences flanking the insert or from the insert itself. The autoradiographs shown represent sequences within the cytochrome oxidase subunit II gene and show the position of the $tRNA^{Asp}$ gene contiguous to the start of the cytochrome oxidase gene. In a comparison with the corresponding beef heart protein amino acid sequence (not shown) it was observed that the triplet UGA is used as a tryptophan codon and not as a termination codon. Two of these are shown in the autoradiograph (i.e. TGA at positions +316 and +487). This was in fact the first demonstration of a different genetic code in human mitochondrial DNA (Barrell et al., 1979).

This method clearly has wide applicability to sequencing cloned double-stranded DNAs. The only prerequisite is a fairly detailed restriction map of the region of interest so that appropriate primers can be selected at each stage to obtain overlapping sequences. In sequencing it is always important, wherever pos-

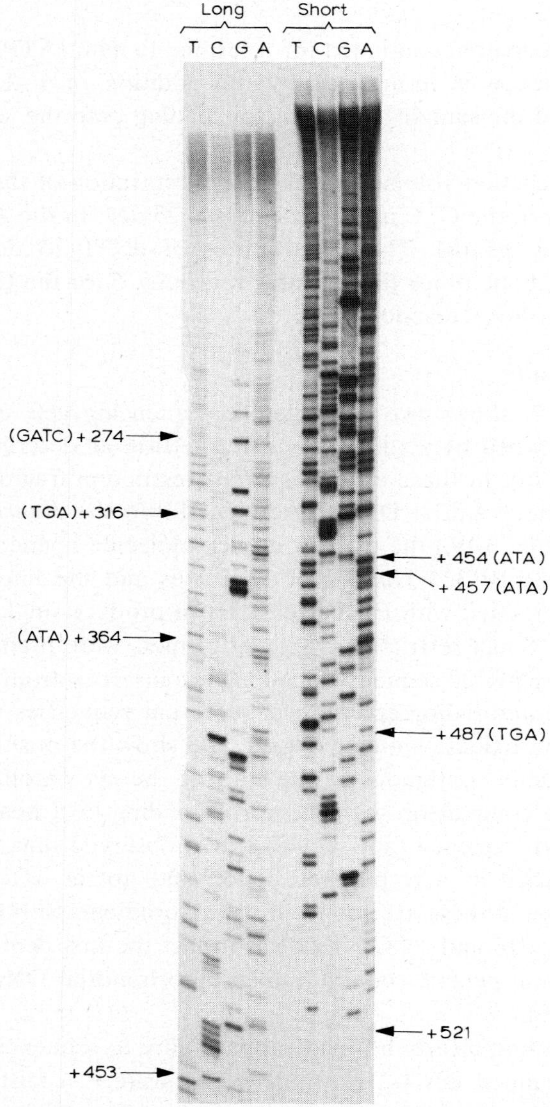

Fig. 3.17. Two examples of chain-termination sequencing gels using single-stranded templates prepared by the exonuclease III method of Smith (1979). In (a) a small *Mbo*I fragment was employed as the primer and in (b) a *Hpa*II fragment. The numbered bands refer to the published human mitochondrial cytochrome oxidase II gene sequence. (Courtesy of Barrell et al., 1979).

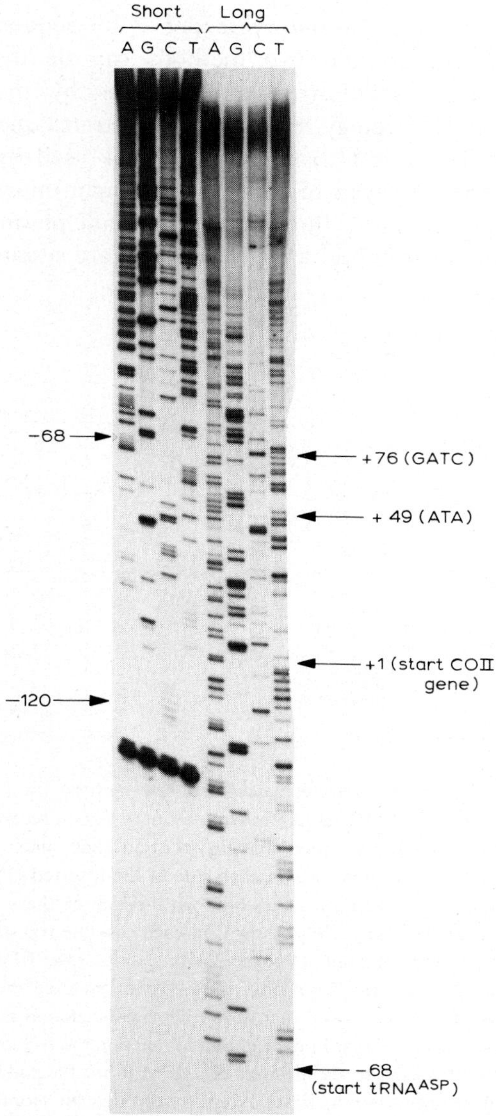
Short
Long
A G C T A G C T
-68
-120
+76 (GATC)
+ 49 (ATA)
+1 (start COII gene)
-68 (start tRNA^ASP)

Fig. 3.17(b).

sible, to obtain confirmatory evidence by sequencing the complementary strand and two methods can be used to produce complementary single-stranded templates by manipulating the exonuclease III strategy. Figure 3.18. illustrates one method which is usually the easiest to employ and was used by Barrell et al. (1979) in their analysis of part of the human mitochondrial DNA genome. In this procedure the recombinant plasmid is linearized by cleaving at two alternative sites which are situated one on each

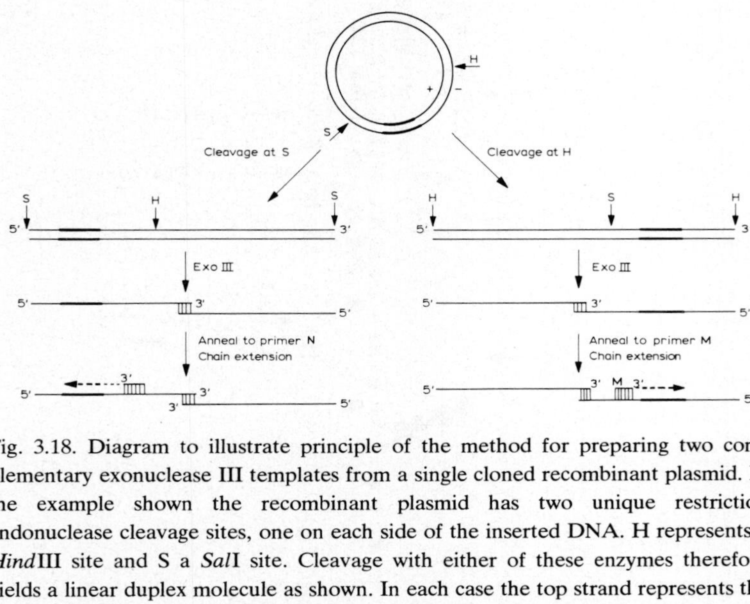

Fig. 3.18. Diagram to illustrate principle of the method for preparing two complementary exonuclease III templates from a single cloned recombinant plasmid. In the example shown the recombinant plasmid has two unique restriction endonuclease cleavage sites, one on each side of the inserted DNA. H represents a *Hind*III site and S a *Sal*I site. Cleavage with either of these enzymes therefore yields a linear duplex molecule as shown. In each case the top strand represents the + strand of the recombinant. Digestion with exonuclease III attacks both linear molecules, at similar rates, from both their 3′-ends yielding the essentially single-stranded products shown. Note that in the left-hand reaction the (+) strand of the insert remains intact whereas in the right hand reaction the (−) strand remains intact. Annealing the products to the primers N and M (after blocking the 3′-ends of the template if necessary) permits chain extension through the cloned sequence in each case. The sequences obtained will be complementary to each other and represent an independent analysis of each strand.

side of the cloned fragment. Digestion with exonuclease III then reduces each duplex to two essentially single-stranded fragments, the difference being that in one case the + (or H) strand of the insert is rendered single-stranded whereas in the other the − (or L) strand is left in a single-stranded form. Annealing these products

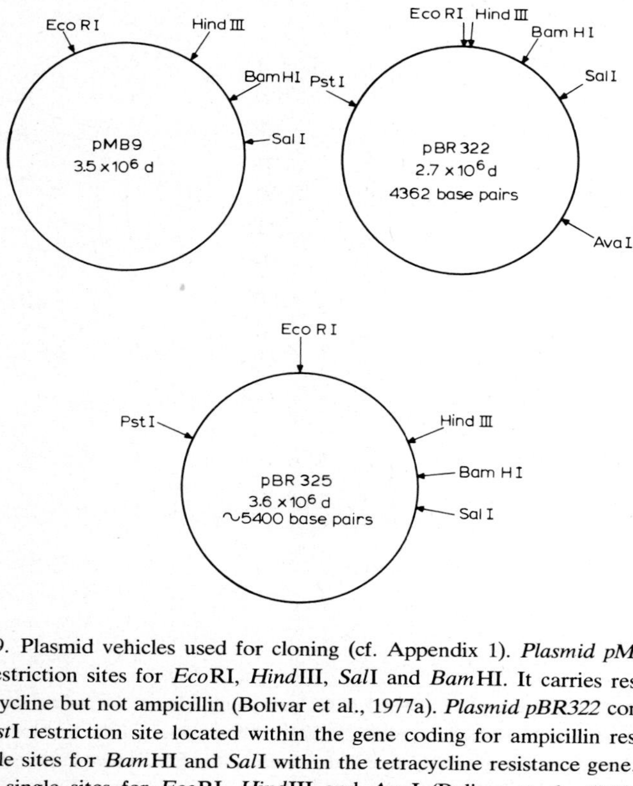

Fig. 3.19. Plasmid vehicles used for cloning (cf. Appendix 1). *Plasmid pMB9* has single restriction sites for *Eco*RI, *Hind*III, *Sal*I and *Bam*HI. It carries resistance to tetracycline but not ampicillin (Bolivar et al., 1977a). *Plasmid pBR322* contains a single *Pst*I restriction site located within the gene coding for ampicillin resistance and single sites for *Bam*HI and *Sal*I within the tetracycline resistance gene. There are also single sites for *Eco*RI, *Hind*III and *Ava*I (Bolivar et al., 1977b). The complete nucleotide sequence has been reported by Sutcliffe (1979). *Plasmid pBR325* is a derivative of pBR322 and carries resistance to three antibiotics: ampicillin, tetracycline and chloramphenicol. The *Pst*I site is located within the Ap^r gene, the *Eco*RI site within the Cm^r gene and the *Bam*HI and *Sal*I sites within the Tc^r gene.

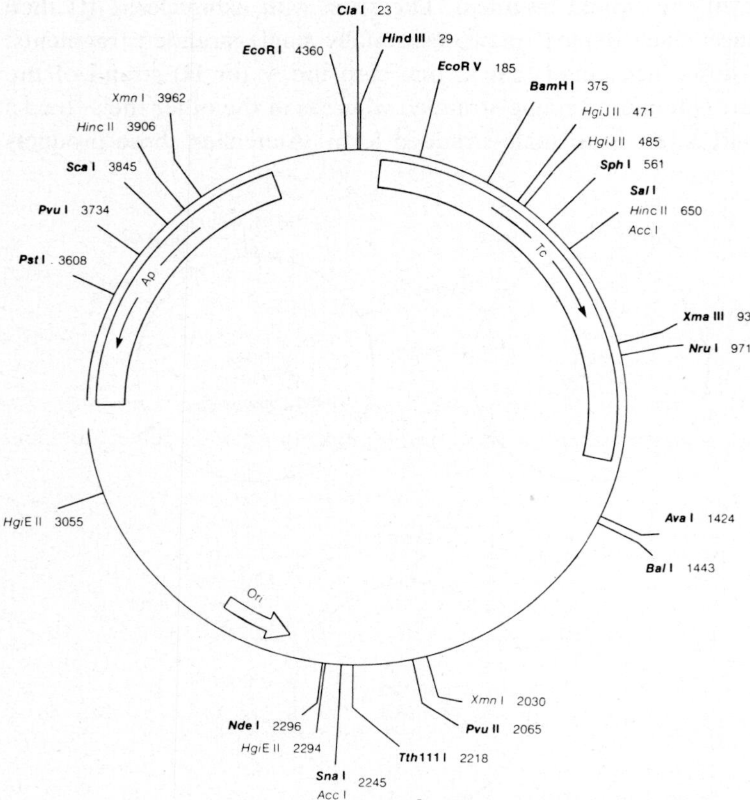

Compiled from 1) Sutcliffe, J.G. (1979) *Cold Spring Harbor Symposium 43*, 77–90
2) Roberts, R.J. (1982) *Nucleic Acids Res. 10*, r117–r144

Fig. 3.20. Restriction map of pBR322 (4362 base pairs). Map of the restriction enzyme sites that occur once or twice in pBR322 DNA. The position of the first base in each recognition sequence is given relative to the centre of the *Eco*RI recognition sequence. Enzymes that cleave the DNA only once are printed in bold type. The positions and orientations of the Ampicillin resistance gene (Ap), the tetracycline resistance gene (Tc), and the origin of DNA replication (Ori) are also shown. Recognition sequences of the following enzymes are not present in pBR322: *Afl* II, *Apa* I, *Asu* II, *Ava* III, *Avr* II, *Bcl* I, *Bgl* II, *Bss*H II, *Bst*E II, *Hpa* I, *Kpn* I, *Mlu* I, *Nco* I, *Sac* I, *Sac* II, *Sau* I, *Sma* I, *Stu* I, *Xba* I, *Xho* I, and *Xma* I.

to suitably positioned primers (conveniently small restriction fragments from the region of interest) allows chain extension through either single-stranded region and thus yields complementary sequence data corresponding to both strands of the insert. This is a generally useful procedure and since most plasmid vectors are constructed to possess single cleavage sites for several different restriction enzymes it should be straight-forward to select a suitable pair. Figure 3.19. shows the position of single cleavage sites on three commonly used plasmid vectors, pMB9, pBR322 and pBR325, all of which are readily propagated in *E. coli.* Figure 3.20 (a and b) shows a detailed restriction map of pBR322 (Sutcliffe, 1979).

The other method for obtaining single-stranded template corresponding to each of the complementary strands relies on iden-

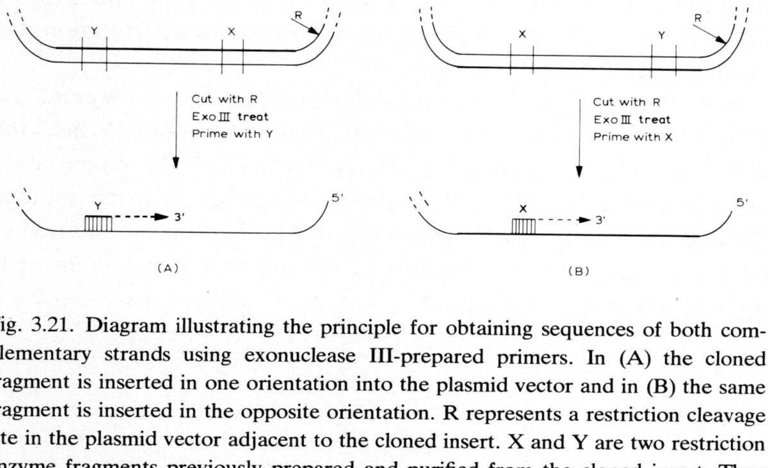

Fig. 3.21. Diagram illustrating the principle for obtaining sequences of both complementary strands using exonuclease III-prepared primers. In (A) the cloned fragment is inserted in one orientation into the plasmid vector and in (B) the same fragment is inserted in the opposite orientation. R represents a restriction cleavage site in the plasmid vector adjacent to the cloned insert. X and Y are two restriction enzyme fragments previously prepared and purified from the cloned insert. They may be the product of the same or different restriction endonucleases. Digestion of the recombinant plasmid with restriction enzyme R yields a linear duplex. Treatment with exonuclease III digests away the top strand in each case leaving, in (A) one strand and in (B), the complementary strand as the template. Priming with fragment Y on (A) and fragment X on (B) allows the sequence of each of the two complementary strands to be determined independently.

tifying and isolating two recombinants in which the cloned sequence has been inserted in opposite orientations. Cleavage of these circular DNAs at a particular site yields a linear duplex which, after digestion with exonuclease III yields two single-stranded templates corresponding to each strand of the inserted fragment. Using appropriate primers, selected from a knowledge of the restriction map, permits the independent sequencing of both strands of the cloned DNA. Figure 3.21. shows the strategy of this approach.

An important advantage of both these procedures is that the sequencing proceeds according to an ordered directed strategy. Potentially useful further restriction sites for generating new primers are revealed as the sequencing progresses and each sequence determined is automatically allocated to one or other of the complementary strands depending on which template preparation is used. The only real drawback is the time spent in the preliminary restriction mapping of the DNA and the isolation and purification of suitable primers.

In summary there is no doubt that this represents a powerful and convenient sequencing strategy of general applicability. Where the decision is to use a primed-synthesis method and the phage M-13 cloning approach (described in the next chapter) is to be avoided for any reason, the *exo*III procedure is likely to be the method of choice. A summarized procedure for chain terminator sequencing is given at the end of Chapter 4.

Primed synthesis methods applied to DNA fragments cloned into phage M13

4.1. General introduction

The various primed-synthesis methods described up till now all have in common the need for reducing large-molecular weight DNA into sets of defined restriction fragments which are then separated, purified and extracted from agarose or acrylamide gels. In order to piece together the sequence of these fragments it is necessary to purify and sequence the fragments from at least one, and usually several other restriction enzyme digests. Where the DNA to be sequenced has a chain length of a few thousand nucleotides the physical isolation of a large number of fragments becomes a major undertaking. For example, in the determination of the ϕX174 DNA sequence the fragments from ten different restriction digests had to be isolated and purified to provide a wide enough spectrum of suitable primers to sequence the whole molecule. These procedures can be both time-consuming and expensive. The DNA needs to be produced in sufficient quantity to provide the wide range of restriction fragments, a number of which often need to be repurified by electrophoresis at different gel concentrations. Repeated purification also increases the hazards of loss or damage to the DNA. Commercial preparations of restriction enzymes are expensive and relatively large quantities are sometimes necessary to provide the desired primers in sufficient quantity.

An approach which gets round these disadvantages is to employ a cloning method in which the restriction fragments derived from cleavage of the DNA with several different enzymes are cloned at random into a suitable vector. The cloning procedure itself usually ensures that the progeny from a single recombinant clone contains a single homogeneous inserted sequence. If the cloning is carried out using a naturally single-stranded DNA phage as the vector (e.g. phages M13, fd) then the recombinants will also contain the cloned sequence in single-stranded form. These phage DNAs can therefore serve directly as single-stranded templates for the primed synthesis reaction. The other essential component of the reaction, i.e. the primer, is either a synthetic oligonucleotide or a restriction fragment, isolated from the parental RF DNA, which hybridizes specifically to a region of the single-stranded DNA within the vector adjacent to the cloned insert (i.e. a flanking primer). The orientation is such that extension by primed synthesis from the 3′-end of the primer leads directly into the cloned sequence which is therefore immediately amenable to sequencing. A 'universal' primer thus serves for sequencing all the recombinant clones and can be used as a standard 'reagent' of known concentration and properties. This random strategy of course requires that the derived sequences are finally collated into a composite sequence by determining the order and polarity in which they are organised in the total DNA sequence. This is usually achieved by cloning and sequencing a further set of fragments derived using another restriction enzyme. The very large volume of sequence data is most effectively stored and processed by computer and a number of different programmes have been written for collating data generated in this way. As the sequence is gradually built up by the acquisition of further data an increasing number of new sequences determined will be redundant since they will be found to be identical in sequence to parts of the DNA for which a sequence has already been established. Some may be useful in providing confirmatory data e.g. the sequence of the complementary strand, but without some discrimination against unwanted sequences an inordinate amount of time would be spent in randomly searching for rare overlaps which might, for example, only clone with a low efficiency into the vector. Computer programmes can search the sequence, as it is built up, for all possible restriction enzyme sites and help predict in the later stages of analysis which restriction enzymes and size lengths of fragments are likely to give the desired overlaps. Such a strategy is continued until the sequence is complete. This is essentially the approach used by Sanger et al. (1980) in their analysis of the human mitochondrial DNA sequence. For less ambitious undertakings the amount of sequence data required to arrive at a final sequence is very much less since fewer restriction fragments need to be selected and cloned to provide the necessary overlaps and problems associated with the organization of long repeated and partially repeated sequences are likely to be less demanding.

This chapter describes in some detail the procedures used for cloning into phage M13 as a way of producing single-stranded template DNA for use in the dideoxynucleotide chain termination procedure. Derivatives of M13 have been constructed containing particular restriction enzyme recognition sequences adjacent to the primer binding region which permit the direct insertion and cloning of cognate restriction fragments. Thus *Eco*RI, *Hind*III and *Sau*3A (or *Bam*HI) generated fragments can be directly ligated into the appropriately modified M13 vector (M13mp2).

The recently introduced M13mp7, derived from M13mp2, possesses an array of restriction endonuclease cleavage sites near to the priming site and has greatly increased the versatility of M13 in this respect. Restriction fragments with blunt ends, e.g. *Alu*I, *Hae*III, *Rsa*I cleavage products, may be cloned directly into the *Hind*II site of M13mp7. Alternatively, such fragments may be modified by the addition of an *Eco*RI linker and cloned into the *Eco*RI site of M13mp2. The following section (4.2.1.) describes these vectors in detail.

The concluding section of this chapter describes an alternative strategy in which the full-length DNA is cloned into an M13 vector thus allowing recombinants to be isolated which contain either one or the other complementary strand in single-stranded form. Sequencing using the flanking primer gives the sequence corresponding to both ends of the cloned fragment and priming with a series of internal primers, derived from a restriction enzyme digest of the double-stranded form of the full-length DNA, serves to determine the remainder of the sequence. The advantages of this procedure are two-fold. Every sequence determined has a known polarity, depending on which of the two single-stranded templates was used in the annealing reaction. Secondly, the relative order of the different sequences determined, in relation to the different restriction enzyme primers used, is automatically revealed as the sequencing progresses. The complete sequence can therefore be determined, at least theoretically, without recourse to searching for fragments from other restriction digests to provide the necessary overlapping data. The author has found this latter procedure to be an efficient and rapid approach for determining sequences of up to 1600 nucleotides. As M13 and its derivatives are filamentous phages there appears to be no packaging problem and cloned fragments of length up to at least 4000 nucleotides can be accommodated and propagated in the recombinant phage.

4.2. Rapid random sequencing by shotgun cloning into single stranded phage vectors (Sanger et al., 1980)

The following sections describe the principles and rationale of the various steps with only brief experimental details. Full experimental details and flow sheets are given in Section 4.3.

4.2.1. Preparation and properties of M13 cloning vectors and primers

Primed synthesis methods require that either the primer or the template DNA is in single-stranded form. Sometimes it is possible to physically separate the complementary strands from a duplex DNA but this is not always a straightforward procedure. The methods developed by Maxam and Gilbert (1977) (Chapter 5 of this volume), while eminently suitable for separating single strands of length up to 200–300 nucleotides rely on gel electrophoretic procedures in which careful attention to loading, gel concentration and cross-linking and running conditions are of paramount importance. Even so some duplex molecules are refractory to strand separation and this procedure is clearly unsuitable for the routine isolation of single-stranded fragments from a large number of duplex restriction fragments. However, the method should be borne in mind since in special situations it may be the only way of isolating a particular single-stranded species.

Sanger et al. (1980) have recently devised an approach to random sequencing which involves the 'shotgun' cloning of fragments of the DNA in question into the single-stranded phage vector M13mp2 (Gronenborn and Messing, 1978). A number of other suitable vectors have also been described (Barnes, 1978, 1979; Herrmann et al., 1978; Ray and Kook, 1978; Boeke et al., 1979) all of which are derived from small single-stranded filamentous bacteriophages. These viruses are secreted by the infected cell into the culture medium at high titres and their isolation is relatively easy.

M13mp2 is a vector constructed by the insertion of a *Hind*II restriction fragment derived from the *lac* operon of *E. coli* into the intragenic space of the wild type M13 phage genome (Messing et al., 1977). The inserted restriction fragment contains a part of the *lac*-I gene, the *lac* promotor, operator and the proximal part of the *lac*-Z gene which codes for a 145 amino acid segment of the enzyme β-galactosidase. Within this gene a single site for the restriction enzyme *Eco*RI was introduced at a position corresponding to the 7th amino acid (Gronenborn and Messing, 1978) yielding the recombinant M13mp2. Despite the introduction of the new *Eco*RI recognition sequence, the 145 amino acid peptide, the α-peptide, is active in α-complementation (Ullman et al., 1967), i.e. it can complement a defective β-galactosidase which contains a deletion corresponding to a sequence within the α-peptide.

The host cells are a derivative of *E. coli* K12 (strain 71-18) in which the *lac pro* region of the chromosome is deleted and instead is carried on an F episome (which is necessary for the horizontal transmission of the phage). As the phage-infected cells are rather unstable and have a tendency to lose the F episome the *pro* marker is used to select for and maintain the F episome in the cells. This is usually achieved by simply maintaining the host cells on minimal agar. Within the *lac* operon contained in the F episome two additional mutations were introduced:

(a) a mutation in the *lac* repressor promotor that causes an over production of *lac* repressor. This avoids the risk of titrating out all the *lac* repressor by the *lac* operator existing on the many copies of the intracellular replicative form of the phage.

(b) a deletion in the structural gene of β-galactosidase corresponding to amino acids 11–41.

The non-infected *E. coli* host is therefore lac^- since it produces a defective β-galactosidase. Phage-infected cells, by contrast are lac^+ since the α-peptide coded by the phage (spanning amino acids 1–145) can complement the deleted segment (amino acids 11–41) in the F episome. Cloning into the *Eco*RI site of M13mp2 produces an inactive α-peptide which fails to complement and thus renders the infected cells lac^-. Clones of these cells are readily distinguished from wild type lac^+ cells since they fail to hydrolyse the chromogenic β-galactoside, 5-bromo-4-chloro-3-indolyl-β-D-galactoside (X-gal) when grown on suitable indicator plates in the presence of the inducer iso-propyl thio-galactoside (IPTG) (Miller, 1972). Recombinant colonies therefore appear as white (colourless) plaques whereas wild type non-recombinant clones are blue. The observed plaques really represent zones of retarded growth of the infected cells which appear as turbid plaques, rather than the clear plaques normally associated with zones of lysis caused by virulent phages. Figure 4.1. outlines this description in diagrammatic form.

Other features of the phage which add to its usefulness as a cloning vehicle are:

(a) small size, about 7200 nucleotides.

(b) high copy number, about 200–300 molecules of supercoiled replicative form (RF) per cell (no amplification with chloramphenicol).

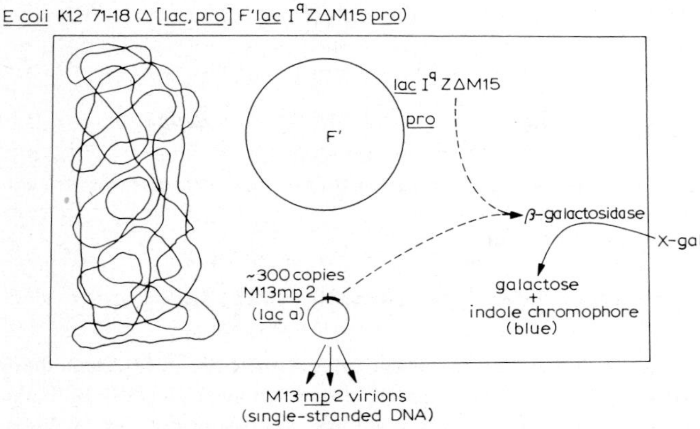

Fig. 4.1. Diagram showing complementation between the α-peptide of β-galactosidase (on M13mp2) and the defective β-galactosidase gene (*lac*I^q ZΔM15) carried on the F′ episome. For discussion see text. (Courtesy of N.L. Brown.)

(c) ease of isolation of the RF of the phage from infected cells (similarity to plasmid DNA).

Phage particles containing single-stranded (the plus strand) DNA are continuously produced and extruded by infected cells without lysis. Titres of phage in the cell-free supernatant range up to 4×10^{12} p.f.u./ml, equivalent to 13 μg of single-stranded DNA (Marvin and Hohn, 1969), which is more than sufficient for ten sequencing experiments. The routine isolation of phage from cell-free supernatants of 1–2 ml cultures is rapid and straightforward. Twenty or more small cultures can be worked up in one day to provide a set of templates ready for sequencing.

Figure 4.2. shows the nucleotide sequence adjacent to the *Eco*RI site in phage M13mp2 into which the cloned sequence is inserted. For the direct cloning of *Bam*HI (or *Sau*3A) and *Hind*III restriction fragments two further derivatives have been constructed (M13mp2/*Bam* and M13mp2/*Hind*III, Rothstein et al., 1979).

The sequences shown in Fig. 4.2. indicate the cloning sites and their relation to the *Eco*RI site.

M13mp2

EcoRI

5'

----GAAACAGCTATGACCATGATTACG↓AATTCACTGGCCGTCGTTTTAC----

M13mp2/Bam

BamHI

5'

----GATTACG**AATTCCCCG↓GATCCGGGG**AATTCACTGGCCGTCGTTTTAC----

M13mp2/Hind III

Hind III

5'

----GATTACG**AATTCCACA↓AGCTTGTGG**AATTCACTGGCCGTCGTTTTAC----

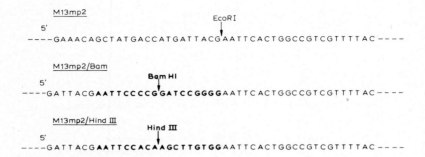

Fig. 4.2. Nucleotide sequences adjacent to the *Eco*RI site in M13mp2, the *Bam*HI site in M13mp2/*Bam* and the *Hind*III site in M13mp2/*Hind*III. Both the latter vectors are derived from the original M13mp2 by the inclusion of an 18-base pair fragment (in bold) which contributes 6 additional codons but does not alter the reading frame. M13mp5 is functionally equivalent to M13mp2/HindIII.

A restriction fragment can usually be inserted into a vector in either of two possible orientations. Ligation of a particular DNA fragment into the double-stranded replicative form of phage M13 therefore gives two different types of recombinant molecules. After transfection into host cells two different types of recombinant phage clones will therefore be obtained depending on which of the two complementary strands of the inserted DNA are included in the progeny phage (Fig. 4.3.). The single-stranded templates derived from these recombinants (the + strands, c.f. Fig. 4.6.), will therefore contain the 3′-end of either of the complementary strands of the original insert adjacent to the priming site (Fig. 4.6.). Sequencing the pair of recombinants therefore gives the sequence of nucleotides from *both* 3′-ends of the original cloned fragment.

In the 'shotgun' sequencing procedure picking recombinant clones is a random procedure and it is often convenient, having sequenced from one end of a single-stranded insert, to turn the sequence around to be able to sequence from the other end. This is done using the 'clone turn-around' procedure described by Winter

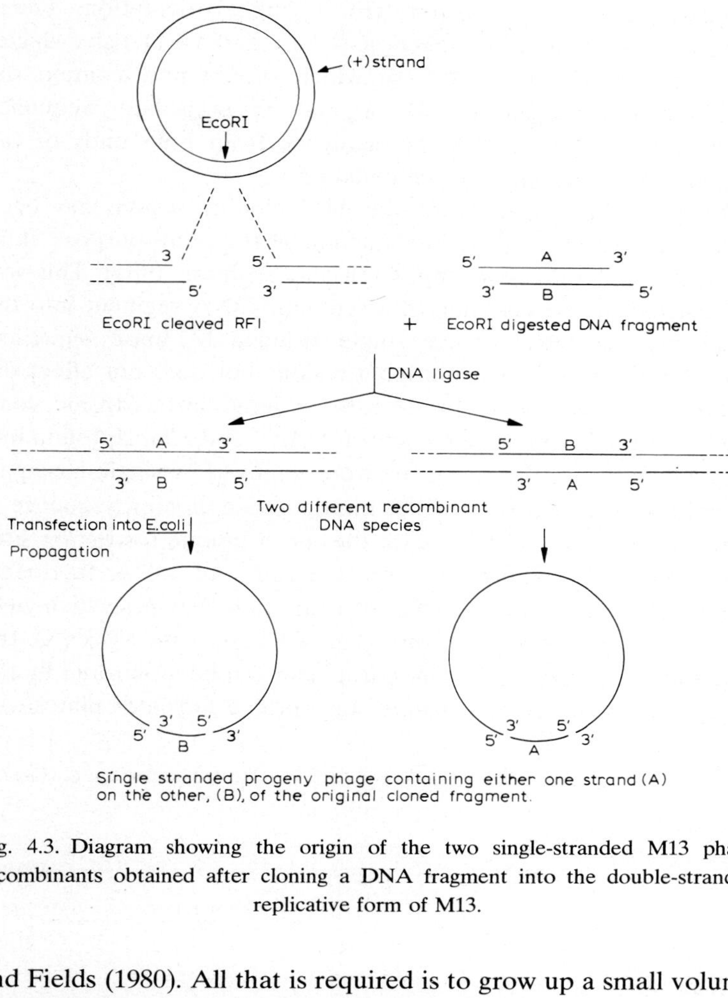

Fig. 4.3. Diagram showing the origin of the two single-stranded M13 phage recombinants obtained after cloning a DNA fragment into the double-stranded replicative form of M13.

and Fields (1980). All that is required is to grow up a small volume (5 ml) of the phage-infected culture (derived from a single plaque, Section 4.3.4.) and isolate the replicative form (RFI) of the phage from the cells using the cleared lysate procedure (Section 4.3.4.). Digestion of the RFI with *Eco*RI (for M13mp2 recombinants) releases the cloned fragment which, on subsequent ligation, is reinserted into the linearised RFI in either orientation. Transformation and plating out (Sections 4.3.10. and 4.3.11.) give single-stranded recombinants, 50% of which should now contain the complementary strand to the original phage isolate. Sequence analysis therefore enables the sequence from both ends of the original duplex fragment to be deduced.

Recently the versatility of the M13 cloning system has been further increased with the introduction of the multi-purpose cloning vector M13mp7 (Messing, Crea and Seeburg, 1981). This was constructed by the insertion of a synthetic DNA segment into the β-galactosidase gene of the single-stranded M13mp2 sequence. This site contributes 14 additional codons but does not affect the ability of the *lac* gene product to undergo intracistronic complementation. Two restriction sites for *Acc*I and *Hinc*II, found in a different region of the viral gene were removed by single-base pair mutations. The structure of the multipurpose cloning sequence is shown in Fig. 4.4., together with the set of unique restriction sites introduced. This allows the direct cloning of DNA fragments generated by a number of different restriction enzymes which yield cohesive or flush-ended termini (Fig. 4.5.). As with M13mp2, the successful integration of a DNA fragment can be monitored by the appearance of colourless plaques on suitable indicator plates. All

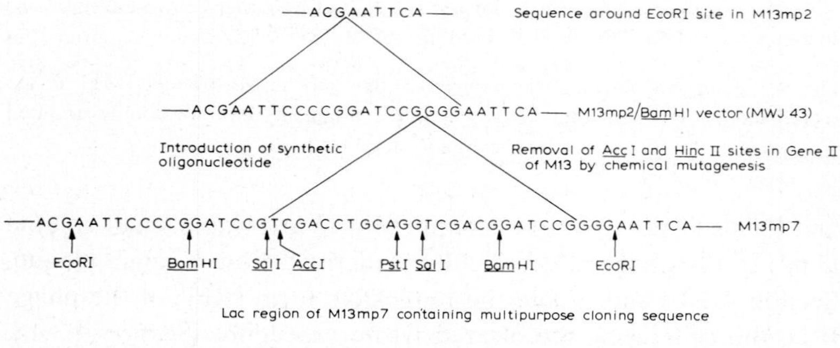

Fig. 4.4. Relationship between multi-purpose cloning vector M13mp7, M13mp2/*Bam*HI vector and the original M13mp2 vector (Messing et al., 1981).

Restriction enzyme used to cut M13mp7 (replicative form)	Sequence cleaved	Potentially clonable fragments	
Bam HI	G↓GATCC	Bam HI Bgl II Bcl I Sau 3A MbO I	G↓GATC C A↓GATC T T↓GATC A N↓GATC N
EcoRI	G↓AATTC	EcoRI EcoRI*	G↓AATT C N↓AATT N
Pst I	CTGCA↓G with subsequent G-tailing	Pst I any fragment with C-tailing	C TGCA↓G
Sal I	G↓TCGAC	SalI XhoI	G↓TCGA C C↓TCGA G
Acc I	GT↓(CT/AG)AC	AccI ClaI HpaII TaqI	GT↓CG AC AT↓CG AT NC↓CG GN NT↓CG AN
Hinc II Hind II	GTPy↓PuAC	Any flush-ended fragment with lower efficiency	

Fig. 4.5. Cloning guide for M13mp7. The enzymes and their recognition sequences are listed on the left hand side. Enzymes which produce DNA fragments which may be cloned at each M13mp7 site are listed on the right. Where this leads to the formation of hybrid sites which are not recognised by either of the enzymes, the flanking restriction sites can be used to cut out the fragment if required. For blunt-ended ligation into M13mp7 the vector is digested with *Hinc*II or *Hind*II. Such vectors, on religation may give a high yield of colourless plaques presumably due to contamination of the enzyme with a nuclease which destroys the reading frame of the α-peptide of β-galactosidase (Section 4.3.1.). Religation of the vector itself in a cloning experiment, however, can be minimized by pretreating the *Hind*II cut M13mp7 with calf-intestinal phosphatase (Worthington) followed by thorough phenol extraction to remove all traces of phosphatase. (cf. Section 4.3.5(b).)

the recombinant phages may be grown on *E. coli* strains 71/18, JM101 or JM103. These three host strains derived from *E. coli* K-12 have the genotype:

71-18: Δ*lacpro*, *sup*E, *thi*, F′*pro*AB, *lac*I^q, ZΔM15.
JM101: as above plus F′*tra*D36.
JM103: as JM101 plus *str*A, *end*A, *sbc*B15, *hsd*R4.

The *tra*D36 mutation (Achtman et al., 1971) permits, under current NIH and GMAG guidelines,* the containment conditions to be dropped by one category. M13mp7 also contains two amber mutations.

The *tra*D36 mutation on the episome in strains JM101 and JM103, however, does increase the risk that in a rich broth the episome may be lost. A strain 71/18 cell when it loses the episome becomes female and readily conjugates with other F$^+$ cells to regain the episome. This horizontal transmission is not possible in the *tra*$^-$ mutant cells and it is important to use fresh inocula from the minimal agar plate (which selects for episome-containing cells) each time a new batch of JM101 or JM103 cells are grown. JM103 is an improvement on JM101 in that it is *E. coli* K12 restriction minus. Thus fragments of DNA which contain an unmodified *Eco*K site cloned into M13 are not restricted and yield recombinant progeny.

All host strains are stored on minimal agar plates (+ glucose) at 4°C and occasionally restreaked.

At the time of writing (1981), M13mp7 is a new arrival on the scene and little information is available regarding its use. Either as a versatile vehicle for preparing banks of DNA fragments or as a vector for the 'shotgun' sequencing approach it is likely to find wide application. The stability of long DNA sequences cloned in this vector have not yet been established. In the related phages M13mp2 and M13mp5, however, DNA sequences of 1000–3000 base pairs have been cloned and sequenced and shown to be identical to the sequences obtained from plasmid clones using the

---

* NIH—National Institutes of Health (U.S.A.), GMAG—Genetic Manipulation Advisory Group (U.K.). Both these bodies offer advice and guidance on the safe handling of recombinant DNA.

Maxam–Gilbert method (Cordell et al., 1979; Buchel et al., 1980; Czernilofsky et al., 1980). With viral DNA up to 6 times the length of M13, however, deletions may occur within the cloned DNA (Messing, unpublished).

Bethesda Research Laboratories Inc., P.O. Box 577, Gaithersburg, MD 20760 in collaboration with Dr. J. Messing is now a central clearing house for all Messing's strains which are obtainable, free of charge, on request to BRL Customer Service Department. BRL also supply M13 cloning/sequencing kits.

Messing et al. (1981, unpublished) have further improved the flexibility of the M13 cloning system by the introduction of two further derivatives, namely M13mp8 and M13mp9. These phages contain an array of restriction endonuclease cleavage sites inserted near the N-terminus of the β-galactosidase gene but in two different orientations. The amino-terminal sequences of M13mp7, mp8 and mp9 are shown below (see pp. 134 and 135). Positions of the different cloning sites are underlined and indicated within their carrier DNA. The *Hae*III site at position 8 serves as a reference.

Cleavage of M13mp8 RF1 with a mixture of, for example, *Eco*RI and *Hind*III restriction endonucleases removes a 30-nucleotide fragment but leaves an asymmetric cloning site in the vector DNA. Cleavage of M13mp9 with the same pair of enzymes also removes a 30-nucleotide fragment, but leaves an asymmetric site in the opposite orientation to that in M13mp8. An *Eco*RI-*Hind*III derived fragment can therefore only be inserted into M13mp8 RFI in one orientation and the single-stranded progeny phage will contain only one of the two complementary strands of the original fragment. Using M13mp9 as the vector the other of the two complementary strands will be inserted, but in the opposite orientation. The same fragment cloned in the replicative form of M13mp8 and M13mp9 will therefore yield two different recombinants; in one case the 3′-end of one of the two complementary strands is put adjacent to the priming site and in the other, the 3′-end of the other strand is put adjacent to the priming site. Primed synthesis, therefore, gives the sequences into both 3′-ends

M13mp7/pUC7						1	2	3	4	5	6	7	8
	1	2	3	4	5	SER	PRO	ASP	PRO	SER	THR	CYS	ARG
	THR	MET	ILE	THR	ASN								
ATG	ACC	ATG	ATT	ACG	AAT	TCC	CCG	GAT	CCG	TCG	ACC	TGC	AGG
					EcoRI			BamHI				PstI	
										SalI, AccI, HincII			

9	10	11	12	13	14			
SER	THR	ASP	PRO	GLY	ASN	6	7	8
TCG	ACG	GAT				SER	LEU	ALA
		BamHI	CCG	GGG	AAT	TCA	CTG	GCC
	SalI, AccI, HindII				EcoRI		HaeIII	

M13mp8/pUC8							1	2	3	4	5	6	7
	1	2	3	4	5	6	ARG	GLY	SER	VAL	ASP	LEU	GLN
	THR	MET	ILE	THR	ASN	SER							
ATG	ACC	ATG	ATT	ACG	AAT	TCC	CGG	GGA	TCC	GTC	GAC	CTG	CAG
					EcoRI			BamHI				PstI	
							SmaI, XmaI			SalI, AccI HindII			

8	9	10	11		
PRO	SER	LEU	ALA	7	8
				LEU	ALA
CCA	AGC	TTG	GCA	CTG	GCC
	HindIII			HaeIII	

M13mp9/pUC9

					1	2	3	4	5	6	7	8	9
	1	2	3	4	PRO	SER	LEU	ALA	ALA	GLY	ARG	ARG	ILE
	THR	MET	ILE	THR									
ATG	ACC	ATG	ATT	ACG	CCA	AGC	TTG	GCT	GCA	GGT	CGA	CGG	ATC
					HindIII			PstI				BamHI	
										SalI, AccI			

10	11				
PRO	GLY	5	6	7	8
		ASN	SER	LEU	ALA
CCC	GGG	AAT	TCA	CTG	GCC
	EcoRI			HaeIII	
SmaI, XmaI					

of the insert. One advantage of the use of these vectors is that fragments inserted in the replicative form of M13mp8 are automatically 'turned-around' when recloned into M13mp9.

Messing et al. (1981, to be published) have also constructed three plasmid vectors, pUC7, pUC8 and pUC9 which contain the Ap® (ampicillin resistance) gene as a selectable marker and the *lac* region of the corresponding M13 counterpart as an insertion detection marker. At the time of writing no details have been published regarding their use.

Hines and Ray (1980) have also described the construction and characterization of new M13 cloning vectors in which additional DNA sequences had been inserted into a *Hae*II restriction site in M13 (RF) DNA. These inserts bring in single restriction sites for *Pst*I, *Xor*II, *Eco*RI, *Sst*I, *Xho*I, *Kpn*I and *Pvu*II and further increase the versatility of the vector for cloning a wide range of fragments.

An alternative to cloning in M13 is to use derivatives of phage fd which have been tailored by Herrmann et al. (1980) to provide an efficient single-stranded DNA vector system. These vectors were constructed *in vitro* by the introduction of various DNA fragments into the intergenic region of phage fd. These inserts introduced into the phage genome unique cleavage sites for restriction enzymes which are suitable for cloning. Since these sites are usually located within genes coding for antibiotic resistance, inactivation of a resistance gene by insertion can be used as a marker for the successful cloning of a DNA fragment. For example in fd 105 an *Eco*RI site is located within the chloramphenicol resistance gene and in fd 106, an additional kanamycin resistance gene permits the selection of $C_m{}^S$, $K_m{}^R$ cell clones by a plaque-assay method. This procedure has not yet been developed into a sequencing tool partly because flanking primers of the sort used for sequencing M13 cloned fragments have not been developed and also because the selection of antibiotic-sensitive clones is a two-step procedure whereas in the M13 cloning system both the identification and selection procedures are carried out on the same plate. However, an alternative to cloning into the β-galactosidase gene of M13 could be important in some circumstances and the use of phage fd derivatives in this context should be kept in mind.

Sequencing by the dideoxy chain termination procedure requires a primer which specifically hybridizes next to the site of insertion of the cloned fragment. Since the phage contains the (+) DNA strand, the polarity of the primer must be such that the 3′-OH end lines up at or near the *Eco*RI site to allow DNA synthesis to proceed along the cloned DNA. One such primer was prepared by Heidecker et al. (1980) by inserting a second *Eco*RI site into the vector M13mp2 to generate the related phage M13mp2962. Cleavage of the double-stranded replicative form of this DNA with *Eco*RI releases a 96-base pair fragment corresponding to amino acids 7 to 35 of the β-galactosidase gene flanking the *Eco*RI site of M13mp2. Denaturation of this fragment followed by annealing to M13mp2 (+) strands will therefore generate a unique hybrid in which the 3′-end of the primer is positioned for extension into the cloned sequence (see Fig. 4.6.). Heidecker et al. (1980) have further described the derivation and isolation of a 92-base pair fragment which hybridizes to the same region (Fig. 4.6.). This

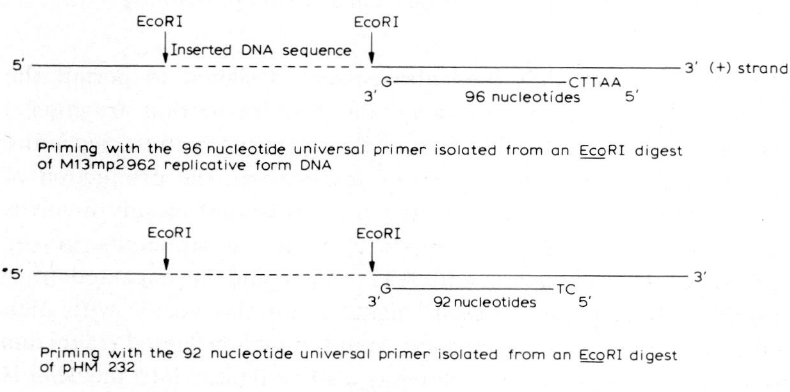

Fig. 4.6. Site of annealing of two long flanking primers for sequencing DNA cloned into M13.

fragment, derived from M13mp2962, was cloned into pBR 325 to yield the recombinant plasmid pHM 232. The advantages of this separate cloning of the primer are firstly that the vehicle, pBR 325 does not contain any long sequences homologous to M13 DNA and the fragment released after *Eco*RI digestion does not need rigorous purification to remove potentially interfering sequences. The results of Heidecker et al. (1980) indicate that no purification at all may be necessary since the pBR 325 sequence is totally unable to prime any chain extension using M13 templates. However, as described later, it is usually advantageous to use Exonuclease III-treated primers for the extension reaction and for this purpose the primer must be separated and purified from the pBR 325 sequence. The second advantage is that it is usually easier to propagate, and obtain high yields, of a plasmid than the intracellular RF of a phage. pHM 232 is readily isolated via the cleared lysate procedure (Section 4.3.4.) followed by CsCl equilibrium centrifugation in the presence of ethidium bromide. The amp^R marker remains intact in pHM 232 and about 350 μg of the crude primer can be isolated from an unamplified 2.5 l-culture containing 20 μg ampicillin/ml.

4.2.2. *Strategies for preparing DNA fragments for cloning into M13 vectors*

The M13 vectors described above were designed to permit the direct insertion of a variety of different restriction fragments. Where the same restriction endonuclease is used both to cleave the vector and produce the fragments for ligation, the production of recombinants is a straightforward procedure and merely involves incubating, at 10–20°C, a mixture of the components in the presence of DNA ligase and ATP. As would be expected fragments with cohesive ends are ligated into the vector with high efficiency but under appropriate conditions blunt-ended fragments (e.g. *Alu*I or *Hae*III products) may also be ligated into the *Hinc*II (*Hind*II) of M13mp7. This latter procedure, however, does have some drawbacks (Section 4.3.6.) and it is often desirable to circumvent these by the addition of a synthetic oligonucleotide linker to the ends of blunt-ended fragments which subsequently permit their ligation into a sticky-ended restriction site. The most widely used linker for this purpose is the oligonucleotide, $^{5'}$GGAATTCC, which contains the *Eco*RI recognition sequence, $^{5'}$GAATTC and thus permits the direct insertion of linker-added fragments into the *Eco*RI site of the vector. Similarly, fragments with 3′- or 5′-sticky ends which cannot be directly ligated into a site in M13mp7 (e.g. *Hha*I, *Hin*fI) can be converted into blunt-ended fragments using the procedures described below and either ligated into the *Hind*II site of M13mp7 or, after the addition of the above linker, inserted into the *Eco*RI site of M13mp2 (or M13mp7).

During the ligation reaction the majority of vector molecules will recircularize without the insertion of an added DNA fragment. Where the frequency of recombinants is low, as in blunt-ended ligation, it is convenient to suppress this non-productive recircularization by removing the 5′-terminal phosphates from the vector. 5′-OH ends are not substrates for DNA ligase and recircularization can only occur where a 5′-phosphorylated DNA fragment is ligated into the site. This procedure is described in Section 4.3.5.

DNA fragments can also be prepared by random cleavage either by using non-specific nucleases such as DNAase I or II, or by mechanical shearing of DNA. Provided the ends of such fragments can be repaired to yield normal base-paired termini these may also be ligated into an M13 vector and used to generate recombinants for sequencing. Suitable procedures are discussed below and Section 4.3.6.(4) gives a detailed protocol for the repair and ligation of DNA fragments obtained by sonicating (shearing) DNA.

In this section these procedures are discussed. The detailed experimental protocols are given in Section 4.3.

4.2.2(a). End-labelling and addition of linkers

The aim of this step is simply to add an *Eco*RI recognition site to both ends of the pool of restriction fragments obtained from the unknown DNA.

The synthetic *Eco*RI linker

$$\begin{matrix} 5' & \mathrm{GGAATTCC} & 3' \\ & \mathrm{\dot{C}\dot{C}\dot{T}\dot{T}\dot{A}\dot{A}\dot{G}\dot{G}} & \end{matrix}$$

is used for this purpose. As purchased, the linker contains free 5′-OH ends and the first step is the phosphorylation of these 5′-ends with polynucleotide kinase using rATP as the phosphorylating agent. In the presence of DNA ligase and rATP these linkers will add on to the ends of flush-ended restriction fragments (e.g. *Alu*I $\begin{matrix}\mathrm{AG{\downarrow}CT}\\ \mathrm{TCGA}\end{matrix}$ or *Hae*III $\begin{matrix}\mathrm{GG{\downarrow}CC}\\ \mathrm{CCGG}\end{matrix}$ ends) to yield a heterogeneous set of products in which a variable number of linker units are polymerized onto the ends of the fragment (see Fig. 4.7.). Many restriction enzymes however yield fragments with staggered 3′- or 5′-ends and it is necessary to first convert these into flush-ended fragments before ligating with the linker. For example, Sanger et al. (1980) in their analysis of part of the human mitochondrial DNA genome needed to add linker to sets of *Hin*fI and *Hpa*II derived fragments, both of which possess 5′-single-stranded extensions (*Hin*fI, G↓ANTC; *Hpa*II C↓CGG). Incubation with DNA polymerase in the presence of the four deoxynucleoside triphosphates results in filling in the ends giving flush-ended fragments. As it is frequently useful to introduce a radioactive label at this stage, in order to follow the subsequent processing of the linker-added fragments, one of the triphosphates may be labelled

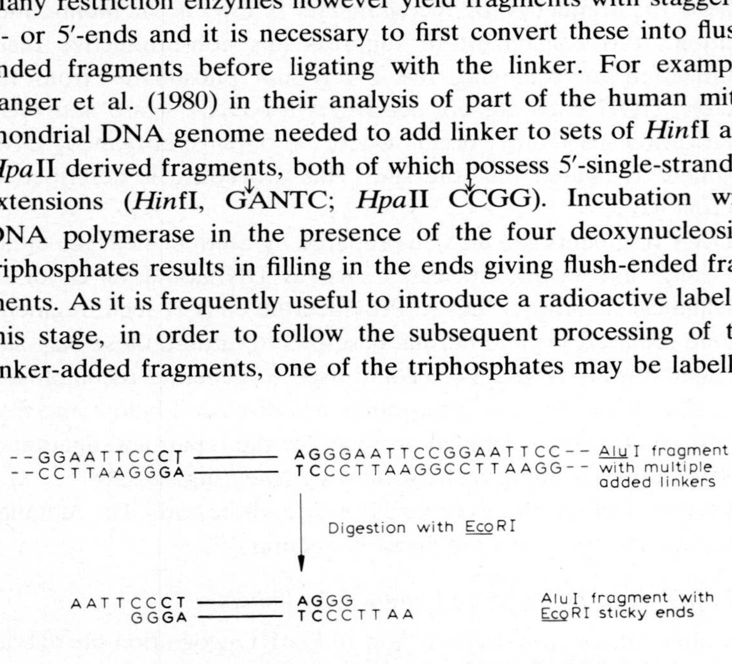

Fig. 4.7. Diagram to illustrate the use of the *Eco*RI linker.

(in the α-position) with [^{32}P]. Thus, in the case of a *Hin*fI fragment, incubation with DNA polymerase in the presence of [α-^{32}P]-dATP, dGTP, dCTP and dTTP yields a flush-ended labelled fragment (Fig. 4.8a). Similarly, the *Hpa*II fragment can be labelled and blunt-ended using [α-^{32}P]dCTP as the labelled substrate. It is usually best to incubate in the presence of all four triphosphates, since though they are not theoretically required, they effectively halt the $3' \rightarrow 5'$ exonuclease activity associated with the polymerase which may damage the molecule.

Naturally flush-ended fragments can also be readily labelled using an exchange reaction catalysed by DNA polymerase (Klenow sub-fragment) or T_4-polymerase. Both these enzymes lack $5' \rightarrow 3'$ exonuclease activity and in the presence of an [α-^{32}P] labelled nucleoside triphosphate, corresponding to the 3′-terminal nucleotide of the fragment, an exchange reaction turns over this residue with the consequent introduction of the labelled residue. The scheme is shown in Fig. 4.8b. As in the previous example, the three unlabelled triphosphates are also added to prevent the $3' \rightarrow 5'$ exonuclease activity from degrading the ends of the DNA. Sometimes (particularly if one is working on a slim research grant) the appropriate [α-^{32}P] triphosphate may not be obtainable. However, it is equally feasible to label, for example, *Alu*I fragments with [α-^{32}P] dATP. In this case the substrate is incubated with DNA-polymerase (Klenow sub-fragment) in the presence of [α-^{32}P] dATP and dCTP and dTTP only. The $3' \rightarrow 5'$ exonuclease removes the terminal G before the enzyme catalyses the turnover of the penultimate A. After incubation for 10 min., dGTP is added and incubation continued for a further 10 min. to fill in the end (Fig. 4.8c).

Whatever labelling scheme is employed the next stage is the polymerization of linker onto the ends of the fragment(s). In order to remove DNA polymerase and the restriction enzyme from the reaction mixture the most straightforward procedure is to shake with phenol (Section 4.3.6.) and precipitate the DNA from the aqueous supernatant with ethanol. After washing with ethanol and

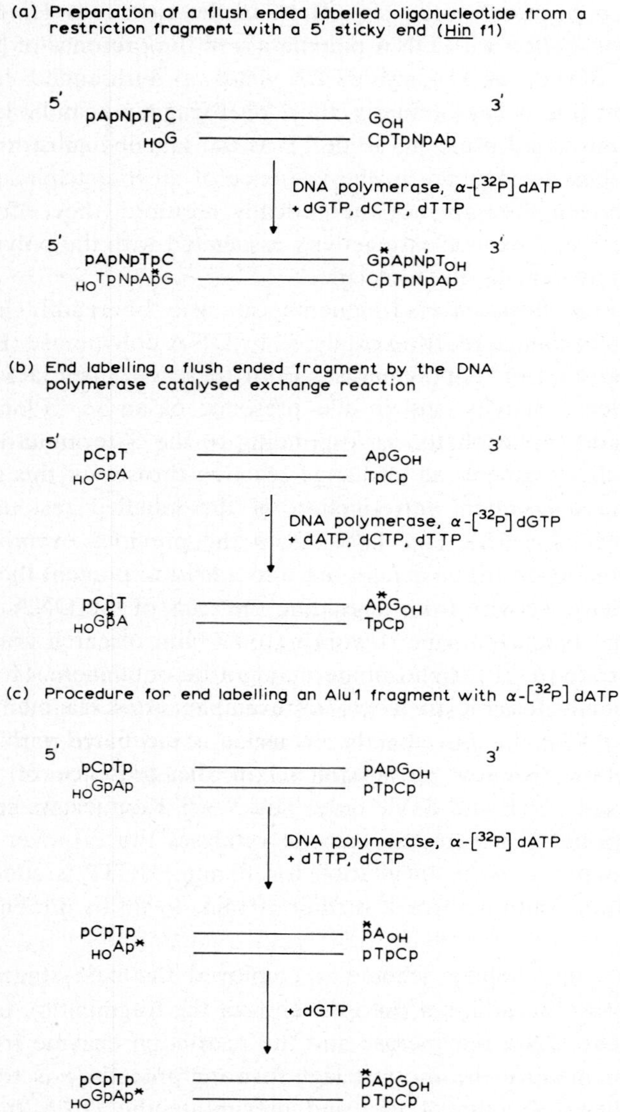
(a) Preparation of a flush ended labelled oligonucleotide from a restriction fragment with a 5′ sticky end (Hin f1)
5′
3′
pApNpTpC
G_OH
HOG
CpTpNpAp
DNA polymerase, α-[32P] dATP
+ dGTP, dCTP, dTTP
5′
3′
pApNpTpC
Gp*ApNpT_OH
HOTpNpAp*G
CpTpNpAp
(b) End labelling a flush ended fragment by the DNA polymerase catalysed exchange reaction
5′
3′
pCpT
ApG_OH
HOGpA
TpCp
DNA polymerase, α-[32P] dGTP
+ dATP, dCTP, dTTP
pCpT
Ap*G_OH
HOGp*A
TpCp
(c) Procedure for end labelling an Alu1 fragment with α-[32P] dATP
5′
3′
pCpTp
pApG_OH
HOGpAp
pTpCp
DNA polymerase, α-[32P] dATP
+ dTTP, dCTP
pCpTp
p*A_OH
HOAp*
pTpCp
+ dGTP
pCpTp
p*ApG_OH
HOGpAp*
pTpCp

Fig. 4.8.

drying, the sample is ready for the addition of linker. The sample is dissolved directly in the phosphorylated linker solution (there is no need to remove the polynucleotide kinase), additional rATP is added to give a final concentration of about 1 mM followed by 0.1–1.0 unit of T_4-DNA ligase. The mixture is incubated at 10–14°C overnight and the enzyme inactivated by heating at 70°C. Considerable differences have been found in different preparations of T_4-ligase and it may be necessary to test several to find a completely satisfactory preparation. Possible contaminants in DNA-ligase preparations include an endonuclease and 5′- and 3′-exonucleases and commercial preparations with minimal (or zero) amounts of these contaminants should be chosen (Appendix IV). Sanger et al. (1980), report an unpublished communication from M. Boguski and W.H. Barnes that the presence of spermidine may be a critical factor in achieving good blunt-ended ligation and it may be advisable to include this compound, at a final concentration of 0.15 mM, in the ligation mixture. The efficiency of the ligation reaction can be tested by electrophoresing a sample of the ligation mixture on a 4–6% acrylamide gel (in the absence of urea) in parallel with a sample of the unligated digest. If the reaction was successful the bands in the digest are replaced with much larger products which are usually not resolved (a smear). In addition a ladder of faster moving bands are usually seen corresponding to polymerized linkers. The ligation mixture should contain a 5–10 fold molar excess of linker over the restriction fragments (for details see Section 4.3.9.).

4.2.2(b). Removal of excess linkers and purification of products for cloning into the M13 vector

This step depolymerizes the chain of linkers yielding restriction fragments containing single *Eco*RI sticky ends at both termini. Before cloning these fragments it is important to remove, as completely as possible, the depolymerized linkers since these would also efficiently ligate into the vector and yield colourless recombinant plaques. Since the vast majority of the restriction fragments are much larger in size than the depolymerized linker, their separation by chromatography on Sepharose or Sephadex gels is usually straightforward.

In a typical experiment the heat-killed ligation mixture is adjusted, by the addition of further Tris HCl, NaCl and $MgCl_2$, to the correct ionic conditions and pH for *Eco*RI digestion, 10–30 units of *Eco*RI added and the mixture incubated for 3 h at 37°C. Excess EDTA (pH 8.0) is added and the mixture extracted with aqueous phenol. The aqueous phase is removed, the phenol layer washed once with water, and the combined aqueous solutions extracted five times with 1 ml ether to remove phenol. Residual ether is evaporated off in a stream of air.

For removing linker fragments Sanger et al. (1980) used a small (1 ml) column of Sepharose 4B equilibrated with 2 mM Tris-HCl (pH 7.4), 0.1 mM-EDTA. The *Eco*RI digest was applied to the column, which was monitored for radioactivity using a hand monitor (Appendix V) and one-drop fractions were collected. A peak of radioactivity is usually found corresponding to the breakthrough volume followed by a long tail of lower activity. 2–3 Drops of the breakthrough peak are collected, taken to dryness in a desiccator and dissolved in a small volume of water. This procedure, in addition to removing linkers, also removes smaller (<100 nucleotides) restriction fragments. This is advantageous since small fragments appear to be more readily incorporated into the vector than larger ones which would result in the majority of random clones picked containing the same small sequences over and over again. Small fragments also contain less sequence information so it is beneficial to exclude them.

In other circumstances the rejection of the smaller fragments may not be desirable and the author has found chromatography on Sephadex G-75 in a high salt buffer will efficiently resolve smaller fragments from the linker. In this procedure the *Eco*RI digest is chromatographed on a 1.4 ml column of G-75 or G-100 (Fine) in a disposable 1 ml plastic pipette using a buffer containing 0.75 M NaCl, 5 mM Tris-HCl (pH 7.5), 0.5 mM-EDTA. One-drop fractions are collected and their radioactivity monitored. A broad smear of radioactivity is found and it is convenient to combine the fractions into two samples comprising the front half and trailing half of the peak. DNA is precipitated by the addition of 2.5–3.0 vol. ethanol, freezing at −70°C and pelleting the DNA. Provided excess ethanol is not added NaCl is not precipitated. This procedure has the advantage of giving a rough separation of restriction fragments so that, in the subsequent cloning, fewer plaques have to be picked to obtain representative clones of all fragments. In addition the concentration of smaller fragments in the trailing part of the peak avoids their interfering with the less efficient cloning of the larger fragments concentrated in the leading part of the peak.

Where random cloning of all restriction fragments is not the aim of the experiment, a preliminary fractionation of the digest by electrophoresis on an agarose gel permits the selection of one, or a few, larger fragments which clone at low efficiencies. Elution of DNA fragments from gel slices is conveniently carried out using the procedure of Yang et al. (1979), in which the gel slice is first dissolved in 5 M sodium perchlorate and the DNA subsequently trapped by filtration through a glass-fibre filter (Section 4.3.8.). The DNA is finally eluted with a small volume of low ionic strength Tris-EDTA buffer. We have found this procedure to give good recoveries of DNA uncontaminated with agarose or other interfering compounds and yield DNA preparations pure enough for further enzymatic manipulation. Alternatively, gel fractionation may be used in place of the Sepharose or Sephadex chromatography. The depolymerized linker migrates far in advance of the restriction fragments and the eluted bands can be ligated directly into the vector ready for cloning.

4.2.2(c). Random cleavage of DNA by methods other than using restriction endonucleases

The set of DNA fragments produced by a restriction enzyme are of course non-random in the sense that the specificity of the enzyme determines the cleavage points in the chain. The final ordering of the different sequences in the DNA in question requires the cloning of another set of restriction fragments from an enzyme of different specificity or the cloning of partial digests in which only a random selection of all the possible cleavage sites are hydrolysed. This latter condition is difficult to attain in practice since cleavage sites vary widely in their relative susceptibility to hydrolysis. Despite the range of restriction enzymes available it is also sometimes found that none of the readily available enzymes cleave the sequence under examination. One way of overcoming these problems would be to cleave the DNA in a truly random, non-specific manner and subsequently clone the fragments. By definition all possible sequences would be represented in the clones and each one examined would yield new data (which adds considerably to the interest of sequencing). Unfortunately no 'non-specific' restriction enzymes seem to exist so one is forced to consider the feasibility of using DNA endonucleases such as DNAases I or II. DNAase I produces predominantly single-stranded nicks in DNA in the presence of Mg^{++} ions. In the presence of Mn^{++}, however, double-stranded cleavages predominate and this has been utilized by Hong (1982) as the basis of an elegant sequencing strategy (note added in 'Update' p. 290). DNAase II (Pig spleen) hydrolysis similarly yields a high proportion of single-stranded breaks but not exclusively so. Under appropriate conditions up to 30% of the cleavages appear to be double-stranded breaks which should produce fragments amenable to cloning. However, DNAase II yields products ending in 3′-phosphates and 5′-hydroxyl ends. These need to be converted into the corresponding 3′-OH and 5′-phosphorylated ends for the subsequent addition of linker and cloning. Only 30% of 30% (9%) of the fragments would be expected to contain flush-ended termini at both ends yielding potentially clonable fragments. If a DNAase II digest of a defined DNA species is run out on an acrylamide gel under non-denaturing conditions a smear of fragments is obtained, the size range of fragments in the smear being determined by the extent of digestion. Under conditions where the centre of the smear corresponds to fragments of length 200–300 nucleotides this region of the gel can be cut out, the fragments eluted and after suitable modification of the ends, used for cloning. The fragments obtained should represent a random selection of all possible sequences in the DNA. Hindley and Pačes (unpublished data) have used this approach to determine the sequence of a *Pst*I fragment derived from pDAM Y8 (Beach et al., 1980) containing a yeast chromosomal origin of replication.

In this experiment the *Pst*I fragment was first digested with DNAase II in sodium acetate buffer, pH 4.7 at room temperature and the reaction halted by chilling and extraction with phenol. After precipitation the DNA was electrophoresed on an 8% polyacrylamide gel and a slice of gel, corresponding to fragments of chain length 150–250 nucleotides cut out and eluted. The 3′-terminal phosphates were removed by treatment with alkaline phosphatase and, after denaturation and removal of the phosphatase with phenol, the DNA was reprecipitated and dissolved in a small volume of water.

The fragments obtained at this stage will predominantly terminate in 3′- or 5′-single stranded extensions of variable length. Products with a 5′-extension can be repaired by filling in the gap using T_4-polymerase in the presence of the four deoxynucleoside triphosphates to produce a blunt end. A 3′-single stranded extension is degraded by the $3' \rightarrow 5'$ exonuclease activity of the enzyme up to the nucleotide opposite the corresponding 5′-terminus, leaving a blunt end. If one of the triphosphates is labelled (e.g. [α-^{32}P]-dATP) the repair synthesis complementary to the 5′-extension will, in a proportion of cases, introduce a radioactive label which is desirable for following the subsequent addition of linker and processing. The final step, which can either be carried out separately, or in the presence of the linker, is the kinase reaction which serves to phosphorylate the 5′-ends. The products at this stage mimic those obtained from a flush-ended restriction enzyme cleavage and can be directly ligated to linker. Figure 4.9. shows schematically the processing stages applied to a DNAase II digest.

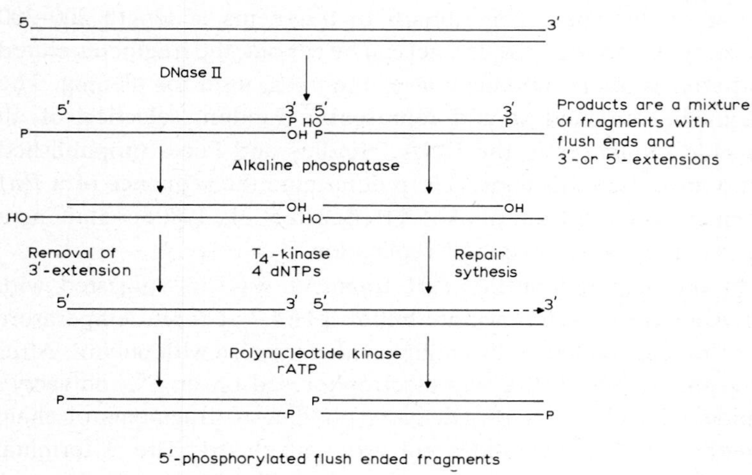

Fig. 4.9. Outline of a procedure for preparing flush-ended fragments from a DNAase II digest. The products may be ligated with RI linker and cloned into M13mp2.

A relatively large excess of linker is required to minimize end to end ligation of the DNAase II fragments yielding false sequences.

An alternative procedure for preparing random fragments for cloning is suggested by the work of Shenk et al. (1975), though this does not yet appear to have been tried. These experiments showed that nuclease S1 cleaved duplex DNA at the site of a single-stranded nick. SV40 DNA which had been nicked once in either of its two strands at the *Hpa*II endonuclease restriction site was shown to be cleaved at or very near that site with nuclease S1. DNAase I, which also introduces single nicks into duplex DNA with the introduction of new 3′-OH and 5′-phosphorylated ends, would be expected to form a similar substrate for nuclease S1 cleavage and yield a population of double-stranded fragments. Though nuclease S1 is known to 'nibble' the ends of duplex DNA, resulting in some shortening of the fragment, this is of no consequence since the aim is only to produce a random set of clonable fragments. Further incubation with DNA polymerase (Klenow sub-fragment) in the presence of the four deoxynucleotide triphosphates should ensure their conversion into flush-ended fragments ready for ligation with the *Eco*RI linker.

In a study on the effect of divalent cations on the mode of action of DNAase I, Campbell and Jackson (1980) found that in the presence of the Mn^{++} ions DNAase I is able to cut both DNA strands within a duplex at or near the same point. This ability to cut both strands is inhibited at lower temperatures and by the addition of a monovalent ion or another divalent cation which is not a transition metal ion. Transition metal ions thus appear to promote the localized unwinding of duplex DNA into a form where DNAase I can introduce breaks into both strands. These observations therefore suggest another route for producing a more or less random set of fragments which, after limited 'polishing' of the ends with DNA polymerase, could be ligated to the RI linker and subsequently cloned into M13mp2.

Frischauf et al. (1980) have described the use of a similar strategy for producing random DNA fragments for cloning into plasmid vectors.

DNA can also be broken down by shearing. To reduce the fragments to sizes suitable for sequencing the shearing forces are generated by sonication. Though it is not clear which bonds in the DNA are most susceptible to breakage, a proportion of the DNA fragments produced may be converted into flush-ends by incubation with T_4 DNA polymerase in the presence of the four deoxynucleotide triphosphates and subsequently cloned into the *Hin*dII (or *Hin*cII) site of M13mp7 (Fig. 4.5.).

4.2.3. Incorporation into vector DNA

The properties of the cloning vector phage M13mp2 and its derivatives have been described previously (4.2.1.). Preparatory to cloning fragments into the linearized RF of the phage the most important check is the purity of the RF. Small quantities of chromosomal DNA, contaminating the RF, will be cleaved into a heterogeneous set of fragments during the linearization with the restriction enzyme which, during the subsequent ligation step will become inserted into the RF and yield unwanted recombinants.

The normal isolation procedure for the intracellular RF involves preparing a cleared lysate from the phage-infected cells and banding the supercoiled RF molecules on a CsCl gradient in the presence of ethidium bromide. Since only very small quantities of RF are required for the cloning step (20–200 ng per experiment) and since the yield of RF reaches 1 μg/ml of growth medium, only a relatively small-scale isolation procedure is required. Starting with a 40–50 ml culture of infected cells and banding the lysate in 30–40 ml CsCl give lightly loaded gradients in which the dense supercoiled RF is clearly separated from the less dense chromosomal DNA. Overloading the gradient gives a poorer resolution with the risk of serious contamination of the isolated RF. Normally the isolation procedure described gives RF pure enough for cloning. If not the equilibrium run has to be repeated or a velocity gradient in 10%–30% sucrose employed.

When starting with a fresh batch of RF it is highly advisable to run a preliminary test to check both the purity of the RF and also the purity of the enzyme involved in the generation of hybrid molecules. Two events may later on cause the unscheduled appearance of colourless plaques. Contamination of the restriction enzyme or DNA ligase with other nucleases can cause the generation of deletions in the *lac* region of the phage and, as mentioned above, contamination of the RF with *E. coli* DNA can result in unspecific insertions into the cloning site. To carry out these checks about 1–2 μg RF from M13mp2 are digested to completion with *Eco*RI, and an aliquot of the product taken for religation using T_4-ligase. Electrophoresis of the products from the two reactions on a 1% agarose gel followed by staining with ethidium bromide and examination under UV illumination should give a good estimate of the quality of the reagents. The initial *Eco*RI digest should yield a single sharp band corresponding to the linearized RF with no traces of other components. After ligation this should be converted to the slower moving circular form which should similarly appear as a sharp discrete band. A more stringent test is to transfect the religated sample into competent cells and assay on an indicator plate as described later. If colourless plaques occur at a frequency of less than 1:100 (relative to blue plaques), the RF and other reagents are pure enough to proceed to the cloning.

One important consideration in setting up the cloning mixture is the relative concentration of vector and *Eco*RI fragments. An excess of vector, which might theoretically be expected to give a maximum yield of recombinants also gives an unacceptably high number of recircularized RF molecules with the result that the occasional colourless plaques are swamped out with non-recombinant blue plaques. An excess of *Eco*RI fragments, on the other hand, can result in the ligation of two, or more, of these sequences into the vector yielding misleading data during the sequencing of that recombinant. Such double-ligation events are not infrequent even under optimal ligation conditions. Usually they can be readily detected since the sequence analysis will reveal the presence of reformed *Eco*RI recognition sequence, –GAATTC– within the cloned sequence. Messing (1979) recommends an optimal concentration of 1–2 μg/ml of linearized vector and DNA fragments with a length between 600–6000 base pairs should be in a 3-fold molar excess. For cloning smaller fragments a larger (up to 10-fold) excess may be used but it is difficult to be precise about this since the number, and size distribution of fragments in a digest is very variable. As a rough guide 20–40 ng vector should be ligated to between 50–200 ng of fragments. The aim is to obtain 100–200 plaques per plate of which 10–30% are colourless (i.e. recombinants). The ligation reaction is carried out in a volume of 2.5–20 μl in the presence of 0.005–0.05 units T_4-ligase at 10–15°C for 2–16 h. Aliquots of the mixture are used directly for the transformation of competent cells.

4.2.4. Transformation

The properties of the *E. coli* K12 derivative (strain 71/18)

(Δ[*lac.pro*], *Sup*E, *thi*, F′, *lac*I^q, ZΔM15, pro^+) constructed by Messing (1979) are described in Section 4.2.1. (above). Sanger (1980) used a tra^- mutant, *E. coli* strain JM101 constructed by Messing (1979) (Δ*lac.pro*), *Sup*E, *thi*, F′, tra D36 *pro* A B, *lac*I^q, ZΔM15). Both strains were constructed to act as host and permit the ready identification of recombinant M13 phage using a chromogenic β-galactoside substrate.

From 2 ng of circular RF molecules about 1000 transformants can be obtained. The transformation efficiency with the same amount of *Eco*RI cleaved RF drops to about zero. Religated *Eco*RI RF gives a transformation efficiency of up to 50% that of the uncleaved molecules. Transfection of the ligation mixture into the *E. coli* host requires that the cells are first made ‘competent’ to take up the recombinant phage DNA. This is usually achieved by harvesting the cells in mid-log phase and suspending them in ice-cold $CaCl_2$ solution. The transfection is carried out in 50–80 mM $CaCl_2$ after which the β-galactosidase inducer, IPTG, and the chromogenic substrate, X-gal are added. Additional fresh exponentially growing strain 71/18 (or JM101) cells are added and sufficient melted agar to give a soft set. The mixture is poured onto a layer of hard agar in a petri dish and allowed to set. After overnight incubation at 37°C, a uniform lawn of bacteria is produced containing a mixture of discrete blue and colourless plaques. Individual plaques can be picked out of the top agar layer using a wooden toothpick or the tip of a pasteur pipette. Only those colourless plaques should be chosen which are discrete from adjacent blue or colourless plaques. Plaques are transferred individually into 1–2 ml medium in 20 ml universal bottles and the phage infected cells grown up in an orbital incubator at 37°C for 12–14 h. The turbid cultures can be stored for short intervals at 4°C but it is advisable to proceed directly to the isolation of recombinant phage DNA.

4.2.5. Preparation of single-stranded templates

This procedure is described in detail, together with the other procedures outlined above, in Sections 4.3.10. and 4.3.11. Briefly, the infected cultures are centrifuged to pellet the cells and the secreted phage precipitated from the supernatant by the addition of polyethylene glycol and salt. The phage pellet is taken up in a small volume of Tris-EDTA buffer and extracted with phenol. The phases are separated by centrifuging, the aqueous layer removed and the DNA precipitated with ethanol. After collecting the precipitate and drying, it is finally dissolved in 50 μl H_2O or 10 mM Tris HCl (pH 7.4) 0.1 mM EDTA. These samples, which constitute the single-stranded templates for sequencing, are stored frozen until required.

4.2.6. Preparation of sequencing primers

See Section 4.3.7.

4.2.7. Preliminary screening of templates with the ddT reaction (Sanger et al., 1980)

Since the recombinant clones contain randomly-inserted sequences many of the isolated templates will contain identical sequences and it is useful to carry out a preliminary screening to identify these. The screening consists of carrying out the chain-extension reaction with a single-chain terminator (ddT) and analysing the fragments in parallel on a sequencing gel. Ten to eighteen samples may be analysed in this way on a single gel and the templates yielding the clearest ladder of fragments representative of each sequence chosen for sequencing.

4.2.8. Sequencing the DNA

The procedures are essentially those described by Sanger et al., 1977; Sanger and Coulson, 1978; Schrier and Cortese, 1979 and Sanger et al., 1980. A full description of the method is given in Section 4.3. Before embarking on a set of sequencing experiments it is worthwhile checking the optimal ratios of template to primer by setting up 4–6 different annealing mixtures in which the template : primer ratio is varied over a ten-fold range. Usually the amount of template DNA isolated from 0.8 ml culture supernatant does not vary very much (about 10–20 pmol), but there is no easy method of measuring this. The yield of primer can also vary significantly in different preparations. Normally one aims at about 3–4 fold molar excess of primer over template, i.e. the annealing mixture contains 0.75–1.5 pmol of primer and 0.25–0.5 pmol template. These quantities yield sufficient primed template for running the four sequencing reactions and, if necessary, analysing the products in triplicate. In order to improve the resolution of the larger oligonucleotides on the gel it is usual to electrophorese two sets of samples in parallel, the second loading being made when the xylene-cyanol FF dye marker (the slow blue) has migrated half way down the gel (cf. Section 4.4.).

When the 96-nucleotide flanking primer (4.3.7.) is used directly for priming a persistent problem is the appearance of a strong artifact band across all four sequencing channels on the gel at a position about 130 nucleotides from the 5′-end of the primer. The sequence preceding this band is usually unreadable (mixed sequence) which means that the first 35–40 nucleotides of the cloned fragment cannot be deduced. This effect is believed to be due to the 3′-end of the primer looping back to base-pair intramolecularly at a position where it can then act as a primer for the DNA polymerase and so extend itself for a further 40 residues before reaching the 5′-end of the strand. This problem can be completely overcome by pretreatment of the primer with Exonuclease III (Schrier and Cortese, 1979; Zain and Roberts, 1979; Smith, 1979), which effectively shortens the primer by half as a result of the removal of nucleotides from the 3′-end of each strand, (Fig. 4.10.). This abbreviated primer is unable to loop-back on itself, priming thus taking place exclusively on the annealed template. This exonuclease III treatment also has the advantage of allowing chain extension to commence before running into the cloned sequence which usually permits the unequivocal identification of the *Eco*RI site generated by the addition of linker. Also, commencing the extension reaction with the shortened

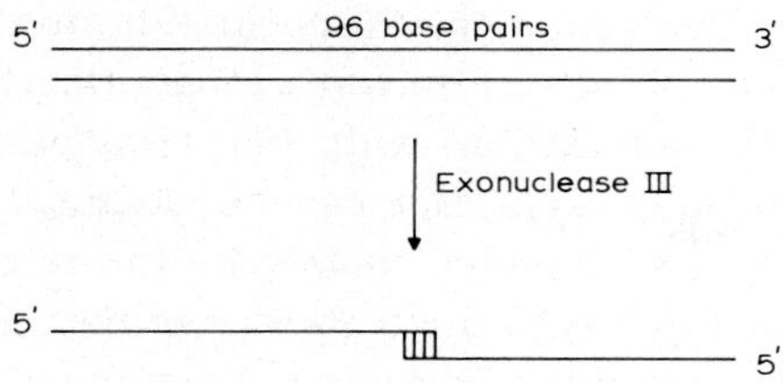

Fig. 4.10. Digestion of the 96 base-pair R_1–R_1 universal primer with exonuclease III. Note that the 5′-ends are left intact thus yielding primers with a fixed defined 5′-end. Since the template used in the annealing reaction is single-stranded (the + strand) only one of the two single-stranded products will subsequently anneal to the template.

primer permits the incorporation of a few labelled 'A' residues before the cloned sequence is encountered thus giving adequately labelled products corresponding to the shortest extension products. This pretreatment of the primer with Exonuclease III is now usually routine and the full-length primer is rarely used.

For reasons which are not altogether clear, excessive exonuclease digestion results in a decrease in priming ability. Theoretically, exonuclease III action should stop when the primer has been reduced to a predominantly single-stranded form since single strands are not attacked by the enzyme. Contaminant endonuclease, or a $3' \rightarrow 5'$ single-stranded exonuclease would effectively reduce the concentration of primer in prolonged incubation. Digestion with Exo III is normally carried out at an enzyme concentration of 0.5–1.0 unit/μl. At room temperature the rate of digestion is about 800–1000 nucleotides per duplex end per hour; for the 96 nucleotide R_1–R_1 primer about 5–10 min. digestion is adequate. Underdigestion of the primer may leave sufficient secondary structure for self-priming to occur on the primer. However, this is rarely a problem. As described above a rough preliminary assay to determine the optimal template : primer ratio is worth doing for each new batch of exonuclease III-treated primer.

The only disadvantage of the 96 nucleotide primer is its excessive length, roughly three or four times longer than is necessary for a stable, specific interaction with the template. Because the amount of sequencing data that can be obtained from a single DNA sequencing gel is limited mainly by the resolution of fragments, it is advantageous to use a shorter primer. Anderson et al. (1980) have described the preparation and cloning of a 30-nucleotide long primer for sequencing M13mp2 clones. Because of its smaller size the primer allows sequences longer than 300 nucleotides in length to be obtained from a single priming. The primer is functionally equivalent to the 3′-terminal sequence (of 24 residues) of the 96-nucleotide primer. Figure 4.11(B). shows the sequence of the primer, as isolated from a cloned *Bam*HI-*Eco*RI fragment in pSP14 (Fig. 4.11(A).). As pointed out above the advantage of this primer is its small size. Pretreatment with exonuclease III is not required and its use allows an additional 70 residues of sequence information to be obtained from a set of sequencing reactions. The T_m of the primer has not been determined but it is recommended (Anderson et al., 1980) that the primer be annealed to the template after heat-denaturation by allowing the temperature to drop slowly (20–30 min.) from 85–90°C down to room temperature. When using this primer it is advantageous to use [α-^{32}P]-dCTP as the labelled nucleotide in the sequencing reactions. With [α-^{32}P]-dATP a strong artifact band may occur which obscures the first three or four residues above the filled-in primer band. The cause of this is not known but it is probably due to hairpin formation and self-priming by the unannealed strand complementary to the primer.

Recently Hindley, Porter and Smith (1981, unpublished results) have used a totally synthetic single-stranded DNA primer of 18 nucleotides equivalent to the 3′-end of the 96-nucleotide primer. This has the sequence:

5′–GTAAAACGACGGCCAGTG–3′

3 pmol of this primer are hybridized to 1–2 pmol of the template in

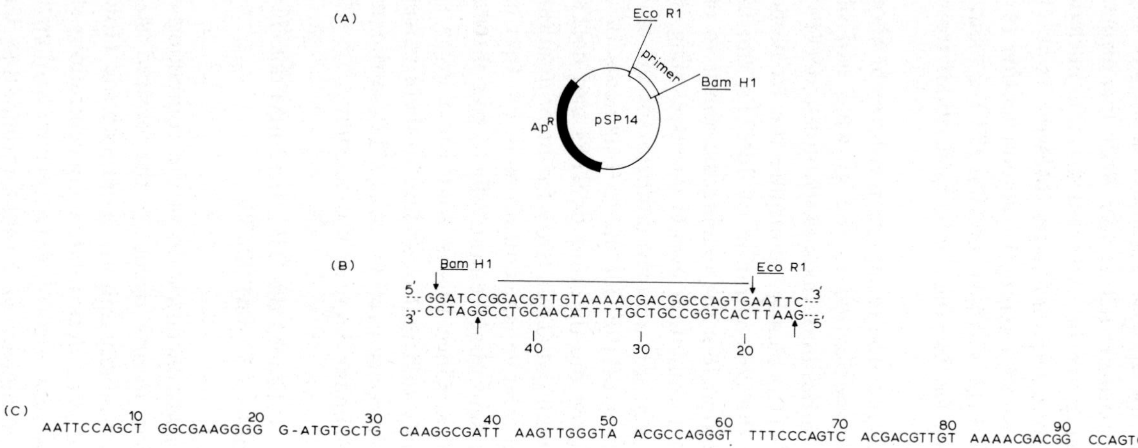

Fig. 4.11. Structure of the cloned primer fragment. A: Plasmid pSP14 consists of the 3980 bp *Eco*RI/*Bam*HI fragment from pBP322 (carrying the ampicillin resistance gene) joined to the primer fragment. B: The sequence of the 26 bp primer fragment is shown. The overlined portion represents that part which is complementary to the M13mp2 template; the numbers indicate base positions in the *lacZ* gene. Some of the 5′-terminal nucleotides of the primer (contributed by the *Bam*HI linker) will be mismatched when it is annealed to the template, but this does not affect its performance as a primer for DNA sequencing. C: Sequence of the M13mp2 universal primer.

the usual way except that the annealing mixture is allowed to cool down slowly to room temperature before use. With this primer the length of readable sequence can be extended to >300 nucleotides using a triple loading on the sequencing gel. Messing et al. (1981) have also described the use of a synthetic oligonucleotide primer. This has the sequence 5′–CCCAGTCACGACGTT–3′ and was designed to prime within the *lac* region 17 nucleotides from the cloning site.

In the primed synthesis reaction termination is brought about by the addition of a dideoxynucleotide to the growing end of the chain. The relative concentration of deoxynucleotides (dNTPs) and dideoxynucleotides (ddNTPs) clearly determine the frequency of termination; the higher the concentration of ddNTPs, the higher the frequency of termination. For sequencing the aim is to obtain the maximum spread of termination events corresponding to the readable region of the gel, i.e. a fairly uniform spread of band intensity from the bottom of the gel up to fragments of 300+ nucleotides in length. Since the different ddNTPs and dNTPs do not appear to compete with equal efficiency for incorporation into the cDNA their optimal relative concentrations in the four extension reactions should be determined empirically to give the desired spread of band intensities. Once this is done for a given set of stock solutions it is usually sufficient to adhere to these values. As a guide the following final concentrations of ddNTPs in the primed synthesis reactions gave similar degrees of extension:

0.15 mM ddATP, 0.12 mM ddCTP, 0.12 mM ddGTP and 0.32 mM ddTTP

In the presence of a 2 μM-concentration of the competing dNTP most of the termination events occur in the desired size-distribution range giving reasonably uniform band patterns. However, it should be emphasized that these relative concentrations should be determined by trial and error before serious sequencing is commenced. Note that in the case of the ddA-terminating reaction mixture the only dATP added is that provided by the added label. If a significantly higher specific activity [α-^{32}P]-dATP is used, it should be diluted with unlabelled dATP to a specific activity of ~300 Ci/mMol. Higher specific activities may be used but care has to be taken that a sufficient molar concentration of dATP is achieved. At the end of the chain-extension reaction, in which dATP is present in limiting amounts, excess unlabelled dATP is added as a 'chase' and the incubation continued for a further 10 min. The reason for this is to allow nascent chains, which have not been terminated by the addition of a dd-nucleotide, to be further extended until a proper termination event occurs. Premature termination needs to be avoided since it results in the appearance of anomalous bands on the sequencing gel which can seriously confuse its interpretation.

4.2.9. Gel electrophoresis

High-resolution thin (0.3 mm) sequencing gels are used according to Sanger and Coulson (1978). Routinely these are 8% or 6% polyacrylamide gels, containing 7 M urea, and run in a continuous buffer system containing 90 mM Tris-Borate, pH 8.3, 1 mM-EDTA.

Prior to electrophoresis the products of the extension reaction are mixed into 15 μl of deionized formamide containing 10 mM-EDTA, 0.2% bromophenol blue and 0.2% xylene cyanol (tracking dyes) and then heated at 100° for 5 min. to denature the DNA. 2–5 μl-samples are loaded into each well of the gel and electrophoresis commenced immediately at a constant current of 30 mA using an LKB 2103 power supply set at constant current. In the initial stages of electrophoresis this corresponds to a voltage of about 1.6–1.7 KV but this drops to 1.2–1.3 KV as the gel heats up and remains constant at this value throughout the run. The gel gets quite hot (~60°C) and this helps to keep the DNA denatured. Normally two or more consecutive loadings are carried out on each gel at 90–100 min.-intervals, the samples being briefly heated to 100°C before each loading. Using 8% gels sequences can usually be read up to bands corresponding to an oligonucleotide chain length

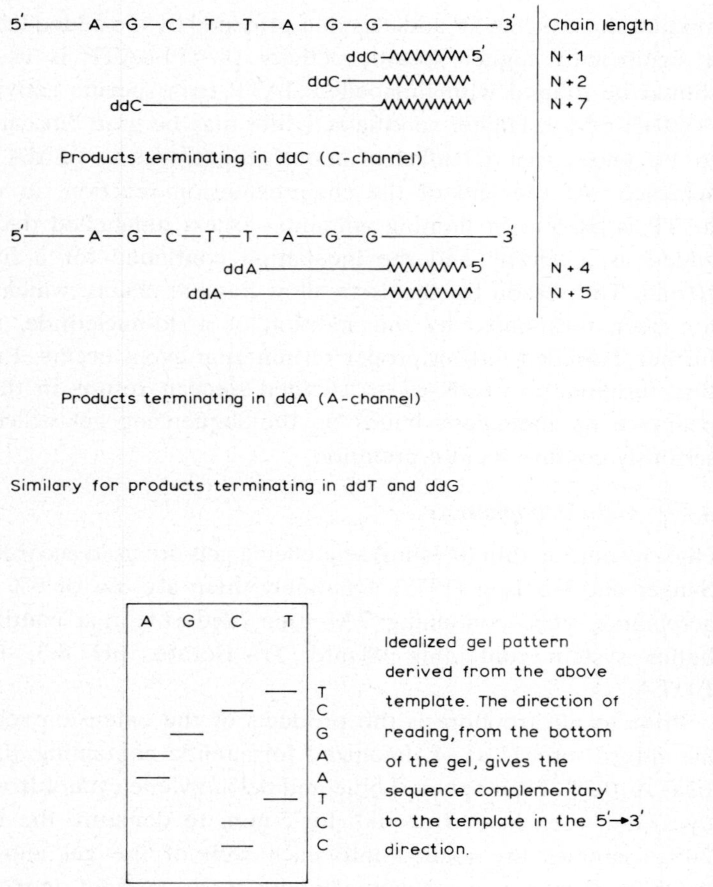

Fig. 4.12. Principle of the dideoxynucleotide chain-termination procedure. Primer (~~) is annealed to the single-stranded template at a site adjacent to the cloned sequence. Chain extension in the presence of the competing dideoxynucleotide results in the random incorporation of that nucleotide at the appropriate sites in the extended product. The mixture of labelled chain-terminated products are fractionated according to size by electrophoresis on a denaturing acrylamide gel and the ladder of products revealed by autoradiography.

of ~250 residues after a $4\frac{1}{2}$ h run. Electrophoresis on 6% gels can be used to push the sequence a further 40–50 nucleotides though the sharpness of the bands tends to decrease somewhat. Detailed procedures are given in Section 4.4.

At the end of the run the gel is autoradiographed either at room temperature after fixing in 10% acetic acid, or directly by autoradiographing the frozen gel at −20°C, or at −70°C when using an intensifying screen (Appendix 4).

Each sequencing channel contains a ladder of fragments all of which terminate in the same dd-nucleotide. The four channels together therefore give a stepped ladder of bands in which every oligonucleotide is represented by a band in one or other of the channels. Under the denaturing conditions of the gel run the mobilities of the oligonucleotides are determined only by their chain length. Reading the sequence therefore merely involves reading up the gel, one mobility unit at a time and noting in which channel the band occurs. The sequence derived represents the sequence complementary to the template, reading in a $5' \rightarrow 3'$ direction (see Fig. 4.12.). As might be anticipated, reading the sequence from the gel is not always entirely straightforward and in Section 4.7. the various problems that can arise are discussed and hints given on the interpretation of gel patterns.

4.3. *Detailed experimental procedures*

This section describes the experimental procedures currently used for cloning into M13 and applying the dideoxy-chain termination procedure for sequencing the cloned sequence. Most of the methods described are those devised by Sanger and his colleagues and are based on their published work and other communications. With the exception of the preparation and use of the pSP14 primer, all the procedures are in current use in the author's laboratory. The Flow Sheet (below) gives in outline the steps for sequencing randomly cloned DNA fragments in M13 vectors. The different steps are described in the appropriate sections.

Flow sheet for sequencing by the M13 cloning-dideoxy chain termination procedure.

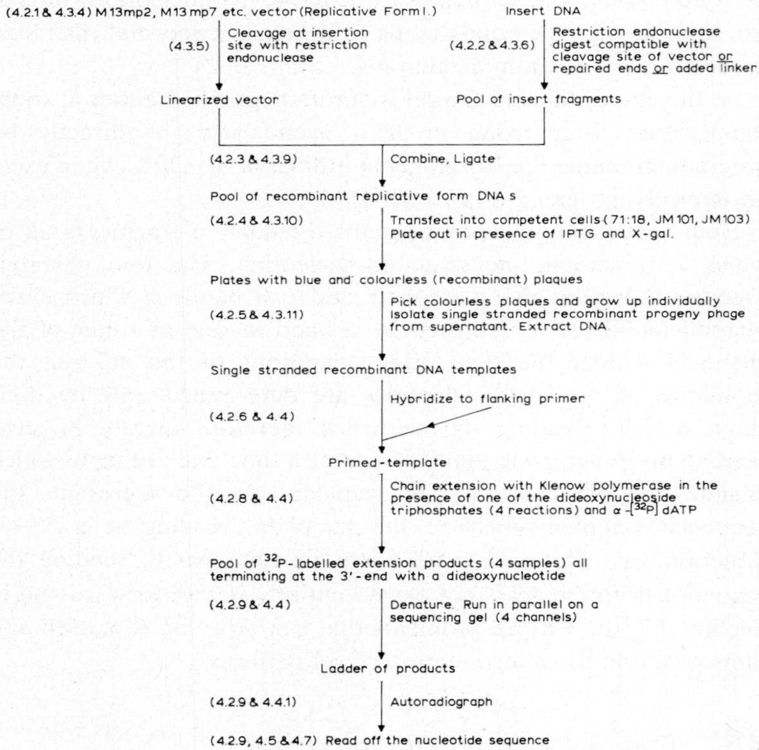

The numbers in brackets (e.g. 4.2.1. etc.) refer to the numbered sections in the text where the procedures are described.

4.3.1. Reagents

DNA Polymerase I, Klenow sub-fragment from Boehringer Mannheim, in a glycerol solution at a concentration of about 1 unit μl^{-1}. Dideoxynucleoside triphosphates are obtained either from P-L-Biochemicals Inc. or from Collaborative Research. The octanucleotide *Eco*RI linker 5′-GGAATTCC was from Collaborative Research. [α-^{32}P]-labelled deoxynucleoside triphosphates (specific activity approximately 400 mCi/μmol) are available at weekly intervals from the Radiochemical Centre, Amersham, Bucks., U.K. Restriction enzymes obtained from New England Biolabs., Boehringer, Mannheim or Bethesda Research Laboratories Inc., were all found to be suitable (see Appendix 5 for full list). A set of automatic adjustable pipettes (such as the Oxford Sampler or Gilson Pipetman) with disposable plastic tips are almost essential. Dade Accupette pipets, especially the 0–5 μl graduated pipet, are generally useful. Suppliers of equipment and reagents are listed in Appendix 5.

4.3.2. Solutions and media used for the cloning procedures

Many of the recipes and details are given in "Experiments in Molecular Genetics" by Jeffrey H. Miller (1972) which is also a valuable source of descriptive detail for many of the bacteriological techniques used:

Low Tris buffer	0.02 M Tris-HCl (pH 7.5), 0.02 M NaCl 0.001 M-EDTA
Triton X-100 buffer	0.05 M Tris-HCl (pH 8.0), 0.01 M-EDTA, 2% Triton X-100
Lysozyme solution	5 mg/ml in 0.05 M Tris-HCl (pH 8.0), 0.01 M-EDTA
Sucrose solution	25% in 0.05 M Tris-HCl (pH 8.0)
0.25 M Tris-EDTA	0.25 M Tris-HCl (pH 8.0), 0.25 M-EDTA
Pancreatic RNAase solution	Dissolve pancreatic RNAase in Low Tris buffer to give a concentration of 10 mg/ml. Boil for 2 min. and store at −20°C.
Ethidium bromide	Dissolve in H_2O to give a concentration of 10 mg/ml. Handle only with gloves.
IPTG	100 mM solution of isopropyl-β-D-thiogalactoside in H_2O. Store frozen.
X-gal	2% solution of 5-chloro-4-bromo-3-indolyl-β-D-galactoside in dimethyl-formamide. Protect from light and store at −20°C.

Concentrated (X10) *Eco*RI buffer	800 mM Tris-HCl (pH 7.5), 800 mM NaCl, 100 mM $MgCl_2$, 10 mM DTT.
YT-medium	Bactotryptone (8 g), Bacto yeast extract (5 g) and NaCl (5 g) per litre H_2O.
2 YT-medium	Bactotryptone (16 g), Bactoyeast extract (10 g) and NaCl (5 g) per litre H_2O.
Minimal agar plates	Minimal agar contains, per litre H_2O: Difco Minimal Agar 15 g Salts: K_2HPO_4 10.5 g KH_2PO_4 4.5 g $(NH_4)_2SO_4$ 1 g Na citrate $2H_2O$ 0.5 g After autoclaving add 1 ml of a 20% solution of $MgSO_4 \cdot 7H_2O$ and 10 ml of a 20% solution of glucose. The salt solution and the agar should be prepared and autoclaved separately at 15 lbs/in^2 for 15 min. It is convenient to autoclave 15 g agar in 800 ml H_2O and the salts (at 5 × normal strength) in 200 ml H_2O. Minimal agar plates are used for the maintenance of the *E. coli* host strains 71/18, JM-101 and JM-103.
Soft agar	6 g minimal agar per litre YT-medium
Hard agar	15 g minimal agar per litre YT-medium

Experimental procedures

4.3.3. Preparation of starter culture of E. coli 71/18/JM101/JM103

Streak cells out on a minimal agar plate and incubate at 37°C until discrete colonies are visible. Pick a single colony for a 2–5 ml overnight culture in YT medium.

4.3.4. Preparation of M13 RFI DNA

RFI DNA, from M13mp2 (Gronenborn and Messing, 1978) and from M13mp2962 (Heidecker et al., 1980), (the source of the 96-nucleotide long universal primer) are prepared from single plaque isolates from a lawn of *E. coli* 71/18.

Prepare a set of ten-fold serial dilutions of the phage, take 0.1 ml from each dilution and mix it with 10 μl IPTG solution, 50 μl X-gal solution, 2 ml of a fresh overnight culture of *E. coli* 71/18 and 3 ml melted soft agar. Pour the mixture onto hard YT agar plates, allow to dry off after the soft agar has hardened and incubate at 37°C overnight. Plaques can be seen after about 4 h but full colour development takes ~12 h. Pick some infected cells from a single blue plaque, put them into 1 ml of 2 YT medium and grow to saturation with shaking at 37°C.

For the preparation of the intracellular RFI of M13mp2 (the cloning vector) a 40 ml culture is sufficient. The procedure given is that of Messing (1979).

Inoculate 0.5 ml infected cell culture into 40 ml 2YT medium and grow to saturation in an orbital incubator at 37°C. Centrifuge the cells in the Sorvall SS 34 rotor at 10,000 rpm for 10 min. The supernatant can be saved to prepare single-stranded DNA if required. The pellet is resuspended in 25% sucrose in 0.5 N Tris-HCl pH 8.0 to make up a volume of 1.5 ml. Keep the suspension in the plastic Sorvall tube which has been used to collect the cells in ice at all times. Add 0.3 ml lysozyme solution and leave for 5 min. All solutions are mixed very gently by slowly revolving and tipping the tube. Then add 0.6 ml of 0.25 M Tris-EDTA followed by 0.025 ml of RNAase solution. After 5 min. incubation lysis is completed by incubation of the mixture with 2.4 ml of Tris-EDTA-Triton solution for a further 10 min. The lysis mixture is cleared by centrifugation at 20,000 rpm in the Sorvall SS 34 rotor for 60 min. at 4°C. The supernatant is diluted with water to a volume of 17 ml containing 0.32 ml ethidium bromide and then used to dissolve 16 g of CsCl. The CsCl solution is centrifuged to equilibrium in two nitrocellulose or polycarbonate tubes in an angle rotor at ~100,000 **g** for 40 h at 15°C. After the run withdraw the denser band illuminated by UV by puncturing the tube from the side with a syringe and a needle. Avoid any contact with the upper, less dense band. The recovered fraction is mixed with an equal amount n-butanol to extract the ethidium bromide. Repeat the extraction twice. Dialyse overnight against 2 l Low Tris buffer. The DNA is precipitated with 1/10 volume 3 M NaAc and 2.5 volumes ethanol, washed once with 70% ethanol, dried under vacuum and redissolved in 200 μl Low Tris buffer. This stock solution should have a concentration of between 50–100 ng DNA/μl. An alternative work-up procedure is to dialyse the DNA band against 2 litres Low Tris buffer before removal of the ethidium bromide and then extract 2–3 times with aqueous phenol. The ethidium bromide is removed into the phenol layer which also removes any traces of protein associated with the DNA. Residual phenol is removed by 5 extractions with diethyl ether and the DNA precipitated as before. Dialysis tubing should be boiled in 2% sodium bicarbonate, 50 mM-EDTA and rinsed thoroughly with distilled H_2O before use.

The M13mp2/(*Bam*HI) and M13mp2/(*Hind*III) and M13mp7 vectors are isolated in exactly the same way starting from a single plaque isolate.

4.3.5. Linearization of vector DNA

(a) *For sticky-end ligation*

Linearization of the M13mp2 RFI DNA is accomplished by opening the molecule at the single *Eco*RI cleavage site.

Mix: 17 μl of RFI stock solution (1–2 μg DNA)

2 μl 10 × *Eco*RI buffer {1 M NaCl, 500 mM Tris-Cl pH 7.5, 50 mM $MgCl_2$}

1 μl *Eco*RI (1–2 units)

20 μl

Incubate 15 min. at 37°C, and inactivate the enzyme by heating at 68°C for 10 min. Store at 4°C, do not freeze.

Electrophorese a 5 μl sample of the above digest in parallel with 4 μl of the untreated RFI solution on a 1% agarose gel in tris-borate buffer (100 mM-Tris base, 100 mM-boric acid, 2 mM-EDTA). Dilute each sample into 15 μl H_2O and add 3 μl loading buffer (3% bromophenol blue in 0.1 M-EDTA, pH 8.3, 50%

glycerol) prior to loading onto the gel. 1 cm wide slots in a 3 mm thick gel are suitable. Electrophorese until the bromophenol blue marker has migrated about 10–15 cm down the gel (2–3 hr at 200 V). Remove the gel, stain by soaking in aqueous ethidium bromide (5 μg/ml) for 10–20 min. and photograph under UV illumination. Complete conversion of the open circular and supercoiled RFI DNA into linear molecules should be obtained as shown by the appearance of a single sharp band of increased mobility relative to the covalently closed circles. Careful inspection of the gel should show no traces of contaminant DNA. Usually the RF DNA prepared as described is pure enough for cloning. If the presence of contaminant *E. coli* DNA is indicated in the above assay the RF should be repurified by banding in another buoyant density centrifugation.

Each new batch of cleaved vector should be checked by transformation (below) to ensure that it gives only a few blue plaques (less than 10/ng), and that on religation it gives only blue plaques (200–400/ng).

The above protocol is essentially unchanged if other enzymes are used to cleave the vector DNA. (e.g. *Acc*I-cleaved M13mp7 DNA used for cloning *Taq*I and *Hpa*II fragments, *Bam*HI-cleaved M13mp2/Bam DNA used for cloning *Sau*3A fragments etc.).

(b) *For blunt-end ligation*

For this purpose the M13mp7 vector is digested with *Hin*dII (or *Hin*cII), both of which cleave the sequence GTPy↓PuAC. However, these enzymes always seem to generate a high (10%) proportion of colourless plaques on religation of the vector in the absence of added restriction fragments. This effect can be minimized by prior phosphatase treatment of the vector to remove the 5′-terminal phosphates since 5′-dephosphorylated ends are not effective substrates for DNA ligase.

Alternatively, if the phosphatase treatment is to be used, the *Eco*RI-cut vector can be used for blunt-ended ligation after repair of the *Eco*RI sticky ends. In practice, this latter procedure appears to give more reproducible results

(i) Cleave RFI DNA (M13mp2, M13mp7) with *Eco*RI and heat to inactivate the enzyme (as above)

(ii) Add to the 20 μl cleaved vector:
1 μl each 5 mM dATP, 5 mM dTTP; 1 unit (1 μl) Klenow polymerase.

- Incubate room temperature for 15 min.
- Extract with phenol (water saturated) (Appendix 4)
- Remove residual phenol by five ether extractions and ethanol precipitate (0.1 vol. 3 M sodium acetate, 2.5 vol. 95% ethanol)
- Collect precipitate by centrifugation (Eppendorf) and dry.

(iii) Redissolve the end-repaired vector in 8 μl H_2O

- Add 10 μl 2 × CIP buffer*
  2 μl calf intestinal phosphatase† (CIP)
  ______
  20 μl
- Incubate at 37°C for 10 min.
- Phenol extract at least *twice*
- Wash with ether four times
- Dilute to about 10 ng/μl final concentration
- Store at 4°C.

The efficiency of the phosphatase treatment is tested by religating the vector and transformation (Sections 4.3.9. and 4.3.10.) and comparing the number of plaques obtained with end-repaired vector *without* phosphatase treatment. The phosphatase-treated vector should give about the same number of plaques as a ligase-minus control.

---

* 2 × CIP buffer: 10 mM glycine, pH 10.5, 1 mM $MgCl_2$, 1 mM zinc acetate.

† CIP is further purified by elution on G75 Sephadex (0.3 cm × 20 cm) in 2 × CIP buffer. The peak fractions are identified using p-nitrophenyl phosphate as substrate, and made 50% in glycerol. Fractions are stored at −20°C. CIP is assayed by digesting *Eco*RI-cleaved vector (not end-repaired) with serial dilutions of CIP. The DNA is then phenol-extracted, ether-extracted and ethanol-precipitated. The DNA is religated and used to transform JM101 (with appropriate controls) (Sections 4.3.10. and 4.3.11.). The correct amount of CIP to use is the lowest which gives the same number of plaques as the ligase-minus control.

4.3.6. *Preparation of 'insert' DNA fragments*

1. Restriction fragments:

(a) Mix DNA (1–3 μg)
10 × restriction buffer
H_2O
Restriction endonuclease (0.5–2 units) in appropriate amounts, and incubate under optimum conditions to obtain complete digestion (manufacturers recommendations) in a final volume of 20 μl.

Heat 70°C for 10 min. to inactivate enzyme (except *Taq*I and other thermostable enzymes).

(b) Phenol-extract, ether-extract, and ethanol-precipitate (0.1 vols 3 M NaOAc, 2.5 vols EtOH) as above and dissolve in a small volume of distilled H_2O (20 μl).

2. Restriction fragments — end-repaired

If the enzyme gives 3′- or 5′-sticky ends which cannot be ligated directly into a site in M13mp7 (e.g. *Hha*I, *Hin*fI), the fragment termini can be converted into blunt ends:

– To 20 μl digest from 1(a) add:
2 μl dNTP mix (1.25 mM in all four dNTPs)
0.5 unit T4 DNA polymerase (or Klenow polymerase, 1 unit/μl)
―――――
22.5 μl.
– Incubate at room temp for 10 min.
– Heat 70°C for 10 min. to inactivate enzyme
– Extract with phenol (as above)
– Ethanol-precipitate (0.1 vol 3 M NaOAc; 2.5 vols EtOH).

Ethanol precipitation will not remove all the dNTPs and these may subsequently inhibit the ligation reaction. If problems are encountered, the DNA should be separated from the dNTPs on a 1 ml sephadex or sepharose column (this section (vi)), precipitated with ethanol (as above) and dissolved in a small volume of distilled H_2O.

3. Restriction fragments—modified by the addition of linker

(i) *Phosphorylation of linker*

The R_1 linker 5′-GGAATTCC as purchased, terminates in 5′-
CCTTAAGG-5′

OH groups. For ligation to restriction fragments 5′-phosphorylated ends are required and the first step is the phosphorylation of these groups using polynucleotide kinase:

– Mix: 1 μg R_1 linker (2 μl) in H_2O
1 μl 500 mM Tris-HCl, pH 7.5, 100 mM $MgCl_2$, 1 M NaCl
1 μl 100 mM DTT
1 μl 5 mM rATP (freshly prepared)
1 μl T_4-kinase (Boehringer, 5 units)
4 μl H_2O
10 μl

– Incubate 37°C, 1 hour

The kinase reaction can be checked by using [γ-^{32}P]rATP and running an aliquot of the mixture on a 20% acrylamide gel. However, this is usually not necessary.

(ii) *End-labelling of restriction fragments*

As discussed in Section 4.2.2. it is convenient to pre-label the restriction fragments to a low specific activity in order to follow their subsequent isolation by gel chromatography. For examples of the procedures used the 3′-end labelling of flush-ended fragments (*Alu*I) and fragments with 5′-protruding ends (*Hin*f) are given.

Flush-end cutters

(a) *Digestion with Alu*I

5 μg DNA (16 μl in H_2O)
4 units *Alu*I (2 μl, 2 units/μl)
2 μl 10 × H buffer*
20 μl

Incubate 37°C, 2 h.

* 10 × H buffer: 66 mM Tris-HCl (pH 7.4), 66 mM-$MgCl_2$, 50 mM DTT, 0.5 M NaCl.

(b) *Labelling by GTP exchange*
Dry down 5 μl (5 μCi) α-[^{32}P]-dGTP (~400 Ci/mMole) in a small siliconized test tube and dissolve in the above *Alu*I digest.
– Add: 2 μl each of 20 mM dATP, 20 mM dCTP and 20 mM dTTP
1 μl 10 × H buffer
2 μl DNA polymerase (Klenow sub-fragment, Boehringer)
– Incubate at room temperature 5–15 min.
– Add 20 μl phenol (redistilled H_2O saturated), vortex, centrifuge and collect aqueous layer. Wash phenol layer with a further 20 μl H_2O. Extract aqueous phase twice with 1 ml ether. Adjust volume to 60 μl with H_2O, add 6 μl 3 M Na acetate and 165 μl ethanol. Chill (−20°C overnight, or −70°C, 10 min.). Centrifuge 10 min. in Eppendorf centrifuge. Wash precipitate with 0.5 ml 95% ethanol. Dry.

5′-protruding ends, e.g. Hinf (*G*↓*ANTC*)
(a) Digestion as above.
(b) Dry down 5 μl (5 μCi) [α-^{32}P]-dATP and dissolve in above restriction digest.
– Add 1 μl DNA polymerase (Klenow sub-fragment).
– Incubate 5 min. room temperature.
– Add 1 μl CTAG mix (equal vols. of 20 mM solutions of the four deoxynucleotide triphosphates).
– Incubate at room temperature, 30 mins. Phenol etc. as above.

The flush-ended labelled fragments from either of the above reactions are now ready for the attachment of linkers.

(iii) *Ligation with linker*
Dissolve dried down fragments directly in the solution of phosphorylated linker from (i) above.
Add: 2 μl T_4-ligase (0.5 unit)
1 μl freshly prepared 5 mM rATP
Mix, incubate at 14°C for 16–20 h.
Inactivate enzyme by heating at 70°C, 10 min.

These procedures can be scaled down for smaller quantities of restriction fragments.

e.g. For 1 μg restriction fragments the dried sample is dissolved in 2 μl of the phosphorylated linker solution in the tip of a drawn out capillary tube.
To this is added 0.5 μl of a solution made from 1.2 μl T_4-ligase (250 units/ml) and 0.3 μl 10 mM rATP.
Seal the tube and incubate 14°C for 20 h.
Heat at 70°C, 10 min. to kill the reaction.

(iv) *EcoRI Digestion*
The excess polymerized linkers attached to the restriction fragments are removed by digestion with *Eco*RI.

Dilute the sample from (iii) above with an equal volume of 2× RI buffer (160 mM Tris-HCl pH 7.5, 160 mM-NaCl, 20 mM-$MgCl_2$, 2 mM-DTT) and digest with an excess (15–20 units) of *Eco*RI, 3 h, 37°C.

The extent of ligation and RI digestion can be checked by electrophoresing samples on an 8% polyacrylamide gel, but the results are often not very encouraging e.g., after ligation and RI digestion the pattern of bands should be similar to the pattern given by the original digest but they rarely are! This should not discourage the experimenter with proceeding with the cloning.

When the scaled-down procedure is used it is sufficient to add 4 μl 1× RI buffer and 0.6 μl *Eco*RI (5 units/μl) to the 4 μl sample from (iii) above and incubate at 37°C for 3 h.

When the digest is complete an excess of EDTA (0.2 M neutralized solution) is added to complex the Mg^{++} and the mixture treated with phenol and ether (as above).

(v) *Purification of linker-added fragments*
At this stage it is important to purify the sample from all traces of depolymerized linker since these latter fragments would become incorporated into the vector and yield colourless plaques.

In the procedure of Sanger et al. (1980) this is achieved by chromatography on a column of Sepharose-4B.

Prepare a column of Sepharose-4B (Pharmacia) in 2 mM-Tris-HCl, pH 7.6, 0.1 mM-EDTA in a 1 ml disposable plastic pipette plugged with siliconized glass wool. Wash thoroughly with buffer and apply the RI digest. The column can be monitored for radioactivity and one drop fractions are collected in Eppendorf tubes. There is usually a radioactive peak followed by a long trail. The front two to four fractions are usually collected.

Example:

Fraction	*Monitor Reading* (*cps*)
1	10
2	120
3	350
4	300
5	160
6	100
7	100

Fractions 2, 3 and 4 are pooled, evaporated to dryness in a desiccator and dissolved in 10–20 μl H_2O.

This procedure selects for the longer restriction fragments which are unretarded on the column. Fragments of <100 nucleotides are eluted later. An alternative procedure which minimizes this selective loss is to chromatograph the sample on Sephadex G-75 or G-100 in high salt:

Prepare a column of G-75 or G-100 Sephadex (Fine) in a buffer containing 0.75 M NaCl, 10 mM Tris-HCl, pH 7.5, 0.5 mM-EDTA in a disposable 1 ml plastic pipette as before. Apply the sample and chromatograph collecting two drop fractions.

Example:

Fraction (2 drops)	*Monitor Reading* (*cps*)
7	25
8	120
9	250
10	250
11	100
12	35
13	17

Fractions 7–9 and 10–12 are pooled separately, 3.0 vol. ethanol added, frozen at −70°C, and the DNA collected by centri-

fugation. Care is taken to remove all traces of supernatant, the precipitates dried and dissolved in 10–20 μl H_2O.

Under these conditions the octanucleotide linker fragments chromatograph at a similar rate to nucleoside triphosphates. There is usually sufficient of the labelled triphosphate, used for the 3′-end labelling, remaining in the sample to act as a marker for the linker peak which elutes later. The first fractions are enriched in the longer DNA fragments—which tend to clone with lower efficiency—and this crude fractionation decreases the number of random plaques which have to be picked to obtain cloned representatives of all the fragments.

4. Sonicated DNA

(i) Take 2–4 μg DNA in 200 μl TE buffer (10 mM Tris-Cl, pH 7.5, 0.5 mM EDTA), chill in ice and sonicate in 7 second bursts at 15 watts output with intermittent cooling. With a 3 mm diameter titanium probe tip between 2 and 5 periods of sonication should be sufficient to fragment most of the DNA into pieces of between 100–500 base pairs in length. The precise conditions will need to be determined by trial and error depending on the equipment used and the molecular weight of the DNA. The easiest way of determining the size distribution of fragments is to electrophorese samples in an 8% acrylamide gel with appropriate markers (Section 4.3.8.). Fragments of about 200 base pairs in length are optimal for sequencing.

(ii) Ethanol precipitate the DNA (0.1 vol. 3 M sodium acetate, 2.5 vols. ethanol), chill, centrifuge, and dry the precipitate.

(iii) End repair of sonicated fragments:

Redissolve the DNA at a concentration of 200 μg/ml in H_2O (i.e. 10–20 μl).

To 10 μl DNA solution (2 μg) add:

1 μl dNTP mix (1.25 mM in all four dNTPs)
2.5 μl 10 × Repair Buffer (500 mM NaCl, 100 mM Tris-Cl, pH 7.5, 100 mM $MgCl_2$, 10 mM DTT)
12 μl H_2O
1 unit T_4 DNA polymerase (0.2 μl)
———
25 μl

Incubate 11°–15°C for 2 h.

(iv) The end-repaired fragments may *either* be fractionated by electrophoresis on an agarose or acrylamide gel (Section 4.3.8.) and desired size range of fragments cut out from the gel and eluted *or* the mixture extracted with phenol, washed with ether and ethanol precipitated. In this latter case it is advisable to further fractionate the mixture to obtain fragments of optimum size range for subsequent cloning. This may be achieved by chromatography on Sepharose 4B or Sephadex (this section, vi) or selective precipitation with polyethylene glycol (PEG6000) (Lis, 1980). The fragments may either be blunt-end ligated into M13mp7 (Section 4.3.9.) or, after addition of the R_1 linker (this section, iii) cloned into the *Eco*RI site of the M13mp2 or M13mp7.

4.3.7. Preparation of the sequencing primers

1. The 96-nucleotide primer

The primer is prepared from an *Eco*RI digest of the RFI DNA, isolated as described above, from M13mp2962 (Heidecker et al., 1980), and separated by polyacrylamide gel electrophoresis or RPC-5 chromatography.

For gel separation 200 μg RFI DNA, isolated from a 500 ml culture of infected cells, is digested to completion with *Eco*RI in a volume of 200 μl and then adjusted to contain 10% sucrose, 0.03% bromophenol blue and 10 mM-EDTA. The sample is loaded into a 10 cm-long slot on an 8% polyacrylamide gel (3 mm × 20 cm × 20 cm) and electrophoresis carried out for 4 h at 150 V in a continuous buffer system containing 0.09 M-Tris-borate (pH 8.3), 2.5 mM-EDTA. The gel is stained with ethidium bromide (1 μg/ml) for 20 min., the DNA visualized under UV light and the rapidly migrating DNA band, moving close to the bromophenol blue marker, excised keeping the gel volume to a minimum. The DNA is eluted using the crush and soak method of Maxam and Gilbert (1977), Section 5.9., modified as follows to deal with a larger volume of gel. The gel slice is placed in the barrel of a 2 ml plastic syringe (without the needle) in which the orifice has been narrowed down by gentle heating to soften the plastic. The volume of gel is roughly estimated and forced through the orifice using maximum thumb pressure into a siliconized glass container. Five volumes of gel extraction buffer are added (0.5 M NH_4 acetate, 10 mM Mg acetate, 0.1% SDS, 0.1 mM-EDTA) and the mixture agitated gently overnight at 37°C. The supernatant is collected by filtration through a plug of siliconized glass wool, the gel washed with one volume of extraction buffer and DNA precipitated by the addition of 3 volumes ethanol. After washing with 95% ethanol and drying, the DNA is taken up in 160 μl H_2O and stored at −20°C.

The 96-nucleotide primer can also be isolated by HPLC on RPC-5 at neutral pH (Eshaghpour and Crothers, 1978, Sanger et al., 1980). For a discussion on the application of this technique to nucleic acid fractionation see Wells et al. (1980).

2. The 92-nucleotide primer from pHM232

Heidecker et al. (1980) have cloned the 92-base pair long *Eco*RI-*Alu*I fragment from M13mp2 into the *Eco*RI site of the plasmid pBR325, yielding the recombinant plasmid pHM232. This plasmid retains the genes for ampicillin (Ap) and tetracycline (Tc) resistance but has lost the chloramphenicol resistance (Cm) gene due to the insertion of the 92-base pair fragment into the *Eco*RI site of the latter. Supercoiled pHM232 DNA can be cleaved with *Eco*RI to release the primer which can be used as an alternative to the 96-base pair primer described above.

An inoculum of *E. coli* strain CSH 26* (*thi ara Δ* [*lac pro*], pHM232) is streaked onto a YT-plate containing 20 μg/ml ampicillin. Colonies are picked into 5 ml 2 YT, 20 μg/ml ampicillin, grown up overnight and the 5 ml inoculum transferred into 700 ml 2YT in a 2 litre flask. The culture is incubated at 37°C with shaking overnight. After harvesting the cells a cleared lysate is prepared (section 4.3.4) and the plasmid isolated by banding on a CsCl equilibrium gradient in the presence of ethidium bromide. The plasmid DNA is isolated in the same way as described for M13mp2962. Digestion with *Eco*RI releases the double-stranded primer which is isolated by preparative gel electrophoresis on an 8% polyacrylamide slab gel, as previously.

3. The 30-nucleotide primer from pSP14 (*Anderson et al., 1980*)
Plasmid pSP14 contains a short segment of DNA complementary to the region adjacent to the *Eco*RI site in the vector M13mp2. This segment is obtained as a small restriction fragment and is an alternative to the 96-nucleotide primer. The advantages of using a shorter primer of this type were discussed in Section 4.2.8.

I *Preparation of pSP14 DNA*
1. Plate out a sample of *E. coli* carrying pSP14 onto a YT plate containing 25 μg/ml ampicillin and grow overnight at 37°C. Resuspend the resultant colonies into 5 ml 2YT and use this to inoculate 1 litre 2YT in a 2 litre flask. Incubate with shaking at 37°C up to a cell density of approximately 0.8 A_{620} units; add 150 mg chloramphenicol and incubate for an additional 16 h at 37°C.
2. Harvest the cells and prepare a cleared lysate as previously (4.3.4.), except that the volumes are increased to give approximately 40 ml cleared lysate. Adjust to 0.1 M NaCl, 0.2% Sarkosyl, and add 4 mg pancreatic RNAase (from a 5 mg/ml stock solution in 0.1 M sodium acetate, pH 5.5, previously boiled for 10 min.). Incubate for 1 hour at 37°C. Add 1.0 ml of a 1 mg/ml solution of proteinase K (BDH Biochemicals) and continue incubation for a further hour at 55°C.
3. Extract crude lysate twice with 0.5 vol. phenol (redistilled equilibrated with 0.2 M Tris-HCl, pH 7.8) plus 0.5 vol. chloroform : isoamyl alcohol (24 : 1). Precipitate with ethanol and dissolve DNA in 15 ml of 10 mM Tris-HCl (pH .78), 1 mM-EDTA.
4. Add 15.4 g CsCl and 0.4 ml of 5 mg/ml ethidium bromide. Centrifuge for 40 h at 150,000 ***g***. Remove the prominent lower band (clearly visible under long-wave UV illumination) preferably

by puncturing the side of the tube with a syringe. Remove the ethidium bromide by four extractions with iso-propanol, and dialyse against 2 l. 10 mM Tris-HCl (pH 7.8) 1 mM-EDTA, 0.3 M NaCl (3 changes over 24 h). The DNA is finally precipitated with 3 vols. 95% ethanol and redissolved to a final concentration of 1 mg/ml in 10 mM Tris-HCl (pH 7.8) 1 mM EDTA, 10 mM NaCl. About 0.5–1.0 mg pSP14 DNA is obtained from a 1 litre culture.

II *Preparation of primer*
200 μg pSP14 DNA in 500 μl of 10 mM Tris-HCl (pH 7.6), 6 mM $MgCl_2$, 50 mM NaCl is digested with 50 units of *Bam*HI and *Eco*RI for 4 h at 37°C. The DNA, after precipitation with ethanol, is dissolved in a small volume of water, mixed with loading buffer and applied to a 10% polyacrylamide gel in just the same way as when preparing the 96-nucleotide primer. A short run (3 h at 200 V) suffices to separate the fragment which appears as a rapidly migrating, rather diffuse band. Extraction from the gel is carried out as described above. Anderson et al. (1980) also describe a fractionation on RPC-5 HPLC, but also suggest that a satisfactory separation should be obtainable using gel filtration (not tried).

4.3.8. Preparation and purification of restriction endonuclease fragments

Restriction digests are carried out under the conditions recommended by the supplier. Reactions are terminated by the addition of EDTA to 10 mM followed by phenol extraction or inactivation of the enzyme by heating at 70°C for 10 min. The DNA is then precipitated by the addition of 1/10th vol. of 3 M sodium acetate and 2.5–3.0 vols. ethanol. After air-drying the precipitate is dissolved in a small volume of loading buffer (20% sucrose, 0.2% bromophenol blue and xylene cyanol tracking dyes in the running buffer) and the sample layered onto a suitable agarose or polyacrylamide gel. Restriction fragments of length up to about 500 nucleotides are most efficiently resolved on 8% polyacrylamide gels. Longer fragments should be fractionated on 1–2% agarose gels. For preparative isolation purposes, e.g. where the cloning of only particular fragments is required, electrophoresis on a 40 cm long acrylamide gel, 1.5 mm thick is recommended. A 5 cm wide slot is compatible with a loading of 10–15 μg DNA without risk of overloading the gel. The DNA should be applied using the minimum volume of loading buffer. A running buffer of 90 mM Tris-borate, pH 8.3, 2.5 mM-EDTA is generally suitable and electrophoresis is carried out at about 20 mA. Bands are visualized by soaking the gel in ethidium bromide (0.5–1.0 μg/ml in water) for 15 min. and examining under UV light. Appropriate bands may be cut out with a razor blade and the DNA extracted using the crush and soak method of Maxam and Gilbert (1977) (Section 5.9., Procedure G). Electroelution methods may also be applied (e.g. Jeppesen, 1980; Galibert et al., 1974) which are reported to give a more quantitative recovery of DNA fragments longer than 200 nucleotides. Excessive grinding of the gel should be avoided in both elution procedures since this releases unpolymerized material which subsequently precipitates with ethanol and is inhibitory to enzymes. Sometimes, where large volumes of gel need to be extracted, this type of contamination is unavoidable and the following procedure is recommended for purifying the DNA (Hindley and Phear, 1979):

> Dilute the DNA extract obtained by the crush and soak method with 4 vols. H_2O and pass through a small (1 ml) column of DEAE-cellulose, DE-52, equilibrated with five-fold diluted gel-extraction buffer. Wash the column with a further 2 ml buffer and elute the absorbed DNA with 1 ml of 1.1 M NaCl in diluted extraction buffer and precipitate by the addition of 3 vols. ethanol. The recovery of DNA is >95% and is free from all traces of gel and acrylamide monomer. DNA extracted by electroelution may be purified in the same manner except that prior dilution is not necessary and the column is equilibrated with the electroelution buffer.

Usually DNA precipitated by the addition of ethanol is recovered by centrifuging in an Eppendorf centrifuge (Model 5412) at

12,000 rpm for 10 min. Where very small quantities of DNA have to be recovered, centrifugation at 45 K rpm for 1 h at 4°C in a SW-60 rotor (or equivalent) is recommended. As much supernatant as possible is drained from the pellet (often invisible) which is finally washed once with 95% ethanol and recentrifuged for the same speed and time. Care is needed to avoid dislodging the pellet after the ethanol wash. After drying, the DNA is dissolved in a small volume of water and stored frozen at −20°C.

Agarose gel electrophoresis
Preparative agarose gel electrophoresis may be carried out in either horizontal or vertical slab gels. The former are often easier to handle as vertical gels sometimes tend to shrink and fall out of the cell. Agarose gels are routinely used for separating the cloned insert in a recombinant plasmid from the plasmid DNA itself. A horizontal gel (20 cm long × 18 cm wide × 0.3 cm depth) made of 1% agarose in a buffer containing 40 mM Tris-acetate, 20 mM Na acetate, 2 mM-EDTA, pH 7.8, is suitable. A 10 cm slot will accommodate >50 μg of a recombinant plasmid DNA digest (e.g. pMB9, pBR322) and resolve an insert up to 2.5 Kbps from the vector. Ethidium bromide may either be incorporated into the gel (0.5 μg/ml) or the gel subsequently stained with 1 μg/ml ethidium bromide in H_2O. The appropriate band(s) are detected under UV-light and excised with a razor blade.

Extraction of the DNA may be achieved by a variety of methods (see e.g. Smith, 1980). Two procedures which are rapid and convenient are described below:

Method 1: takes advantage of the properties of a low melting temperature agarose (Bethesda Research Laboratories, Rockville, Md., USA). The gel slice is melted at 70°C and an equal volume of 10 mM Tris-HCl, (pH 7.6), 0.1 mM-EDTA added and the solution extracted with the same volume of water-saturated phenol in one or more 1.5 ml Eppendorf centrifuge tubes. After brief centrifugation the aqueous layer is removed and the phenol extraction repeated a further 2–3 times. Finally the aqueous phase is extracted with ether and the DNA precipitated at −70°C following the addition of 1/10th vol. 3 M Na acetate, pH 5.5 and 2.5 vols. ethanol. The DNA is pelleted by centrifugation, dried and dissolved in H_2O (Smith, 1980).

Method 2: takes advantage of the solubility of agarose gels in high concentrations of chaiotropic salts such as $NaClO_4$ or NaI and the selective retention of DNA on glass-fibre filters in the presence of these salts (Vogelstein and Gillespie, 1979; Yang et al., 1980).

The gel segment is dissolved in 4 vols. of 6 M $NaClO_4$, 10 mM-Tris-HCl (pH 7.5) at 40–45°C for 30 min. The final concentration of $NaClO_4$ is about 5 M and the agarose approximately 0.25%. The solution is then filtered dropwise through a single 8 mm diameter GF/C glass-fibre filter placed on a sintered-glass filtration apparatus connected to a water pump. A filter of this size can accommodate up to 10 μg DNA. If desired a pad of up to four filters stacked on top of each other can be used. The filter is washed with 0.5 ml 6 M $NaClO_4$ and then with 0.5 ml propan-2-ol followed by 1 ml ethanol. The filter disc(s) are removed and air-dried. The ethanol wash effectively removes all $NaClO_4$ from the filter. DNA is eluted from the filter in the following way. The cap of a closed 1.5 ml Eppendorf microfuge tube is pierced several times with a 26-gauge needle and the filter placed into the exterior indentation of the cap. 25 μl Tris-EDTA buffer (5 mM Tris-HCl, pH 7.4, 0.5 mM-EDTA) is added to the filter which is then covered over with Parafilm. After standing a few minutes the tube is centrifuged for 10 s. in the microfuge. The elution step is repeated twice more. The DNA solution, about 60 μl, collects in the bottom of the centrifuge tube. If more filters are used, the volume of eluting buffer is increased. DNA fragments less than ~200 base pairs are not efficiently retained by the filter but for longer fragments recoveries of 90% or more are obtained. The author has found this to be an extremely convenient method. The DNA is recovered in a small volume of buffer suitable for further enzymatic reactions.

4.3.9. Ligation into M13mp2 or M13mp7 vector DNA

The tests described under 4.3.5. "Linearization of vector" indicate whether the DNA and T_4-ligase are of sufficient quality to proceed with the ligation. The experimental procedure described below includes a biological control which also gives a more stringent test of the reagents. However, before proceeding to the ligation the optimal ligase concentration should be established.

1. Optimum ligase concentration

(a) Make a series of 10-fold serial dilutions of T4-ligase in ligase dilution buffer (50% glycerol, 50 mM KCl, 10 mM pot. phosphate, pH 7.6, 6 mM 2-mercaptoethanol, 100 μg/ml autoclaved gelatin).

(b) Mix: 1 μl (10 ng) vector
1 μl 10 × ligase buffer (500 mM Tris-Cl, pH 7.5, 100 mM $MgCl_2$, 10 mM DTT)
6 μl H_2O
1 μl 10 mM rATP
1 μl T4 DNA ligase
10 μl

Seal in a capillary and incubate at 14°C for 6 h. Use mixture to transform competent cells (e.g. JM101) and plate out (see Section 4.3.11. below). Count the number of plaques for each ligase concentration and choose the optimum concentration. NOTE: Blunt-end ligation requires more ligase than sticky-end ligation of *Acc*I-cleaved vector (2-base overlap, Fig. 4.5.) which requires more than ligation of *Eco*RI, *Pst*I or *Bam*HI-cleaved vector (4-base overlap).

(c) The presence of white plaques on religation of the vector indicates that the vector may be contaminated with foreign (e.g. *E. coli* chromosome) DNA, or that the ends may have been damaged by a contaminating nuclease (Section 4.3.11. below).

2. Ligation into vector DNA

– Mix: 4 μl linearized vector DNA (20–40 ng) prepared by diluting the stock vector digest with water
4 μl appropriate DNA fragments (50–100 ng)
2 μl × 10 ligase buffer (1(b) above)
2 μl autoclaved gelatin (2 mg/ml)
5 μl 5 mM-rATP
7 μl H_2O
1 μl T_4-ligase of appropriate concentration
20 μl

Incubate overnight in bath of water at 15°C placed in a cold-room so that temperature drops slowly to about 8°C. The ligation mixture is now ready to proceed to the transformation of competent cells. For blunt-ended ligations incubate at 14°C for 24 h. Always include a control with no DNA fragments.

Figure 4.13. shows, as an example, the ligation of a *Sau*3A fragment into the *Bam*H1 site of the M13mp2/*Bam*H1 vector.

4.3.10. Transformation

The number of competent cells required depends on the number of transformation mixtures to be assayed. For 2 ng of vector DNA, 0.3 ml competent cells and one agar plate are required. The competent cells are concentrated ten-fold out of the growth volume and extra cells are needed for controls.

Experimental Procedure:

(a) Dilute a fresh overnight culture of *E. coli* 71/18 (or JM101 or 103) 1/100 into YT medium (Section 4.3.2.). Grow cells with shaking until a density of 0.6–0.7 A_{660} is reached and chill the flask on ice.
(b) Collect the cells by centrifuging in a Sorvall SS 34 rotor at 9000 rpm, 5 min. at 4°C and resuspend the cells gently in half the growth volume of 200 mM $CaCl_2$. Keep in ice for 20 min.
(c) Collect the cells again by centrifuging and resuspend in 1/10th growth volume of ice-cold 80 mM $CaCl_2$.
(d) Mix the amount of DNA solution from 4.3.9. (above)

equivalent to 2 ng vector (0.5–1 μl) with 0.3 ml competent cells and leave in ice for 40 min.

(e) Heat shock for 2 min. at 45°C.

(f) Return to room temperature and add directly:
10 μl IPTG (100 mM)
50 μl X-Gal (2% in dimethylformamide)
0.2 ml fresh exponentially growing 71/18 (JM101, JM103) cells
3 ml melted soft agar (at 45°C)

(g) Pour directly onto hard YT agar plates, swirl gently to cover the agar layer and allow the plates to dry off and harden. Incubate at 37°C overnight. The fresh exponential 71/18 cells may be obtained by adding fresh YT medium to the culture flask which still contains a few drops from the harvested cells. These will grow up to a sufficient density while making the cells competent and during the transformation.

This procedure yields about 100–200 plaques per plate, 15–30% of which are recombinants as distinguished by their colourless appearance compared to the blue plaques produced by phage with no inserted fragment at the cloning site.

Each transformation experiment should be accompanied by at least one control from a ligation mixture containing the vector DNA only. When starting with a fresh batch of vector it is also recommended that control plates are set up using the non-linearized vector and the linearized vector before ligation.

These assays will indicate the quality of one's reagents and that the procedures are working properly. The following approximate results should be obtained:

1. Original (non-linearized vector) —100–300 blue plaques only
2. Linearized vector —0–10 blue plaques only
3. Linearized vector (after ligation) —100–200 blue plaques
0–2 colourless plaques
4. Linearized vector ligated to fragments —mixture of blue and colourless plaques (15–30% colourless)

The most important control is No. 3. If colourless plaques occur with a frequency of >1% the purity of the vector or ligase are suspect.

Small-scale transformation procedure (*Sanger et al., 1980*)

– Mix: 1 μl containing 10 ng vector digested with *Eco*RI
 1.2 μl 2× ligase buffer (as above but diluted 1:5 and containing 0.5 mM-rATP)
 0.2 μl linker-added fragments (as prepared above), ~12 ng
 0.25 μl T_4-ligase (25 units/ml)

Seal in capillary, at 15°C for 2 h

Add 20 μl 10 mM Tris-HCl, (pH 7.5), 0.1 mM-EDTA

Use 5 μl for transformation of 0.2 ml competent cells for one plate

Should give about 20 colourless plaques

4.3.11. Preparation of single-stranded M13 DNA templates

Colourless plaques from the transformation experiment are picked individually from the plate and transferred into 2 ml 2YT medium in 50 ml screw-capped universal bottles. Wooden toothpicks are convenient or the plaques may be stabbed out using pasteur pipettes. The cultures are incubated with shaking for 8–14 hours and 1 ml samples of the turbid cultures transferred to 1.5 ml Eppendorf microtubes. The cells are pelleted by centrifugation in an Eppendorf microfuge for 10 min. and 0.8 ml of the clear supernatant withdrawn into another Eppendorf tube. Care should be taken not to disturb the cell pellet. 0.2 ml of a solution containing 2.5 M NaCl and 20% PEG (polyethyleneglycol 6000, BDH), is added and mixed by inverting the tube a few times. The mixture is allowed to stand for 15–30 min. either at room temperature or in ice and centrifuged for 10 min. The supernatant is decanted as completely as possible and the tubes recentrifuged briefly to collect the remaining supernatant at the bottom of the tube. This is then removed using a drawn-out capillary without disturbing the just visible precipitate.

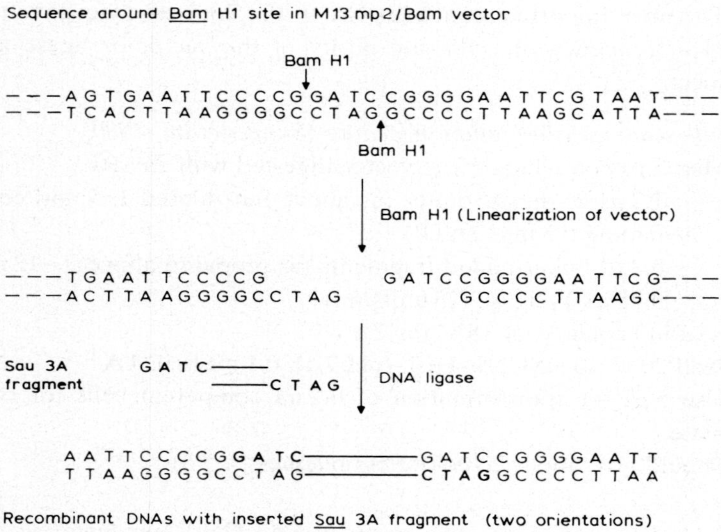

Fig. 4.13. Procedure for cloning *Sau*3A derived fragments into the *Bam*HI site of M13mp2/Bam.

50 μl 10 mM Tris-HCl, pH 7.9, 1 mM-EDTA are added followed by 50 μl water-saturated phenol equilibrated with the above buffer. The mixture is vortexed briefly, allowed to stand for 30 min., revortexed, and centrifuged for 1 min. The upper aqueous phase is removed (about 45 μl) into another Eppendorf tube and extracted three times with ether. Finally, 10 μl 1 M sodium acetate pH 5.5 and 140 μl 95% ethanol are added and the mixture left at −20°C overnight. The DNA is collected by centrifuging for 10 min. and the supernatant removed as previously. The precipitate is barely visible. The pellet may be washed with 0.5 ml 95% ethanol but great care has to be taken not to lose the pellet.

The precipitate is air-dried, dissolved in 50 μl H_2O and stored frozen at −20°C. These samples provide the single-stranded templates for sequencing.

4.4. *Dideoxy-nucleotide sequencing procedures*

(i) *Exonuclease III digestion of primer from M13mp2962 or pHM232* (*Smith, 1979*)

– Mix, in an Eppendorf tube:

40 μl primer (derived from 50 μg M13mp2962 RF or pHM232) (Section 4.3.7.)
8 μl 10 × Exonuclease III buffer (1 M Tris-HCl (pH 8.0), 10 mM $MgCl_2$, 10 mM DTT)
28 μl H_2O
4 μl Exonuclease III (P.L. Biochemicals, 10,000 units/ml)
80 μl

Incubate at 37°C for 10 min.
Heat to 75°C for 10 min. to kill the enzyme

This gives sufficient primer for 20–30 sequencing experiments.

(ii) *Annealing primer to template*

The single-stranded templates from 4.3.11. (above) and primer are denatured by boiling and then annealed together.

To a drawn-out capillary add:

5 μl M13 cloned template DNA
3 μl exonuclease III-treated primer
1 μl 10 × H buffer (66 mM Tris-HCl, pH 7.4, 66 mM-$MgCl_2$, 0.5 M-NaCl, 10 mM-DTT)
1 μl H_2O
10 μl

Seal the capillary, submerge in boiling water for 3 min.
Anneal at 67°C for 15 min. (or longer)

(iii) *Chain-extension reaction*

– Dry down 5 μl (5 μCi) [α-^{32}P]-dATP sp. activity 350 Ci/mMole, in a well-siliconized glass tube. Dissolve residue in 5 μl H_2O and keep covered on ice.
– Set up four drawn-out capillaries, labelled T, C, G, A held horizontally in plasticine. Introduce into the tip of each capillary the following components using, for example, an Oxford 1 μl Sampler.

	T	C	G	A
Sample (from annealing reaction)	2 μl	2 μl	2 μl	2 μl
T° mix	1 μl			
C° mix		1 μl		
G° mix			1 μl	
A° mix				1 μl
ddT	1 μl			
ddC		1 μl		
ddG			1 μl	
ddA				1 μl
[α-^{32}P]-dATP	1 μl	1 μl	1 μl	1 μl
0.2 unit DNA polymerase Klenow sub-fragment	1 μl	1 μl	1 μl	1 μl

- Mix by expelling the contents of each capillary 2–3 times onto the bottom of a siliconized glass tube held on ice (a separate tube for each sample). Draw up into the tip of the capillary and incubate for 15 min. at room temperature.
- Add 1 μl 0.5 mM dATP to each sample
- Mix as before (the same set of 4 tubes can be used)
- Incubate for a further 15 min. at room temperature
- Expel contents of each capillary into 15–20 μl Dye-Formamide-EDTA mixture (see below).
- Denature in a boiling water bath 3–5 min.

The samples are now ready for application to the sequencing gel.

Composition of dXTP mixes

Store frozen at −20°C

	T°	C°	G°	A°
0.5 mM dTTP	1 μl	15 μl	15 μl	20 μl
0.5 mM dCTP	15 μl	1 μl	15 μl	20 μl
0.5 mM dGTP	15 μl	15 μl	1 μl	20 μl
50 mM Tris-HCl pH 8.0, 1 mM EDTA	5μl	5 μl	5 μl	5 μl

Composition of dd-XTP solutions

Stocks of 4 mM or 10 mM dideoxy-nucleotide triphosphates (P.L. Biochemicals or Collaborative Research) are kept frozen in water.

Working solutions are made by appropriate dilutions into H_2O (determined by trial and error) to give a similar spread of band intensities in each of the sequencing channels. The approximate concentrations are:

1 mM dd-TTP
0.4 mM dd-CTP
0.4 mM dd-GTP
0.2 mM dd-ATP

Keep working stocks frozen at −20°C.

Formamide-dye-EDTA mix
Gently stir 100 ml formamide with 5 g Amberlite MB1 (Mixed Bed resin) for 30 min. Remove resin by filtration. Add 0.3 g xylene cyanol FF, 0.3 g bromophenol blue and Na_2 EDTA to 10 mM. Store at 4°C.

(iv) *Gel electrophoresis*
(a) *Preparation of thin sequencing gel*
The following stock solutions are required:
40% acrylamide (1/20 acrylamide : bis). Make up 190 g acrylamide (electrophoresis grade) and 10 g bis-acrylamide (NN′-methylene-bisacrylamide) to 500 ml with distilled water and stir gently with 5 g Amberlite MB-1 resin for 30 min. Remove resin by filtration and keep stock at 4°C. *10 × TBE* (Tris-borate-EDTA buffer). Make up 108 g Tris base, 55 g Boric acid and 9.5 g Na_2 EDTA.$2H_2O$ to 1 litre with distilled water. The resultant pH should be about 8.3. Store at room temperature.
10% ammonium persulphate a 10% solution in water will keep at 4°C for several weeks.
The gels are conveniently made during the annealing step and set up in the gel apparatus during the boiling of the samples in formamide-dyes.
For two gels: Weigh out 42 g urea (BDH AR or Aristar), add 15 ml

40% acrylamide stock solution and 10 ml 10 × TBE. Add water to less than 100 ml and leave urea to dissolve on magnetic stirrer.

Make the acrylamide/urea solution up to 99 ml with H_2O, add 0.8 ml 10% ammonium persulphate and 50 μl TEMED (NNN′N′-Tetramethylethylene-diamine). Mix briefly by swirling and pour into the electrophoresis cell (Section 3.1.2.) making sure that no bubbles are trapped in the gel mixture. Insert the slot former (12–20 slots) made from a piece of Plasticard (Fig. 3.3.), place the gel raised slightly from the horizontal and put a weight over the top of the gel to ensure a tight fit of the slot former.

When set (about 30 min.) remove the tape from the bottom of the gel, remove the slot former and clamp the gel plates vertically into the electrophoresis apparatus. Fill the reservoirs with 1 × TBE and flush out the wells with a pasteur pipette to wash out unpolymerised acrylamide.

Load 3–5 μl of each of the four samples into adjacent wells, connect to the power supply and electrophorese at a constant current of 30 mA. The initial voltage is about 1.7 KV falling to about 1.2 KV as the gel heats up.

A second loading is usually made after 90–100 min. by which time the slower moving dye band (xylene cyanol) should have migrated half way down the gel. Before the second loading the samples should again be briefly boiled and the wells rinsed out with buffer to remove urea that has leached out of the gel. Electrophoresis is terminated when the xylene-cyanol band from the second loading reaches the bottom of the gel.

4.4.1. Autoradiography See Section 3.1.2. and Appendix III

4.5. *Results*

Examples of sequencing gels prepared as described are shown in Plates I–IV. The procedure used in reading the gel is described in Section 4.2.9. Hints on reading the sequence and trouble-shooting are given in Section 4.7.

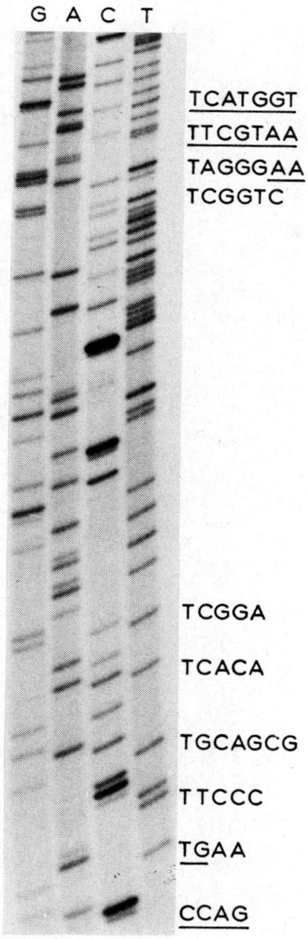

Plate I. Autoradiograph of a sequencing gel from M13mp2 containing an 80-nucleotide long inserted *Alu*I fragment. For discussion see text.

Plate I shows part of a sequencing gel which gives the sequence of an 80-nucleotide *Alu*I fragment derived from the yeast 2 μ plasmid (Elleman and Hindley, 1980). The sequences underlined correspond to the M13mp2 sequences around the site of insertion

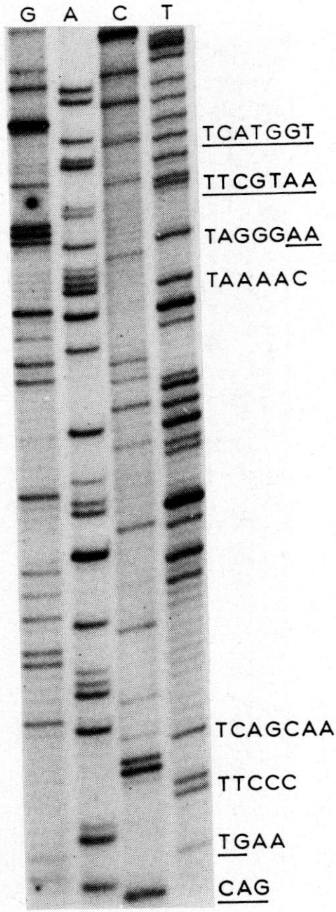

Plate II. Autoradiograph of a sequencing gel showing an increased background of all possible chain lengths. For discussion see text.

and define the ends of the cloned fragment. The basis for this is shown in Fig. 4.14. Part of the sequence of the *Alu*I fragment, as deduced from the gel, is shown in Plate I. One feature which can be noted is the variation of intensity within the three bands

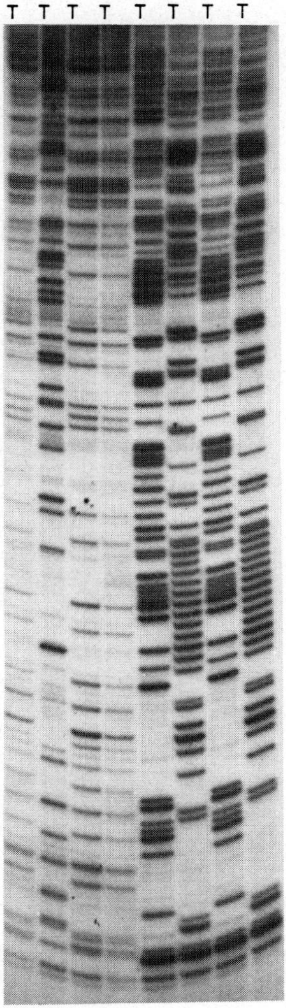

Plate III. A sequencing gel showing the patterns obtained from eight M13mp2 recombinant phage DNAs picked at random. Chain extension was carried out in the presence of ddT only to identify recombinants containing identical cloned sequences.

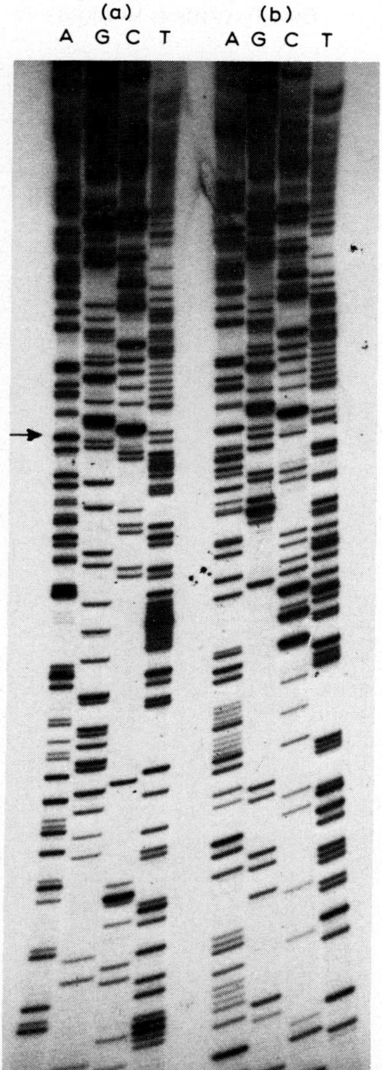

Plate IV. Autoradiograph from two templates produced by cloning *Sau*3A fragments into the M13mp2/Bam vector. In (b) the sequence reading from the bottom of the gel is: –TTCCCCGGATC– which identifies the beginning of the cloned fragment (see Fig. 4.15.). In (a), the sequence from the arrowed position is –GATCCGGGGAATT– which identifies the end of the cloned fragment (see Fig. 4.15.).

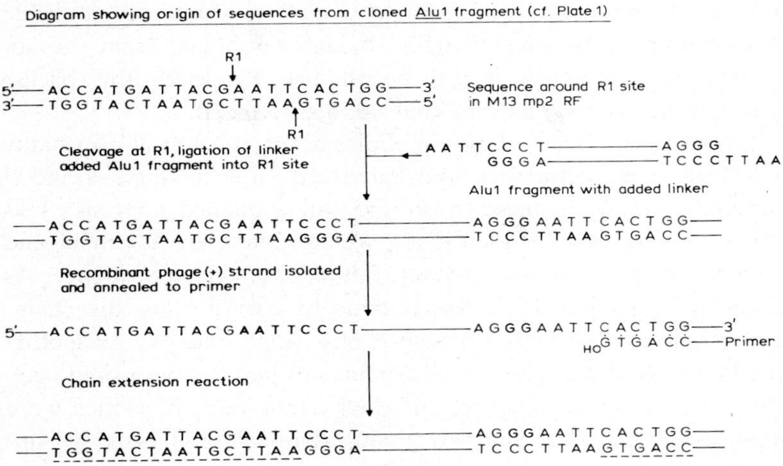

Fig. 4.14. Diagram showing origin of sequences from cloned *Alu*I fragment (cf. Plate I). The dashed underlined sequence is seen to correspond to the underlined bands in Plate I and represents the sequence around the *Eco*RI site in the vector. The third C of the sequence –CCC– defines the start of the cloned *Alu*I fragment and the first G of the –GGG– sequence defines its end.

corresponding to the sequence 5′–CCC–3′. The first band is very weak whereas the second is disproportionately strong. The same effect is seen in the –CC– sequence at the bottom of the gel. This is characteristic of runs of C's and is discussed further in Section 4.7. (Hints on reading sequences).

Plate II shows another autoradiograph of a sequencing gel corresponding to a different cloned *Alu*I fragment. In this gel a much higher background of all possible chain lengths is observed. The reasons for this are discussed in the next section. In the example shown this background does not significantly interfere with the reading of the sequence but a more intense background would lead to problems of interpretation. Note the distribution of intensities in the bands corresponding to runs of A's. The 5′–A is the strongest and successive bands are progressively weaker—in contrast to the situation where runs of C's are encountered. A visible background of all chain lengths, though detracting from the appearance of the gel, is seen to be useful in estimating the spacing between bands and ensuring that none are missed.

Since recombinant clones are picked at random it is usually worthwhile characterizing the sequence from each single-stranded template before embarking on the full sequence analysis. This saves time in quickly identifying which clones are identical and which, for one reason or another, fail to give a clean sequence. As detailed in Section 4.2.7. this is done by carrying out the chain-extension reaction in the presence of a single dideoxy-nucleotide (ddT) and analysing the set of products in parallel on a single gel. Plate III shows an example in which eight random clones were analysed in parallel. Samples giving identical sequences are immediately obvious.

Plate IV shows an autoradiograph from two templates produced

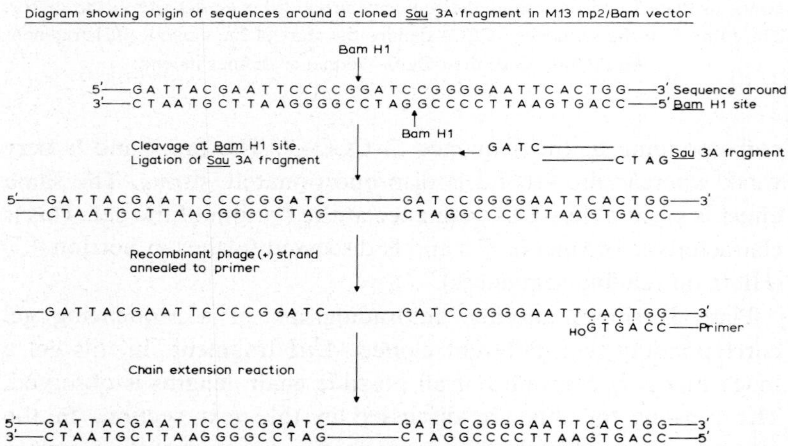

Fig. 4.15. Diagram showing origin of sequences around a cloned *Sau*3A fragment in M13mp2/Bam vector. The beginning of the inserted *Sau*3A fragment is defined by the sequence –CCCCG– followed by the *Sau*3A recognition sequence –GATC–. The end of the cloned fragment is recognised by another –GATC– sequence followed by –CGGGG–. Plate IV shows both these sequences on a sequencing gel.

by cloning *Sau*3A fragments into the M13mp2/*Bam* vector. In this case linkers are not used and Fig. 4.15. shows how the beginning and end of the *Sau*3A fragment is identified.

4.6. *Discussion*

Among the primed synthesis methods for sequencing DNA the procedures described in this chapter are, at present, the most powerful and versatile we have. The speed and simplicity with which sequence information can be obtained is unmatched by any other procedure and with the development of new derivatives of M13 (e.g.: M13mp7) the range of restriction fragments that can be cloned directly, without recourse to the use of linkers, is rapidly increasing. The use of short synthetic primers which anneal to regions within the *lac*Z sequence adjacent to the cloned insert also speeds up the procedure by avoiding the time-consuming preparation of primer from the replicative form of M13mp2962 or recombinant plasmids containing the cloned primer. It is to be hoped that synthetic primers of the type described in Section 4.2.8. will soon be available commercially. (See note added in Update, p. 286.)

A sequence of about 250 nucleotides can usually be determined from the priming site. Occasionally on particularly clear gels the readable sequence can be extended to ~300 nucleotides, especially when the shorter synthetic primers are used, but it is doubtful if much is in general to be gained by resorting to lower gel concentrations, reducing the relative concentrations of the dideoxynucleotides in the chain-extension reaction or increasing the running times above about $4\frac{1}{2}$ h in order to push the sequencing a little further. In some circumstances, e.g. where sequencing a particular cloned fragment to the limit will take the sequence through a repeated region or through a restriction site subsequently used to generate an overlap this may be justified and yield valuable data but in general, excessive zeal in extracting the last few nucleotides in a sequence is not recommended. The aim when using the random cloning approach is first to reduce the DNA into a set of fragments each of which can be sequenced from end to end on a single gel. One approach to this ideal is to successively digest the DNA with a set of restriction enzymes of different specificities and at each stage to fractionate the products by gel chromatography so that the large products which are unretarded on the gel are collected separately and cleaved with the next enzyme in the set. Ultimately, most if not all of the DNA is reduced into sequence-able-sized fragments which are then cloned into the vector. This procedure, however, does mean that a large volume of sequence data is collected since no control can be exercised over the frequency with which a particular sequence is cloned or picked up as a recombinant. For large DNA's (> about 4000–5000 nucleotides) the use of computer programmes is almost essential for the storage and processing of the data. Building up the final sequence requires the cloning and analysis of further sets of fragments to determine the overlaps and also to ensure that no fragments have been overlooked. Ultimately every sequence can be unequivocally positioned and related to the surrounding sequence. When this stage is reached and the sequences of both strands have been determined in fragments covering the entire molecule, the sequence is considered to be established. Computer programmes which can scan, identify and tabulate the data as they become available and continuously correct and upgrade the sequences have been written by Staden (1980). This is essentially the approach used by Sanger et al. (1980) which additionally indicated a number of refinements that can be used to increase the efficiency of the procedure. For example, as the sequencing proceeds, computer scanning of the data can indicate which restriction enzyme would most suitably provide overlaps. Where a sequence larger than about 250 nucleotides has been cloned, the sequence can be 'turned around' by isolating the RF of the phage from the clone, cleaving out the insert and recloning. About 50% of the progeny from this recloning will contain the complementary strand and sequencing will yield the first 250 nucleotides or so from the 3′-end of this strand. Thus cloned fragments of up to 500–600 nucleotides can be sequenced (e.g.: Winter and Fields, 1980).

One problem which does sometimes cause difficulty springs from the selection technique used to identify the recombinant clones. A colourless plaque merely indicates some interruption in that part of the *lac* gene contained in the M13 vector, however caused. If, for example, the restriction enzyme used to linearize the vector contained an exonuclease activity, resealing of the vector itself can give rise to deletions and result in an out of phase reading of the *lac* gene. This will give a colourless plaque which on analysis gives only M13 sequence. This problem is not usually troublesome when using cloning sites with three or four nucleotide 5′-sticky ends but it could be serious when attempting blunt-ended ligation into, for example, the *Hin*dII (*Hin*cII) site of M13mp7 unless a rigorously purified enzyme is used to cleave the vector (Brown, personal communication). In general linearisation of the vector is best carried out using the minimum amount of digestion (it is not necessary to achieve 100% ring opening) with a high-quality enzyme. Sometimes depending on the history of a particular set of fragments to be cloned, a marked reduction in the number of blue plaques may be observed. This may often be attributed to damage to the ends of the fragments such that the majority only contain linker, or an intact cloning sequence, at one end of the fragment. After ligation to the vector at the intact end, such molecules fail to circularize and neither blue nor colourless plaques are formed. A large excess in the ratio of fragments to vector in the ligation mixture may also have a similar effect since the continued ligation of fragments into the vector may effectively compete with its circularization. Whatever the reason it is sometimes found that the number of blue plaques on particular plates falls by up to 90% and this is often associated with extended manipulative procedures in preparing the fragments for cloning.

One other problem which could cause difficulty in rare cases is the apparent failure of a particular sequence to clone into M13. Signals within the insert may terminate the replication of the vector and so fail to produce any progeny.

4.7. *Hints for reading sequencing gels*

The accurate derivation of a sequence from a sequencing gel is not always a straightforward procedure and the purpose of this section is to set out some general guidelines which should be followed when interpreting the results of a sequencing experiment.

The commonest source of error is due to the presence of artifact bands which may occur in one or more of the four sequencing channels. These may be due to a variety of causes and a careful examination of the pattern will often point to the source of the problem. The DNA template for example, may be contaminated with traces of its complementary strand which can in turn result in priming taking place from the 3′-end of this strand. The pattern of bands will consequently be superimposed on a fainter background pattern which in places can obscure or confuse the reading of the sequence. A similar situation arises if the primer is contaminated with traces of other restriction fragments. Since the isolation of restriction fragments is usually achieved by gel electrophoretic methods, small fragments, which tend to give more diffuse bands, may often be contaminated with fragments of similar size and give rise to an anomalous background pattern. Careful attention to the purity of both template and primer DNA will usually serve to overcome problems of this sort. Another problem, more rarely encountered, is heterogeneity in size at the 5′-end of the primer DNA. This will give rise to multiple bands, rather than a single band, corresponding to each position of a nucleotide in a sequence. The most likely source of this problem is contamination of the restriction enzymes used to generate the primers with a $5' \rightarrow 3'$ exonuclease or an impure exonuclease III preparation. While it is still often possible to deduce a sequence provided these anomalies are not too serious, it is always preferable to repurify or retreat the DNA in such a way as to eliminate the problem.

Two types of primer DNA (Sections 4.2.1. and 4.8.1.) may be employed for the chain-extension reaction. When the desired sequence, of chain length up to ~400 nucleotides, is cloned into M13mp2, a flanking primer which hybridizes to a region of the

M13 DNA sequence adjacent to the cloned fragment is usually employed. This has the advantage of providing a unique priming site outside the sequence to be determined and since the primer is not derived from any part of the unknown sequence there is little risk of any significant homology between the sequences which might result in hybridization to regions other than the correct priming site. On the other hand, where the primers are restriction fragments corresponding to regions of a longer fragment whose sequence is to be determined the formation of a unique template-primed hybrid in the annealing reaction cannot always be taken for granted. DNA sequences, particularly those derived from eukaryotic sources, often contain blocks of repeated or partially repeated sequences and under the annealing conditions used non-specific hybrid structures are produced in which the primer is annealed to two or more different regions of template. Often, the repeated sequences take the form of identical blocks of nucleotides and improving the stringency of the annealing reaction does not improve matters. When this situation arises the result is an unreadable pattern of bands in the sequencing gel caused by the superimposition of two or more overlapping sequences of similar intensity. If there is no reason to suspect the purity of one's primer or template DNA, an unreadable sequence of this sort is almost diagnostic of a set of repeated sequences in the template. This is confirmed when a similar result is obtained using the complementary DNA strand as the template. One solution is to rely on primers derived from regions of the sequence outside the repeated region and sequence right through the latter. Where this is not possible the only solution is to reclone subsets of fragments from the repeated region into the vector and sequence these using a flanking primer.

In the dideoxy chain-termination procedure all the bands produced on the sequencing gel should ideally result from specific chain termination caused by the incorporation of a dideoxynucleotide. However, spurious chain termination can also occur and while the reasons for this are not altogether clear, it is supposed that this may be the result of particular features of template which the polymerase finds difficulty in copying. The extending chains consequently 'pile up' at these regions yielding bands which can be mistaken for authentic chain-termination events. Other features, however, generally serve to distinguish such spurious termination events. Usually not all four sequencing channels are similarly affected and a band, preceding an authentic band in that channel, which has the same mobility as a band in one of the other sequencing channels, is often indicative of spurious termination. Less frequently unscheduled chain termination may occur at particular positions on the template in all four reactions resulting in an artifact band running across all the sequencing channels. Often, the most intense band corresponding to a particular chain length is the authentic band but as other factors can also influence band intensity caution is needed in making this assumption.

The relative intensities of bands in the different channels depend on the relative concentrations of the limiting deoxynucleotide and corresponding dideoxynucleotide in each extension reaction. A significant increase in the ddN/dN concentration ratio will cause a greater proportion of the chains to terminate early in the reaction and will cause a rapid fall-off in the intensity of bands higher up the sequencing gel. This limits the number of nucleotides that can be read from the gel as well as confusing the identification of artifact bands due to spurious termination. For these reasons it is important to carefully control the relative deoxy- and dideoxynucleotide concentrations in the reaction mixtures to obtain a similar relative frequency of chain termination in each of the four reactions. Where the template is particularly asymmetric in its base composition, e.g. rich in A residues, due allowance may need to be made for the rapid fall off of intensity of bands in the corresponding complementary (T) channel by adjusting the concentrations of the deoxynucleotide in question.

When an over-exposed autoradiograph of a sequencing gel is examined it is seen that the desired bands are in fact, superimposed on a much fainter background of bands corresponding to every oligonucleotide chain length in each sequencing channel. This is thought to be due to the presence of traces of a contaminating endonuclease present in most commercial preparations of the DNA polymerase (Klenow sub-fragment) used for the chain-extension reaction. This contaminating activity cleaves the labelled extension products more or less at random and gives a weak background of labelled fragments in all positions. Usually this does not interfere with the reading of the pattern. At low ddNTP concentrations, such as one might use when pushing a primed synthesis reaction to the limit, a considerable amount of label is incorporated into the products and their partial breakdown by the endonuclease can give a high background. When this background is variable between the different channels allowance needs to be made in distinguishing between authentic and artifactual bands.

A feature of the enzymology of the chain-termination procedure is that there does seem to be some discrimination as to whether a dNTP, or its corresponding dd-counterpart, is incorporated onto the end of a growing chain. Ideally there should be no discrimination and the frequency of termination should reflect only the relative concentrations of dNTP : ddNTP in the reaction. Deviations from this random behaviour are particularly noticeable where a run of the same nucleotide is encountered in a sequence and a characteristic pattern of band intensities is often found. This is a valuable feature in reading a sequence since it can help to establish the authenticity or otherwise of companion bands in other channels depending on whether they disturb the characteristic pattern or not. For example, in runs of A residues the band at the bottom (5′-end) of the run is the most intense, the others all having a lower and either equal or gradually diminishing intensity toward the 3′-end of sequence (Fig. 4.16.(a)). The sequence 5′AAAGAA3′ will be represented in the 'A' channel by a run of six consecutive bands. That the fourth A band is an artifact can immediately be recognised since the fifth A band is comparable in intensity to the first A of a run of A's (Fig. 4.16.(b)). A failure to disturb the A pattern (Fig. 4.16.(c)) would strongly suggest that the

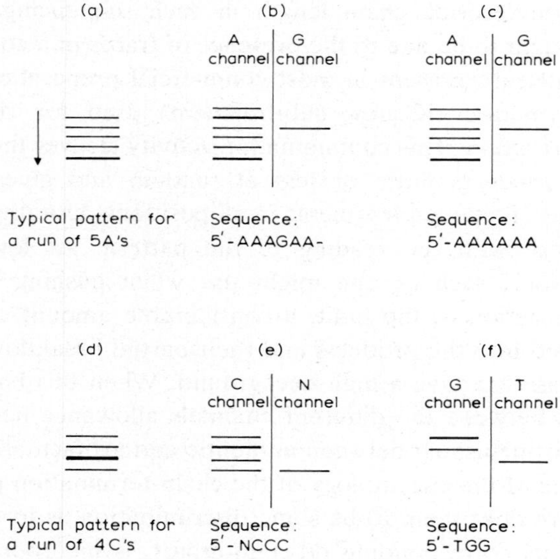

Fig. 4.16. Some characteristics of the patterns of band intensities found on sequencing gels. For discussion see text.

G band is an artifact. In the T reaction consecutive residues usually give a band pattern of equal intensities except where the run of T's follow a C. In this case the pattern is often reminiscent of a run of A's. A run of consecutive C residues also has a characteristic pattern in which the 5′-C is much weaker than the second band (Fig. 4.16.(d)), while the third and subsequent bands fall off in intensity towards the 3′-end. Thus for example, the sequence 5′XCCC3′ (where X is G, A or T) is represented by four consecutive bands in the C channel and whatever the intensity of the first band it can be recognised as an artifact since it will be followed by a characteristic weak band and not the intense band expected if it was the second C in a run of consecutive C's (Fig. 4.16.(e)). Runs of G's, where they follow T residues, give a similar pattern of intensities to runs of C (Fig. 4.16.(f)).

The last cause of error, which is often the most intractable and is shared by all the rapid DNA-sequencing techniques, is due to the anomalous mobilities exhibited by some oligonucleotides during the gel electrophoresis. The main reason for running sequencing gels under as strongly denaturing conditions as possible is to maintain all the oligonucleotides in a single-stranded, nonhydrogen bonded configuration where their mobility is only a function of their chain length. As residues are successively added in the polymerising reaction sequences may be formed which are capable of yielding base-paired looped structures which survive the denaturing conditions of the gel and will exhibit anomalous mobilities; this confuses the interpretation of the gel. In general, secondary structure will result in an increased mobility relative to that of the non-looped structure and this will be reflected in the gel as a disturbance (compression) of band spacings. In the extreme form the band spacings may collapse to zero giving rise to a single intense band which might be read as a single oligonucleotide but actually represents several oligonucleotides of different chain length. The order of bands can also be inverted such that the observed order is the reverse of the order of residues in the sequence. Compression is often associated with regions of sequence rich in G and C residues but it is difficult to correlate or predict these compression zones with the sequence. Compression can usually though not always, be detected by careful examination of the band spacings and it is also observed that a region of compression is often followed by an expansion of the band spacing. The only reliable way of determining the sequence in these regions is to sequence independently the complementary strand. Since the sequence causing the anomalous behaviour will in this case be approached from the opposite direction there is little chance that the same pattern of secondary structure will be assumed by the extending oligonucleotides.

While, at first sight, this may seem a rather daunting list of problems associated with the method, in practice most of them can be recognised and either allowed for or eliminated. The speed and simplicity of the method permits the experimenter to either modify the procedure to overcome a particular problem or to obtain confirmatory data from an independent analysis.

As with any method, procedures and techniques can be improved and the author is always glad to hear of such developments.

4.8. Sequencing long single-stranded DNAs cloned in bacteriophage M13 using internal primers

4.8.1. Introduction

Sequencing DNA by cloning sets of restriction fragments into M13 involves the random selection and sequencing of a large number of recombinants picked from agar plates. With increasing size of the DNA the number of recombinants that need to be analysed to obtain at least one representative clone of all the possible restriction fragments becomes a major exercise. Frequently a detailed restriction map of the sequence to be analysed is not available and the additional problem arises of knowing when all the possible fragments have been accounted for. In order to put the different fragments together in the correct sequence at least one, and usually several other sets of restriction fragments have to be cloned and sequenced and difficulties may even then be encountered in finding particular overlaps due to the large number of clones which need to be scanned for a particular piece of sequence. These methods also generate a large volume of data and without suitable computing facilities for keeping track of data and comparing each new sequence with all the previous sequences the work becomes increasingly time-consuming. Many of these problems are counterbalanced by the ease and rapidity of the sequencing method itself but frequently a more ordered strategy would be desirable in which the sequencing can be carried out in a way that minimizes the random elements in the above approach.

In their earlier work Sanger et al. (1977) and Smith (1979) used restriction fragments as primers for sequencing either naturally occurring single-stranded DNA or single strands produced by the action of exonuclease III on a duplex DNA. Another approach is to clone the original full-length DNA segment into M13 and use a set of restriction fragments, derived from the original duplex DNA, as a series of internal primers for priming at predetermined sites on the single-stranded template. We have used this method (Hindley and Phear, 1981) for sequencing DNA fragments up to 1600 nucleotides long and have found the method to offer a number of advantages not the least being the saving in the number of clones that have to be picked and the number of sequencing gels that have to be run. The method should be applicable to any length of DNA which can be successfully cloned into M13, the upper limit being at least in excess of 4000 nucleotides. For larger sequences preliminary cleavage into fragments of 1000–2000 nucleotides before cloning would seem to offer the best approach.

4.8.2. *Principle of the method*

The starting point is a duplex DNA which has been cloned and amplified in a suitable vector e.g. pBR325. A few hundred micrograms of recombinant plasmid may be purified from a 1 litre culture and this provides ample material for the whole sequence analysis. The fragment of interest is cleaved out of the recombinant and is usually isolated and purified by electrophoresis on an agarose gel.

A small aliquot ($\sim$0.5%) of this double-stranded fragment is ligated into a suitable M13 RF vector, which is then used to transform competent cells and a few recombinant plaques are picked from a single plate. Since only one fragment was cloned, the screening of 6–10 clones (picked at random) using the M13 flanking primer and a single dideoxynucleotide and then running in parallel on a single sequencing gel usually serves to identify the two sets of recombinants corresponding to the cloning of one or other of the two complementary strands of the original DNA. This procedure thus yields two full-length cloned templates corresponding to each of the original complementary strands.

The primers are prepared separately by digesting a sample of the original duplex DNA with a suitable restriction enzyme and fractionating the products by electrophoresis on an acrylamide gel. After elution, the cleanly resolved fragments are digested to completion with exonuclease III and used as primers on the prepared M13 templates. The primers are therefore all internal and anneal to different regions within the cloned sequences. Extension in either direction from each primer is obtained by using the two individually cloned single strands as templates. We have found that using sets of primers prepared from one or two restriction enzyme digests allowed the sequence of two DNA fragments of 600 and 1600 nucleotides respectively, to be deduced without recourse to any other approach. There are three features of this procedure which makes this possible:

(i) The sequences derived can be immediately ascribed to one or other of the complementary strands since this depends only on which of the two templates is used for the primed synthesis. In the random cloning method there is no immediate way of telling which strand of the cloned fragment has been sequenced.

(ii) The ability to sequence in both directions from a particular priming site (by annealing the same primer to the two complementary templates) gives the sequences adjacent to each restriction site thus automatically ordering the different restriction fragments within the cloned sequence.

(iii) The use of exonuclease III-treated primers usually permits the entire sequence of the primer itself to be deduced from the sequences obtained using the two complementary templates. This is because exonuclease III is not a processive enzyme and the single strands generated by its action are sufficiently variable in length to give overlapping sequences corresponding to the centre of the primer when used to prime the two different templates.

Figure 4.17. shows the principle of this approach and Fig. 4.18. illustrates the argument in (iii).

This procedure is therefore in theory capable of independently sequencing all of each strand of a duplex DNA. To obtain com-

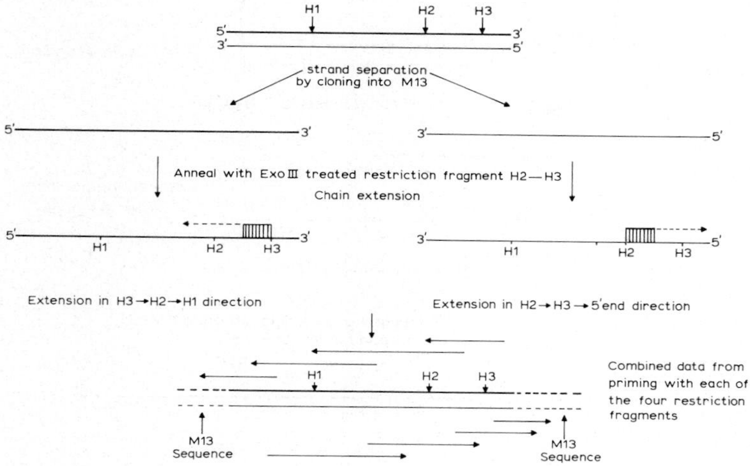

Fig. 4.17. Principle of the primed synthesis method applied to single-stranded templates prepared by cloning into M13 and primed with exonuclease III digested restriction fragments. The sequence to be determined is first cloned into M13 and recombinants picked and identified containing one or the other of the two complementary strands. This is conveniently done using the M13 flanking primer to sequence the ends of 6–10 recombinants picked at random. The sequence shown contains three restriction enzyme cleavage sites, H_1, H_2, and H_3. Cleavage at these sites produces four fragments which are separated by electrophoresis on an acrylamide gel and after elution are digested with exonuclease III to reduce the duplex fragments to two essentially single-stranded half molecules. A sample of each product is annealed to the two complementary cloned templates and the primers extended in the usual way using the dideoxynucleotides as chain terminators. In the example shown the H_2–H_3 primer is extended in the $H_3 \rightarrow H_2 \rightarrow H_1$ direction on one template and in the opposite direction, $H_2 \rightarrow H_3 \rightarrow 5'$-end, on the other template. Repeating the priming with the other exonuclease III treated restriction fragments in turn rapidly permits the deduction of the entire sequence.

plete coverage of both strands does mean that most, if not all of the restriction fragments are used as primers. This creates one difficulty since a single-gel electrophoresis will not always clearly separate all the fragments, and primers prepared from a second

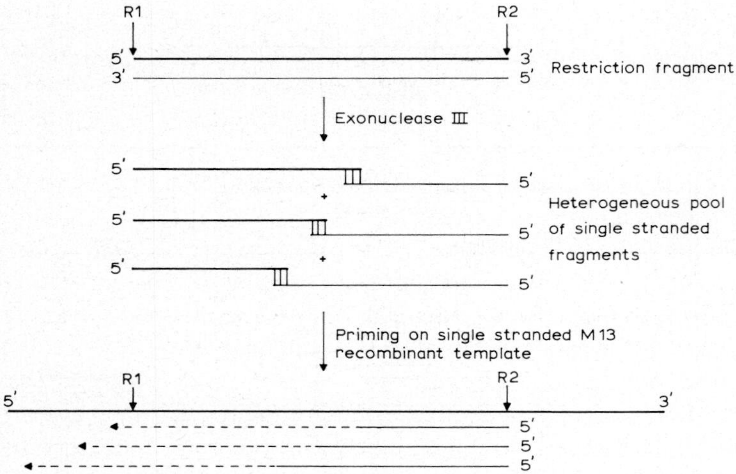

Fig. 4.18. Priming with exonuclease III treated restriction fragments. Exonuclease III digestion of a restriction fragment gives a heterogeneous pool of predominantly single-stranded fragments. When annealed to the template chain extension can, in some cases, commence at points further back than the mid-point of the restriction fragment. When the same pool of primers is used for chain extension on the complementary strand, the sequences overlap at the centre of the restriction fragment thus allowing the entire R_1–R_2 sequence to be deduced in addition to the sequences flanking R_1 and R_2.

restriction endonuclease digest may often need to be employed to obtain sequence data from all of each strand.

In many instances some of the restriction fragments will be of such a length that even after exonuclease III digestion, the priming strand is too long to give much readable sequence after chain extension. In such cases suitable shorter primers from a different set of restriction fragments will need to be employed. The sequence data obtained from the first set of primings will usually indicate which further restriction digests would yield suitable primers.

In addition to using internal primers it is also convenient to sequence the ends of the cloned single strands using the M13 flanking primer. This quickly permits the sequencing of the first 200–250 nucleotides from the 3′-end of each of the two templates and consequently aids the initial ordering of the sequence by identifying which primers subsequently extend the sequence into the known terminal regions.

One problem that can cause difficulty is when the sequence to be determined contains extended regions of repeated sequences. Primers derived from such regions can hybridize to more than one site on the template and therefore give a mixture of different primed products. On a sequencing gel these give rise to a complex pattern of bands due to the superimposition of two or more patterns which cannot be distinguished from each other and therefore fail to give any sequence data. When a similar pattern of mixed sequences is also obtained using the complementary template, this can be taken as confirmation that the template contains regions of repeated sequences. This can be turned to advantage, however, since the position and length of the repeated region can thus be approximately established and if necessary a different strategy used to sequence this region. Provided the region is not too extensive, primers which anneal to unique sequences flanking the repeated region may be used to sequence right through the repeat. Where this is not possible restriction fragments containing the repeated sequences may be cloned into M13 and sequenced using the M13 flanking primer. The results from the internal priming experiments are therefore used to estimate the length, position and complexity of the repeat and indicate which restriction fragments will be most useful for subsequent cloning and sequencing. Eukaryotic DNA in particular abounds in repeats and inverted repeats and a feature of this method is that they are detected early in the sequencing strategy and the techniques can be modified to deal with them. In the random cloning of restriction fragments the occurrence of identical or nearly identical fragments from a repeated region can lead to difficulty in identifying the existence of such structures during the early stages of the sequencing work and increase the labour involved in finally defining their extent and location.

For several reasons therefore this more structured approach may offer advantages. Provided the DNA sequence to be analysed can be cloned intact into M13 it may be worthwhile seeing how far one can go with one or two sets of internal primers. Frequently this will give the entire sequence but where it does not it will indicate which fragments need to be cloned separately or which additional primers need to be employed to complete the sequence.

4.8.3. Results

Figure 4.19. shows a line diagram of a 1578 nucleotide sequence derived from a *Sau*IIIA digest of the recombinant plasmid pDAM Y14 (Beach et al., 1980) containing a yeast chromosomal repli-

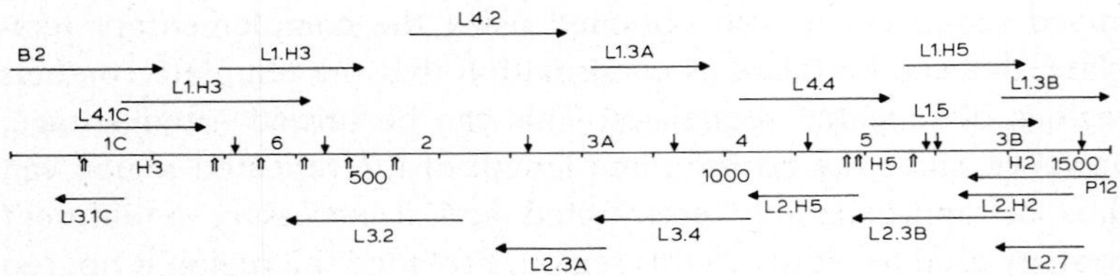

Fig. 4.19. Diagram of 1578 base pair *Sau*3A fragment showing sequences determined using a set of seven exonuclease III digested *Alu*I fragments and three *Hae*III fragments as primers. ↓-*Alu*I cleavage sites, ⇑-*Hae*III cleavage sites. Numbers immediately above the line show the *Alu*I fragments and below, the *Hae*III fragments. The arrowed sequences are designated such that, for example, L3.4 is the sequence determined by priming template L3 with exonuclease III digested *Alu*I fragment 4. The terminal sequences B2 and P12 were determined using a flanking primer. The advantages of this procedure (see text) are: (a) the automatic assignment of each sequence to one or other strand depending only on which template is used. The identical templates L1 and L4 give sequences reading left to right and correspond to one of the complementary strands of the *Sau*3A fragment. Templates L2 and L3 similarly give sequence reading right to left corresponding to the other strand. (b) each pair of primings, e.g. L4.4 and L3.4, L4.2 and L3.2 etc., give both the sequence of the primer itself, due to the non-processive action of exonuclease III, and extended sequences into both adjacent fragments. The order of fragments is thus automatically determined. (c) the sequence of region 228–575, which contains $5\frac{1}{2}$ direct repeats (Fig. 4.20.) is determined directly by read through and its boundaries defined.

cation origin. The entire sequence was determined using a set of seven *Alu*I fragments and three *Hae*III fragments as primers (derived from the *Sau*3A fragment) on the individually cloned full-length single strands of the *Sau*IIIA fragment. The sequences at the 3′-ends of the M13 cloned single strands were initially determined using the M13 flanking primer. The validity of this sequencing approach was confirmed by subsequently cloning the *Alu*I fragments into M13mp2 using the RI linker and sequencing with the M13 flanking primer.

Fig. 4.20. shows a part of the sequence, from residues No. 228–575 which is composed of $5\frac{1}{2}$ repeated blocks of 62–63 nucleotides. The *Alu* fragment, stretching from 315–440, is entirely situated within the repeat and all attempts to use this fragment as a primer, after exonuclease III digestion, gave unreadable mixed sequences due to priming at multiple sites. This result in fact originally suggested the presence of the repeat which was confirmed by sequencing through the region using flanking primers. For confirmation, the three *Alu*I fragments involved in this region were cloned into M13 and their sequences determined. As can be seen from Fig. 4.19., the sequences generated from priming with six different *Alu* fragments gave overlapping data for all the *Alu* sites in the DNA. While additional primings were needed to confirm the detailed sequence, this approach can be seen to rapidly generate an overall structure for the fragment. One can be sure that no fragments are missing (since all the overlaps are revealed and both ends of the sequence are known) and the

```
                            228 TATCTTTGTTAACGAAGCATCTGTGCTTCA TTTTGT
                                                        ↓
264 AGAACAAAAATGCAACGCGAGAGCGCTAATTTTTCAAACAAAGAATCTGAGCTGCATTTTTAC
327 AGAACAGAAATGCAACGCGAAAGCGCTA TTTTACCAACGAAGAATCTGTGCTTCATTTTTGT
                                                        ↓
389 AAAACAAAAATGCAACGCGAGAGCGCTAATTTTTCAAACAAAGAATCTGAGCTGCATTTTTAC
452 AGAACAGAAATGCAACGCGAGAGCGCTA TTTTACCAACAAAGAATCTATACTTCTTTTTTGT
514 TCTACAAAAATGCATCCCGAGAGCGCTA TTTTTCTAACAAAGCATCTTAGATTACTTTTTTT 575
```

Fig. 4.20. A part of the sequence shown diagrammatically in Fig. 4.19. aligned to show the $5\frac{1}{2}$ blocks of repeated sequence.

data are detailed enough to make a precise assessment of which additional restriction fragments are required to complete and confirm the sequence.

4.9. Sequencing cDNA from reverse transcripts of RNA

The ability to clone nanogram quantities of DNA obtained by whatever route, into M13 DNA clearly has implications for the sequencing of cDNAs produced by the reverse transcription of RNA species. The cloning procedure both segregates and subsequently amplifies individual DNA fragments and so makes feasible the sequencing of any RNA which can be reverse transcribed. Thus messenger RNAs, purified by immunoprecipitation from a polysome preparation (e.g. Shapiro and Schimke, 1975) or viral RNAs (e.g. from poliomyelitis or influenza virions) can in theory be sequenced by this approach. In order to illustrate the potential of this method it is instructive to review briefly the existing methods for cDNA sequencing.

These essentially fall into two groups depending on whether the single-stranded cDNA transcript itself is sequenced using either a 5′-end labelled primer or by using the dideoxynucleotide chain-termination procedure, or alternatively by producing a full-length double-stranded DNA transcript, cloning into a suitable vector, and subsequently sequencing the cloned insert.

The first group of methods are exemplified by the earlier work of Hamlyn et al. (1978) in their sequencing of the 'constant' region of an immunoglobulin light chain mRNA and more recently by the work of Houghton et al. (1980) on the sequence of human interferon mRNA. Fig. 4.21. illustrates Hamlyn's approach. A hexanucleotide, 5′T–T–G–G–G–T3′ was synthesized chemically and shown to give specific priming at a single site on the mRNA. Four separate dideoxy-chain termination reactions were set up in the presence of the mRNA, the primer, the appropriate dNTP/ddNTP mixtures and reverse transcriptase. At the end of the reaction the products were denatured and fractionated on a

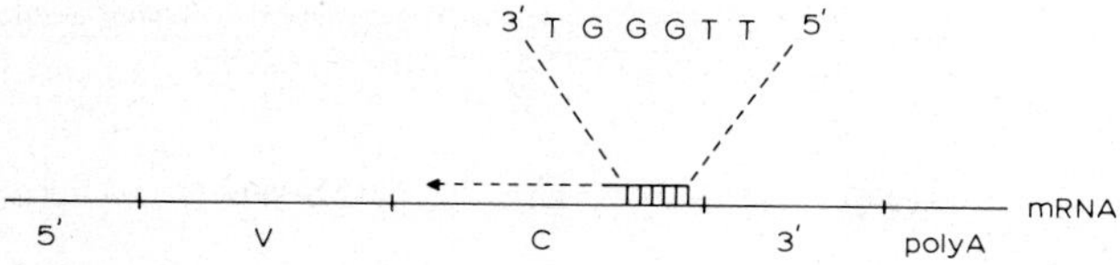

Fig. 4.21. Site of priming of the hexanucleotide TTGGGT on the constant region of MOPC21 immunoglobulin light-chain mRNA. The dotted arrow shows the direction of chain extension in the dideoxynucleotide chain-termination reaction. 5′ and 3′ refer to the 5′ and 3′ non-coding regions and V and C the 'variable' and 'constant' regions of the gene.

thin sequencing gel. In this way a sequence of 300 residues was determined from the site of priming. It was, in retrospect, fortunate that a hexanucleotide primer gave such specific and high yield priming and in other work, e.g. Proudfoot (1977), a phased primer, $(pT)_{10}$-G-C, which hybridizes to the poly(A) tail of β-globin mRNA was used and allowed the entire 3′-non-coding region to be determined.

This type of approach was taken a stage further by Houghton et al. (1980) in their analysis of the interferon mRNA sequence. A chemically synthesized primer was given sufficient specificity to anneal with a single mRNA in a complex mixture of polyadenylated mRNA species isolated from human fibroblasts. This 15-nucleotide long primer was designed from a knowledge of the amino terminal sequence of fibroblast interferon and the observed codon usage preferences in human genes (Fig. 4.22.). One such primer, after 5′-end labelling with α-[^{32}P]ATP and polynucleotide kinase, was hybridized to poly(A)-mRNA isolated from induced fibroblasts and the primer extended with AMV reverse transcriptase. The products were fractionated on a denaturing acrylamide gel and the predominant cDNA band eluted and sequenced according to the methodology of Maxam and Gilbert (cf. Chapter 5). From this experiment the nucleotide sequence between the methionine codon at the amino terminus of the mature protein and the 5′-terminus of the mRNA was determined. In subsequent work

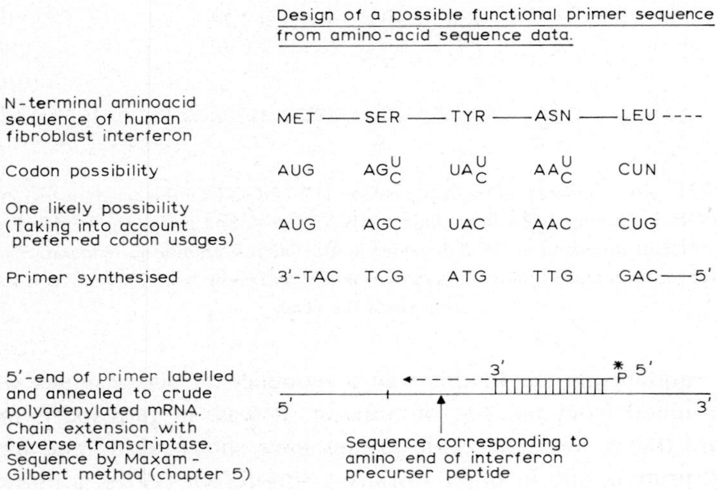

Fig. 4.22. Design of a possible functional primer sequence from amino-acid sequence data. (Houghton et al., 1980.)

using an additional chemically synthesized primer the entire sequence of fibroblast interferon mRNA was determined.

These priming methods therefore offer a powerful way of sequencing rare mRNA species that are present in a heterogeneous population of mRNAs without resort to either purification methods (e.g. immunoprecipitation) or cloning methods. The principal drawback is the length of sequence that can be determined from a single 5′-end labelled primer (about 250 nucleotides) and the fact that the data obtained relies on the sequencing of only one DNA strand.

The second approach to sequencing RNAs, via their cDNA transcripts, is exemplified by the sequence analysis of the influenza virus haemagglutin gene (Sleigh et al., 1979). In this method a double-stranded (dsDNA) copy of the RNA genome was inserted, after digestion with nuclease S_1, into the double-stranded bacterial plasmid pBR322 which was subsequently used to transform a suitable strain of *E. coli.* Tetracycline-resistant bacterial colonies were screened for the presence of plasmid containing the haemagglutinin (HA) gene transcript by testing their ability to hybridize with a specific [^{32}P]-labelled single-stranded DNA probe. Hybrid plasmids containing the cloned gene were finally analysed by restriction enzyme mapping and the sequence of the insert determined using the method of Maxam and Gilbert. This rather complex sequence of reactions is shown in Fig. 4.23. Various modifications of the basic pathway can be employed, for example cloning the ds-cDNA into the plasmid vector by the addition of the *Hind*III linker (d[CCAAGCTTGG]) and subsequent cloning into the *Hind*III site of pBR322 (Roberts et al., 1979; Porter et al., 1979; Bedbrook et al., 1980); improvement of the translational efficiency of mRNAs by pretreatment with methyl-mercury hydroxide CH_3HgOH (Payvar and Shimke, 1979) or prior fractionation of the RNA by a variety of means. This procedure is currently the most widely used reaction scheme and a variety of RNA molecules have been copied and cloned using these basic steps. The final cloning step is essential since the yield of full-length dsDNA is only a few nanograms and the heterogeneous ends make it unsuitable for direct sequencing by end-labelling.

Bearing in mind the above, the advantages of sequencing by cloning into an M13 vector may be considered. In the first place the initial requirement for a defined RNA may, in some circumstances be relaxed. For example the eight RNA species constituting the influenza virion genes, which together account for about 14,000 nucleotides should be no more difficult to sequence than a single DNA species of this length, and in fact would probably be considerably easier. In the hypothetical scheme shown in Fig. 4.24., the double-stranded cDNA preparation obtained after nicking with nuclease S_1 could be digested with a restriction enzyme and after blunt-ending the fragments (if necessary) directly ligated with the *Eco*RI linker and the fragments cloned into M13mp2 for sequencing. Since the cloning procedure itself segregates the different fragments into unique clones, the sequencing of all eight genes

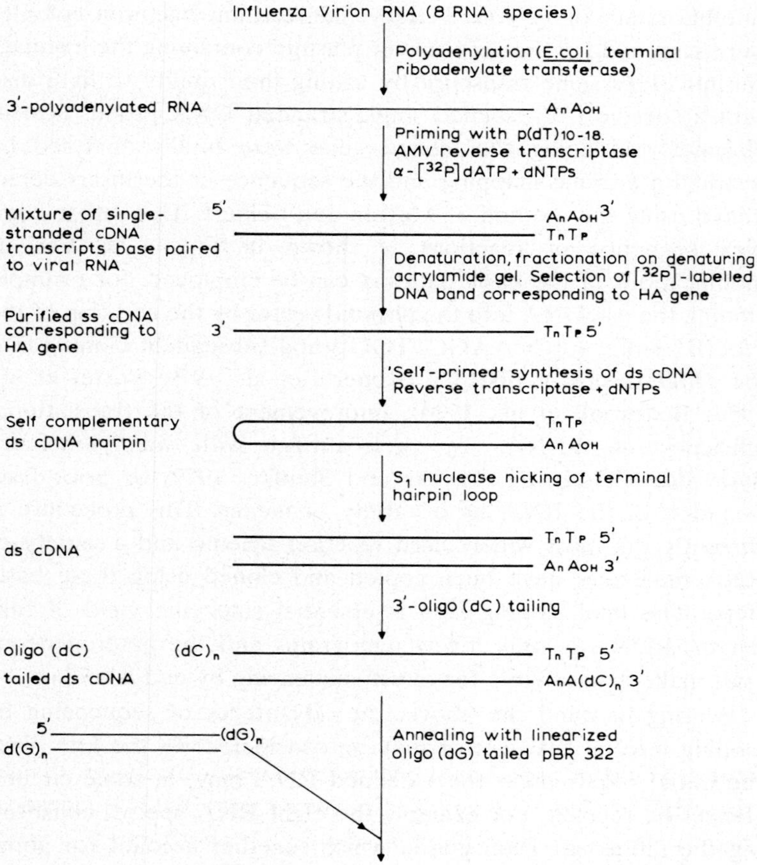

Fig. 4.23. Route for cloning double-stranded cDNA from reverse transcripts of viral RNA and subsequent sequencing (not using M13 cloning procedure).

proceeds simultaneously without recourse to any fractionation of either the RNAs or the cDNA transcripts. In the same way the purification of an mRNA species could be less rigorous. Fractionation on a denaturing formamide gel followed by identification of the required component by translation in a suitable cell-free

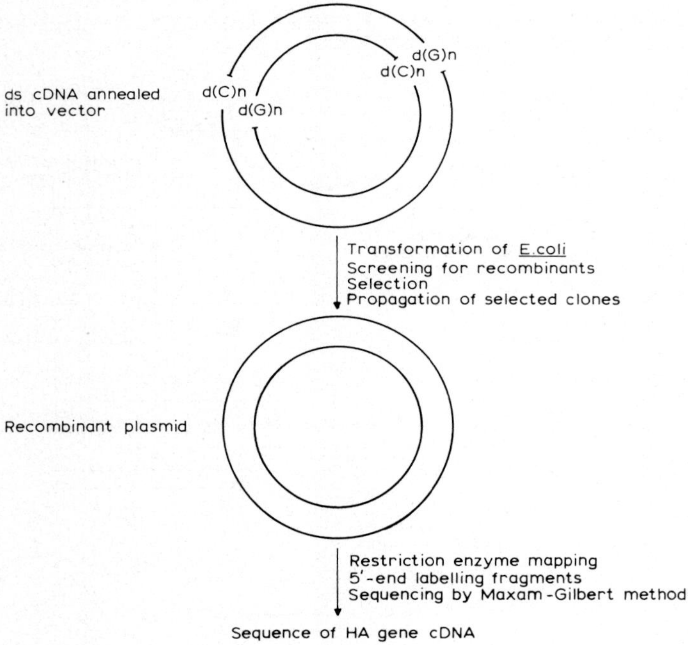

Fig. 4.23, continued.

system would be expected to give pure enough material for reverse transcription and cloning the fragments into M13. Extraneous fragments will be recognised as the sequencing proceeds, particularly if a partial amino acid sequence of the coded protein is known. The advantages of possibly avoiding rigorous purification by immunoprecipitation of polysomes (which can be technically difficult) and the need for cloning into a plasmid vector and identification of recombinants with a hybridization probe should considerably simplify the procedure. Alternatively, priming with a unique synthetic oligonucleotide in a pool of mRNAs as described above and subsequent conversion into the ds cDNA can yield a few nanograms of product, sufficient for restriction and cloning

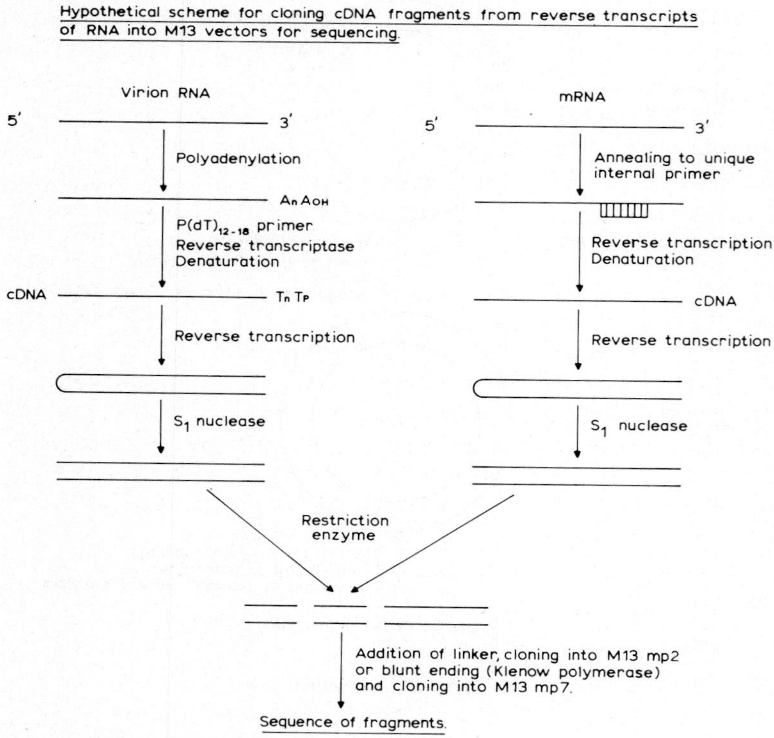

Fig. 4.24. Hypothetical scheme for cloning cDNA fragments from reverse transcripts of RNA into M13 vectors for sequencing.

into M13. This contrasts with sequencing using a 5′-end labelled primer where, at most, about 250 nucleotides can be sequenced from the labelled terminus. Longer sequences would require either the successive use of different end-labelled primers or the cloning and amplification of the ds cDNA in a plasmid to obtain sufficient material. With the development of methods for synthesizing oligonucleotides by the phosphotriester approach (Narang et al., 1980) the use of defined single-stranded primers for the transcription of particular RNAs seems likely to assume an increased importance. Preliminary knowledge of short sequences at the 3′- and 5′-ends of the RNA would enable such defined primers to be tailored and avoid the oligo(dT) priming. In addition, the back-copying of the single-stranded cDNA could be specifically primed and carried out with DNA polymerase (Klenow sub-fragment). This is probably a more efficient, and accurate, procedure for obtaining full-length molecules than the self-priming reaction with reverse transcriptase and also avoids the need for subsequent S_1 cleavage.

These latter approaches to RNA sequencing are still very much in the developmental stage but with appropriate variation to suit a particular problem they do seem to offer a powerful new strategy.

*Since the time of writing Winter et al. (1981) have described the complete double-stranded copying of the influenza virus RNA genes using two synthetic oligonucleotide primers. The RNA species were not fractionated before reverse transcription and the random cloning of Taq*I, *Alu*I *and Sau*3A *fragments into M13 followed by plaque isolation and sequencing using a flanking primer permitted the complete analysis of the genes coding for two non-structural viral proteins.*

Key references

A full list of references is given at the end of this volume. The following list deals exclusively with the background and procedures to the M13 sequencing method.

Anderson, S., M.J. Gait, L. Mayol and I.G. Young, 1980. A short primer for sequencing DNA cloned in the single-stranded phage vector M13mp2. Nucleic Acids Res. *8,* 1731.

Gronenborn, B. and J. Messing, 1978. Methylation of single-stranded DNA in vitro introduces new restriction endonuclease cleavage sites. Nature *272,* 375.

Heidecker, G., J. Messing and B. Gronenborn, 1980. A versatile primer for DNA sequencing in the M13mp2 cloning systems. Gene *10,* 69.

Hu, N-t and J.B. Messing, 1982. The making of strand specific M13 probes. Gene *17,* 271–277.

Messing, J.B., 1981. M13mp2 and derivatives: a molecular cloning system for DNA sequencing, strand specific hybridization and in vitro mutagenesis, in: Third Cleveland Symposium on Macromolecular Recombinant DNA, Ed. A. Walton (Elsevier, Amsterdam) 143–153.

Messing, J.B., B. Gronenborn, B. Muller-Hill and P.H. Hofschneider, 1977. Filamentous coliphage M13 as a cloning vehicle: insertion of a *Hin*dII fragment of the *lac* regulatory region in M13 replicative form in vitro. Proc. Natl. Acad. Sci. USA *74*, 3642.

Messing, J., R. Crea and P.H. Seeburg, 1981. A system for shotgun DNA sequencing. Nucleic Acids Res. *9*, 309.

Mills, D.R. and F.R. Kramer, 1979. Structure-independent nucleotide sequence analysis. Proc. Natl. Acad. Sci. USA *76*, 2232.

Poncz, M., D. Solowiejczyk, M. Ballantine, E. Schwartz and S. Surrey, 1982. Nonrandom DNA sequence analysis in bacteriophage M13 by the dideoxy chain-termination method. Proc. Natl. Acad. Sci. USA *79*, 4298–4302.

Sanger, F. and A.R. Coulson, 1978. The use of thin acrylamide gels for DNA sequencing. FEBS Lett. *87*, 107.

Sanger, F., A.R. Coulson, B.F. Barrell, A.J.H. Smith and B. Roe, 1980. Cloning in single-stranded bacteriophage as an aid to rapid DNA sequencing. J. Mol. Biol. *143*, 161.

Sanger, F., S. Nicklen and A.R. Coulson, 1977. DNA sequencing with chain terminating inhibitors. Proc. Natl. Acad. Sci. USA *74*, 5463.

Schreier, P.H. and R. Cortese, 1979. A fast and simple method for sequencing DNA cloned in the single-stranded bacteriophage M13. J. Mol. Biol. *129*, 169.

Winter, G. and S. Fields, 1980. Cloning of influenza cDNA into M13: the sequence of the RNA segment encoding the A/PR/8/34 matrix protein. Nucleic Acids Res. *8*, 1965.

Summarised procedure for chain terminator sequencing

This section is essentially a summarised account of the procedures described in the preceding sections. It includes the most recent procedural variations as practised in Fred Sanger's laboratory.

1. *Materials*

1. High voltage power supply capable of delivering 2 KV at 40 mA output e.g. LKB 2103.
2. 20 × 40 cm Slab gel apparatus (at least two) e.g. Raven Scientific RGA 505.
3. Tempered glass plates and clips for above.
4. 'Plasticard' (0.35 mm thick) for cutting out spacers and combs.
5. Vinyl insulating tape.
6. Small bunsen burner.

7. Capillary tubes, e.g. Gallenkamp YC-650-R.
8. 1–5 microlitre graduated disposable micropipettes, e.g. Dade® Accupettes, Supracaps No. 7090 07.
9. Pushbutton type propipettes with disposable tips e.g. Oxford sampler, Gilson 'Pipetman'.
10. Glass test tubes 1×5 cm.
11. Cling film, e.g. Sarenwrap.
12. X-ray film for autoradiography e.g. Fuji.
13. Miscellaneous items i.e. desiccator, test tube racks, hot plate/stirrer, Parafilm, glass cutting pencil, Repelcote.

2. *Solutions and buffers*

A. *For preparing gels*

1. *10 × TBE*

Make up 162 g Tris base, 27.5 g boric acid and 9.5 g Na_2 EDTA, $2H_2O$ to 1 litre (pH 8.8). Store at room temperature. This is reported to give a better resolution, particularly on longer gel runs, than the 10 × TBE, pH 8.3 buffer (Section 4.4.). For shorter runs the latter buffer is marginally preferable.

2. *Acrylamide stock solutions*

Make up 288 g urea (preferably ultrapure), 34.2 g acrylamide and 1.8 g bis-acrylamide to about 500 ml with distilled water. Dissolve with gentle warming (do not exceed 40°C). Add 20 g Amberlite MB-1 resin and stir gently for 30 min. Remove resin by filtration through a sintered glass funnel (No. 2 porosity) with suction. Add 60 ml 10 × TBE and make up to 600 ml with distilled water. Degas on a water pump and store at 4°C. Used to prepare 6% sequencing gels.

A 4% mix is prepared using 22.8 g acrylamide and 1.2 g bis-acrylamide. Alternatively, a 40% acrylamide/bis-acrylamide (38/2) stock solution can be prepared, and the urea and buffer solution added as required.

B. *For chain termination reactions*

(1) *10 × Hin buffer*:

Mix 9 microlitres (0.1 M Tris-Cl pH 7.4, 0.1 M $MgCl_2$ 0.5 M NaCl) + 1 microlitre (0.1 M DTT) (dithiothreitol). Keep stock solutions frozen, prepare mixture immediately before use.

(2) *dNTP*-mix buffer*:

50 mM Tris-Cl, pH 8.0

1 mM EDTA.

(3) *Dideoxytriphosphates (ddNTP) solutions*

Stocks of 10 mM dideoxynucleoside triphosphates are kept frozen in water (or 5 mM Tris-Cl pH 8.0, 0.1 mM EDTA). Working stocks are prepared by appropriate dilution (with water). As a first approximation prepare:

0.5 mM ddTTP
0.3 mM ddCTP
0.3 mM ddGTP
0.1 mM ddATP

Store frozen.

(4) *Deoxynucleoside triposphate (dNTP) solutions*

Stocks of 20 mM dNTPs and 0.5 mM working solutions in water. Keep frozen.

(5) *dNTP* mixes*

	dNTP mix*			
	dTTP	dCTP	dGTP	dATP
0.5 mM dTTP	1 μl	20 μl	20 μl	20 μl
0.5 mM dCTP	20 μl	1 μl	20 μl	20 μl
0.5 mM dGTP	20 μl	20 μl	1 μl	20 μl
dNTP* mix buffer	5 μl	5 μl	5 μl	5 μl

The above keep indefinitely when frozen (μl = microlitres). Make up 200 microlitres 0.5 mM dATP in 5 mM Tris-Cl, pH 8.0, 0.1 mM EDTA for the chase.

(6) *Formamide dye mix*

Deionize 100 ml formamide with 5 g Amberlite MB1 resin. Remove resin by filtration. Add 0.03 g xylene cyanol FF, 0.03 g bromophenol blue and Na_2EDTA to 20 mM. Store at room temperature.

(7) *[α-^{32}P]-deoxyadenosine triphosphate*

e.g. PB 164, Specific Activity ~ 400 Ci/mmole in 50% ethanol, store at −20°C.

(8) *DNA-polymerase (Klenow subfragment)*

0.5–1.0 units per microlitre in glycerol solution. Store at −20°C.

3. *Preparations for sequencing*

(1) All reactions take place in drawn out capillary tubes resting in siliconised 1 × 5 cm test tubes, or in the test tubes themselves. The required number of tubes should be siliconised, washed and dried at 100°C.

The capillaries are flamed and drawn out about three-quarters of their length from one end. They should be scored with a diamond pencil and broken cleanly to avoid a ragged end. The drawn out end should not be excessively long or have a too narrow bore.

(2) Gel plates must be thoroughly cleaned to prevent the formation of bubbles when pouring the gel. This is best done by scrubbing with detergent, rinsing thoroughly with distilled water followed by ethanol and drying. The inner side of the notched plate is siliconised to encourage the gel to adhere to the other plate when the gel is dismantled.

(3) The slot formers and side spacers are cut from the 'Plasticard'. Slot formers

should have 16×0.5 cm wide slots. The side spacers should be about 1 cm wide. New 'Plasticard' contains a coating of material which inhibits polymerization of the gel. This can be removed by *briefly* wiping with tissue moistened with acetone or by washing with undiluted detergent followed by a thorough water wash.

4. *Gels*

For sequencing uncharacterised inserts it is convenient to run two 6% gels for 1.5 and 4 hours respectively. After 1.5 hours, the bromophenol blue marker should be at the bottom of the gel and corresponds to an oligonucleotide of about 25 residues, i.e. the 3′ end of a 16–20 nucleotide primer and therefore the beginning of the unknown DNA sequence. The two runs together should give a sequence of about 300–350 nucleotides. The sequence can be pushed further using a 4 or 5% gel with up to 6 hours electrophoresis.

When the gel is prepared using stock solutions stored at 4°C the mixture should be warmed to room temperature by swirling under the hot tap to prevent irregular polymerization around the slot former. Add 40 μl TEMED and 300 μl of 10% ammonium persulphate to 50 ml of gel solution and mix by swirling.

Pour the gel solutions down an inside edge of the gel former using a pipette or large syringe with the plates sloped to about 30° from the horizontal. Bubbles should not form but if they do they can often be dislodged by tapping the plates or rocking from side to side.

Insert the slot former and leave to set tilted slightly from the horizontal. The gel should be left for at least one hour before use. Keeping a gel overnight before use does not detract from the resolution.

5. *Primers*

(1) The original 96 base-pair primer is cloned in the *Eco*RI site of pBR325 to form pHM232. The plasmid can be prepared from cultures of the infected host cells, *E. coli* C600 (pHM232) by the standard cleared lysis-ethidium bromide-CsCl centrifugation procedure. The primer is released from the plasmid DNA by *Eco*RI digestion. As pBR325 shares no extensive sequence homology with M13mp2 or its derivatives, the primer fragment does not require separation from the vector DNA before use. In practice however it is often found that clearer results are obtained if the primer fragment is first purified by gel electrophoresis. Though not obligatory it is usual to pretreat the primer with Exonuclease III to yield an essentially single stranded product. This prevents the formation of strong artefact bands appearing on the gel.

(2) Shorter cloned primers can be prepared from plasmids:

pSP14 (*Eco*RI/BamHI digestion)

pSP16 (*Eco*RI digest only)

The latter is also available in M13mp2 and are obtainable from Biolabs or

Bethesda Research Laboratories. These primers do not require prior Exonuclease III digestion but should be separated from the vector after release.

(3) Synthetic primers are currently being developed and are available from Collaborative Research and New England Biolabs.

In general primer solutions are stored frozen at −20°C at a concentration of approx. 0.2–0.5 pmol/μl. As a guide, the primer obtained from 100 μg of plasmid dissolved in a final volume of 100 μl water is sufficient for 100 primings, one microlitre of primer being annealed to 5 microlitres of the clone DNA.

6. *The sequencing reaction*

A. *Hybridization*

A mixture of the single stranded M13 template and primer are denatured at 100°C and annealed.

(i) To a drawn out capillary add:

5 microlitres M13 cloned DNA
1 microlitre 10 × Hin buffer
1 microlitre primer
3 microlitres water
10 microlitres

(ii) Seal the capillary and immerse in a test tube of water placed in a boiling water bath. After 3 minutes remove the test tube and allow to cool down to room temperature (15–30 min.). This procedure should be used for the short synthetic primers and the primers from pSP14 and pSP16. For longer primers (from pHM232) the mixture should be annealed at 67°C for 30 min. after denaturation by boiling.

(iii) Dry down five microlitres of α-^{32}P dATP (specific activity ~ 400 Ci/mmole) in a siliconized glass test tube in a vacuum desiccator. Take up in 5 microlitres water and keep on ice.

(iv) Draw out four capillaries and rest each in a separate siliconized glass tube.

(v) Distribute 2 microlitres of the hybridized DNA solution (from (i)) into each of the four capillaries. These are conveniently labelled A, G, C and T.

B. *Polymerization reaction*

(i) To the first capillary, labelled A, add successively through the drawn out tip: 1 microlitre dATP* mix and 1 microlitre ddATP (see 2.B (3) and (5) above). To the second capillary, G, add 1 microlitre dGTP* and 1 microlitre ddGTP. To the third capillary, C, add 1 microlitre dCTP* and 1 microlitre ddCTP. To the fourth capillary, T, add 1 microlitre dTTP* and 1 microlitre ddTTP. These additions are conveniently made using a 1 microlitre push button propipette (e.g. Oxford Sampler) with disposable plastic tips.

(ii) Into the tip of each capillary add 1 microlitre α-^{32}P dATP solution. Expel the mixture into the appropriate siliconized glass tube, take up into the capillary and repeat two or three times to ensure thorough mixing.

(iii) Add 1 microlitre DNA polymerase (Klenow subunit), previously diluted 1:5 with 50% glycerol, 20 mM potassium phosphate, pH 7.4, into the first capillary. Mix by gently expelling into the siliconized test tube and drawing up into the capillary two or three times. Leave the capillary standing in the tube, start the stopwatch (t = 0) and incubate at room temperature. At convenient intervals (approx. 2 min.) add 1 μl polymerase to each of the other samples, mix and incubate at room temperature.

(iv) At t = 15 min. add 1 microlitre 0.5 mM dATP chase to the first capillary, A. Expel into the glass tube, mix, and discard the capillary. At 2 min. intervals repeat the dATP chase with each of the other three samples.

(v) At t = 30 min. add 4 microlitres of formamide-dyes mixture to the A sample and mix in by rotating the tube. At 2 min. intervals repeat the procedure with the other three samples. Ensure that each sample has a full 15 min. chase. A 20% error in timing is not important. The ratio of ddNTP : dNTP is much more critical.

If the sequencing gels are not to be run immediately the reaction mixtures are frozen at −20°C, only adding the formamide-dyes just before electrophoresis.

7. *Gel electrophoresis*

(i) Stand the open tubes in a boiling water bath for 5 min. and load the samples onto a 6% acrylamide-7 M urea thin gel with a long drawn out capillary. The gel slots should be flushed out *immediately* before loading and not more than 2 microlitres of sample added to each slot. Commence electrophoresis at a constant current of 30 ma (~1500 v).

After approx. $1\frac{1}{2}$–2 hours, when the bromophenol blue marker dye is just emerging from the bottom of the gel, switch off the current, change the electrophoresis buffer in the reservoirs, and load a further set of four samples into adjacent slots. Recommence electrophoresis at 30 ma. and continue until the bromophenol blue from the second set of samples is near the bottom of the gel.

In general two DNA clones are sequenced simultaneously and with a little practice this is easily achieved. In this case it is convenient to premix the label with the set of four dNTP* : ddNTP mixes, e.g. dTTP* : ddTTP; dCTP* : ddCTP, etc. in order to minimize the number of pipetting operations.

Two 6% gels are prepared. The first is loaded with eight samples (from clone 1 and clone 2) and electrophoresed for $1\frac{1}{2}$ hours.

The second gel is similarly loaded and run for approx. 4 hours, with one buffer change after approx. 2 hours.

(ii) Dismantle the apparatus and remove the siliconized glass plate leaving the gel adhering to the other plate.

The gel can be autoradiographed either after freezing (a), fixing in acetic acid (b), or drying (c).

(a) Freezing

Dry the surface of the gel by placing several layers of tissue paper on top and rolling gently with a pipette. Carefully peel off the paper and invert the gel onto a piece of used X-ray film. Lift off the glass plate leaving the gel adhering to the film, cover with sarenwrap, place a film on top and autoradiograph at −20°C in a film cassette or under a heavy steel plate.

(b) Acid fixing

Immerse the gel, still attached to the plate, in 10% acetic acid for 10 min. To dry the gel ensure that all wrinkles or bubbles are removed and blot with tissue paper as above. Cover with sarenwrap and autoradiograph at room temperature.

(c) Drying

Gels can be dried either using a proprietary machine or simply by placing in a 70°C oven overnight after air-drying for ~2 hr. at room temperature. The film is then placed directly onto the gel without sarenwrap. If the gels stick to the film this can be avoided by spraying the dried gel with a water repellant (e.g. PTFE). Drying the gel improves the resolution and can be useful for longer electrophoretic runs where the slower moving bands are closely spaced.

8. *Additional reactions*

(i) Arabino-C

The use of araCTP instead of ddCTP results in a different pattern of relative band intensities in a run of C residues. Use 1 microlitre of a 25 mM solution per reaction in place of ddCTP.

(ii) Deoxyinosine triphosphate

This is an additional ddG terminated reaction in which dITP replaces dGTP.

dGTP* (I) mix:

20 microlitres	0.5 mM dCTP
20 microlitres	0.5 mM dTTP
1 microlitres	2.0 mM dITP
5 microlitres	50 mM Tris-HCl, pH 8.0, 1 mM EDTA

Reaction mixture:

2 microlitres	annealed primer-template
1 microlitre	dGTP* (I) mix
1 microlitre	0.05 mM ddGTP
1 microlitre	α-^{32}P dATP
1 microlitre	DNA polymerase, Klenow subunit (0.2 units)

This reaction is frequently useful in resolving 'pile-ups' caused by regions of local secondary structure in the copied DNA (cf section 4.7).

9. *Time course for an experienced sequencer*
Providing all the reagents, solutions and apparatus are at hand a set of 2–4 M13 recombinant DNA samples can be sequenced in an 8 hour working day. When using fresh α-^{32}P dATP label overnight autoradiography is often sufficient and if all is well sequences totalling 500–1000 nucleotides may be determined daily.

CHAPTER 5

DNA sequencing by the Maxam-Gilbert chemical procedure

5.1. Introduction

Since its introduction in 1977 this has become the most widely used sequencing procedure for DNA. In the intervening period the method has been refined and developed to the point where it has become the method of choice in laboratories all over the world. Though not as fast as the primed synthesis methods, it is, up to the time of writing the most widely tried and tested procedure and the majority of DNA sequences reported in the literature to date have been determined by the Maxam-Gilbert procedure.

In the chemical DNA sequencing method end-labelled DNA is partially cleaved at each of the four nucleotide bases in four different reactions, the products ordered by size by gel electrophoresis and the sequence read-off an autoradiograph by noting which base-specific agent cleaved at each successive nucleotide along the strand. No *in vitro* enzymatic copying is required and either single- or double-stranded DNA, labelled at either the 3′- or 5′-end, can be sequenced. Given a DNA and a restriction endonuclease which cuts the molecule this method can display the sequence of at least 250 nucleotides in both directions from the site of cleavage. In addition by using a different end labelling strategy (3′-end labelling as opposed to 5′-end labelling) the sequence of the complementary strands can be determined thus giving a valuable check on the deduced sequences. As with the primed synthesis methods, the main limitation imposed on the method is the resolving ability of the sequencing gels.

The crucial step in the Maxam-Gilbert procedure is the specific chemically-induced cleavage at one or two of the four nucleotide bases. This is a three-stage process involving modification of the base, removal of the modified base from its sugar and finally strand scission at that sugar. Figure 5.1. illustrates these steps with the initial methylation of the N7 position in guanine with dimethylsulphate, release of 7-methyl-guanine by cleavage of the weakened glycosidic bond, and finally strand scission by an alkali-catalysed β-elimination reaction. The second and third reactions in such a scheme are contingent on the one that preceded it. Thus only an appropriately modified base will be released from its sugar and the DNA will only break at a sugar lacking a base. Since the specificity lies in the first reaction, it is clear that the subsequent strand-cleaving reactions must be quantitative and free from side reactions.

The two reagents employed to attack and subsequently open the nucleotide bases are dimethylsulphate and hydrazine. Hydrazine, under alkaline conditions, attacks the two pyrimidine bases thymine and cytosine at C4 and/or C6 and subsequently cyclizes with C4–C5–C6 to form a new five membered ring. Further hydrazine reactions release this pyrazolone ring leaving the N1–C2–N3 fragment of the pyrimidine base still attached to the sugar. In the presence of piperdine both phosphates are released from the sugar fragment by a β-elimination reaction leading to breakage of the DNA chain at that nucleotide. The chemistry of these reactions are shown in Fig. 5.2. In the presence of 1 M salt the reaction of hydrazine with thymine is slowed down sufficiently to provide a cytosine-specific cleavage reaction and thus distinguish between the C and C + T degradation products.

Dimethylsulphate is an alkylating agent which methylates ring nitrogens in DNA bases. As far as sequencing is concerned the relevant methylation sites are the N7 position in guanine and the N3 in adenine. Methylation of guanosine N7 in DNA is 4–10 times faster than the N3 methylation of adenine and in both cases the effect is to render the glycosidic bond labile to hydrolysis at neutral

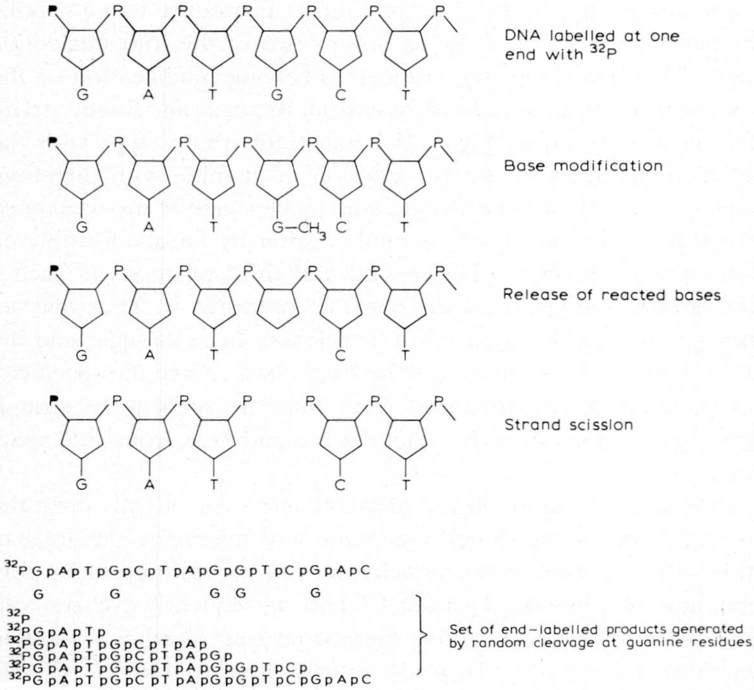

Fig. 5.1. Sequencing end-labelled DNA by limited, base-specific chemical cleavage. Fig. 5.1.(a) shows three consecutive reactions in which the DNA chain is cleaved at one guanine residue. In the first step a limited methylation reaction is carried out in which some of the guanine residues are methylated in the N7 position. In the second reaction this modified base is removed from the sugar thus weakening the chain at this point. The third reaction eliminates the phosphates from the sugar resulting in cleavage of the DNA chain. Figure 5.1.(b) shows how partial chemical cleavage at random guanine residues produces a set of end-labelled fragments. Each radioactive product extends from the ^{32}P-labelled end to the phosphate adjacent to the guanine which was methylated in the initial reaction. Electrophoresis on an acrylamide gel resolves these labelled fragments giving a series of bands in which their mobility is directly related to their chain length. If the DNA is cleaved in four such reactions, each specific for a particular base, and the four sets of products are electrophoresed in parallel and autoradiographed, the sequence can be read off the gel by noting in which channel each degradation product occurs.

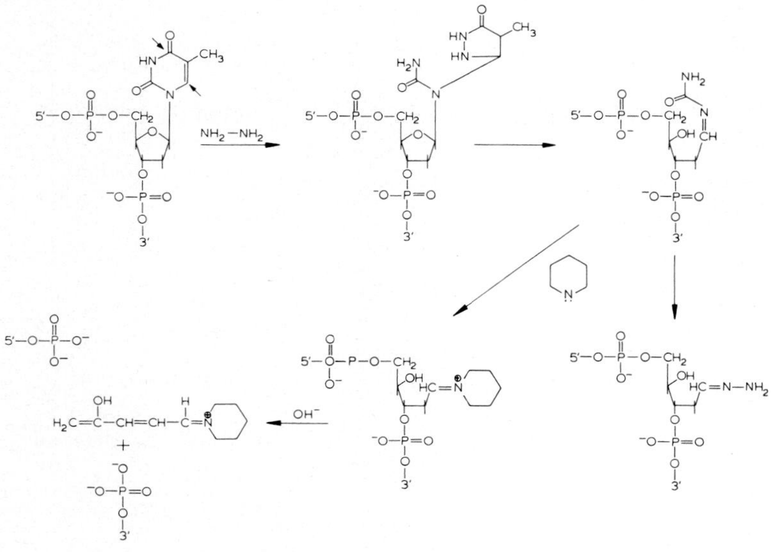

Fig. 5.2.(a). Sequential reactions with hydrazine and piperidine which break DNA at thymine residues. Hydrazine attacks thymine in DNA at C4 and C6, opens the pyrimidine ring and cyclizes with C4–C5–C6 to form a new five-membered ring. Further reaction with hydrazine releases this pyrazolone ring and also the N1–C2–N3 as a urea derivative. This results in the loss of the thymine leaving the ring-opened sugar as a hydrazone as the only connecting link in the DNA backbone. Piperidine catalyses the β-elimination of both phosphates from this sugar which results in the DNA being cleaved at that point. The reactions leading to strand breakage at cytosine residues follow essentially the same mechanism. In the presence of salt the reaction of hydrazine with thymine is suppressed giving a cytosine-specific reaction only.

pH. Rupture of this bond leaves the sugar (minus the base) as the link in the sugar-phosphate backbone. This linkage, however, is weakened sufficiently to permit ready alkali-catalysed β-elimination of the phosphates flanking the sugar and thus breakage of the DNA at that point. While the methylation of adenine N3 is considerably slower than that of guanine N7, the glycosidic bond in 3-methyl adenosine is considerably weaker than the corresponding

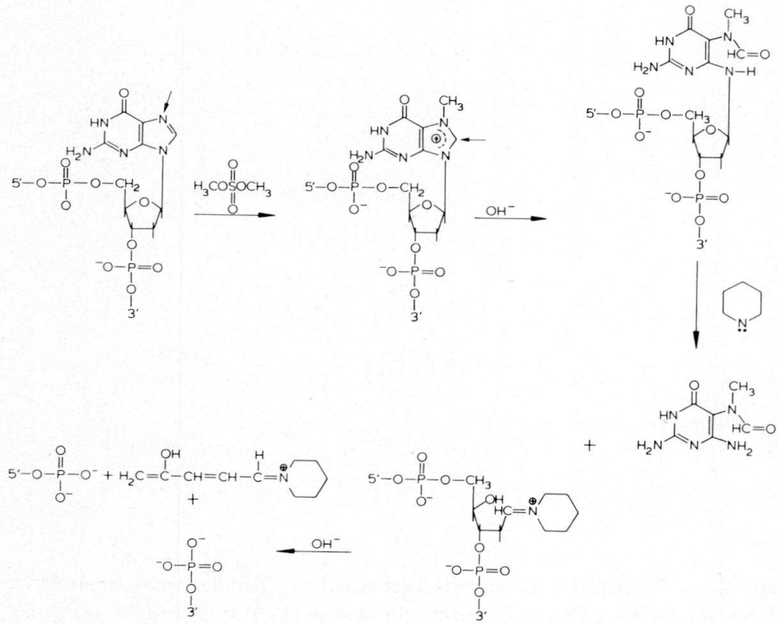

Fig. 5.2.(b). Sequential reactions with dimethylsulphate and piperidine which break DNA at guanine residues. Dimethylsulphate methylates the N7 of guanine fixing a positive charge in the imidazole ring. The base then breaks the bond between C8 and N9. Piperidine then displaces the ring-opened 7-methyl guanine and catalyses the β-elimination of both phosphates from the modified deoxyribose thereby breaking the DNA chain at that point.

bond in 7-methyl guanosine and is hydrolysed 4–6 time faster at neutral pH. These differential rates of purine methylation and methylated purine release formed the basis in the original Maxam-Gilbert method as the means for distinguishing between the adenines and guanines in DNA. More recently an alternative reaction has been employed in which the methylated DNA is reacted with piperdine (Fig. 5.2.(b)). This reaction is specific for methylated guanine and therefore results in strand scission uniquely at guanine residues.

TABLE 5.I.
Base-specific cleavage reactions for sequencing DNA

Reaction	Cleavage	Base modification	Modified base displacement	Strand scission	Experimental details
RI	G > A	Dimethylsulfate	Heat at pH 7	Sodium hydroxide	Maxam and Gilbert (1977)
R2	A > G	Dimethylsulfate	Acid	Sodium hydroxide	
R3	C + T	Hydrazine	Piperidine	Piperidine	
R4	C	Hydrazine + salt	Piperidine	Piperidine	
R5	G	Dimethylsulfate	Piperidine	Piperidine	Maxam and Gilbert (1980)
R6	G + A	Acid	Acid	Piperidine	
R7	C + T	Hydrazine	Piperidine	Piperidine	
R8	C	Hydrazine + salt	Piperidine	Piperidine	
R9	A > C	Sodium hydroxide	Piperidine	Piperidine	
R10	G > A	Dimethylsulfate	Heat at pH 7	Piperidine	
R11	G	Methylene blue	Piperidine	Piperidine	Friedman and Brown (1978)
R12	T	Osmium tetroxide	Piperidine	Piperidine	

A summary of reactions which have been successfully used to partially cleave end-labelled DNA at one or two of the four bases. Reactions R1–R4 are those originally recommended by Maxam and Gilbert (1977) while R5–R8 are new versions, which are faster and provide salt-free cleavage products for thin sequencing gels. Reaction R9 is an alternative to R6 for adenine cleavage.

Acid depurination will also lead to DNA cleavage at guanine and adenine though it does not distinguish between the two. Protonation of the purines at pH 2 results in depurination as a consequence of the hydrolysis of the glycosidic bond and a piperidine catalysed β-elimination of the phosphates then serves to break the DNA at that point.

Alkali will open adenine and cytosine rings at positions which can ultimately lead to DNA backbone cleavage. Incubation with 1 N NaOH splits some deoxyadenosine residues between C8 and N9 and a lower proportion of deoxycytidine rings, probably between C2 and N3. Treatment of the product with piperdine gives fairly uniform cleavage at the modified bases. The ratio of adenine:cytosine breaks occur with a frequency of about 5:1.

All the base-specific cleavage reactions described have been used successfully for sequencing DNA. Other novel base-specific modifications have been described e.g. the piperdine-catalysed cleavage at guanine bases after U.V. irradiation in the presence of methylene blue or rose bengal and a specific thymine cleavage by piperidine treatment after reaction of the DNA with osmium tetroxide (Friedmann and Brown, 1978). These latter procedures however, have not, as yet, been widely adopted as sequencing tools.

With this armoury of cleavage reactions a variety of different strategies can be designed for producing sets of specific cleavage products and Table 5.I lists the most widely used and recommended (Maxam and Gilbert, 1980) sets of cleavage reactions. Experimental details for each of these procedures are given in Section 5.9.

5.2. End-labelling DNA segments

The starting material for sequence analysis is usually a defined duplex DNA fragment purified from a restriction enzyme digest or obtained by a cleavage, at a unique site, of a defined circular DNA from a plasmid, phage or virus. Unique ends may also originate *in vivo* and on double-stranded DNA with a singular gap, nick or tail (cf. Chapter 1, Fig. 1.2.). These procedures give linear DNA molecules with (usually) unique sequences at both ends which can be labelled at either the 5′- or 3′-termini. In each case the product, after end-labelling, will yield a doubly labelled fragment and since a singly end-labelled DNA is required in order to generate a unique sequence the next step requires either the separation of the labelled strands or cleavage of the fragment with another restriction enzyme to yield separate singly labelled fragments. The principle is shown in Fig. 5.3. Modern cloning methods using plasmid and phage vectors have made accessible for DNA sequencing any fragment for which a suitable selection procedure can be devised and it is usually a straightforward undertaking to propagate and amplify the sequence in question as part of a recombinant plasmid or phage. In this way a few hundred picomoles of the DNA can be prepared which is sufficient to sequence fragments up to 5000 b.p's in length. The requirement for unique ends, though desirable, is not absolute. NucleaseS_1 resistant or ligand-protected fragments can be sequenced if fragments of different chain length can be resolved after labelling. For frag-

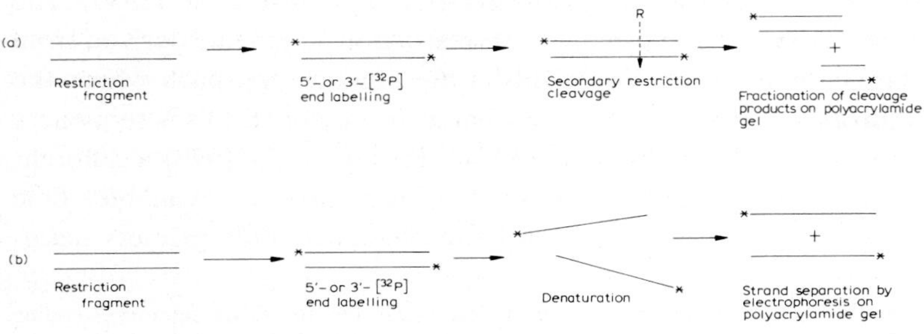

Fig. 5.3. Two routes for preparing singly-labelled DNA fragments. In scheme (a) the duplex fragment, labelled at both ends is cleaved with a secondary restriction enzyme at R to yield two singly-labelled duplex fragments. In scheme (b) the doubly-labelled fragment is denatured and the separated strands resolved by polyacrylamide gel electrophoresis.

ments of 100 nucleotides or less, electrophoresis on an 8% sequencing gel will often do this.

However, not all sequences of potential interest can be readily isolated as defined DNA fragments. In some cases difficulties in preparing suitable hybridization probes for detecting particular genomic fragments or in devising suitable selection procedures can prove a major stumbling block in the identification and isolation of particular sequences. In a different context the determination of coding sequences in an mRNA species may be the objective rather than that of the complex genomic DNA, with its pattern of interspersed intron and exon sequences,* and it is clearly desirable to be able to sequence directly single-stranded DNAs corresponding to a particular RNA species. Methods for achieving this have been described and are currently assuming an increased importance. In principle the methods are similar and depend on the primed synthesis of DNA transcripts of the RNA, usually with the enzyme reverse transcriptase, followed by analysis of the resultant complementary DNAs (cDNAs). If the DNA primer (usually 15–100 nucleotides) is 5′-end labelled prior to the annealing reaction, the resultant cDNA will be similarly labelled and amenable to sequence analysis by the Maxam-Gilbert procedure. A suitable procedure has been described in detail by Ghosh et al. (1980). The DNA primers have usually been restriction fragments derived from the genomic DNA. While this is the obvious approach where the appropriate primers can be prepared from cloned DNA sequences or from small genomes (e.g. SV40, Reddy et al., 1979) a different strategy is required where such primers are not available. One novel approach is to use synthetic oligonucleotide primers desig-

---

* In 1977 it was discovered that, in higher organisms, the DNA sequences coding for most mature mRNA transcripts are discontinuous. The DNA of the gene appears to be a mosaic of sequences, some of which are conserved in the mRNA of that gene (the exon sequences) whilst others (the intron, or intervening sequences) are discarded during the processing of the primary transcript to form the mature mRNA. The sequence of the mRNA is therefore not continuous with that of the gene (e.g. Breathnach et al., 1977; Jeffreys and Flavell, 1977).

ned from a knowledge of the amino acid sequence of the coded protein. This approach, for example, was successfully used in the determination of the nucleotide sequence of human fibroblast interferon mRNA using a synthetic 15 nucleotide primer which hybridized uniquely to a sequence within the interferon mRNA (Houghton et al. 1980) (Chapter 4, Section 4.9.). Though the interferon mRNA accounted for only about 0.03% of the total polyadenylated mRNA extracted from fibroblasts this method permitted the selective reverse transcription and sequencing of the cDNA. These techniques are therefore likely to find increasing application particularly as improved methods for the synthesis of defined deoxynucleotide oligomers (e.g. Hsiung et al., 1979) become available thus increasing the specificity and stringency of hybridization to a single unique sequence in a complex mixture of mRNAs. It has long been a dream of molecular biologists to be able to identify and isolate gene sequences corresponding to any defined polypeptide and these methods provide an important step in that direction. The procedures which may be used to sequence RNA molecules by analysis of the corresponding cDNAs were outlined in Chapter 4.

The method used for 5′-end labelling DNAs, be they single- or double-stranded depends on the phosphorylation of the free 5′-OH ends with T_4-polynucleotide kinase using high specific activity γ-[^{32}P]rATP as the phosphate donor. Since DNA fragments produced by restriction endonuclease cleavage usually possess phosphorylated 5′-ends it is necessary to first dephosphorylate the fragments using alkaline phosphatase. This sequence of reactions is shown diagramatically in Fig. 5.4.(a). After dephosphorylation, the alkaline phosphatase is denatured and removed by EDTA and phenol, the DNA recovered by precipitation and end-labelled with γ-ATP of specific activity 1000 Ci/mmol (or greater). Protruding 5′-ends are usually easier to phosphorylate than flush or recessed ends though, using the recent protocols described by Maxam and Gilbert (1980) labelling of these latter ends offers no real difficulty. Studies on the optimal conditions for kinase label-

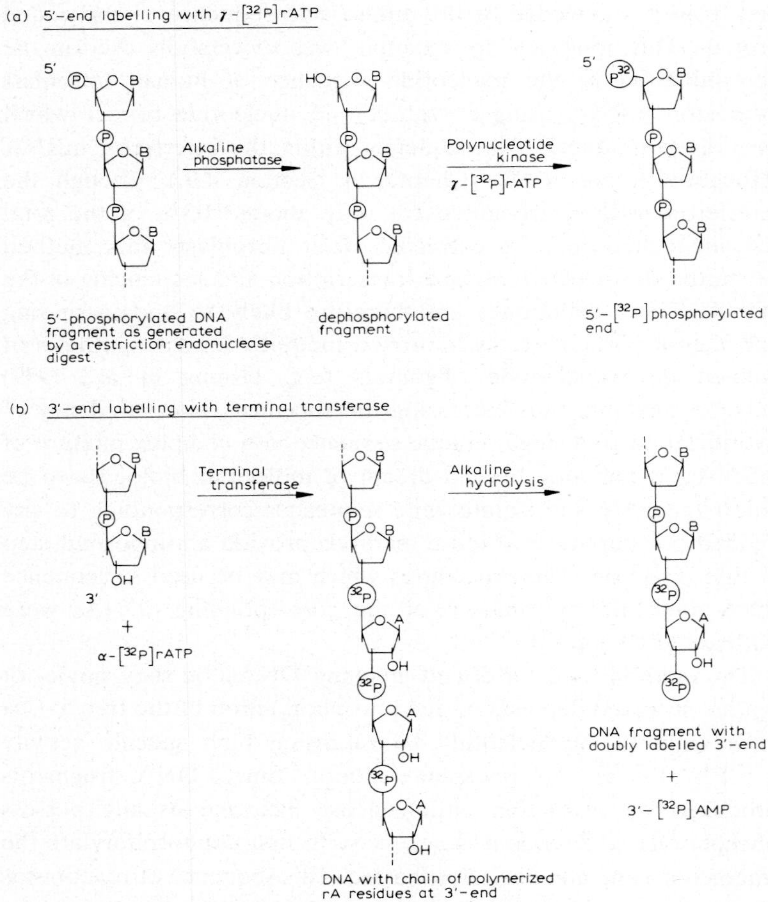

Fig. 5.4. Methods for end-labelling DNA.

ling are described in van de Sande et al. (1973) and Lillehaug et al. (1976). Polynucleotide kinase will also catalyse the exchange of existing 5′-phosphorylated ends and procedures have been described for using this reaction to label directly phosphorylated

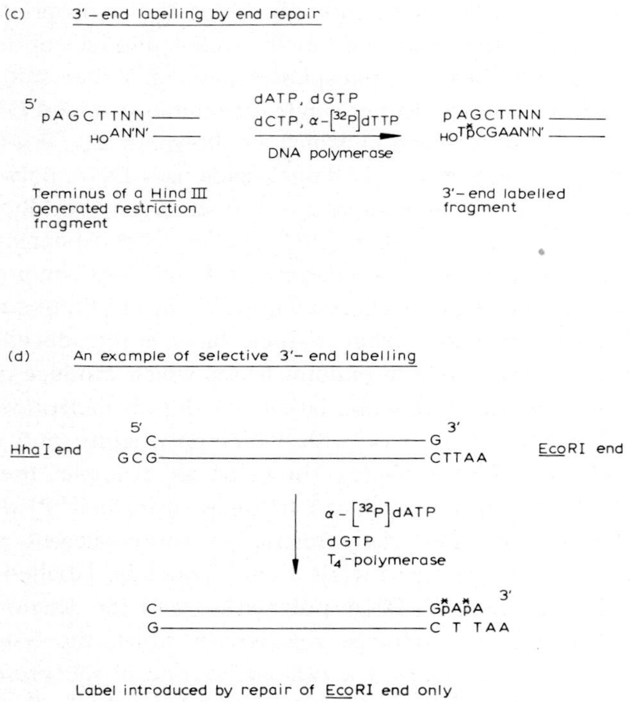

Fig. 5.4 (continued).

termini e.g. Chaconas et al. (1975), Berkner and Folk (1977). A suitable procedure, developed by Maxam and Gilbert and based on the Berkner and Folk method, is described in Section 5.9, Procedure D.

3′-end labelling can be carried out using two different reaction schemes. In the presence of calf thymus terminal transferase and a ribonucleoside triphosphate a variable number of ribonucleotide residues are polymerized onto the 3′-ends of DNA strands in a template-independent reaction. If the triphosphate is labelled with [^{32}P] in the alpha position, they will label the 3′-ends of the DNA.

Treatment of the product with alkali, or the appropriate ribonuclease, serves to remove all but the initial ribonucleotide which is retained with two [^{32}P]-phosphates, one on either side of the ribonucleotide residue. Either single- or double-stranded DNA can be labelled in this reaction which is shown in Fig. 5.4.(b). An alternative procedure for labelling 3′-ends uses DNA polymerase. A restriction fragment with recessed 3′-ends can be labelled by filling out the ends using α-[^{32}P] labelled triphosphates in the presence of *E. coli* DNA polymerase or T_4-DNA polymerase. The labelled triphosphates are chosen with reference to the sequence of the 5′-protruding end so that a single label is introduced. Table 5.II. lists fifteen restriction endonucleases which produce protruding 5′-ends and are readily end-labelled with polynucleotide kinase and γ-[^{32}P]ATP (5′-end) or with DNA polymerase and α-[^{32}P]-dNTP (3′-end). Figure 5.4.(c), shows as an example, the 3′-end labelling of a *Hind*III produced fragment using α-[^{32}P] dTTP as the labelled substrate. Many restriction endonucleases produce flush-ended fragments and while these cannot be labelled by the "filling-in" reaction, T_4-DNA polymerase and the Klenow polymerase catalyse an exchange reaction in which the 3′-terminal nucleotide is exchanged for a radioactive one in the presence of the appropriate α-[^{32}P] deoxynucleoside triphosphate. Both these methods have been successfully used to produce gel sequencing patterns and further details can be found in Soeda et al., 1978 and Schwarz et al., 1978. In some cases, using a DNA fragment generated by the combined action of two restriction endonucleases of different specificity, it is possible to label selectively only one 3′-end by an appropriate choice of labelled α-[^{32}P] deoxynucleotide triphosphate. Figure 5.4.(d) shows an example in which a fragment, produced from a combined *Eco*RI and *Hha*I digest, is singly labelled at the 3′ *Eco*RI end (Hindley and Phear, 1979).

Maxam and Gilbert (1980) have described detailed protocols for the preparation and isolation of dephosphorylated DNA fragments, and reaction schemes for the selective 5′-end and 3′-end labelling of these products either individually or *en masse*. Procedures A–D (Experimental Section 5.9.) recapitulate, in an abbreviated form, their recommended methods.

Procedure A describes the preparative scale isolation of DNA restriction fragments and their dephosphorylation with alkaline phosphatase.

Procedure B describes methods for the 5′-end labelling of dephosphorylated fragments with either protruding 5′-ends or with flush or recessed 5′-ends.

Procedure C gives the recommended procedure for labelling 3′-ends with terminal transferase and α-[^{32}P] rATP.

Procedure D is a method for 5′-end labelling of a set of fragments, not previously dephosphorylated, using the polynucleotide kinase catalysed exchange reaction.

5.3. *Separating the two labelled ends of complementary strands*

Whatever labelling procedure is utilized for end-labelling of double-stranded DNA the product of the reaction is almost always a doubly labelled fragment and since the gel sequencing methods cannot analyse two different sequences simultaneously it is necessary to resolve the two labelled ends from each other. This problem does not arise if single-stranded DNA is labelled and provided the population of single-stranded molecules is a mixture of different but discrete lengths, electrophoresing the products of an end-labelling reaction through an appropriate denaturing gel will usually resolve the different unique fragments. For a double-stranded DNA restriction fragment, labelled at either both 5′- or 3′-termini there are two choices. One is to cleave the fragment into two or more subfragments with another restriction enzyme and separate these by electrophoresis on a non-denaturing acrylamide gel (Procedure F, Experimental Procedures). The other choice is to 'melt' the DNA and separate the individual complementary strands by gel electrophoresis. This latter procedure is mandatory where no restriction site exists in the DNA fragment to be analysed and is usually the method of choice for small DNA fragments.

5.4. *Strand separation*

The separation of dissociated DNA strands can be achieved, with varying degrees of success, by several methods. Under defined conditions agarose gel electrophoresis (Hayward, 1972, Flint et al., 1975; Perlman and Huberman, 1977), electrophoresis on cellulose acetate membranes (Maniatis et al., 1975), homochromatography and reversed-phase column chromatography are all capable of resolving denatured DNA into its complementary strands. Maxam and Gilbert (1977) and Szalay et al. (1977) used electrophoresis on acrylamide gels for strand separation and with subsequent improvements this latter method has become the most widely adopted. Why complementary strands separate in this way is not known but it is presumably related to differences in their size, charge or shape. The most likely explanation is that when double-stranded DNA is dissociated each strand folds on itself giving a particular semi-stable conformation which is determined by its primary structure. Thus strands of the same size, but different in sequence, may acquire different conformations and be differentially retarded by the gel matrix during electrophoresis. Whatever the explanation strand separation offers the major advantage that the need for cleavage with a secondary restriction enzyme is obviated and the entire sequence of both strands can often be determined without recourse to the necessity of obtaining overlapping sequences across internal restriction sites. Acrylamide gels of between 5–8% acrylamide concentration and 1:60 to 1:30 cross-linked with bis-acrylamide are widely used. Initial denaturation of the DNA sample may be achieved by alkali (Maxam and Gilbert, 1977), heat (Szalay et al., 1977) or by heating in dimethylsulphoxide (Maxam and Gilbert, 1980). The optimal conditions may vary with the size of the DNA fragment in question. For restriction fragments up to 200 nucleotides alkali denaturation followed by electrophoresis on a

1:30 cross-linked, 8% acrylamide gel often gives good strand separation while for longer DNA fragments of several hundred nucleotides heat denaturation in dimethylsulphoxide followed by electrophoresis in a 1:60 cross-linked 5% gel is preferable (Maxam and Gilbert, 1980). A suitable procedure is described in Procedure E (Experimental procedures).

5.5. Elution of DNA from polyacrylamide gels

For sequencing it is necessary that the DNA fragments resolved on one or other of the above gels, are extracted from the gel matrix and concentrated into a small volume. The process needs to be as quantitative as possible, convenient and rapid, and yield undegraded DNA free of contaminants from the gel matrix. Two methods have been developed which meet these criteria. The first, developed by Maxam and Gilbert (1977), depends on diffusion of the DNA fragments out of the crushed gel into a medium of high ionic strength while the second method uses electroelution to promote the migration of the DNA out of the gel segment. (McDonell et al., 1977, Smith, H.O., 1980). (See also Section 4.3.8. for a description of alternative methods.)

5.5.1. (*i*) *Elution by diffusion*

The essential steps in this procedure are crushing the gel, soaking in a salt solution to allow the DNA fragments to diffuse out, removal of the gel fragments by filtration and precipitation of the DNA with ethanol. This is the most used procedure and efficiently recovers DNA fragments up to several thousand base pairs in length. Maxam and Gilbert (1980) describe the procedure in detail and Section G (Experimental Procedures) summarises their method. The extraction buffer contains (i) ammonium acetate, which promotes diffusion of the DNA out of the gel matrix, and is readily soluble in ethanol in the subsequent precipitation step, (ii) sodium dodecyl sulphate, a detergent which effectively denatures any contaminating DNAase and (iii) magnesium ions which aids the precipitation of DNA from dilute solutions. We have found that concentration of the aqueous gel extract with n-butanol prior to the ethanol-precipitation step is a convenient way of reducing the volume of the gel-extraction medium where large gel slices have been eluted. Six volumes of n-butanol will take up approximately one volume of water and two or three extractions serve to concentrate the DNA ten-fold prior to the precipitation step. In general 80–95% of the DNA is recoverable despite the unfavourable ratio of gel to elution buffer volume. Apparently the configurations available to the DNA in gel are limited compared to free solution and partitioning into the buffer is highly favourable.

5.5.2. (ii) Electroelution

This method has been recommended for the elution of high-molecular weight DNA (Smith, H.O., 1980). Different versions of the electroelution technique have been used, the most straightforward being that described by McDonell et al. (1977). In this procedure the gel slice is placed inside a dialysis bag together with a small amount of low-conductivity buffer, e.g. 5 mM Tris, 2.5 mM acetic acid (pH 8.0). A shallow box with platinum electrodes running along opposite sides serves as the electroelution chamber. The bag is tied at both ends and laid in the box parallel to the electrodes and sufficient buffer added to establish electrical contact with the bag but not to cover it as this only adds to the electrical heating. A voltage gradient of 10–15 V/cm for 15–30 min. is usually sufficient to drive the DNA out of the gel and into the buffer in the bag. The buffer is then drained from the bag and concentrated or precipitated as required. Excessive electrophoresis can result in the DNA adhering to the dialysis bag and it may be advantageous to reverse the polarity of the electrodes briefly to drive adhering DNA back into the buffer.

An alternative method is to electroelute from a crushed gel. A convenient way of doing this is to introduce the crushed gel as a slurry in a low-conductivity buffer (as above) into a wide bore plastic pipette plugged with siliconized glass wool in the tip to which is attached a small dialysis bag filled with buffer (Fig. 5.5.). Electrical contact is made by immersing the bag into a buffer reservoir (anode) and a bridge connecting the buffer overlaying the gel to another reservoir (cathode). With radioactively labelled DNA the progress of elution can be readily followed by a hand monitor. Ethidium-bromide stained DNA elution can also be monitored by inspecting under long wave U.V. After elution the DNA may be recovered by precipitation with ethanol in the presence of 0.3 M sodium acetate either before or after concentration.

Problems are sometimes encountered in contamination of the DNA with acrylamide monomers or other impurities from the gel matrix. Such contamination is usually apparent from the formation of an excessive turbidity during the ethanol-precipitation step. Pure DNA preparations in the microgram amounts eluted rarely give visible turbidity. A convenient purification procedure is to absorb the DNA onto a small (1 ml) column of DEAE-cellulose (DE-52, Whatman) in a low ionic strength buffer. Non-ionic im-

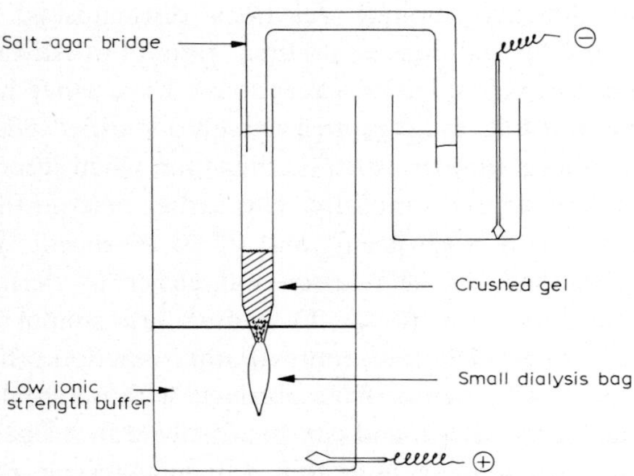

Fig. 5.5. Schematic diagram of apparatus for electroelution of DNA fragments from a crushed acrylamide gel.

purities are removed by washing with 4–5 volumes of the loading buffer and the DNA subsequently eluted with 1.1 M NaCl and precipitated by the addition of 2.5 volumes of ethanol. With radioactive DNA absorption and desorption to the DEAE-cellulose can be followed with a hand monitor (Appendix V). We have found this procedure to be useful in conjunction with the Maxam-Gilbert elution procedure (see above). Dilution of the filtered gel extract with water (1:5) and passage through 1 ml bed volume DE-52 quantitatively retains DNA. Elution with 1.1 M NaCl in 1:5 dilution of gel extraction buffer (0.7–1.0 ml) gives virtually quantitative recovery of DNA freed from all interfering contaminants (Hindley and Phear, 1979). This procedure is applicable to double-stranded DNA fragments but may result in losses of single-stranded DNAs.

5.6. Base-specific chemical cleavage reactions

In their original description of the base-specific chemical cleavage reactions Maxam and Gilbert presented a set of reaction schemes which partially cleaved DNA at either one or two of the four bases. The dimethyl-sulphate reactions distinguished adenines from guanines by differences in their rate of methylation and depurination (A > G and G > A reactions). Experience has shown that G and A residues are more precisely identified when bands are dependably present or absent rather than when interpretation is based on alterations in intensity. The earlier reactions have now been superseded by a G-specific and (G & A)-specific reactions. The new reactions are also faster and easier to perform. The hydrazine reactions for (C & T) and C are similar to those originally described. The four contemporary reaction schemes (G, G & A, C & T, C) yield salt-free products with minimal amounts of carrier DNA and RNA and can be dissolved in a small volume of formamide for loading onto thin sequencing gels. These procedures, as recommended by Maxam and Gilbert (1980) are set out in the Experimental Section 5.9., Procedures H, I, J and K.

In following these reaction schemes a number of factors need to be borne in mind which can markedly affect the successful outcome of a sequencing experiment. Some of these concern the proper care, quality and handling of reagents while others are related to the type of chemistry and reaction conditions and aim of the particular sequencing experiment. In the following discussion these different points are considered in more detail.

Reagents

Some of the sequencing reagents are labile. Anhydrous dimethylsulphate may pick up moisture from the air and hydrolyse to methanol and sulphuric acid; hydrazine can also pick up moisture and oxidize to diimine. Piperidine can also undergo oxidation. To avoid anomalous side reactions with these products, reagents of high purity should be obtained and kept sealed from the atmosphere. It is advisable to keep the reagents in two lots: the bulk in the original tightly-capped container and a small aliquot as a working solution in a screw-capped glass tube. Hydrazine in particular is unstable and can give an anomalous reaction with thymine that is not salt-suppressed. Working solutions of hydrazine are replaced every day. At various stages the DNA solution is heated to 90°C and all the buffers and salts used should be of high quality to avoid side reactions. Carrier DNA should be sheared by sonication and thoroughly extracted with phenol. After removal of phenol with ether, the DNA is dialysed against distilled water and diluted with water to 1 mg/ml. The stock solution is kept frozen.

It should be noted that dimethylsulphate and hydrazine are poisonous volatile reagents and should be stored and handled in a fume-hood. Disposal of dimethylsulphate into sodium hydroxide solution and hydrazine into 3 M ferric chloride inactivates these compounds.

All the initial base-modification reactions are conveniently carried out in 1.5 ml conical Eppendorf polypropylene tubes with snap-caps. It is recommended that these are siliconized with a 2% solution of dimethyldichlorosilane in 1.1.1. trichloroethane (cf.

Section 2.3.), dried and rinsed with distilled water before use. The advantage of these reaction vessels is that many of the reactions can be carried out successively in the same tube. Centrifugation of ethanol precipitates is rapid in the Eppendorf microfuge (12,000 ***g***) and precipitates are collected as barely visible pellets at the bottom of the tube.

5.6.1. Base modification reactions

The DNA sequencing chemistry begins with a base-modification reaction, the extent of which determines the frequency of DNA cleavage in the subsequent phosphate-elimination reaction. The number of bases modified in each molecule depends on the concentration of dimethylsulphate (G and G & A reactions) and hydrazine (C & T reactions) as well as the temperature and duration of the reaction. For speed and convenience the Maxam-Gilbert procedure makes use of temperature shifts and dilution to control the rate and extent of these reactions. The reagents are mixed at 0°C, incubated at 20°C for the required time and the DNA precipitated with cold sodium acetate and ethanol to slow down or halt the reaction. A fixed concentration of the different reagents is usually used so the main factor determining the extent of reaction is the time of incubation at 20°C.

Two factors need to be considered in choosing a reaction time. Long reaction times yield on average smaller fragments so where the objective is a limited sequence adjacent to the labelled end, or where only a short fragment is to be sequenced, it is advantageous to obtain the maximum yield of labelled fragments in this size range. On the other hand long range sequencing, which is the more usual objective, requires that the base-modification reactions are limited so as to yield fragments up to the maximum length that can be resolved on the gel. For the 8% acrylamide thin sequencing gels, fragments up to 300 nucleotides can be resolved so one aims to obtain a spread of chain lengths at least up to that figure. In this latter case the actual length of the DNA is irrelevant. Whether it is 500 or 5000 nucleotides in length the conditions for base modification are identical. The only consideration is how much of it can be sequenced from one end and this is determined by the resolving power of the gel. Maxam and Gilbert (1980) make the point that there is no virtue in carrying out several identical reactions for different times and then pooling the products. The times and temperatures given are those appropriate for sequencing 250 nucleotides from a labelled end. They can be adjusted if either too much label is left at the top of any of the sequencing channels or if the bands in a channel fade out near the top. It is not advisable to change the concentrations of the base-modification reagents; vary the time or temperature instead.

After the base-modification reaction cold (0°C) sodium acetate and three volumes of cold (0°C) ethanol are added to dilute and chill the reaction mixture and precipitate the DNA. A five-minute spin (12,000 **g**) quantitatively pellets the DNA leaving most of the excess dimethylsulphate or hydrazine in the supernatant. A second precipitation followed by an ethanol wash removes the last traces of reagents and sodium acetate.

5.6.2. Strand scission reactions

The DNA from the above reactions contain a few methylated bases (dimethylsulphate reaction), missing purine bases (acid) or ring-opened piperidine bases (hydrazine). Treatment with 1 M piperidine will open the 7-methylguanine rings, displace ring-opened bases from sugars and catalyse the β-elimination of phosphates from the empty sugars. The result is the cleavage of the DNA at these points. 1 M piperidine should be made fresh by dilution of the concentrated reagent into water. For a 1 M solution dilute 100 μl piperidine into 0.9 ml distilled water taking care to wash out the micro-pipette repeatedly to prevent loss of piperidine which tends to adhere to glass. The reaction with piperidine at 90°C was originally carried out in sealed glass capillary tubes but provided precautions against loss of piperidine by evaporation are taken the reaction can be carried out in the original Eppendorf tubes. The dried-down DNA is taken up in 100 μl 1.0 M piperidine and a layer of conformable tape (Teflon or polyvinyl) (Appendix 5) is stuck to the underside of the cap, which is then closed. This provides a vapour-proof gasket. The tubes are set in a rack in a 90°C water bath and a lead weight placed on top so that the caps do not pop open as pressure builds up inside. After 30 min. at 90°C the top of the tube is punctured and the samples lyophilized three times. This leaves four salt-free samples, one cleaved at random guanine residues (G), one at both guanine and adenine residues (G & A), another at cytosine and thymidine (C & T) and one at cytosines only (C). These four sets of products are dissolved in formamide, heat-denatured, and are then ready for electrophoresis on the sequencing gel. Detailed procedures for the above reactions are set out in 5.9. Procedures, H, I, J and K (Experimental procedures).

5.7. *Gel electrophoresis* (cf. also Sections 2.2., 3.1.2., 4.2.9. and 4.4.)

The thin-layer gel system used for resolving the different fragments produced by the chemical degradation method is essentially that of Sanger, Nicklen and Coulson (1977) as described in Chapter 4. An 8% polyacrylamide gel is usually used though for special purposes a 10 or 12% gel can give an improved resolution of the smallest fragments. For the standard 8% gel the polymerization mixture contains 7.6% (wt/vol) acrylamide, 0.4% (wt/vol) bisacrylamide, 50% (wt/vol) urea, 100 mM Tris-borate, pH 8.3, 2 mM EDTA, 0.07% (wt/vol) ammonium persulphate and TEMED catalyst. This solution is poured or injected into a $0.4 \times 200 \times 400$ mm mold to form a gel slab.

If less than about 100 nucleotides are to be sequenced the whole of each sample can be loaded onto the gel and electrophoresed until the xylene-cyanol dye marker (turquoise) has moved about a third of the way down the gel, (xylene-cyanol runs with fragments about 70 nucleotides long on an 8% gel). If sequences of 200 or more nucleotides are to be determined multiple loadings of the sample should be made after different times of electrophoresis. As a rough guide two loadings, the second being made when the xylene-cyanol marker from the first loading has moved to the bottom of the gel, and then continuing electrophoresis until the second xylene-cyanol band has moved about half-way, gives a sequence from about nucleotide 25 (beginning from the labelled end) up to about residue 250. Double loadings can sometimes result in a disturbance of the pH gradient between the electrodes with the formation of a discontinuity which moves down the gel and interferes with the resolution of DNA bands. Gel systems which employ paper wicks are better avoided and should not be used for double loadings. The sequence of the first 30 nucleotides is best determined by running a third sample on a thin 20% gel until the bromophenol blue has moved only a third of the way down the gel. The sequence of the first two or three nucleotides from the labelled end are often the most difficult to obtain. Since however, this restriction site will usually be included within a different set of labelled restriction fragments, the sequence will emerge from this latter data. When the aim of the experiment is to push the sequencing as far as possible from the labelled end, the third sample may be electrophoresed on a 6% acrylamide gel. This speeds up the running time and can improve the resolution of the slower moving bands.

5.8. *Reading the gel*

The autoradiograph of a sequencing gel shows four ladders of staggered horizontal bands. The bands at the bottom of the gel represent the smallest end-labelled fragments, increasing in size as one reads up the gel until either the pattern fades out due to the absence of any full-length labelled molecules left in the sample, or the pattern terminates abruptly in a strongly labelled band, in all four channels, corresponding to the uncleaved full-length labelled fragment. Of the four ladders in each pattern, bands in the first, (G) derive from breaks at guanine residues, bands in the (G & A) channel from breaks at both guanines and adenines, in the third, (C & T) from breaks at cytosines and thymidines and in the fourth channel (C) from breaks at only cytosines. Figure 5.6. shows a diagram of a hypothetical sequencing pattern and its interpretation. As one reads up a gel the band spacings gradually decrease until the number and order of the different bands becomes difficult to see. One then turns to the autoradiograph of the longer run on the same sample and reads this as far as possible. If the procedure has worked properly a sequence of up to 250 or more nucleotides can be deduced.

Sequencing patterns should be continuous, i.e. free from breaks

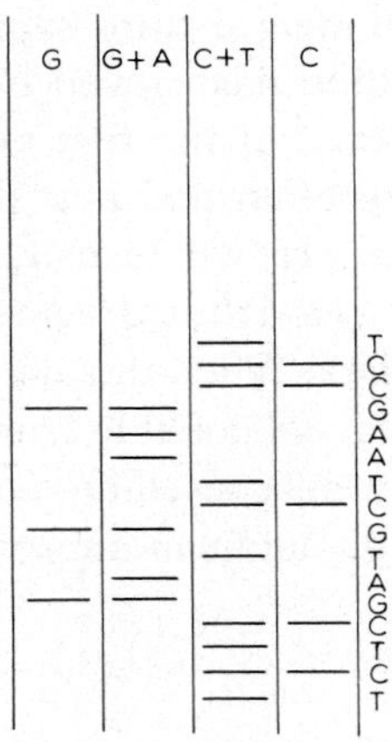

Fig. 5.6. Diagram of a sequencing gel and its interpretation. The first band reading from the bottom occurs in the (C & T) channel but not in the (C) channel, therefore it represents a fragment produced by cleavage at a T residue. The next band is found in both the (C & T) and (C) channels. This means the fragment arose by cleavage at a C residue. The change in mobility between these two fragments is equivalent to the addition of a single nucleotide therefore the sequence starts T–C. The next band occurs only in the (C & T) channel indicating a T cleavage thus extending the sequence to T–C–T. The same argument applies to the (G) and (G & A) channels and the hypothetical sequence shown can be deduced from the pattern of bands. Note that the two centre channels (G & A) and (C & T) between them give bands corresponding to every nucleotide.

in the pattern, and the bands should be sharply resolved and consistent with a unique DNA sequence. In practice however, difficulties may be encountered due to departures from this ideal pattern and bands may be doubled up, blurred, extraneous, retarded, or even missing. When such aberrations are encountered it is important to try and pinpoint the cause so that it can be corrected. In general difficulties may arise due to four causes:

(i) Contamination or heterogeneity of the labelled DNA. Also the presence of nicks or heterogeneity of the labelled end due to impure or contaminated restriction enzymes.

(ii) Faults in the chemical reactions such as loss of base specificity during modification or incomplete β-elimination.

(iii) Compression and the pile-up of bands within a narrow region of the gel due to persistent secondary structure within a set of particular fragment lengths.

(iv) Faults in the acrylamide gel due to impure reagents, uneven polymerization or uneven glass plates.

To diagnose a particular aberration is not always straightforward but a careful examination of the gel pattern will often suggest the nature of the problem. One can ask:

(1) Is the same fault found consistently on separate gels?

(2) Does it appear in all or only one or two of the sequencing channels?

(3) Does it appear in a particular position on the gel or throughout a given ladder?

Section 5.11. lists some of the more common problems encountered and suggests appropriate action. For an authoritative discussion the reader should consult Maxam and Gilbert (1980).

5.9. *Experimental procedures*

Procedure A

Large-scale preparation of unique DNA segments

1 mg DNA (phage, plasmid, virus)

120 μl distilled water
20 μl × 10 restriction enzyme buffer
50 μl restriction enzyme (100 units)

(a) Combine in 1.5 ml Eppendorf tube. Incubate 16 hours
(b) Electrophorese 2 μl on a gel and stain with ethidium bromide. If digestion is complete proceed with the next step. If not, add more enzyme and reincubate.
(c) Add: 2 μl 2.0 M Tris HCl, pH 8.0–8.5
5 μl alkaline phosphatase, 0.1–0.5 unit
(d) Incubate at 37°C for 30 min.
(e) Add: 2 μl 1.0 M EDTA
200 μl redistilled phenol equilibrated with 50 mM Tris-HCl, pH 8.0, 1 mM EDTA
(f) Cap the tube and vortex thoroughly
(g) Centrifuge briefly to separate the phases
(h) Remove bottom phase with a micropipette and discard
(i) Repeat phenol extraction
(j) Wash aqueous layer twice with 500 μl ether
(k) Add: 25 μl 3 M Sodium acetate
750 μl 95% ethanol
(l) Cap the tube, mix, chill at −70°C for 5 min.
(m) Centrifuge at 12,000 × g for 5 min.
(n) Remove supernatant
(o) Wash pellet with 1 ml 70% ethanol. Recentrifuge for 15 sec. Dry in vacuum for a few min.
(p) Dissolve pellet in 100 μl gel-loading buffer, 10% (vol/vol) glycerol in 50 mM Tris-borate, pH 8.3, 1 mM EDTA, 0.05% (wt/vol) xylene cyanol and bromophenol blue.
(q) Load onto a long, thick non-denaturing gel:
5% acrylamide, 0.17% bis-acrylamide
50 mM Tris-borate pH 8.3, 1 mM EDTA
Slab gel: 6 × 200 × 400 mm, or 3 × 200 × 400 mm
Sample wells: two 6 × 30 mm or four 3 × 30 mm
(r) Electrophorese until the smallest fragment wanted has traversed most of the gel

(s) Transfer gel onto a thin sheet of transparent plastic wrap and lay on a fluorescent thin-layer chromatography plate (Appendix V) and examine under short-wave U.V. light in a dark room. The DNA appears as dark bands against a light fluorescent background. Cut out the appropriate bands with a razor blade and crush by squeezing through the narrowed-down orifice of a disposable plastic syringe into a siliconized 12 ml centrifuge tube.

(t) Add: 8 ml 0.5 M Ammonium acetate, 10 mM Magnesium acetate, 1 mM EDTA, 0.1% sodium dodecyl sulphate (Gel extraction buffer)

(u) Seal tube, mix by inversion and incubate at 37°C for 10–16 h. Centrifuge at 10,000 × *g* for 15 min. Remove supernatant and save. Re-extract gel with a further 4 ml gel-extraction buffer. Combine supernatants and filter through a Millipore 'Millex' unit under gravity. Collect filtrate in a siliconized 50 ml centrifuge tube.

(v) Add 35 ml 95% ethanol

(w) Seal, mix and freeze at −70°C (ethanol–dry ice bath)

(x) Centrifuge at 10,000 × *g* for 15 min., 4°C. Remove supernatant with a pipette

(y) Add 0.25 ml 0.3 M sodium acetate to redissolve the DNA Transfer solution to 1.5 ml Eppendorf tube

(z) Add 0.75 ml 95% ethanol
- Close the cap, mix and freeze as before. Centrifuge at 12,000 × *g* for 5 min. (Eppendorf, 4°C). Remove supernatant with pasteur pipette
- Add 1 ml 95% ethanol
- Centrifuge for 15 sec. and remove supernatant. Dry under vacuum for a few min. The DNA residue appears as a small pellet or film and can be dissolved in a small volume of water.

Notes

1. The DNA bands in the gel can also be vizualized by staining

with ethidium bromide. This is not recommended, however, since this may cause nicking when illuminated with UV light and residual ethidium bromide can contaminate the final DNA preparation. If no other choice is available this method can be used and the gel inspected under long-wave UV.

2. The volume of gel extract (12 ml) can be reduced before precipitation by several extractions with n-butanol. One volume of water is taken up by 6 volumes of butanol. A ten-fold concentration does not appear to affect the purity of the DNA recovered after ethanol precipitation.

Procedure B

Labelling 5′-ends with polynucleotide kinase and [γ-^{32}P]-rATP

(*i*) *For protruding 5′-ends*

(a) Combine, in a siliconised 1.5 ml Eppendorf tube:
5 μl dephosphorylated DNA, equivalent to 1–50 pmol 5′-ends
35 μl distilled water
5 μl 500 mM Tris-HCl, pH 7.6, 100 mM $MgCl_2$, 50 mM DTT, 1 mM spermidine, 1 mM EDTA
5 μl γ-[^{32}P] rATP, >1000 Ci/mmol, at least 50 pmol
1 μl T_4-polynucleotide kinase, 20 units

(b) Mix and incubate at 37°C for 30 min.

(*ii*) *For flush or recessed 5′-ends*

(a) Combine, in a siliconized 1.5 ml Eppendorf tube:
5 μl dephosphorylated DNA, equivalent to 1–50 pmol. 5′-ends
32 μl 20 mM Tris-HCl, pH 9.5, 1 mM spermidine, 0.1 mM EDTA

(b) Mix, heat at 90°C for 2 min., chill rapidly in ice-water and immediately add:
5 μl 500 mM Tris-HCl, pH 9.5, 100 mM $MgCl_2$, 50 mM DTT in 50% glycerol
5 μl γ-[^{32}P] rATP > 1000 Ci/mmol, at least 50 pmol
1 μl T_4 polynucleotide kinase

(c) Mix and incubate at 37°C for 15 min.

Work-up of (i) and (ii) samples

(a) Add: 200 μl 2.5 M ammonium acetate
1 μl t-RNA, 1 mg/ml
750 μl 95% ethanol

(b) Add reagents in the above order to prevent the tRNA from becoming labelled

(c) Close the tube, mix by inversion, chill at −70°C for 5 min.

(d) Centrifuge at 12,000 × **g** for 5 min. (Eppendorf, 4°C)

(e) Remove supernatant
Add: 250 μl 0.3 M sodium acetate to dissolve DNA

(f) Mix
Add: 750 μl 95% ethanol

(g) Mix, chill and centrifuge as before

(h) Wash with 1 ml 95% ethanol, centrifuge (15 s)

(i) Remove supernatant.

(j) Dry in vacuo for a few minutes.

The product in each case is 5′-[^{32}P] labelled DNA plus unlabelled carrier RNA. The preparation is essentially salt-free and ready for either secondary restriction enzyme cleavage or strand separation.

Procedure C

Labelling 3′-ends with terminal transferase and α-[^{32}P]rATP

(a) Combine in a siliconized 1.5 ml Eppendorf tube:
5 μl DNA, equivalent to 1–50 pmol 3′-ends
10 μl distilled water

(b) Mix, heat at 90°C for 1 min., quick-chill and centrifuge for a few seconds

(c) Add: 2.5 μl 1.0 M potassium cacodylate, pH 7.6, 10 mM $CoCl_2$, 2 mM dithiothreitol
5 μl α-[^{32}P] rATP, 300–3000 Ci/mmol, at least 50 pmol
1 μl terminal deoxynucleotidyl transferase, 10 units

(d) Mix and incubate at 37°C for 30 min.

(e) Add: a further 1 μl terminal deoxynucleotidyl transferase, 10 units

(f) Mix and incubate at 37°C for 30 min.

(g) Add: 1 μl 100 mM rATP
(h) Mix, incubate at 37°C for 10 min., chill in ice
(i) Add: 200 μl 2.5 M ammonium acetate (0°C)
  1 μl tRNA, 1 mg/ml
  750 μl 95% ethanol
(j) Close the cap, mix by inversion, chill at −70°, 5 min.
(k) Centrifuge and remove supernatant
(l) Redissolve in 250 μl 0.3 M sodium acetate precipitate and wash with ethanol as described in Procedure B
(m) To dried residue add: 100 μl 1.0 M piperidine (freshly prepared)
(n) Close the cap and agitate until DNA is dissolved
(o) Centrifuge briefly. Seal cap with stretchable tape (e.g. polyvinyl insulating tape)
(p) Heat at 90°C for 30 min.
(q) Centrifuge briefly, punch holes in cap, freeze and lyophilize
(r) Redissolve DNA in 10 μl water, freeze and lyophilize
(s) Repeat once more.

The product is 3′-[^{32}P]-labelled DNA essentially free of salt and ready for strand separation. Where secondary cleavage with a restriction enzyme is to follow it may be necessary to renature the sample. This may be achieved as follows:

Dissolve the sample in 20 μl distilled water
 Add 2 μl 200 mM Tris-HCl, pH 7.4, 50 mM $MgCl_2$, 10 mM DTT, ±500 mM NaCl.

Anneal at 67°C for 15 min. Cool to 37°C.

The sample is now ready for secondary restriction cleavage.

Procedure D

Three reactions and one gel for preparing 5′-[^{32}P]-labelled DNA fragments

This procedure describes a way for producing sets of 5′-end labelled fragments, ready for sequencing, from a phage, plasmid or viral DNA, or a large restriction fragment. The DNA is first cleaved with a restriction enzyme (A) to produce a collection of fragments which are phosphorylated using the polynucleotide-kinase exchange reactions. After the fragments have all been labelled, at both ends, they are either all denatured and electrophoresed to separate single strands, or the mixture is digested with a second restriction enzyme (B) and electrophoresed to separate out the secondary cleavage products from the uncut fragments. In both cases some undenatured, or uncut material is run in parallel to distinguish the strand-separated or split products from the intact fragments. The experiment is then repeated using the same pair of enzymes in the reverse order (Fig. 5.7.). The pair of enzymes chosen should preferably be ones which yield 5′-protruding ends.

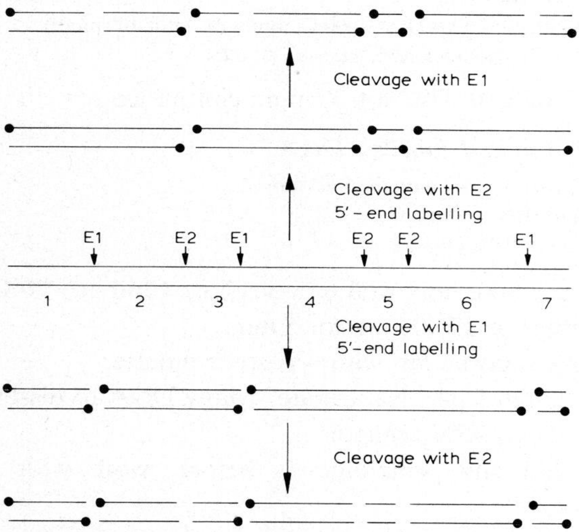

Fig. 5.7. Pattern of singly-labelled fragments produced using two restriction enzymes in the reverse order. The hypothetical fragment contains three E1 and three E2 restriction sites as shown. Cleavage with E1 + E2 yields seven fragments (numbered 1–7). In the scheme shown, singly-labelled fragments are produced from 1, 2, 3, 4, 6 and 7. Fragment 5 is either doubly-labelled or unlabelled. Fragments 1 and 7 are singly-labelled in one reaction sequence (E2 followed by E1). Fragments 2, 3, 4 and 6 are all singly-labelled at opposite ends depending on the reaction sequence.

Primary restriction cleavage

10 μg DNA
15 μl distilled water
2 μl 200 mM Tris-HCl, pH 7.4, 50 mM $MgCl_2$, 10 mM DTT, ±500 mM NaCl (depending on restriction enzyme, see manufacturers recommendations)
5 μl restriction enzyme A (10 units)

– Combine above in a siliconized 1.5 ml Eppendorf tube. Incubate at 37°C for 1 hour

End-labelling

Add: 10 μl distilled water
5 μl 500 mM imidazole-HCl, pH 6.6, 100 mM $MgCl_2$, 50 mM DTT, 1 mM spermidine
3 μl 5 mM rADP
10 μl γ-[^{32}P]-rATP, >1000 Ci/mmol (at least 100 pmol)
1 μl T_4-polynucleotide kinase, 20 units

– Mix, incubate at 37°C for 30 min., chill in ice.

Precipitation of end-labelled DNA

Add: 200 μl 2.5 M ammonium acetate
1 μl tRNA, 1 mg/ml
750 μl 95% ethanol

(a) Close the tube, mix well by inversion. Chill at −70°C.
(b) Centrifuge at 12,000 × ***g*** for 5 min.
(c) Remove supernatant with a pasteur pipette.
(d) Add: 250 μl 0.3 M Na acetate. Allow DNA to dissolve.
750 μl 95% ethanol
(e) Mix, chill and centrifuge as before, wash with 1 ml 95% ethanol.
(f) Dry in vacuo for a few minutes.
(g) Dissolve in 20 μl distilled water.
(h) Proceed with strand separation or secondary restriction enzyme cleavage, or both.

Strand separation

Divide the end-labelled DNA into two portions:

18 μl (for strand separation)	2 μl (for undenatured control)
20 μl 60% dimethylsulphoxide,	20 μl 60% dimethylsulphoxide,
1 mM EDTA, 0.05% xylene cyanol, 0.05% bromophenol blue.	1 mM EDTA, + dyes as opp.
	15 μl distilled water
Mix, heat at 90°C for 2 min. quick-chill in ice-water	Mix and keep at 0°C

(a) Immediately load both samples in adjacent slots on a strand-separation gel (procedure E) using 1.5 × 20 mm sample wells.
(b) Electrophorese, autoradiograph and identify strand-separated fragments (Procedure E).
(c) Elute and sequence.

Secondary restriction cleavage
Divide end-labelled DNA into two portions:

18 μl (for secondary digestion)	2 μl (undigested marker)
2 μl 200 mM Tris-HCl, pH 7.4 50 mM $MgCl_2$, 10 mM DTT, ±500 mM NaCl	2 μl buffer (as opposite)
	20 μl distilled water
5 μl restriction enzyme B, (10 units)	
Mix and incubate at 37°C, 1 h.	Mix and keep at 0°C

(a) Add concentrated gel loading solution to each sample, 5 μl (50% glycerol, 25 mM EDTA, 0.25% xylene-cyanol and bromophenol blue).
(b) Load samples into adjacent slots of a non-denaturing gel (Procedure F).
(c) Electrophorese, autoradiograph and identify singly-labelled cleavage products.
(d) Elute and sequence.

Procedure E

DNA strand separation by denaturation and gel electrophoresis
(a) Mix: Approximately 1 μg DNA restriction fragment, [^{32}P]-labelled at both ends, in an Eppendorf tube

40 μl of 30% (vol/vol) dimethylsulphoxide,
1 mM EDTA, and 0.05% (wt/vol) xylene-cyanol
and bromophenol blue.

(b) Close cap and redissolve the DNA. Centrifuge for a few seconds, heat at 90°C for 2 min. and quick-chill in ice-water.

(c) Immediately load onto a pre-electrophoresed (30 min.) strand-separation gel.
5% (wt/vol) acrylamide, 0.1% (wt/vol), bis-acrylamide
50 mM Tris-borate, pH 8.3, 1 mM EDTA
Slab gel: 1.5 × 200 × 400 mm
Sample well: 1.5 × 20 mm

(d) Electrophorese at ~300 V (8 V/cm) or less. The gel should not get appreciably warm.

(e) Stop electrophoresis when the complementary strands show adequate resolution (determined empirically).

(f) Autoradiograph.

(g) Cut required bands out of the X-ray film and use this as a template to excise the gel bands. Elute the DNA (5.9., G).
Note: The most important parameter in achieving satisfactory strand separation appears to be the concentration of the sample. Where larger amounts of DNA are used a thicker gel and longer sample wells should be employed.

Procedure F

DNA fragment division by secondary restriction cleavage and gel electrophoresis

(a) Mix: 1 μg DNA restriction fragment, [^{32}P]-labelled at both ends (salt-free pellet in Eppendorf tube)
20 μl distilled water

(b) Allow sample to dissolve (agitation), centrifuge briefly

(c) Add: 2 μl 200 mM Tris HCl, pH 7.4, 50 mM $MgCl_2$, 10 mM DTT, ±500 mM NaCl (cf. procedure D above, primary restriction cleavage)
0.5–1.0 μl restriction enzyme (1 unit)
Incubate at 37°C for 60 min.

(d) Add: 5 μl 50% (wt/vol) glycerol, 25 mM EDTA, 0.25% xylene-cyanol and bromophenol blue
(e) Mix and load onto pre-electrophoresed (30 min.) non-denaturing gel:
5–10% (wt/vol) acrylamide, 0.17–0.33% bisacrylamide
50 mM Tris-borate, pH 8.3, 1 mM EDTA
Slab gel: 1.5 × 200 × 200 mm or 1.5 × 200 × 400 mm
Sample well: 1.5 × 15–20 mm
(f) Electrophorese at 15–20 v/cm (constant voltage) until the position of the tracking dyes indicate adequate resolution of fragments (determined empirically)
(g) Autoradiograph
(h) Cut required bands out of the film and use this as a template to excise the gel bands. Elute the DNA (5.9., G).

Maxam and Gilbert (1980) recommend chromatographing small (0.2 μl) aliquots of the digest on a PEI-cellulose thin layer to check that the restriction enzyme is free of phosphatase activity which might remove the terminally labelled nucleotide. Samples are taken at the beginning (0 min.) and end (60 min.) of the digest, chromatographed on a PEI (polyethylene-imine) thin layer using 0.75 M potassium phosphate pH 3.5, and autoradiographed. A loss of label from the origin spot indicates contamination with phosphatase. This can sometimes be minimized by including 1 mM orthophosphate in the restriction enzyme buffer.

Procedure G

Elution of DNA from polyacrylamide gel

Maxam and Gilbert (1980) recommend the following procedure for small segments of gel. For longer pieces of gel the method described under Procedure A may be used:
(a) Plug a 100 μl plastic pipette tip with siliconized glass wool and seal the tip in a small flame. Drop in the gel segment and grind to a paste with a siliconized glass rod
(b) Add: 0.6 ml 500 mM ammonium acetate
10 mM $MgCl_2$, 1 mM EDTA, 0.1% sodium dodecyl sul-

phate, 10 μg/ml tRNA (omit carrier RNA if labelling is to follow)

(c) Stir the paste into this solution, seal the top of the tip with Parafilm and incubate at 37°C for 10 h. or more. Remove parafilm, cut off the sealed point and support the tip in a siliconised centrifuge tube. Centrifuge at ~3000 rpm (2 min.) until all the supernatant has collected into the centrifuge tube.

(d) Add: 0.2 ml above gel elution buffer to tip and recentrifuge. Combine eluates, then add 2 ml 95% ethanol.

(e) Mix by inversion, chill at −70°C, centrifuge at 10,000 × ***g*** for 15 min. (4°C).

(f) Remove supernatant with pasteur pipette. Dissolve precipitate in 0.5 ml 300 mM sodium acetate. Reprecipitate with 1.5 ml 95% ethanol etc. Centrifuge and remove supernatant.

(g) Dissolve precipitate in 25 μl distilled water. Reprecipitate with 1 ml 95% ethanol, chill and centrifuge. Remove supernatant and dry under vacuum for a few minutes. The DNA/RNA is obtained as a small pellet on film at the bottom of the tube. It is essentially salt-free and can be dissolved in a small volume of water (20 μl) ready for the sequencing reactions.

5.10. Sequencing reactions (Maxam and Gilbert, 1980)

Procedure H

Limited DNA cleavage at guanines (G)

– 200 μl 50 mM sodium cacodylate, pH 8.0, 10 mM $MgCl_2$, 1 mM EDTA, 5 μl end-labelled DNA in water

– Combine the above in a 1.5 ml Eppendorf snap-cap tube
  1 μl dimethylsulphate

– Close the cap and mix (Vortex)

– Let stand at 23°C for 5 min.
  50 μl 1.5 M sodium acetate pH 7.0, 1.0 M mercaptoethanol, 100 μg/ml tRNA (0°C)
  750 μl 95% ethanol (0°C)

- Close the cap and mix well (invert the tube four times)
- Chill at −70°C for 5 min.
- Centrifuge at 12,000 × *g* for 5 min. (Eppendorf, 4°C)
- Remove the supernatant with a Pasteur pipette

  250 μl 0.3 M sodium acetate (0°C)
- Close the cap and redissolve the DNA (Vortex)

  750 μl 95% ethanol (0°C)
- Invert to mix, chill, centrifuge, and remove supernatant

  1 ml 95% ethanol
- Centrifuge at 12,000 × *g* 15 sec. and remove supernatant
- Place the tube under vacuum for a few minutes

  100 μl 1.0 M piperidine (freshly diluted)
- Close cap and redissolve DNA (Vortex and manual agitation)
- Centrifuge for a few seconds (Eppendorf)
- Place conformable tape under cap and close tightly
- Heat at 90°C for 30 min. (under weight in water bath)
- Remove the tape and centrifuge for a few seconds
- Punch holes in the tube cap with a needle
- Freeze the sample, and lyophilize
- Redissolve the DNA in 10 μl water, freeze, and lyophilize
- Redissolve DNA in 10 μl water, freeze, and lyophilize again.*

  10 μl 80% (vol/vol) deionized formamide†, 10 mM NaOH, 1 mM EDTA, 0.1% (wt/vol) xylene cyanol, 0.1% (wt/vol) bromophenol blue
- Close cap and redissolve DNA (Vortex and manual agitation)
- Centrifuge for a few sec. (Eppendorf)
- Heat at 90°C for 1 min. and quick-chill in ice-water
- Load on sequencing gel(s) immediately (Procedure L)

---

* The repeated lyophilizations in this and the following procedures are to remove all traces of piperidine which otherwise may interfere with the resolution of the gel.

† 100 ml formamide stirred for 30 min. with 5 g Amberlite MB-1 (mixed bed resin). The resin is removed by filtration.

Procedure I

Limited DNA cleavage at guanines and adenines $(G + A)$

10 μl distilled water

10 μl end-labelled DNA, in water

– Combine the above in a 1.5 ml Eppendorf snap-cap tube

2 μl 1.0 M piperidinium formate [4%(vol/vol) formic acid, adjusted to pH 2.0 with piperidine]

– Mix, and let stand at 23°C for 60–80 min.

– Punch holes in the tube cap with a needle

– Freeze the sample, and lyophilize

– Redissolve the DNA in 20 μl water, freeze, and lyophilize again

100 μl 1.0 M piperidine (freshly diluted)

– Close cap and redissolve DNA (Vortex and manual agitation)

– Centrifuge for a few sec. (Eppendorf)

– Place conformable tape under cap and close tightly

– Heat at 90°C for 30 min. (under weight in water bath)

– Remove the tape and centrifuge for a few sec. (Eppendorf)

– Freeze the sample, and lyophilize

– Redissolve the DNA in 10 μl water, freeze, and lyophilize

– Redissolve DNA in 10 μl water, freeze, and lyophilize again

10 μl 80% (vol/vol) deionized formamide, 10 mM NaOH, 1 mM EDTA, 0.1% (wt/vol) xylene-cyanol, 0.1% (wt/vol) bromophenol blue

– Close cap and redissolve DNA (Vortex and manual agitation)

– Centrifuge for a few sec. (Eppendorf)

– Heat at 90°C for 1 min. and quick-chill-in ice-water

– Load on sequencing gel(s) immediately

In this procedure two reactions partially cleave DNA at purines, i.e., limited acid depurination followed by piperidine-catalyzed β-elimination. The extent of cleavage will reflect the extent of depurination, the rate of which is pH-dependent. The use of piperidinium formate (rather than free acid) provides some control over the pH and ultimately the cleavage. More extensive cleavage can be obtained by increasing the reaction temperature (to 37°C) on increasing the reaction time. The piperidinium formate should be stored in aliquots at −20°C.

Procedures J and K

Limited DNA cleavage at pyrimidines

J. Cytosines and thymines (C + T)	*K. Cytosines (C)*
10 μl distilled water	10 μl 5 M sodium chloride
10 μl end-labelled DNA	10 μl end-labelled DNA

- Prepare the above in two 1.5 ml Eppendorf snap-cap tubes
  30 μl hydrazine, 95% reagent grade (0°C)
- Close cap on tube and mix gently (manual agitation)
- Let stand at 23°C for 5 min.
  200 μl 0.3 M sodium acetate, 0.1 mM EDTA, 25 μg/ml tRNA (0°C)
  750 μl 95% ethanol (0°C)
- Close the cap and mix well (invert the tube four times)
- Chill at −70°C for 5 min.
- Centrifuge at 12,000 × *g* for 5 min. (Eppendorf, 4°C)
- Remove the supernatant with a Pasteur pipette and transfer into a waste bottle containing 2 M ferric chloride
  250 μl 0.3 M sodium acetate (0°C)
- Close the cap and redissolve the DNA (Vortex)
  750 μl 95% ethanol (0°C)
- Invert to mix, chill, centrifuge, and remove supernatant
  1 ml 95% ethanol
- Centrifuge at 12,000 × *g* 15 sec. and remove supernatant
- Place the tube under vacuum for a few min. (desiccator)
  100 μl 1.0 M piperidine (freshly diluted)
- Close cap and redissolve DNA (Vortex and manual agitation)
- Place conformable tape under cap and close tightly
- Heat at 90°C for 30 min. (under weight in water bath)
- Remove the tape and centrifuge for a few sec. (Eppendorf)
- Punch holes in the tube cap with a needle
- Freeze the sample, and lyophilize
- Redissolve the DNA in 10 μl water, freeze and lyophilize
- Redissolve DNA in 10 μl water, freeze, and lyophilize again
  10 μl 80% (vol/vol) deionized formamide, 10 mM NaOH, 1 mM

EDTA, 0.1% (wt/vol) xylene-cyanol, 0.1% (wt/vol) bromophenol blue
- Close cap and redissolve DNA (Vortex and manual agitation)
- Centrifuge for a few sec. (Eppendorf)
- Heat at 90°C for 1 min. and quick-chill in ice-water
- Load on sequencing gel(s) immediately

L. Preparation and running of sequencing gels
The procedures employed for resolving end-labelled fragments prepared by the Maxam and Gilbert method are identical to those described in Chapter 4 for fractionating primed synthesis fragments. Full and detailed descriptions are also given in Maxam and Gilbert (1980).

5.11. Trouble-shooting guide (Maxam and Gilbert, 1980)

Fault	Probable cause	Solution
Extraneous bands in all positions throughout a sequencing pattern	Terminal heterogeneity at the labelled end due to action of contaminating nuclease in the restriction enzyme or in the labelling procedure	Replace or purify the enzymes involved
	Unavoidable terminal heterogeneity as a consequence of using DNAase I or Sl nuclease	After labelling ragged ends cleave with a restriction enzyme to yield end-labelled fragments of less than 100 nucleotides. Denature and electrophorese on a sequencing gel (1.5 mm thick). Select a well-resolved single band, elute, and use for sequence analysis

Fault	Probable cause	Solution
	Contamination of end-labelled fragment with other end-labelled DNAs	Try to improve the resolution of the restriction fragment by altering the running conditions on the gel (e.g. longer running times, lower loading, altering the acrylamide concentration). Alternatively use a different restriction enzyme
Any or some of the following:		
Smearing of pyrimidine cleavage products	Incomplete cleavage of hydrazinolysis products with piperidine. Hydrazine has reacted with C's or T's giving products with an altered charge and/or mass which have resisted subsequent cleavage with piperidine	Ensure that sufficient piperidine has been used and that tubes are well sealed in the 90°C heating step. Use an efficient vacuum pump in the lyophilizations to remove all traces of hydrazine and piperidine
Loss of electrophoresis dye colour in (C) and (C & T) channels		
Poor suppression of T's in the (C) ladder		
Bands in (C) and/or (C & T) channels at positions where purines follow pyrimidines	Hydrazine has undergone oxidation	Replace with a fresh aliquot or new bottle
	Residual sodium acetate from the second ethanol precipitation buffering the piperidine reaction leading to incomplete β-elimination of phosphate	Redissolve the DNA, at the appropriate stage, in 25 μl H_2O. Reprecipitate with 1 ml 95% ethanol, centrifuge and dry the pellet before proceeding. Sodium acetate is soluble in 95% ethanol
T's are weak in the (C & T) ladder	Residual salt from ethanol-precipitation step suppressing reaction of T with hydrazine	

Fault	Probable cause	Solution
Anomalous bands are present in (C) and/or (C & T) channels corresponding to position of bands in (G) channel	A reaction between hydrazine and guanine at lower pH followed by piperidine cleavage at guanines	Chill after hydrazinolysis step. Use chilled (0°C) hydrazine stop solution, and ethanol. Centrifuge in fridge or cold room
Uneven running and retardation of electrophoresis dyes in some channels	Sample contaminated with residual piperidine	Employ a good vacuum for the final lyophilization steps. Use conc. H_2SO_4 in desiccator
All ladders show correct cleavage products but the bands are not sharp enough to permit reading far into the sequence	Sample wells damaged or uneven loading of sample	Check that wells are free from pieces of gel adhering to sides, load with a fine pointed capillary sweeping the sample evenly from side to side
	Diffusion or secondary structural effects during electrophoresis	Increase the current so that the gel is run at >50°C
	Scatter during autoradiography	Use thin gels. Ensure even pressure of X-ray film against the gel
Localised loss of band spacings in all four channels. Bands sometimes superimposed on each other as spacing drops to zero. The 'compression' effect	Secondary structure forming stable hairpin loops in fragments of a particular length class	Thorough denaturation of the DNA before loading and running the gel as hot as possible (>50°C). Prerun the gel to heat it up before loading the sample
		Obtain the sequence from the complementary strand

Fault	Probable cause	Solution
An extraneous band across all four channels at one position in the gel when a double-stranded fragment is being sequenced	A proportion of the DNA strands being sequenced have been nicked at a unique position, probably due to some variable specificity or contaminant in the restriction enzyme. Such breaks are not revealed when the double-stranded fragment is electrophoresed on a non-denaturing gel	Avoid over-digestion with the restriction enzyme. If the sequence around the site is similar to that of the cleavage site of the restriction enzyme, assume nicking due to variable specificity. Alternatively use a different restriction enzyme (or isoschizomer) or use strand separation
More than one strong band terminates the sequencing patterns at the top of the gel	Incomplete denaturation of sample, one band is single-stranded, the other double-stranded	Increase time of reaction with dimethylsulphate or hydrazine. Denature vigorously before applying sample to gel
	Labelled fragment is heterogeneous in length	Provided the length heterogeneity does not involve the labelled end (i.e. a unique sequence can be read up to the positions of the first strong band in all four channels) the sequence can be taken as valid

5.12. Results

Figure 5.8. shows two examples of sequencing gels. Autoradiograph (a) shows a short electrophoretic run in which the sequence of the first 29 nucleotides from the labelled 5′-end can be deduced. The sequencing procedures used were those originally proposed by Maxam and Gilbert (1977). The discrimination between the G's

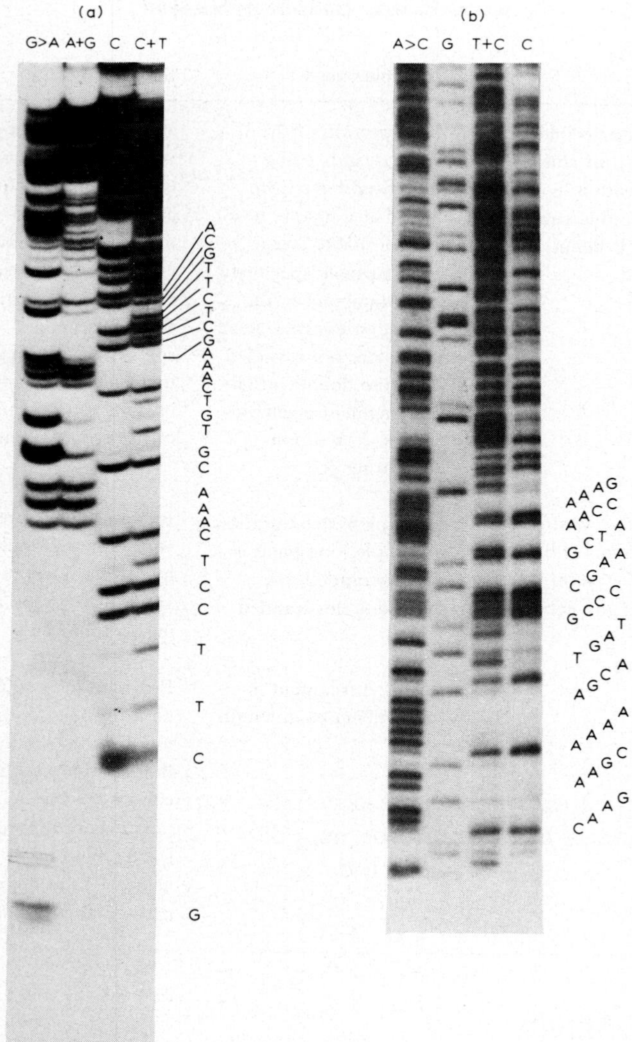

Fig. 5.8. (a) Autoradiograph of a sequencing gel prepared using the original Maxam–Gilbert (1977) procedure (reactions R_1–R_4, Table 5.I.). In this method the discrimination between the G and A residues is not absolute and depends on a comparison between a G > A reaction and an A + G reaction. (b) Autoradiograph of a sequencing gel prepared using reactions R_5 (G-specific), R_9 (A > C) and R_3 and R_4 (for C + T and C, cf. Table 5.I.). The sequence shown is part of the human fibroblast-interferon gene sequence obtained by reverse transcription of the interferon mRNA using a 5′-end labelled synthetic primer. (Courtesy of Dr. A.G. Porter).

and A's depends on a comparison between the G > A reaction and A + G reaction. Autoradiograph (b) shows part of a sequencing gel using the alternative G-specific and A > C reactions (R5 and R9 respectively, Table 5.I). Inspection shows that each of the labelled bands may be clearly and unequivocally interpreted.

Both these examples show the result of short, single loading gel runs. In practice, with multiple loadings made at different times, the patterns are expanded and sequences of 200–300 nucleotides may be routinely determined. Published examples of sequencing gels abound in the literature and Maxam and Gilbert (1980) show the degree of resolution that can be achieved.

5.13. *Strategies for sequencing double-stranded DNA*

The main consideration in deciding one's approach to a particular sequencing problem is simply the size of the DNA in question. Is the aim to sequence the entire genome of a virus or plasmid or some other large genetic element previously cloned in a suitable vector, or is the purpose of the experiment to sequence a particular defined region within a much larger sequence? In the former, the strategy will be to accumulate rapidly as much sequence data as possible and using this data select the appropriate restriction enzymes which will give the necessary overlaps and confirmatory sequences. While a restriction map of the sequence is useful it is not a necessary prerequisite and as this emerges with the sequence it is not a particularly rewarding exercise to start by constructing such a map. On the other hand where the targeted sequencing of a particular region is the aim one wants ideally to be able to identify and isolate a single restriction fragment containing the sequence in question and for this it is usually important to have a suitable restriction map. In many instances the fragment will have been identified and cloned into a suitable double-stranded plasmid vector as a means of amplifying the amount of material available. An appropriate choice of vector and cloning site will often permit the intact fragment to be cleaved out leaving the vector as a much larger linear d.s. molecule. Gel electrophoresis, or sucrose gradient centrifugation, serves to isolate the fragment in pure form. Procedure *A* gives a scheme essentially as recommended by Maxam and Gilbert (1981), for the large-scale preparation of unique DNA segments using a restriction enzyme and a polyacrylamide gel.

For sequencing an entire plasmid or virus DNA one can simply begin by cleaving a sample of the DNA into fragments, preferably using a restriction enzyme which yields protruding 5′-ends for ease of labelling. Table 5.II. lists fifteen restriction enzymes which make

TABLE 5.II.

Restriction endonucleases which cleave DNA into fragments readily end-labelled with T_4-polynucleotide kinase and γ[^{32}P]-ATP or with DNA polymerase and α-[^{32}P]-dNTP

Restriction enzyme	Sequence recognized	Fragment end-structure after cleavage
HinfI	G↓ANTC	pANTCNN GNN
HpaII	C↓CGG	pCGGNN CNN
Sau3A	↓GATC	pGATCNN NN
TaqI	T↓CGA	pCGANN TNN
EcoRII	↓CC(A/T)GG	pCCXGGNN NN
AvaII	G↓G(A/T)CC	pGXCCNN GNN
AvaI	C↓YCGRG	pYCGRGNN CNN
BamHI	G↓GATCC	pGATCCNN GNN
BglII	A↓GATCT	pGATCTNN ANN
EcoRI	G↓AATTC	pAATTCNN GNN
HindIII	A↓AGCTT	pAGCTTNN ANN

TABLE 5.II (continued)

Restriction enzyme	Sequence recognized	Fragment end-structure after cleavage
SalI	G↓TCGAC	pTCGACNN GNN
XbaI	T↓CTAGA	pCTAGANN TNN
XhoI	C↓TCGAG	pTCGAGNN CNN
XmaI	C↓CCGGG	pCCGGGNN CNN

Restriction enzymes which cleave double-stranded DNA, making staggered cuts which leave the 5′-end of each strand extended. Double-stranded DNA with protruding 5′-ends is more efficiently phosphorylated by polynucleotide kinase than DNA with flush or recessed 5′-ends. The fore-shortened 3′-ends can also be labelled by extending them with ^{32}P-labelled nucleoside triphosphates complementary to bases in the extended template strand using DNA polymerase. Y—pyrimidine nucleotide, R—purine nucleotide, N—any nucleotide. A complete list of commercially obtainable restriction endonucleases is given in Appendix III.

these staggered cuts. The whole collection of fragments are then labelled at either their 5′- or 3′-ends and the mixture of fragments either digested with a second restriction enzyme or denatured for strand separation before electrophoresis on an acrylamide gel. By running a small sample of the uncleaved (or denatured) sample in parallel, the singly labelled products produced can be readily identified and cut out of the gel. If two restriction enzymes are used in the reverse order, they will yield the same fragments but labelled at their opposite ends. The usefulness of this approach is shown in Fig. 5.7. Procedure B gives the scheme recommended by Maxam and Gilbert (1981) for preparing a set of single 5′-[^{32}P]-end labelled fragments ready for sequencing. This sequencing routine is then repeated with another pair of restriction enzymes thus producing sequences which start to overlap those already determined. An examination of the sequences obtained from the first pair of restriction enzymes for potential restriction sites can often help in the selection of the second pair of cutters. Searching for overlaps can be a time-consuming operation as the amount of data increases and the identification of potential restriction enzyme recognition sequences is as good a way as any for discovering matches between the different sequences. Other easily spotted markers are runs of a particular nucleotide. Where a large volume of data needs to be scanned a computer programme is the most efficient procedure and has the added advantage that sequences can be entered, stored and amended as the work proceeds. Modern programmes also include uncertainty codes for doubtful

TABLE 5.III.
Summary of four base-specific reactions for sequencing end-labelled DNA

G *See Procedure H*	G+A *See Procedure I*	C+T *See Procedure J*	C *See Procedure K*
200 μl DMS buffer	10 μl water	10 μl water	10 μl 5 M NaCl
5 μl ^{32}P-DNA	10 μl ^{32}P-DNA	10 μl ^{32}P-DNA	10 μl ^{32}P-DNA
1 μl DMS	2 μl pip-for, pH2	30 μl hydrazine	30 μl hydrazine
206 μl	22 μl	50 μl	50 μl
23°C, 5 min.	23°C, 60–80 min.	23°C, 5 min.	23°C, 5 min.
50 μl DMS Stop	Freeze	200 μl HZ Stop	200 μl HZ Stop
750 μl ethanol	Lyophilize	750 μl ethanol	750 μl ethanol
Chill 5 min.	—	Chill 5 min.	Chill 5 min.
Centrifuge 5 min.	—	Centrifuge 5 min.	Centrifuge 5 min.
Pellet	—	Pellet	Pellet
250 μl 0.3 M NaAc	—	250 μl 0.3 M NaAc	250 μl 0.3 M NaAc
750 μl ethanol	—	750 μl ethanol	750 μl ethanol
Chill 5 min.	—	Chill 5 min.	Chill 5 min.
Centrifuge 5 min.	—	Centrifuge 5 min.	Centrifuge 5 min.
Pellet	—	Pellet	Pellet
Ethanol rinse	—	Ethanol rinse	Ethanol rinse

TABLE 5.III. (continued)

G *See Procedure H*	G + A *See Procedure I*	C + T *See Procedure J*	C *See Procedure K*
Vacuum dry	—	Vacuum dry	Vacuum dry
100 μl 1.0 M piperidine	100 μl 1.0 M piperidine	100 μl 1.0 M piperdine	100 μl 1.0 M piperidine
90°C, 30 min.	90°C, 30 min.	90°C, 30 min.	90°C, 30 min.
Lyophilize	Lyophilize	Lyophilize	Lyophilize
10 μl water	10 μl water	10 μl water	10 μl water
Lyophilize	Lyophilize	Lyophilize	Lyophilize
10 μl water	10 μl water	10 μl water	10 μl water
Lyophilize	Lyophilize	Lyophilize	Lyophilize
10 μl formamide-NaOH-dyes	10 μl formamide-NaOH-dyes	10 μl formamide-NaOH-dyes	10 μl formamide-NaOH-dyes
90°C, 1 min.	90°C,1 min.	90°C, 1 min.	90°C, 1 min.
Quick-chill	Quick-chill	Quick-chill	Quick-chill
Load on gel	Load on gel	Load on gel	Load on gel

The abbreviated protocols above indicate how four base-specific cleavage reactions are coordinated to sequence one end-labelled DNA fragment. They are given only as a chronological guide, and initially the more detailed descriptions in Procedures H–K should be followed. DMS is dimethylsulphate, HZ is hydrazine, and pip-for is piperidinium formate.

residues and the scanning of sequences for all possible restriction sequences. Ultimately most of the sequences will be known, overlapped and placed in order and to complete the sequence it may be necessary to use particular restriction enzymes to confirm just one questionable overlap or sequence. In all cases it is important to sequence right through the restriction enzyme sites used to generate the different fragments, particularly since the first two or three residues from the labelled end can sometimes be difficult to interpret from the gel.

The length of sequence that can be analysed from a particular labelled end (about 250 nucleotides) implies that this is the optimal size to aim for in producing restriction fragments. Eight different tetranucleotide-recognising restriction enzymes are now available (Table 5.IV.) each of which will, on average cut double-stranded DNA once every 256 base-pairs. Table 5.V. lists restriction enzymes which will cut subsets of these sequences within a hexanucleotide recognition site. Given these, the strand separation techniques and the ability to label either the 5′- or 3′-ends it is difficult to avoid getting the sequence from both strands of the majority, if not all, of the different fragments. Maxam and Gilbert (1980) strongly advocate deliberate two-stranded sequencing throughout the DNA. If, after having sequenced both strands and across all restriction fragment sites it is found that the sequence of one strand is complementary to the sequence of the other at all positions, the structure is considered to be established.

These general approaches of course may be modified or restructured to suit a particular problem. In probably the majority of sequencing undertakings the DNA of interest will have been cloned and amplified in a suitable vector and while in some instances the cloned fragment can be precisely cleaved out from the recombinant vector and isolated by gel electrophoresis, in other cases this may not be possible. The 'shotgun' method described above will yield a complex array of end-labelled fragments and the problem immediately arises of distinguishing which ones

TABLE 5.IV.
Eight restriction endonucleases with tetranucleotide recognition sites

Restriction endonuclease	Cleavage site
EcoRI*	↓AATT
AluI	AG↓CT
TaqI	T↓CGA
Sau3A	↓GATC
HpaII	C↓CGG
ThaI	CG↓CG
HhaI	GCG↓C
HaeIII	GG↓CC

TABLE 5.V.
Sequences at which restriction endonucleases cleave DNA

	NNNN	ANNNNT	TNNNNA	CNNNNG	GNNNNC
AATT	EcoRI	—	—	—	EcoRI
ATAT	—	—	—	—	—
TATA	—	—	—	—	—
TTAA	—	—	—	—	HpaI, HindII
ACGT	—	—	—	—	AcyI
AGCT	AluI	HindIII	—	PvyII	SacI
TCGA	TaqI	ClaI	—	XhoI, AvaI	SalI, HindII
TGCA	—	AvaIII	—	PstI	—
CATG	—	—	—	—	—
CTAG	—	—	XbaI	AvrII	—
GATC	Sau3A	BglII	CpeI	PvuI	BamHI
GTAC	—	—	—	—	KpnI
CCGG	HpaII	—	—	XmaI, AvaI	—
CGCG	ThaI	—	—	BacII	—
GCGC	HhaI	HaeII	MstI	—	HaeII, AcyI
GGCC	HaeIII	HaeI	BalI, HaeI	—	—

Listed on the left are the 16 self-complementary tetranucleotides while flanking nucleotides which expand these to the 64 self-complementary hexanucleotides head the next four columns. This arrangement is useful in looking for restriction enzymes which increase the specificity of cutting at a particular tetranucleotide. Enzymes that do not have symmetrical recognition sequences are not included.

come from the inserted sequence. One approach to this, elaborated by Maxam (1979) uses a strategy in which the pattern of fragments obtained from the recombinant vector (RV) by cutting, labelling and sub-cutting, is displayed in parallel with two control patterns, one from the RV DNA before digestion with the second cutter and the other from the vector (V) DNA itself—without any insert—using the same set of enzymes to cut, label and sub-cut as for the RV DNA. The two control and one preparative restriction digest are interpreted as follows (Fig. 5.9.). Any fragment in the preparative channel which is also present in the right-hand channel (RVA) has not been cut with enzyme B and therefore has both ends labelled and cannot be sequenced. Similarly, any frag-

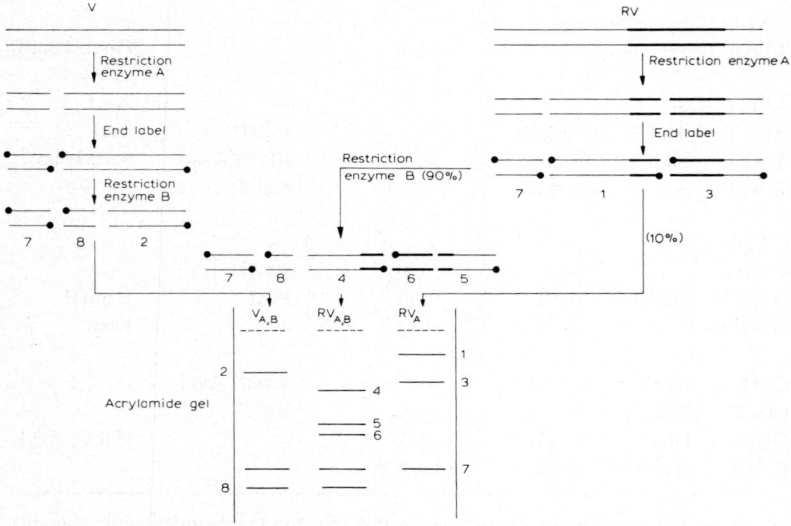

Fig. 5.9. Strategy for sequencing a cloned gene. In two parallel reactions a restriction enzyme A cleaves the vector V and the recombinant vector RV at the sites shown. Cleavage may occur at one or several sites within the vector sequence. The fragments produced are then end-labelled (either at the 3′- or 5′-ends). All of the vector digest, and 90% of the RV digest are then cleaved with a secondary restriction enzyme B of different specificity and the products electrophoresed in parallel on an acrylamide gel together with the 10% of the cleaved RV not digested with enzyme B. Any band in the recombinant vector lane $RV_{A,B}$ which is also in the once cut vector lane $V_{A,B}$ comes from the vector sequence and is ignored (bands 7 and 8). Similarly any band in the $RV_{A,B}$ which also occurs in the once cut lane RV_A, was not cut by enzyme B and is therefore labelled at both ends (Band 7). All other bands in the centre preparative lane (bands 4, 5 and 6) therefore arise from the inserted sequence and are labelled at one end only. These may be eluted and sequenced. This strategy is not limited to cloned sequences; V and RV may be any two related DNAs differing in a block of non-homologous sequence. (Redrawn from Maxam, A.M., 1979).

ment in the preparative channel which is also present in the vector-only channel ($V_{A,B}$) cannot come from the insert. Those bands which appear *only* in the centre channel are *single-end labelled* fragments from the cloned insert and can be eluted from the gel and sequenced. The same strategy applies using strand separation except that the heat-denaturation step substitutes for digestion with restriction enzyme B. An unheated portion of RV_A indicates which fragments are undenatured and therefore double-labelled.

The approach is not limited to a particular cloned insert. V and RV may be two related DNAs differing by a block of non-homologous sequence. Studies on gene rearrangement and viral integration often focus on two sets of sequences which differ only in a discrete region of non-homology and this procedure offers one route for identifying and sequencing these regions as well as defining the sequence around the site of the non-homology.

For problems of this sort it is highly advantageous to have a restriction enzyme cleavage map which can initially locate the divergent region within the sequence and help choose suitable pairs of restriction enzymes for producing sequenceable fragments. In this connection Smith and Birnstiel (1976) originally developed a rapid method for mapping restriction sites on DNA using long, singly-labelled restriction fragments which are subjected to partial digestion with a second enzyme (of lower specificity) and subsequently ordering all the labelled fragments by chain length on an acrylamide gel. If this procedure is applied in parallel to two DNAs, one of which contains an added or deleted region the pattern of bands, which are identical to begin with, will diverge at a position corresponding to the start of the modified region. This therefore established the positions of one end of the insertion or deletion in relation to a particular restriction enzyme map. Repeating the experiment with other enzymes will gradually narrow down the position of the site at which the sequences diverge and additionally suggest which restriction enzymes are most useful for the targeted sequencing of that region. Figure 5.10. illustrates the

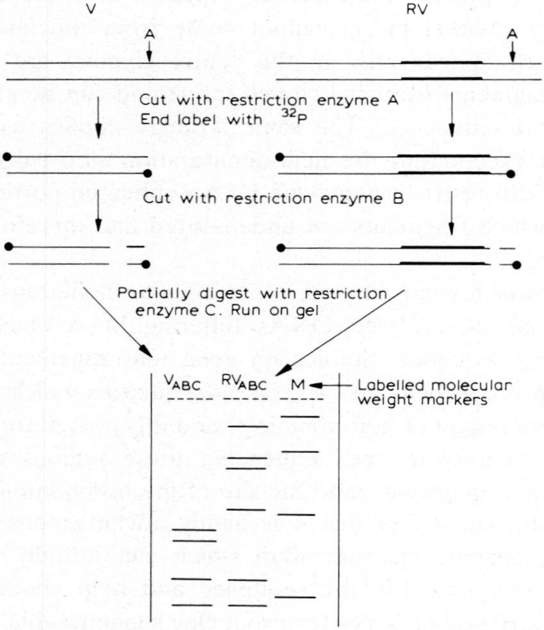

Fig. 5.10. Two DNAs, related to one another by an insertion, deletion, or region of rearranged or altered sequence are digested with a restriction enzyme 'A' which cuts outside the region of interest. The product is end-labelled with ^{32}P and cut with a second restriction enzyme 'B' to yield singly end-labelled fragments. The products are then partially digested with restriction enzyme 'C' and the two sets of fragments electrophoresed on an acrylamide gel in parallel with labelled marker fragments of known size. From a graph of marker chain length versus mobility the lengths of the end-labelled partial products are deduced and restriction maps for the two DNAs, V and RV, deduced. The start of the non-homologous region in RV is defined as lying within the first partial fragment in the RV_{ABC} channel which is not also found in the V_{ABC} channel. Whether and where the two sequences converge again can be determined by reversing the order of the restriction enzymes 'A' and 'B' and so running the analysis from the other labelled end. This procedure is essentially an extension of the Smith-Birnstiel (1976) mapping method structured to differentially locate a major sequence divergence in two otherwise identical DNAs. (Redrawn from Maxam, A.M., 1979).

logic of this argument. Such an approach can also be extended to sequencing across the boundaries of recombination events where the recombination involves any gross sequence rearrangement including insertion, deletion, inversion and unequal crossing-over. Provided the unmodified form of the chromosomal segment is available simultaneous mapping by the Smith-Birnstiel procedure will indicate where in the DNA the event has occurred. Repeating the mapping with different restriction enzymes will pinpoint that region with increasing accuracy and suggest the most useful pair of enzymes for sequencing that region using the simultaneous shotgun approach shown in Fig. 5.9.

The Maxam-Gilbert method clearly forms the basis of a highly flexible approach to sequencing problems. The procedures can be structured to suit particular sequencing aims and the chemical sequencing reactions themselves are convenient, rapid and reproducible. For the majority of sequencing undertakings the choice of approach therefore will either be those described in this chapter or the primed-synthesis methods elaborated by Sanger and his colleages as described in Chapter 4.

Most of the descriptions and procedures set out in this chapter are inevitably taken from, or based on, the published works of Maxam and Gilbert (1977, 1980) and Maxam (1979). The author is much indebted to Dr. Allan Maxam for permission to make extensive use of their protocols. Sufficient background and detail is included to undertake a sequencing project but it is strongly recommended that Maxam and Gilbert (1980) is consulted for further refinements and details.

Update

Update to chapter 4

Since the time of writing a number of advances have been reported which further extend the scope and usefulness of the M13 cloning-sequencing procedures described in Chapter 4. These are described under the separate headings below and are cross-referenced to Chapter 4.

1. M13 sequencing primers (cf. Sections 4.2.8 and 4.6)

The universal M13 sequencing primer, d(TCCCAGTCAC-GACGT) is now available from New England Biolabs. This anneals to the + strand of all M13 *lac* phages which carry the β-galactosidase gene (M13mp2, mp5, mp7, mp8, mp9 and M13mWJ22). The length and position of annealing of the primer permits the reading of the first nucleotide of the cloned fragment. The 3′-terminal hexanucleotide sequence of the primer was selected since it only occurs once in the M13 genome and assures negligable non-specific priming at other sites in the M13 template.

The 26 base pair universal primer isolated from phage pSP14 (Anderson et al. Nucleic Acids Res. *8,* 1731–43 (1980), cf. Section 4.3.7, Fig. 4.11), is available from Bethesda Research Laboratories.

BRL also supply, free of charge, the following phage/plasmid strains: M13mp2, M13mp7, M13mp8/pUC8 and M13mp9/pUC9 and the bacterial strains JM103, JM101, CSH34/pHM232 (a derivative of pBR 325 containing the 96 b.p. primer insert (Fig. 4.6, Section 4.2.1), and CSH 26/pHM 325 (a pBR 325 derivative containing a 21 base pair primer cloned as an *Eco*RI insert). This latter primer is complementary to the *lac* region on the opposite side of the *Eco*RI site from the 96 b.p. primer annealing position and is used to generate radioactive probes for screening M13 recombinant libraries (see Section 2 below). Strain JM83 is the host for pUC8 and pUC9. The 17 nucleotide primer used for (−) strand priming in the Hong procedure (see 3 below) is not yet commercially available. Its synthesis is described in Duckworth et al. Nucleic Acids Res. (1981) *9*, 1691–1706.

2. A system for making strand-specific M13 hybridization probes

Hu and Messing (Gene, *17* (1982) 271–277) have recently described a novel approach for the preparation of highly radioactive, strand-specific M13 probes. While this is not strictly a sequencing procedure it is a potentially valuable tool for screening M13 recombinant libraries as well as searching for insert sequences complementary to very small quantities of specific sequences as, for example in the identification of single copy genes. This is facilitated by the very high specific activities attainable in these probes (1.5–3×10^8 cpm/μg DNA) and possibly of even greater value is the opportunity it provides to probe RNA transcripts with single-stranded DNA probes and thus define the genomic DNA 'sense' strand and the polarity of transcription.

In the conventional M13 sequencing strategy a primer is annealed adjacent to the cloning site so that elongation of the primer occurs through the cloned sequence yielding a radioactive strand complementary to the insert. In this new procedure a universal primer, complementary to the region 5′ to the multiple cloning site of M13mp7, is used to initiate DNA synthesis of the complementary strand of M13 downstream from the inserted sequence. The synthesis of the (−) strand, labelled with α-[^{32}P]dATP (or preferably α-[^{32}P]dCTP, since this appears to give greater incorporation of label) is not allowed to proceed to completion so that the inserted sequence remains single-stranded. A partially double stranded probe is thus obtained which has the specificity of the single-stranded insert but which may incorporate up to 7000 nucleotides of 'carrier' M13(−) strand DNA labelled to the desired specific activity. The primer is a 21 b.p. fragment which has been cloned into pBR325 as an *Eco*R1 insert. The new plasmid is designated pHM235 and is contained and propagated in *E. coli*

CSH26/pHM235 (obtainable from BRL). The cells are grown in the presence of ampicillin and the plasmid may be amplified by chloramphenicol. The primer is released from the plasmid by digestion with *Eco*RI and a yield of 1–2 μg (sufficient for 500–1000 primings) is obtained from a 500 ml amplified culture. The labelling of M13 templates with the 21 b.p. primer is carried out by using a limiting concentration of an α-[^{32}P]dNTP, specific activity 2000–3000 Ci/mmol so that the DNA synthesis reaction terminates before the insert DNA sequence is encountered. The labelled product is not denatured before use and experiments have shown that the labelled M13(−) strand remains stably associated with the unlabelled (+) strand containing the single-stranded insert.

Full experimental details are given in Hu and Messing's paper and are not repeated here.

The stability of the partially double stranded probe means that, under the conditions described in their paper it can be used to probe other M13 recombinants. Since the viral strand in all M13 clones is the (+) strand and since the labelled (−) strand does not dissociate from its template, hybrids are only formed when the separate M13-derivatives contain regions of complementary single stranded insertions. M13 recombinants containing long (3 kb) inserts derived from genomic DNA can therefore be probed with M13 recombinants containing, for example, cDNA sequences obtained by reverse transcription of a mRNA.

A further suggested application of this procedure is the selective, strand-specific probing of mRNA sequences using the 'Northern' blot procedure of Alwine et al., Proc. Natl. Acad. Sci. USA *74* (1977) 5350–5354. For a given single stranded M13 insert the labelling scheme described would give a probe specific for RNA sequences complementary to the insert. If a radioactive copy is made using a conventional primer the labelled DNA product would be specific for sequences identical to the insert and would, in theory, permit the polarity of any RNA transcript to be determined. This seems likely to be developed into a very useful method.

3. A method for sequencing single stranded cloned DNA in both directions by the dideoxynucleotide-chain termination procedure (Hong, G.F., Bioscience Reports 1, 243–252 (1981))

In sequencing randomly chosen M13 recombinant clones a sequence of about 300 nucleotides can usually be determined from one end of the insert. However the insert is frequently longer than this and it is clearly desirable to have a means of sequencing from both ends of these longer sequences. In addition the ability to obtain complementary sequence data from short inserts is a valuable means of checking sequence data. One way of achieving this result is to use the clone 'turn-around' procedure of Winter and Fields (1980) (Sections 4.1 and 4.6). In Hong's procedure a different strategy is employed in which an unlabelled copy of the insert sequence is first made with DNA polymerase and the four deoxynucleotide triphosphates by priming with a normal flanking primer. This (–) strand copy is then used as a template for the usual terminator sequencing method using a new flanking primer corresponding to a portion of the vector DNA adjacent to the other side of the insert. The 17 nucleotide primer used (CAGGAAACAGCTATGAC) is identical to residues 6204–6220 of M13mp7 and therefore primes the (−) strand at a position ten nucleotides upstream from the *Eco*RI site. Extension of this primer in the $5' \rightarrow 3'$ direction therefore permits the sequence of the complement of the (−) strand, i.e. the sequence of the original insert, to be determined from its 5′-end.

Two difficulties had to be overcome to develop this into a satisfactory method. After the first priming reaction there remained some unreacted primer which had to be removed before the second priming reaction. Selective precipitation with polyethyleneglycol (PEG) was found to be a rapid and convenient way of doing this. In the second priming reaction a short single stranded template has to be annealed to a double stranded template. Conditions had therefore to be found wherein sufficient annealing of the primer and template could occur without completely reforming the original double strand. Rapid chilling of the annealing mixture in ethanol-dry ice favoured the selective annealing of the short primer and under these conditions satisfactory priming on the (−) strand was found to occur.

Full experimental details are found in Hong's paper which also shows the application of this method to the sequencing of an 806 nucleotide insert cloned into M13mp2.

4. A systematic DNA sequencing strategy (Hong, G.F. (1982) J. Mol. Biol. (in the press)

Among the procedures developed for a more systematic approach to DNA sequencing the most recent is the one described by Hong (1982). Procedures such as clone 'turn-around' and priming with exonuclease III digested restriction fragments were already discussed (Sections 4.2.1 and 4.8). For the systematic sequencing of long (>1000) nucleotide inserts cloned in M13 two different strategies may be envisaged; either priming at different sites along the sequence using a set of internal primers derived from restriction fragments (cf. Section 4.8) or devising some means for bringing different regions of the insert adjacent to the 'universal' priming site on M13. The Hong procedure develops this latter approach.

The length of DNA along which the sequence data can be progressively read appears to be limited only by the insertion capacity of the vector and in the example given by Hong, an M13 recombinant RF with a 2327 nucleotide long *Hind*III insert was used to illustrate the method. The replicative form of the recombinant was first partially digested with DNAase I in the presence of Mn^{++} ions. By carefully controlling the time of digestion the circular RF was converted to the corresponding double stranded linear form. In the presence of Mn^{++}, DNAase I appears to give predominantly double stranded cuts with only limited single stranded regions at the cleavage sites. These cleavages will occur at random in the RF DNA but the sections ultimately sequenced will only be those that are derived from breaks within the insert. Since the origin of replication of the vector is close to the left-hand end of the insert (Schaller et al., 1978, in 'The Single-stranded DNA

Phages' pp. 139–163, Cold Spring Harbor Laboratory, N.Y.), breaks on this side will subsequently yield molecules without an origin which do not give viable phage. Cleavages on the other side of the insert will give phage which, in the subsequent stages, will yield progeny with no priming site and therefore cannot be sequenced.

In the example given by Hong, the insert was cloned into the *Hind*III site of M13mp9. After linearizing with DNAase I, the DNA was cleaved at the single *Sma*I site which occurs 27 nucleotides upstream from the cloning site and is immediately adjacent to the priming site. A size fractionation based on the selective precipitation of longer DNA fragments with polyethylene glycol removes the smaller fragments (<2000 nucleotides) including those arising from DNAase I cuts within the insert (since this is close to the *Sma*I site). After polishing the DNAase I end (with DNA polymerase (Klenow sub-unit) and the 4 dNTPs), religation yields circular recombinants in which the DNAase I cut site is fused to the *Sma*I site. Transfection into competent cells and plating out yield progeny clones which are picked and worked up as usual (Sections 4.3.11 and 4.4). The single stranded phage progeny therefore contain inserts created by the fusing of random DNAase I cut sites to the priming site.

Phages with longer insert DNA sequences will diffuse more slowly on the agar plate, thus giving rise to smaller plaques. Plaques size is therefore a rough, but useful, guide to the identification of the length of DNA. This, coupled to the ddT screening reaction (Sections 4.2.7 and 4.5) allows a set of recombinants to be chosen for sequencing in which each insert partially overlaps the next longest insert. In other words the length of the insert is related only to the size of the DNAase I-*Sma*I fragment that was cut out of the full length insert before ligation. The insert size is therefore a function of which region of the original sequence was fused to the priming site. The determination of sequences of 200–250 nucleotides from 24 clones originally selected from the agar plate sufficed to sequence the entire 2327 nucleotide fragment.

Full details are given in Hong (1982). In our laboratory the sequence of a 2356 nucleotide fragment containing the yeast cell division cycle 2 (cdc 2) gene was rapidly determined by this procedure. Additionally we used the RF from two clones in which the original insert was cloned in either orientation. This enabled complementary sequence data to be obtained for each region of the insert. The only disadvantage of this procedure is the rather lengthy series of manipulations required to obtain the final recombinants for sequencing. Once that stage is reached progress is rapid.

5. Shotgun cloning into M13mp7

With the increasing experience of shotgun cloning procedures and the development of computer programmes to handle the data there is a marked drift to the increasing use of random degradation methods to obtain a pool of DNA fragments for cloning into M13 derivatives. Either digestion with DNAase I in the presence of Mn^{++} (Section 4.2.2) or shearing by sonication (4.3.6) yield fragments which can be end-repaired and cloned either with or without the use of linkers. Experience in this and other laboratories has tended to show that addition of the *Eco*RI linker (Section 4.2.2) followed by cloning into the *Eco*RI site of M13mp7, or mp2, is the more reliable procedure. DNA fragments of any desired size range can be produced by these methods and further fractionated by gel electrophoresis to obtain a fairly defined size class, ideally of 250–300 residues. Inserts of this size can be completely sequenced by double loading on a single sequencing gel. The random nature of the cleavage means that representatives of many different sequences are found in the clones isolated and sequencing continues until all the overlaps are joined into a single 'contig' (Chapter 6) and their sequences confirmed. In this way the drawbacks inherent in using restriction digest fragments, i.e. variability in size and the necessity for using a second enzyme to provide the overlaps, are avoided. Though this is probably the most straightforward approach the investment in time and materials for sequencing DNAs greater than 2-3000 nucleotides can become excessive and computing facilities become almost essential to keep track of, and assemble the data. For these reasons the use of one or more of the structured approaches discussed earlier should be entertained for longer sequences. For sequences of less than about 2000 nucleotides, the random approach is the fastest though precautions do have to be taken if blocks of repeated sequences are suspected in the DNA. Knowledge of a restriction map of the DNA will often indicate such regions and provide a valuable check on the correct disposition of restriction sites in the finished sequence.

The final point to note is that sequencing by the primed synthesis methods will not reveal the presence of methylated or other modified nucleotides in the original DNA sequence. This is also true of the Maxam–Gilbert method where the DNA to be sequenced has been cloned into a plasmid vector.

6. *Effect of temperature on primed DNA synthesis in the presence of chain terminating inhibitors*

One recurrent problem in the interpretation of DNA sequencing gels is the presence of artifact bands caused by spurious chain termination events during the elongation reaction (cf. pp. 108, 112, 200–206). These are often, though not always associated with short regions of symmetrical sequences which can fold back on each other yielding hair-pin loops at which the advancing DNA polymerase molecules can 'pile-up' and thus slow down or even terminate the nascent DNA chain. These events subsequently cause artifact bands which can be mistaken for authentic bands brought about by incorporation of dideoxy-nucleotide.

Recent experience in this laboratory has shown that such problems can be largely overcome by carrying out the elongation reaction and subsequent chase at 29°C rather than at 'room-temperature' as previously described.

In one experiment (Hindley, J., unpublished data) a set of dideoxy-nucleotide chain extension reactions, using an identical primed-template, was carried out at 14, 20 and 29°C respectively and the reaction products analysed on a standard thin sequencing gel. Numerous pile-ups and regions of ambiguity occurred in the 14°C product. At 20°C one pile-up remained but at 29°C the entire sequence was free from ambiguity. In this laboratory the chain extension reactions are now carried out routinely at 26–29°C. Under these conditions we have not encountered any examples of spurious termination.

7. *Direct determination of sequences cloned into plasmid pBR322* (Wallace, R.B., Johnson, M.J., Suggs, S.V., Ken-ichi, M., Bhatt, R. and Itakura, K., Gene 16 (1981) 21–26

Plasmid pBR322 (pp. 119–120) is widely used as a vector for preparing recombinant DNAs by cloning the desired sequence into one or another of the unique restriction sites situated within the Amp^R or Tc^R genes. In order to sequence the DNA insert it is usually necessary to excise the fragment from the recombinant plasmid using a restriction endonuclease and then either end label (for Maxam–Gilbert sequencing, cf. Chapter 5) or reclone into one of the M13 phage RFs to prepare the single stranded template (Chapter 4).

A third route for sequencing is to take advantage of the observation that a small single stranded primer can anneal to its target sequence in a denatured double stranded DNA and support primed synthesis from that site (cf. 3. above, Hong, G.F.). Wallace et al. (1981) have taken advantage of this and synthesized a set of seven oligonucleotide primers complementary to the plasmid vector pBR322 at positions adjacent to five of the unique restriction endonuclease recognition sites (*Eco*RI, *Hind* II, *Bam*H I, *Sal* I and *Pst* I). The polarity of the primers is such that any DNA inserted at one or a combination of two of the above restriction sites may be sequenced by the chain termination method using one of the synthetic DNA primers. The plasmid DNA is first linearized by digestion with a restriction endonuclease which will not excise the inserted DNA and then mixed with an approximately 20-fold molar excess of the oligonucleotide primer (e.g. 0.3 pmol linearized plasmid : 6 pmol primer). Following denaturation (100°C, 3 min) the mixture is rapidly chilled in ice-water and apportioned out for the four base-specific chain termination reactions (Chapter 4). Since the plasmid DNA is double stranded both strands of the insert may be sequenced independently by choosing an appropriate pair of primers. Full details of the experimental procedure are given in Wallace et al. (1981).

This method can clearly be extrapolated to other plasmids. Aves, S.J. (unpublished data, this laboratory) has used a formally similar approach to sequence double stranded DNAs cloned into pUC8 and pUC9 (pp. 134–135) using the synthetic primer 5′-GTAAAACGACGGCCAGTG-3′. One advantage of this latter procedure is that host cells (*E. coli* JM83, which are constitutive for the defective β-galactosidase, cf. Update 1) and which contain recombinant plasmids fail to produce β-galactosidase and give white colonies on indicator plates containing X-gal (p. 126). Colonies containing unmodified plasmid assume a blue colour. White colonies are picked, grown up in a small volume of medium and plasmid DNA extracted using either the cleared lysate procedure (cf. p. 165) or the Birnboim-Doly procedure (Nucleic Acids Res. (1979) 7, 1813–1823). Banding on CsCl-EtBr gradients (p. 165) is not necessary and the plasmid DNAs are pure enough for sequencing as described above.

Update to chapter 5

Another addition to the range of cloning vectors is the multi-purpose plasmid, pUR222 (Rüther, U., Koenen, M., Otto, K. and Müller-Hill, B. (1981). Nucleic Acids Res. *16*, 4087–4098). This contains six unique cloning sites (*Pst* I, *Sal* I, *Acc* I, *Hind* II, *Bam*H I and *Eco*RI) in a small region of part of the *lac Z* gene. In general, the chemical sequencing of DNA fragments by the Maxam–Gilbert procedure (Chapter 5) requires the isolation of fragments which, after end labelling, need either to be cut with a second restriction endonuclease or strand separated and the fragments reisolated from an acrylamide gel. DNA fragments cloned in pUR222 can be sequenced very readily by the Maxam–Gilbert method since a number of these steps can be simplified or omitted.

The polylinker region of plasmid pUR222 has the structure:

Hind II
Acc I

ACG AAT TGC TGC AGG TCG ACG GAT CCG GGG AAT TCA

PstI SalI BamHI ECoRI

4 5 6

The codon numbers of the original *lac Z* gene are given under the sequence.

Shotgun cloning into any of these restriction sites yield progeny which usually fail to produce an active β-galactosidase and produce colourless colonies on indicator plates (cf. Section 4.2). A DNA fragment ligated, for example, into the *Pst* I of the plasmid will yield a recombinant which can be subsequently cleaved at the *Bam*H I site and the ends labelled and filled in by incubation with DNA polymerase (Klenow sub-fragment) and the appropriate α-[^{32}P]dNTP and unlabelled nucleoside triphosphates (Section 4.2.2.). Cleavage with *Eco*RI releases the small (9 nucleotide) end-labelled *Bam*H I–*Eco*RI fragment from the polylinker which may be removed if desired by gel chromatography or selective precipitation. If this fragment is not removed the only penalty is that the first 9 residues on the sequencing gel will represent a region of mixed sequence and will not be readable.

The singly end-labelled insert may be sequenced directly by the Maxam–Gilbert method. Experimental details are given in Rüther et al. (1981). The variety of cloning and restriction sites in pUR222 should permit the method to be adapted to sequencing any set of fragments via the use of appropriate linkers. As the method does not require highly purified plasmid for sequencing the method of Birnboim and Doly (1979) Nucleic Acids Res. 7, 1813–1523, for the rapid lysis of the host cells and selective precipitation of circular plasmid DNA after alkaline denaturation yields sufficiently pure starting material. This avoids the time consuming and expensive CsCl–EtBr banding procedure for plasmid isolation.

Appendices

APPENDIX 1

List of contractions and special terms

Nucleotides

dATP—Deoxyadenosine triphosphate	dTTP—Deoxythymidine triphosphate (Same as Thymidine triphosphate)
dGTP—Deoxyguanosine triphosphate	
dCTP—Deoxycytidine triphosphate	dNTP—collective abbreviation for Deoxynucleoside triphosphates

A-residue, G-residue etc.: or more simply dA, dG (or A, G etc. where there is no possibility of confusion with the ribonucleotides, for example as in a DNA sequence)	A deoxyadenosine (deoxyguanosine) monophosphate either at one end, or within, a polynucleotide chain and linked to the adjacent nucleotide(s) by phosphodiester bond(s)
pdT etc.	deoxythymidine 5′-monophosphate
dTMP, dAMP etc.	the deoxynucleoside 3′-monophosphate
rATP, rCTP etc.	the ribonucleoside triphosphates, adenosine triphosphate, cytidine triphosphate etc.
ribo C (rC) etc.	a ribocytidine monophosphate residue
ddATP	2′,3′-dideoxyadenosine triphosphate
ddGTP	2′,3′-dideoxyguanosine triphosphate
ddCTP	2′,3′-dideoxycytidine triphosphate
ddTTP	2′,3′-dideoxythymidine triphosphate
ddA, etc.	2′,3′-dideoxyadenosine monophosphate residue
ddNTP	collective abbreviation for dideoxynucleotide triphosphates
ara-CTP	an analogue of CTP in which the deoxyribose sugar is replaced by arabinose (structural formula, Fig. 3.1.)

Reagents

EDTA	Disodium salt of ethylenediamine-tetra-acetic acid
Tris	Trihydoxymethylaminomethane
DTT	Dithiothreitol (Cleland's reagent)
bis	Bis-acrylamide
IPTG	iso-propyl-β-D-thiogalactopyranoside

X-gal	5-bromo-4-chloro-3-indolyl-β-galactoside
XCFF	the light blue tracking dye, xylene-cyanol FF
BPB	the dark blue tracking dye, bromophenol blue
TEMED	NNN′N′-tetramethylethylene-diamine

Enzymes

Restriction endonucleases are named in accordance with the proposals of Smith and Nathans (1973). Their names consist of a three letter abbreviation for the host organism (e.g. *Eco* for *E. coli*, *Hin* for *Haemophilus influenzae*, *Bam* for *Bacillus amyloliquefaciens*) followed by a strain designation and a Roman numeral. Thus *Haemophilus influenzae*, strain R_d contains at least three distinct restriction endonucleases and these are called *Hin*dI, *Hin*dII and *Hin*dIII respectively. Restriction enzymes are generally known and described in the catalogues of suppliers by this code.

DNA polI or *E. coli* polI	DNA polymerase I from *E. coli*. Possesses both $3' \rightarrow 5'$ and $5' \rightarrow 3'$ exonuclease activities in addition to its polymerase activity. Usually used for nick translation (Section 1.3.1.)
T_4-polymerase	A DNA polymerase isolated from phage T_4 infected *E. coli*. Possesses a $3' \rightarrow 5'$ exonuclease activity but lacks the $5' \rightarrow 3'$ activity found in *E. coli* DNA polI. Used for synthesis of DNA on a single-stranded template (Section 1.3.4.)
DNA polymerase (nach Klenow). Klenow polymerase. Large fragment *E. coli* DNA polymerase I. DNA polymerase (Klenow sub-fragment).	Variety of names for the large fragment produced by proteolytic cleavage of *E. coli* DNA polI. This enzyme lacks the $5' \rightarrow 3'$ exonuclease activity of *E. coli* DNA polI. Generally used for sequencing DNA
T_4-kinase	T_4-DNA kinase. Enzyme isolated from phage T_4 infected *E. coli* which catalyses the transfer of the γ-phosphate group of rATP to the 5′-hydroxyl terminus of DNA and RNA
T_4-ligase	T_4-DNA ligase. Enzyme isolated from phage T_4 infected *E. coli* which catalyses its formation of a phosphodiester bond between adjacent 5′-phosphorylated and 3′-hydroxy terminated DNA strands
AMV reverse	Enzyme isolated from Avian Myeloblastosis virus

transcriptase	particles which catalyses the synthesis of DNA from an RNA template. The DNA product, whether single-stranded or double-stranded, is known as cDNA (complementary DNA). Used for making cDNA transcripts from mRNA (messenger RNA) for subsequent cloning into a plasmid vector and sequencing
Exo III	Exonuclease III from *E. coli.* The enzyme is a double-stranded specific 3′-exonuclease and is used for degrading double-stranded DNA into an essentially single-stranded form (Fig. 1.3.)
Terminal transferase	Terminal deoxynucleotidyl transferase. Enzyme isolated from calf thymus which catalyses the limited addition of ribonucleotides to the 3′-end of a DNA fragment. Used for 3′-end labelling of DNA for sequencing by the chemical procedure (Chapter 5.)

Buffers

It is often convenient to make up buffer solutions at ten times normal strength (10 ×) which can then be diluted accordingly. For example 10 × TBE is ten times normal strength Tris-borate-EDTA buffer. × 1 TBE is a ten-fold dilution with distilled water.

Terminology describing types of ends on double-stranded DNA molecules

Blunt- or flush-ended fragments	Both terms are synonymous and refer to ends which both terminate precisely opposite each other and are fully base-paired
Staggered, protruding or recessed ends	Fragments in which the ends of the two complementary strands do not coincide so that either the 5′-ends or the 3′-ends protrude in a single-stranded extension of variable length
Sticky or cohesive ends	A class of staggered ends usually generated by restriction endonuclease cleavage such that the single-stranded protruding ends at each end of the fragment, or on other fragments generated by the same enzyme are complementary to each other. At low temperatures these ends can base-pair and the gaps may be sealed with T_4-DNA ligase (Fig. 1.5.)

Plasmids

The naming of plasmids is generally haphazard. Some, e.g. plasmid Col E1 (pCol E1) carries, as the name implies, immunity to colicin E1. The majority, e.g. pMB9, pBR322, pACYC184 etc., are usually named after the initials of their discoverer (or engineer).

Miscellaneous

RFI	Replicative form 1. This refers to the double-stranded closed circular form of a plasmid or phage DNA which can be isolated from infected cells by virtue of its increased buoyant density on caesium chloride-ethidium bromide density gradients
Tm	'Melting' temperature of double-stranded DNA in aqueous solution. The Tm corresponds to the point of inflection on the curve relating hyperchromicity (measured at 260 μm) to temperature
Vector or vehicle DNA	The genome of a phage of plasmid into which a piece of foreign DNA is inserted

APPENDIX 2

Cloning vectors

Some commonly used plasmid vectors of E. coli

Plasmid	MW ($\times 10^{-6}$)	*Pertinent antibiotic markers*	*Sites at which insertion inactivates a marker*
pACYC184	2.65	Cm^R, Tc^R	*Bam*HI, *Eco*RI, *Hind*III, *Sal*I
pBR322	2.8	Ap^R, Tc^R	*Bam*HI, *Hinc*II, (two sites) *Hind*III, *Pst*I, *Sal*I, *Pvu*I, *Ava*I
pBR325	3.6	Cm^R, Ap^R, Tc^R	*Ban*HI, *Eco*RI, *Hinc*II (two sites) *Hind*III, *Pst*I, *Sal*I, *Pvu*I, *Ava*I
pMB9	3.5	Tc^R, ImmEI	*Bam*HI, *Hind*III, *Sal*I
pACYC177	2.46	Ap^R, Km^R	*Pvu*I, *Pst*I, *Hinc*II, *Hind*III, *Sma*I, *Xho*I

ImmEI—immune to colicin EI, Ap^R—ampicillin-resistant, Cm^R—chloramphenicol resistance, Km^R—kanamycin-resistant, Tc^R—tetracycline-resistant.
Plasmids pACYC184 and pBR325 and their derivatives which retain an intact chloramphenicol gene can be amplified with chloramphenicol or spectinomycin. For detailed descriptions see Bolivar and Backman (1979), Kahn et al. (1979) and references therein.

Some commonly used bacteriophage vectors of E. coli

Bacteriophage	*Mol. Wt. (K. base pairs)*	*Cleavage sites at which foreign DNA may be inserted*	*Comments*
Charon 4 and	45	*Eco*RI	Can replicate more than

4A			22 kb of inserted DNA. Insertion at *Eco*RI site inactivates *lac*Z gene and allows selection on indicator plates
Charon 28	39.4	*Bam*HI, *Eco*RI, *Hinc*II, *Sal*I, *Xho*I	
λgtWES-λB	49.4	*Eco*RI, *Sal*I, *Sst*I, *Xho*I	Useful for replicating DNA of warm-blooded vertebrates
M13mp2(RF)	6.4	*Eco*RI	Insertional inactivation of *lac*Z gene. Used for preparing single-stranded templates for sequencing
M13mp5(RF)	6.4	*Hind*III	
M13mp7(RF)	6.4	*Acc*I, *Bam*HI, *Eco*RI, *Hind*II, *Pst*I, *Sal*I	

For further description of Charon or λ-phages see Morrow (1979) and references therein. The single-stranded M13 phages are discussed in Chapter 4 of this volume.

APPENDIX 3

Commercially obtainable restriction endonucleases (1981)

Endonuclease	*Sequence*	*Endonuclease*	*Sequence*
AccI	GT↓(AT/CG)AC		
AluI	AG↓CT	HinfI	G↓ANTC
		HgaI	GACGC*
AvaI	C↓PyCGPuG	HpaI	GTT↓AAC
AvaII	G↓G(A/T)CC	HpaII	C↓CGG
BalI	TGG↓CCA	KpnI	GGTAC↓C
BamHI	G↓GATCC	MboI	↓GATC
		MboII	GAAGA*
BclI	T↓GATCA	MnlI	CCTC*
		MspI	C↓CGG, C↓^{Me}CGG
BglI	GCC(N)$_4$↓NGGC	PstI	CTGCA G
		PvuI	CGAT↓CG
BglII	A↓GATCT	PvuII	CAG↓CTG
		SalI	G↓TCGAC
BstEII	G↓GTNACC	Sau3A	↓GATC, ↓G^{Me}ATC
CfoI	GCG↓C	Sau961	G↓GNCC

APPENDIX 3 (continued)

Endonuclease	*Sequence*	*Endonuclease*	*Sequence*
ClaI	AT↓CGAT		
DdeI	C↓TNAG	SmaI	CCC↓GGG
		SphI	GCATG↓C
DpnI	G^{Me}A↓TC	SstI	GAGCT↓C
EcoRI	G↓AATTC	SstII	CCGC↓GG
EcoRII	↓CC(A/T)GG	TaqI	T↓CGA
HaeII	PuGCGC↓Py	ThaI	CG↓CG
HaeIII	GG↓CC	XbaI	T↓CTAGA
HhaI	GCG↓C	XhoI	C↓TCGAG
		XmaIII	C↓GGCCG
HincII	GTPy↓PuAC	XorII	CGAT↓CG
HindIII	A↓AGCTT		

* Enzymes which do not cleave within the listed recognition sequence. Bethesda Research Laboratories, Inc. and Boehringer Mannheim provide useful lists of enzymes and their properties with extensive references. Available free on request.

APPENDIX 4

Autoradiography

Autoradiography is a technique for detecting radioactively labelled species in a gel, paper or thin layer by sandwiching a sheet of photographic film next to the support medium and allowing the radioactive emissions to expose the film. After developing the film, the positions and intensities of the radioactive sources are shown by areas of blackening on the film.

In sequencing work the isotope used exclusively is [^{32}P] which decays with the production of high-energy β-particles (1.71 Mev) and has a half life of 14.3 days.

The energy of the β-particles is sufficient to penetrate several millimetres thickness of hydrated gels on paper without significant absorption and thus allows the direct autoradiographic detection of ^{32}P in these materials. All that is required is a waterproof sheet of material between the gel and the film and storage in a light-proof container which exerts an even gentle pressure on the sandwich. For autoradiographing wet acrylamide or agarose gels at room temperature, a thin sheet of polythene (polyethylene) should be used between the film and gel.

For sequencing gels, which are usually autoradiographed frozen, a sheet of 'Saren-wrap' or similar flexible film is usually used. Freezing is simply a convenient method of preventing loss of resolution by diffusion of the radioactive molecules in the gel matrix during exposure of the film.

Autoradiography can be carried out in boxes in which the sandwich is enclosed between steel sheets weighted with lead bricks. It is more convenient however, to use X-ray film cassettes (holders) for autoradiographing sequencing gels in which the film (or gel) can be cut to make a snug fit. A felt-lined spring-loaded lid to the cassette gives an even pressure on the sandwich. After closing, the cassette is light-proof and can be stored in a deep-freeze for the period of exposure.

The high energy β-particles pass through and beyond the film. However, their excess energy can be trapped and returned to the film as visible light by placing a high-density fluorescent 'intensifying screen' behind the film. This enhances the latent image 7–10 fold, with a correspondingly shorter exposure time to obtain the same intensity of blackening after development. As expected this causes some loss of resolution of the image but where the incorporation of [^{32}P] is low this is a valuable way of speeding up the time required for autoradiography and sometimes permits the autoradiography of sequencing gels which otherwise would be too weak to develop. For the maximum efficiency of this method it is necessary to bypass the reversible first stage of latent image forrmation in the film. This is a technique used by astronomers to photograph faint stars and is most easily achieved by pre-exposing the film to an instantaneous flash of light ($\leqslant$1 msec). An electronic flash-gun, masked with an orange filter (e.g. Kodak Wratten number 21 or 22) or alternatively covered with 2–3 thicknesses of yellow paper (as used in packaging Kodak X-ray film), is first calibrated by varying the distance between the flash-gun and the film so that a single flash increases the absorbance of the film by 0.15 (A_{540}) above the absorbance of the unexposed film. The distance should be $\geqslant$80 cm in order to obtain uniform illumination of the film.

Having calibrated the system the usual procedure is to clamp the flash-gun at the correct distance pointing vertically downwards onto the darkroom bench. The gel, in the cassette, is covered with the film and, in *total darkness*, exposed to a single flash. The film is overlayed with the intensifying screen and the cassette closed. The cassette is finally stored at −70°C (for optimum quantative accuracy) for the required time. The film should subsequently be developed in total darkness.

This procedure is fully described in Laskey (1980) and references therein.

Radioactive ink

Ordinary ink supplemented with [^{35}S]-sulphate at a concentration of about 100 μCi/ml. [^{32}P] can be used but it gives a blurred image if overloaded and needs changing frequently on account of its short half-life. It is convenient to attach small sticky labels to the saren-wrap covering the gel and mark these with the radioactive ink. When radioactive bands have to be cut out of the gel for further analysis it is important to mark all four corners of the gel to permit accurate subsequent alignment of the film on the gel.

Preparation and running of sequencing gels

The final stage in each of the different sequencing procedures is the separation of the labelled DNA fragments by electrophoresis on denaturing gels. The composition, dimensions and running conditions of the gels vary slightly for the different protocols and details are given in the appropriate experimental sections.

Section 2.2. Step 5 (Chapter 2) describes the setting up and running of 1–1.5 mm thick gels.

Section 3.1.2. (Chapter 3) shows the apparatus and describes the assembly and running of thin sequencing gels and a slightly modified protocol is given in Sections 4.2.9. and 4.4. Section 5.7. (Chapter 5) describes the thin-layer gel system used in connection with the Maxam-Gilbert procedure.

Phenol

Phenol is redistilled under nitrogen, saturated with water or 10 mM Tris-HCl, 0.5 mM EDTA, pH 8.0, and stored frozen in small aliquots at −20°C. The addition of 0.1 % hydroxyquinoline to the phenol inhibits atmospheric oxidation. Also, the strong yellow colour helps in defining the interface between the phenol and aqueous layers after small-scale extractions.

APPENDIX 5

Equipment

Slab gel apparatus for gel electrophoresis: Raven Scientific Ltd. (Cat. No. IN/97). Sturmer End, Haverhill, Suffolk CB9 7VV, U.K. or BRL sequencing gel electrophoresis system (Cat. No. 1030), Uniscience House, 8, Jesus Lane, Cambridge, CB5 8BA, U.K.

Power supply for high-voltage electrophoresis: LKB Biochrom 2103 power supply unit, LKB Instruments Ltd., 232 Addington Road, Selsdon, South Croydon, Surrey CR2 8YD, U.K. This instrument may be used to run gels at constant voltage, constant current or constant power.

Micropipettes: Dade® Accupette pipets, disposable, Dade Diagnostics, Inc., Headquarters: 1851 Delaware Parkway, Miami, FL 33152, and local suppliers. Offers a range of glass capillary micropipettes for volumes of 1 μl up to 50 μl.

Automatic adjustable pipettes with disposable tips: Gilson pipetman, Models P20, delivers 2 μl–20 μl, P200 delivers 20 μl–200 μl, P1000 delivers 200 μl–1000 μl. Obtainable from Anachem Ltd., 15 Power Court, Luton, Beds, LU1 3JJ, U.K. Similar adjustable pipetting instruments from Anderman (Eppendorf Multipette), Laboratory Supplies Division, Central Avenue, East Molesey, Surrey KT8 OQ2, U.K. Oxford Sampler micropipetting system (range includes a 1 μl pipetter with disposable plastic tips) from Oxford Laboratories, BCL, Bell Lane, Lewes, East Sussex BN7 1LG, U.K.

Eppendorf Centrifuge Models 5412 and 5413: Obtainable from Anderman Laboratory Supplies Division, Central Avenue, East Molesey, Surrey KT8 OQZ, U.K.

These are small bench-top centrifuges for use with Eppendorf 1.5 ml snap-cap microfuge tubes. An essential piece of equipment for all sequencing work.

Plastic snap-cap centrifuge tubes: 1.5 ml, 0.5 ml. W. Sarstedt Ltd., (Cat. No. 39/10A) U.K. address: 47 Highmeres Road, Leicester LE4 7LZ. Used as reaction tubes for sequencing procedures. Essential.

'*Plastikard*' (0.35 mm thick) from Slaters' Ltd., Matlock Bath, Matlock, Derbyshire, U.K. Used for cutting spacers and well-formers for thin sequencing gels.

Vinyl tape for gels: from Sellotape Ltd., (Tape No. 1607) or Universal Scientific, 232 Plashet Road, London E13 OQU, U.K.

Plastic 1 ml disposable pipettes for columns: Sterilin Ltd., (Cat. No. 106) 43–45 Broad Street, Teddington, Middlesex, TW11 8QZ, U.K.

'*Sarenwrap*' cling film from Dow Chemical Co. Ltd. Many other brands of locally obtainable 'clingfllm' are suitable though some may stick to X-ray film and cause fogging.

'*Toothpicks*': obtainable from Boots Chemists (U.K.) but may be locally difficult to obtain. Ensure that they do not contain added bacteriocidal agents. (Sterilized pasteur pipettes are equally suitable as a way of stabbing out plaques from culture plates).

Autoradiography film: Kodak X-Omat R or Fuji Rx X-ray film (for hand developing). These are the most sensitive films for recording the visible light produced using intensifying screens. If intensifying screens are not used, Kodak 'No-screen' (for hand developing) or Kodak XH-1 (for autoprocess developing) are recommended.

Intensifying Screens: Du Pont Cronex Lighting Plus or Fuji Mach 2 medical X-ray screens. Size approx. 40 cm × 15 cm. These are among the most sensitive calcium tungstate screens. Obtainable from Hospital X-ray equipment suppliers.

X-ray film cassettes: Kodak Ltd. or X-ograph manufacture suitable X-ray cassettes. Generally obtainable from Hospital X-ray equipment suppliers.

Hand monitor (*Radiation and Contamination Monitor*): obtainable from Mini-instruments Ltd., 8 Station Industrial Estate, Burnham-on-Crouch, Essex CH0 8RN, U.K. (g–m meter type 5.10). A useful instrument for scanning chromatography columns, tubes, gels, etc. for ^{32}P-radioactivity. Specially useful for checking against loss of small (or invisible) precipitates in the bottom of Eppendorf tubes. The instrument has a calibrated range of 0–2000 cps and incorporates a small speaker for aural indication of radioactivity.

Fluorescent thin-layer plates: Eastman Kodak Co. product no. 13254. Plates coated with a fluorescent compound. Under short-wave UV illumination the plate fluoresces with a light blue colour. DNA, in gels, absorbs some of the exciting wavelengths and therefore shows up as dark bands on a light background.

PEI–thin layer plates: Plastic sheets coated with PEI (polyethylene-imine)-cellulose for thin-layer chromatography. 20 × 20 cm sheets Polygram CEL 300 PEI/UV_{254} obtainable from Macherey–Nagel & Co., 516 Düren, Werkstrasse, 6–8, Postfach 307, West Germany or from laboratory suppliers.

Conformable tape: (for Eppendorf tube gaskets) is plastic film tape No. 471, from 3M Company and distributors. Other types of soft Teflon or polyvinyl film are equally suitable.

Other items are from standard laboratory suppliers.

Enzymes, chemicals

A number of firms are now providing reagents for sequencing work. The following list gives the names of suppliers we have used and does not necessarily mean that other firms' products are unsuitable.

Enzymes

T4-DNA ligase	B.R.L., Miles, P.-L.
T4-kinase	Boehringer
Klenow polymerase	Boehringer, B.R.L.
T4-DNA polymerase	Miles, P.-L.
Exonuclease III	P.-L.
Restriction enzymes	any supplier or home-made but test by cleavage, religation and second cleavage.
Alkaline phosphatase from calf intestine	Boehringer, Worthington.
Alkaline phosphatase from *E. coli*	P.-L.

B.R.L.	Bethesda Research Laboratories, Inc. Md., U.S.A. U.K. distributors: Uniscience Ltd., 8, Jesus Lane, Cambridge CG5 8BA.
P.-L.	P.-L. Biochemicals Inc. Wisconsin, U.S.A. U.K. Distributors: International Enzymes Ltd., Hanover Way, Vale Road, Windsor, Berks SL4 5NJ.
Boehringer	The Boehringer Corporation (London) Ltd., Bell Lane, Lewes, East Sussex BN7 1LG.
Miles	Miles Laboratories Ltd., P.O. Box 37, Stoke Poges, Slough SL2 4LY.
New England Biolabs, Inc.	U.K. distribution: CP Laboratories Ltd., P.O. Box 22, Bishop's Stortford, Herts.

Chemicals

α-[^{32}P]-labelled nucleoside triphosphates—Radiochemical Centre, Amersham, Bucks. U.K. or New England Nuclear, 2, New Road, Southampton, U.K.
Deoxyribonucleoside triphosphates—Boehringer, P.-L.
Ribonucleoside triphosphates—Boehringer, P.-L.
Dideoxyribonucleoside triphosphates—P.-L.
X-Gal and IPTG (see list of abbreviations)—Sigma
Acrylamide, bis-acrylamide (High-purity grades)—BDH, Sigma
Urea (High-purity grades)—BRL, BDH.
*Eco*RI linkers—Collaborative Research, Inc. (U.K. distributors: Uniscience Ltd., 8, Jesus Lane, Cambridge CB5 8BA).
Anhydrous Hydrazine—Pierce Chemical Company, Rockville, Illinois, U.S.A.
Other chemicals—usually 'Analar' grade from BDH or other supplier.
Sigma: Sigma (London) Chemical Co. Ltd., Fancy Road, Poole, Dorset BH17 7NH, U.K.
BDH: BDH Chemicals Ltd., Poole, Dorset BH12 4NN, U.K.

For sequencing by the chemical procedure, Maxam and Gilbert (1980) give a detailed list of approved chemicals and suppliers.

APPENDIX 6

New England Biolabs, Inc. catalogue 1982–83 contains up to date restriction endonuclease maps of pBR322, M13mp8, phage λ and SV40 DNA; updated lists of restriction endonucleases, M13 primers, linkers and adapters as well as cross indexes of palindromic recognition sequences and much other useful data. A valuable up to date compendium obtainable from:

Europe	*USA*
New England Biolabs GmbH	New England Biolabs Inc
Postfach 2750	32 Tozer Road
6231 Schwalbach/Taunus	Beverly, MA 01915
West Germany	USA

APPENDIX 7

Nucleotide sequence of M13mp7; 7238 residues

The array of restriction endonuclease cleavage sites which has been inserted near the N-terminus of the β-galactosidase gene occupies positions 6231 (*Eco*RI site) to 6238 (centre of the *Hae*III site) cf. page 134, chapter 4). A restriction map of the closely related M13mp8 is given in Appendix 8.

```
        10         20         30         40         50         60         70         80         90        100        110        120
AATGCTACTA CTATTAGTAG AATTGATGCC ACCTTTTCAG CTCGCGCCCC AAATGAAAAT ATAGCTAAAC AGGTTATTGA CCATTTGCGA AATGTATCTA ATGGTCAAAC TAAATCTACT
TTACGATGAT GATAATCATC TTAACTACGG TGGAAAAGTC GAGCGCGGGG TTTACTTTTA TATCGATTTG TCCAATAACT GGTAAACGCT TTACATAGAT TACCAGTTTG ATTTAGATGA

       130        140        150        160        170        180        190        200        210        220        230        240
CGTTCGCAGA ATTGGGAATC AACTGTTACA TGGAATGAAA CTTCCAGACA CCGTACTTTA GTTGCATATT TAAAACATGT TGAGCTACAG CACCAGATTC AGCAATTAAG CTCTAAGCCA
GCAAGCGTCT TAACCCTTAG TTGACAATGT ACCTTACTTT GAAGGTCTGT GGCATGAAAT CAACGTATAA ATTTTGTACA ACTCGATGTC GTGGTCTAAG TCGTTAATTC GAGATTCGGT

       250        260        270        280        290        300        310        320        330        340        350        360
TCCGCAAAAA TGACCTCTTA TCAAAAGGAG CAATTAAAGG TACTCTCTAA TCCTGACCTG TTGGAGTTTG CTTCCGGTCT GGTTCGCTTT GAAGCTCGAA TTAAAACGCG ATATTTGAAG
AGGCGTTTTT ACTGGAGAAT AGTTTTCCTC GTTAATTTCC ATGAGAGATT AGGACTGGAC AACCTCAAAC GAAGGCCAGA CCAAGCGAAA CTTCGAGCTT AATTTTGCGC TATAAACTTC

       370        380        390        400        410        420        430        440        450        460        470        480
TCTTTCGGGC TTCCTCTTAA TCTTTTTGAT GCAATCCGCT TTGCTTCTGA CTATAATAGT CAGGGTAAAG ACCTGATTTT TGATTTATGG TCATTCTCGT TTTCTGAACT GTTTAAAGCA
AGAAAGCCCG AAGGAGAATT AGAAAAACTA CGTTAGGCGA AACGAAGACT GATATTATCA GTCCCATTTC TGGACTAAAA ACTAAATACC AGTAAGAGCA AAAGACTTGA CAAATTTCGT

       490        500        510        520        530        540        550        560        570        580        590        600
TTTGAGGGGG ATTCAATGAA TATTTATGAC GATTCCGCAG TATTGGACGC TATCCAGTCT AAACATTTTA CTATTACCCC CTCTGGCAAA ACTTCTTTTG CAAAAGCCTC TCGCTATTTT
AAACTCCCCC TAAGTTACTT ATAAATACTG CTAAGGCGTC ATAACCTGCG ATAGGTCAGA TTTGTAAAAT GATAATGGGG GAGACCGTTT TGAAGAAAAC GTTTTCGGAG AGCGATAAAA

       610        620        630        640        650        660        670        680        690        700        710        720
GGTTTTTATC GTCGTCTGGT AAACGAGGGT TATGATAGTG TTGCTCTTAC TATGCCTCGT AATTCCTTTT GGCGTTATGT ATCTGCATTA GTTGAATGTG GTATTCCTAA ATCTCAACTG
CCAAAAATAG CAGCAGACCA TTTGCTCCCA ATACTATCAC AACGAGAATG ATACGGAGCA TTAAGGAAAA CCGCAATACA TAGACGTAAT CAACTTACAC CATAAGGATT TAGAGTTGAC

       730        740        750        760        770        780        790        800        810        820        830        840
ATGAATCTTT CTACCTGTAA TAATGTTGTT CCGTTAGTTC GTTTTATTAA CGTAGATTTT TCTTCCCAAC GTCCTGACTG GTATAATGAG CCAGTTCTTA AAATCGCATA AGGTAATTCA
TACTTAGAAA GATGGACATT ATTACAACAA GGCAATCAAG CAAAATAATT GCATCTAAAA AGAAGGGTTG CAGGACTGAC CATATTACTC GGTCAAGAAT TTTAGCGTAT TCCATTAAGT

       850        860        870        880        890        900        910        920        930        940        950        960
CAATGATTAA AGTTGAAATT AAACCATCTC AAGCCCAATT TACTACTCGT TCTGGTGTTT CTCGTCAGGG CAAGCCTTAT TCACTGAATG AGCAGCTTTG TTACGTTGAT TTGGGTAATG
GTTACTAATT TCAACTTTAA TTTGGTAGAG TTCGGGTTAA ATGATGAGCA AGACCACAAA GAGCAGTCCC GTTCGGAATA AGTGACTTAC TCGTCGAAAC AATGCAACTA AACCCATTAC

       970        980        990       1000       1010       1020       1030       1040       1050       1060       1070       1080
AATATCCGGT TCTTGTCAAG ATTACTCTTG ATGAAGGTCA GCCAGCCTAT GCGCCTGGTC TGTACACCGT TCATCTGTCC TCTTTCAAAG TTGGTCAGTT CGGTTCCCTT ATGATTGACC
TTATAGGCCA AGAACAGTTC TAATGAGAAC TACTTCCAGT CGGTCGGATA CGCGGACCAG ACATGTGGCA AGTAGACAGG AGAAAGTTTC AACCAGTCAA GCCAAGGGAA TACTAACTGG

      1090       1100       1110       1120       1130       1140       1150       1160       1170       1180       1190       1200
GTCTGCGCCT CGTTCCGGCT AAGTAACATG GAGCAGGTCG CGGATTTCGA CACAATTTAT CAGGCGATGA TACAAATCTC CGTTGTACTT TGTTTCGCGC TTGGTATAAT CGCTGGGGGT
CAGACGCGGA GCAAGGCCGA TTCATTGTAC CTCGTCCAGC GCCTAAAGCT GTGTTAAATA GTCCGCTACT ATGTTTAGAG GCAACATGAA ACAAAGCGCG AACCATATTA GCGACCCCCA

      1210       1220       1230       1240       1250       1260       1270       1280       1290       1300       1310       1320
CAAAGATGAG TGTTTTAGTG TATTCTTTCG CCTCTTTCGT TTTAGGTTGG TGCCTTCGTA GTGGCATTAC GTATTTTACC CGTTTAATGG AAACTTCCTC ATGAAAAAGT CTTTAGTCCT
GTTTCTACTC ACAAAATCAC ATAAGAAAGC GGAGAAAGCA AAATCCAACC ACGGAAGCAT CACCGTAATG CATAAAATGG GCAAATTACC TTTGAAGGAG TACTTTTTCA GAAATCAGGA

      1330       1340       1350       1360       1370       1380       1390       1400       1410       1420       1430       1440
CAAAGCCTCT GTAGCCGTTG CTACCCTCGT TCCGATGCTG TCTTTCGCTG CTGAGGGTGA CGATCCCGCA AAAGCGGCCT TTAACTCCCT GCAAGCCTCA GCGACCGAAT ATATCGGTTA
GTTTCGGAGA CATCGGCAAC GATGGGAGCA AGGCTACGAC AGAAAGCGAC GACTCCCACT GCTAGGGCGT TTTCGCCGGA AATTGAGGGA CGTTCGGAGT CGCTGGCTTA TATAGCCAAT

      1450       1460       1470       1480       1490       1500       1510       1520       1530       1540       1550       1560
TGCGTGGGCG ATGGTTGTTG TCATTGTCGG CGCAACTATC GGTATCAAGC TGTTTAAGAA ATTCACCTCG AAAGCAAGCT GATAAACCGA TACAATTAAA GGCTCCTTTT GGAGCCTTTT
ACGCACCCGC TACCAACAAC AGTAACAGCC GCGTTGATAG CCATAGTTCG ACAAATTCTT TAAGTGGAGC TTTCGTTCGA CTATTTGGCT ATGTTAATTT CCGAGGAAAA CCTCGGAAAA

      1570       1580       1590       1600       1610       1620       1630       1640       1650       1660       1670       1680
TTTTTGGAGA TTTTCAACGT GAAAAAATTA TTATTCGCAA TTCCTTTAGT TGTTCCTTTC TATTCTCACT CCGCTGAAAC TGTTGAAAGT TGTTTAGCAA AACCCCATAC AGAAAATTCA
AAAAACCTCT AAAAGTTGCA CTTTTTTAAT AATAAGCGTT AAGGAAATCA ACAAGGAAAG ATAAGAGTGA GGCGACTTTG ACAACTTTCA ACAAATCGTT TTGGGGTATG TCTTTTAAGT

      1690       1700       1710       1720       1730       1740       1750       1760       1770       1780       1790       1800
TTTACTAACG TCTGGAAAGA CGACAAAACT TCAGATCGTT ACGCTAACTA TCAGGGTTGT CTGTGGAATG CTACAGGCGT TGTAGTTTGT ACTGGTGACG AAACTCAGTG TTACGGTACA
AAATGATTGC AGACCTTTCT GCTGTTTTGA AATCTAGCAA TGCGATTGAT ACTCCCAACA GACACCTTAC GATGTCCGCA ACATCAAACA TGACCACTGC TTTGAGTCAC AATGCCATGT

      1810       1820       1830       1840       1850       1860       1870       1880       1890       1900       1910       1920
TGGGTTCCTA TTGGGCTTGC TATCCCTGAA AATGAGGGTG GTGGCTCTGA GGGTGGCGGT TCTGAGGGTG GCGGTTCTGA GGGTGGCGGT ACTAAACCTC CTGAGTACGG TGATACACCT
ACCCAAGGAT AACCCGAACG ATAGGGACTT TTACTCCCAC CACCGAGACT CCCACCGCCA AGACTCCCAC CGCCAAGACT CCCACCGCCA TGATTTGGAG GACTCATGCC ACTATGTGGA

      1930       1940       1950       1960       1970       1980       1990       2000       2010       2020       2030       2040
ATTCCGGGCT ATACTTATAT CAACCCTCTC GACGGCACTT ATCCGCCTGG TACTGAGCAA AACCCCGCTA ATCCTAATCC TTCTCTTGAG GAGTCTCAGC CTCTTAATAC TTTCATGTTT
TAAGGCCCGA TATGAATATA GTTGGGAGAG CTGCCGTGAA TAGGCGGACC ATGACTCGTT TTGGGGCGAT TAGGATTAGG AAGAGAACTC CTCAGAGTCG GAGAATTATG AAAGTACAAA

      2050       2060       2070       2080       2090       2100       2110       2120       2130       2140       2150       2160
CAGAATAATA GGTTCCGAAA TAGGCAGGGG GCATTAACTG TTTATACGGG CACTGTTACT CAAGGCACTG ACCCCGTTAA AACTTATTAC CAGTACACTC CTGTATCATC AAAAGCCATG
GTCTTATTAT CCAAGGCTTT ATCCGTCCCC CGTAATTGAC AAATATGCCC GTGACAATGA GTTCCGTGAC TGGGGCAATT TTGAATAATG GTCATGTGAG GACATAGTAG TTTTCGGTAC

      2170       2180       2190       2200       2210       2220       2230       2240       2250       2260       2270       2280
TATGACGCTT ACTGGAACGG TAAATTCAGA GACTGCGCTT TCCATTCTGG CTTTAATGAA GATCCATTCG TTTGTGAATA TCAAGGCCAA TCGTCTGACC TGCCTCAACC TCCTGTCAAT
ATACTGCGAA TGACCTTGCC ATTTAAGTCT CTGACGCGAA AGGTAAGACC GAAATTACTT CTAGGTAAGC AAACACTTAT AGTTCCGGTT AGCAGACTGG ACGGAGTTGG AGGACAGTTA

      2290       2300       2310       2320       2330       2340       2350       2360       2370       2380       2390       2400
GCTGGCGGCG GCTCTGGTGG TGGTTCTGGT GGCGGCTCTG AGGGTGGTGG CTCTGAGGGT GGCGGTTCTG AGGGTGGCGG CTCTGAGGGA GGCGGTTCCG GTGGTGGCTC TGGTTCCGGT
CGACCGCCGC CGAGACCACC ACCAAGACCA CCGCCGAGAC TCCCACCACC GAGACTCCCA CCGCCAAGAC TCCCACCGCC GAGACTCCCT CCGCCAAGGC CACCACCGAG ACCAAGGCCA

      2410       2420       2430       2440       2450       2460       2470       2480       2490       2500       2510       2520
GATTTTGATT ATGAAAAGAT GGCAAACGCT AATAAGGGGG CTATGACCGA AAATGCCGAT GAAAACGCGC TACAGTCTGA CGCTAAAGGC AAACTTGATT CTGTCGCTAC TGATTACGGT
CTAAAACTAA TACTTTTCTA CCGTTTGCGA TTATTCCCCC GATACTGGCT TTTACGGCTA CTTTTGCGCG ATGTCAGACT GCGATTTCCG TTTGAACTAA GACAGCGATG ACTAATGCCA

      2530       2540       2550       2560       2570       2580       2590       2600       2610       2620       2630       2640
GCTGCTATCG ATGGTTTCAT TGGTGACGTT TCCGGCCTTG CTAATGGTAA TGGTGCTACT GGTGATTTTG CTGGCTCTAA TTCCCAAATG GCTCAAGTCG GTGACGGTGA TAATTCACCT
CGACGATAGC TACCAAAGTA ACCACTGCAA AGGCCGGAAC GATTACCATT ACCACGATGA CCACTAAAAC GACCGAGATT AAGGGTTTAC CGAGTTCAGC CACTGCCACT ATTAAGTGGA

      2650       2660       2670       2680       2690       2700       2710       2720       2730       2740       2750       2760
TTAATGAATA ATTTCCGTCA ATATTTACCT TCCCTCCCTC AATCGGTTGA ATGTCGCCCT TTTGTCTTTA GCGCTGGTAA ACCATATGAA TTTTCTATTG ATTGTGACAA AATAAACTTA
AATTACTTAT TAAAGGCAGT TATAAATGGA AGGGAGGGAG TTAGCCAACT TACAGCGGGA AAACAGAAAT CGCGACCATT TGGTATACTT AAAAGATAAC TAACACTGTT TTATTTGAAT

      2770       2780       2790       2800       2810       2820       2830       2840       2850       2860       2870       2880
TTCCGTGGTG TCTTTGCGTT TCTTTTATAT GTTGCCACCT TTATGTATGT ATTTTCTACG TTTGCTAACA TACTGCGTAA TAAGGAGTCT TAATCATGCC AGTTCTTTTG GGTATTCCGT
AAGGCACCAC AGAAACGCAA AGAAAATATA CAACGGTGGA AATACATACA TAAAAGATGC AAACGATTGT ATGACGCATT ATTCCTCAGA ATTAGTACGG TCAAGAAAAC CCATAAGGCA

      2890       2900       2910       2920       2930       2940       2950       2960       2970       2980       2990       3000
TATTATTGCG TTTCCTCGGT TTCCTTCTGG TAACTTTGTT CGGCTATCTG CTTACTTTTC TTAAAAAGGG CTTCGGTAAG ATAGCTATTG CTATTTCATT GTTTCTTGCT CTTATTATTG
ATAATAACGC AAAGGAGCCA AAGGAAGACC ATTGAAACAA GCCGATAGAC GAATGAAAAG AATTTTTCCC GAAGCCATTC TATCGATAAC GATAAAGTAA CAAAGAACGA GAATAATAAC

      3010       3020       3030       3040       3050       3060       3070       3080       3090       3100       3110       3120
GGCTTAACTC AATTCTTGTG GGTTATCTCT CTGATATTAG CGCTCAATTA CCCTCTGACT TTGTTCAGGG TGTTCAGTTA ATTCTCCCGT CTAATGCGCT TCCCTGTTTT TATGTTATTC
CCGAATTGAG TTAAGAACAC CCAATAGAGA GACTATAATC GCGAGTTAAT GGGAGACTGA AACAAGTCCC ACAAGTCAAT TAAGAGGGCA GATTACGCGA AGGGACAAAA ATACAATAAG

      3130       3140       3150       3160       3170       3180       3190       3200       3210       3220       3230       3240
TCTCTGTAAA GGCTGCTATT TTCATTTTTG ACGTTAAACA AAAAATCGTT TCTTATTTGG ATTGGGATAA ATAATATGGC TGTTTATTTT GTAACTGGCA AATTAGGCTC TGGAAAGACG
AGAGACATTT CCGACGATAA AAGTAAAAAC TGCAATTTGT TTTTTAGCAA AGAATAAACC TAACCCTATT TATTATACCG ACAAATAAAA CATTGACCGT TTAATCCGAG ACCTTTCTGC

      3250       3260       3270       3280       3290       3300       3310       3320       3330       3340       3350       3360
CTCGTTAGCG TTGGTAAGAT TCAGGATAAA ATTGTAGCTG GGTGCAAAAT AGCAACTAAT CTTGATTTAA GGCTTCAAAA CCTCCCGCAA GTCGGGAGGT TCGCTAAAAC GCCTCGCGTT
GAGCAATCGC AACCATTCTA AGTCCTATTT TAACATCGAC CCACGTTTTA TCGTTGATTA GAACTAAATT CCGAAGTTTT GGAGGGCGTT CAGCCCTCCA AGCGATTTTG CGGAGCGCAA

      3370       3380       3390       3400       3410       3420       3430       3440       3450       3460       3470       3480
CTTAGAATAC CGGATAAGCC TTCTATATCT GATTTGCTTG CTATTGGGCG CGGTAATGAT TCCTACGATG AAAATAAAAA CGGCTTGCTT GTTCTCGATG AGTGCGGTAC TTGGTTTAAT
GAATCTTATG GCCTATTCGG AAGATATAGA CTAAACGAAC GATAACCCGC GCCATTACTA AGGATGCTAC TTTTATTTTT GCCGAACGAA CAAGAGCTAC TCACGCCATG AACCAAATTA

      3490       3500       3510       3520       3530       3540       3550       3560       3570       3580       3590       3600
ACCCGTTCTT GGAATGATAA GGAAAGACAG CCGATTATTG ATTGGTTTCT ACATGCTCGT AAATTAGGAT GGGATATTAT TTTTCTTGTT CAGGACTTAT CTATTGTTGA TAAACAGGCG
TGGGCAAGAA CCTTACTATT CCTTTCTGTC GGCTAATAAC TAACCAAAGA TGTACGAGCA TTTAATCCTA CCCTATAATA AAAAGAACAA GTCCTGAATA GATAACAACT ATTTGTCCGC
```

```
      3610       3620       3630       3640       3650       3660       3670       3680       3690       3700       3710       3720
CGTTCTGCAT TAGCTGAACA TGTTGTTTAT TGTCGTCGTC TGGACAGAAT TACTTTACCT TTTGTCGGTA CTTTATATTC TCTTATTACT GGCTCGAAAT TGCCTCTGCC TAAATTACAT
GCAAGACGTA ATCGACTTGT ACAACAAATA ACAGCAGCAG ACCTGTCTTA ATGAAATGGA AAACAGCCAT GAAATATAAG AGAATAATGA CCGAGCTTTA ACGGAGACGG ATTTAATGTA

      3730       3740       3750       3760       3770       3780       3790       3800       3810       3820       3830       3840
GTTGGCGTTG TTAAATATGG CGATTCTCAA TTAAGCCCTA CTGTTGAGCG TTGGCTTTAT ACTGGTAAGA ATTTGTATAA CGCATATGAT ACTAAACAGG CTTTTTCTAG TAATTATGAT
CAACCGCAAC AATTTATACC GCTAAGAGTT AATTCGGGAT GACAACTCGC AACCGAAATA TGACCATTCT TAAACATATT GCGTATACTA TGATTTGTCC GAAAAAGATC ATTAATACTA

      3850       3860       3870       3880       3890       3900       3910       3920       3930       3940       3950       3960
TCCGGTGTTT ATTCTTATTT AACGCCTTAT TTATCACACG GTCGGTATTT CAAACCATTA AATTTAGGTC AGAAGATGAA ATTAACTAAA ATATATTTGA AAAAGTTTTC TCGCGTTCTT
AGGCCACAAA TAAGAATAAA TTGCGGAATA AATAGTGTGC CAGCCATAAA GTTTGGTAAT TTAAATCCAG TCTTCTACTT TAATTGATTT TATATAAACT TTTTCAAAAG AGCGCAAGAA

      3970       3980       3990       4000       4010       4020       4030       4040       4050       4060       4070       4080
TGTCTTGCGA TTGGATTTGC ATCAGCATTT ACATATAGTT ATATAACCCA ACCTAAGCCG GAGGTTAAAA AGGTAGTCTC TCAGACCTAT GATTTTGATA AATTCACTAT TGACTCTTCT
ACAGAACGCT AACCTAAACG TAGTCGTAAA TGTATATCAA TATATTGGGT TGGATTCGGC CTCCAATTTT TCCATCAGAG AGTCTGGATA CTAAAACTAT TTAAGTGATA ACTGAGAAGA

      4090       4100       4110       4120       4130       4140       4150       4160       4170       4180       4190       4200
CAGCGTCTTA ATCTAAGCTA TCGCTATGTT TTCAAGGATT CTAAGGGAAA ATTAATTAAT AGCGACGATT TACAGAAGCA AGGTTATTCA CTCACATATA TTGATTTATG TACTGTTTCC
GTCGCAGAAT TAGATTCGAT AGCGATACAA AAGTTCCTAA GATTCCCTTT TAATTAATTA TCGCTGCTAA ATGTCTTCGT TCCAATAAGT GAGTGTATAT AACTAAATAC ATGACAAAGG

      4210       4220       4230       4240       4250       4260       4270       4280       4290       4300       4310       4320
ATTAAAAAAG GTAATTCAAA TGAAATTGTT AAATGTAATT AATTTTGTTT TCTTGATGTT TGTTTCATCA TCTTCTTTTG CTCAGGTAAT TGAAATGAAT AATTCGCCTC TGCGCGATTT
TAATTTTTTC CATTAAGTTT ACTTTAACAA TTTACATTAA TTAAAACAAA AGAACTACAA ACAAAGTAGT AGAAGAAAAC GAGTCCATTA ACTTTACTTA TTAAGCGGAG ACGCGCTAAA

      4330       4340       4350       4360       4370       4380       4390       4400       4410       4420       4430       4440
TGTAACTTGG TATTCAAAGC AATCAGGCGA ATCCGTTATT GTTTCTCCCG ATGTAAAAGG TACTGTTACT GTATATTCAT CTGACGTTAA ACCTGAAAAT CTACGCAATT TCTTTATTTC
ACATTGAACC ATAAGTTTCG TTAGTCCGCT TAGGCAATAA CAAAGAGGGC TACATTTTCC ATGACAATGA CATATAAGTA GACTGCAATT TGGACTTTTA GATGCGTTAA AGAAATAAAG

      4450       4460       4470       4480       4490       4500       4510       4520       4530       4540       4550       4560
TGTTTTACGT GCTAATAATT TTGATATGGT TGGTTCAATT CCTTCCATAA TTCAGAAGTA TAATCCAAAC AATCAGGATT ATATTGATGA ATTGCCATCA TCTGATAATC AGGAATATGA
ACAAAATGCA CGATTATTAA AACTATACCA ACCAAGTTAA GGAAGGTATT AAGTCTTCAT ATTAGGTTTG TTAGTCCTAA TATAACTACT TAACGGTAGT AGACTATTAG TCCTTATACT

      4570       4580       4590       4600       4610       4620       4630       4640       4650       4660       4670       4680
TGATAATTCC GCTCCTTCTG GTGGTTTCTT TGTTCCGCAA AATGATAATG TTACTCAAAC TTTTAAAATT AATAACGTTC GGGCAAAGGA TTTAATACGA GTTGTCGAAT TGTTTGTAAA
ACTATTAAGG CGAGGAAGAC CACCAAAGAA ACAAGGCGTT TTACTATTAC AATGAGTTTG AAAATTTTAA TTATTGCAAG CCCGTTTCCT AAATTATGCT CAACAGCTTA ACAAACATTT

      4690       4700       4710       4720       4730       4740       4750       4760       4770       4780       4790       4800
GTCTAATACT TCTAAATCCT CAAATGTATT ATCTATTGAC GGCTCTAATC TATTAGTTGT TAGTGCACCT AAAGATATTT TAGATAACCT TCCTCAATTC CTTTCTACTG TTGATTTGCC
CAGATTATGA AGATTTAGGA GTTTACATAA TAGATAACTG CCGAGATTAG ATAATCAACA ATCACGTGGA TTTCTATAAA ATCTATTGGA AGGAGTTAAG GAAAGATGAC AACTAAACGG

      4810       4820       4830       4840       4850       4860       4870       4880       4890       4900       4910       4920
AACTGACCAG ATATTGATTG AGGGTTTGAT ATTTGAGGTT CAGCAAGGTG ATGCTTTAGA TTTTTCATTT GCTGCTGGCT CTCAGCGTGG CACTGTTGCA GGCGGTGTTA ATACTGACCG
TTGACTGGTC TATAACTAAC TCCCAAACTA TAAACTCCAA GTCGTTCCAC TACGAAATCT AAAAAGTAAA CGACGACCGA GAGTCGCACC GTGACAACGT CCGCCACAAT TATGACTGGC

      4930       4940       4950       4960       4970       4980       4990       5000       5010       5020       5030       5040
CCTCACCTCT GTTTTATCTT CTGCTGGTGG TTCGTTCGGT ATTTTTAATG GCGATGTTTT AGGGCTATCA GTTCGCGCAT TAAAGACTAA TAGCCATTCA AAAATATTGT CTGTGCCACG
GGAGTGGAGA CAAAATAGAA GACGACCACC AAGCAAGCCA TAAAAATTAC CGCTACAAAA TCCCGATAGT CAAGCGCGTA ATTTCTGATT ATCGGTAAGT TTTTATAACA GACACGGTGC

      5050       5060       5070       5080       5090       5100       5110       5120       5130       5140       5150       5160
TATTCTTACG CTTTCAGGTC AGAAGGGTTC TATCTCTGTT GGCCAGAATG TCCCTTTTAT TACTGGTCGT GTGACTGGTG AATCTGCCAA TGTAAATAAT CCATTTCAGA CGATTGAGCG
ATAAGAATGC GAAAGTCCAG TCTTCCCAAG ATAGAGACAA CCGGTCTTAC AGGGAAAATA ATGACCAGCA CACTGACCAC TTAGACGGTT ACATTTATTA GGTAAAGTCT GCTAACTCGC

      5170       5180       5190       5200       5210       5220       5230       5240       5250       5260       5270       5280
TCAAAATGTA GGTATTTCCA TGAGCGTTTT TCCTGTTGCA ATGGCTGGCG GTAATATTGT TCTGGATATT ACCAGCAAGG CCGATAGTTT GAGTTCTTCT ACTCAGGCAA GTGATGTTAT
AGTTTTACAT CCATAAAGGT ACTCGCAAAA AGGACAACGT TACCGACCGC CATTATAACA AGACCTATAA TGGTCGTTCC GGCTATCAAA CTCAAGAAGA TGAGTCCGTT CACTACAATA

      5290       5300       5310       5320       5330       5340       5350       5360       5370       5380       5390       5400
TACTAATCAA AGAAGTATTG CTACAACGGT TAATTTGCGT GATGGACAGA CTCTTTTACT CGGTGGCCTC ACTGATTATA AAAACACTTC TCAAGATTCT GGCGTACCGT TCCTGTCTAA
ATGATTAGTT TCTTCATAAC GATGTTGCCA ATTAAACGCA CTACCTGTCT GAGAAAATGA GCCACCGGAG TGACTAATAT TTTTGTGAAG AGTTCTAAGA CCGCATGGCA AGGACAGATT

      5410       5420       5430       5440       5450       5460       5470       5480       5490       5500       5510       5520
AATCCCTTTA ATCGGCCTCC TGTTTAGCTC CCGCTCTGAT TCCAACGAGG AAAGCACGTT ATACGTGCTC GTCAAAGCAA CCATAGTACG CGCCCTGTAG CGGCGCATTA AGCGCGGCGG
TTAGGGAAAT TAGCCGGAGG ACAAATCGAG GGCGAGACTA AGGTTGCTCC TTTCGTGCAA TATGCACGAG CAGTTTCGTT GGTATCATGC GCGGGACATC GCCGCGTAAT TCGCGCCGCC

      5530       5540       5550       5560       5570       5580       5590       5600       5610       5620       5630       5640
GTGTGGTGGT TACGCGCAGC GTGACCGCTA CACTTGCCAG CGCCCTAGCG CCCGCTCCTT TCGCTTTCTT CCCTTCCTTT CTCGCCACGT TCGCCGGCTT TCCCCGTCAA GCTCTAAATC
CACACCACCA ATGCGCGTCG CACTGGCGAT GTGAACGGTC GCGGGATCGC GGGCGAGGAA AGCGAAAGAA GGGAAGGAAA GAGCGGTGCA AGCGGCCGAA AGGGGCAGTT CGAGATTTAG

      5650       5660       5670       5680       5690       5700       5710       5720       5730       5740       5750       5760
GGGGGCTCCC TTTAGGGTTC CGATTTAGTG CTTTACGGCA CCTCGACCCC AAAAAACTTG ATTTGGGTGA TGGTTCACGT AGTGGGCCAT CGCCCTGATA GACGGTTTTT CGCCCTTTGA
CCCCCGAGGG AAATCCCAAG GCTAAATCAC GAAATGCCGT GGAGCTGGGG TTTTTTGAAC TAAACCCACT ACCAAGTGCA TCACCCGGTA GCGGGACTAT CTGCCAAAAA GCGGGAAACT

      5770       5780       5790       5800       5810       5820       5830       5840       5850       5860       5870       5880
CGTTGGAGTC CACGTTCTTT AATAGTGGAC TCTTGTTCCA AACTTGAACA ACACTCAACC CTATCTCGGG CTATTCTTTT GATTTATAAG GGATTTTGCC GATTTCGGAA CCACCATCAA
GCAACCTCAG GTGCAAGAAA TTATCACCTG AGAACAAGGT TTGAACTTGT TGTGAGTTGG GATAGAGCCC GATAAGAAAA CTAAATATTC CCTAAAACGG CTAAAGCCTT GGTGGTAGTT

      5890       5900       5910       5920       5930       5940       5950       5960       5970       5980       5990       6000
ACAGGATTTT CGCCTGCTGG GGCAAACCAG CGTGGACCGC TTGCTGCAAC TCTCTCAGGG CCAGGCGGTG AAGGGCAATC AGCTGTTGCC CGTCTCGCTG GTGAAAAGAA AAACCACCCT
TGTCCTAAAA GCGGACGACC CCGTTTGGTC GCACCTGGCG AACGACGTTG AGAGAGTCCC GGTCCGCCAC TTCCCGTTAG TCGACAACGG GCAGAGCGAC CACTTTTCTT TTTGGTGGGA

      6010       6020       6030       6040       6050       6060       6070       6080       6090       6100       6110       6120
GGCGCCCAAT ACGCAAACCG CCTCTCCCCG CGCGTTGGCC GATTCATTAA TCCAGCTGGC ACGACAGGTT TCCCGACTGG AAAGCGGGCA GTGAGCGCAA CGCAATTAAT GTGAGTTACC
CCGCGGGTTA TGCGTTTGGC GGAGAGGGGC GCGCAACCGG CTAAGTAATT AGGTCGACCG TGCTGTCCAA AGGGCTGACC TTTCGCCCGT CACTCGCGTT GCGTTAATTA CACTCAATGG

      6130       6140       6150       6160       6170       6180       6190       6200       6210       6220       6230       6240
TCACTCATTA GGCACCCCAG GCTTTACACT TTATGCTTCC GGCTCGTATG TTGTGTGGAA TTGTGAGCGG ATAACAATTT CACACAGGAA ACAGCTATGA CCATGATTAC GAATTCCCCG
AGTGAGTAAT CCGTGGGGTC CGAAATGTGA AATACGAAGG CCGAGCATAC AACACACCTT AACACTCGCC TATTGTTAAA GTGTGTCCTT TGTCGATACT GGTACTAATG CTTAAGGGGC

      6250       6260       6270       6280       6290       6300       6310       6320       6330       6340       6350       6360
GATCCGTCGA CCTGCAGGTC GACGGATCCG GGGAATTCAC TGGCCGTCGT TTTACAACGT CGTGACTGGG AAAACCCTGG CGTTACCCAA CTTAATCGCC TTGCAGCACA TCCCCCCTTC
CTAGGCAGCT GGACGTCCAG CTGCCTAGGC CCCTTAAGTG ACCGGCAGCA AAATGTTGCA GCACTGACCC TTTTGGGACC GCAATGGGTT GAATTAGCGG AACGTCGTGT AGGGGGGAAG

      6370       6380       6390       6400       6410       6420       6430       6440       6450       6460       6470       6480
GCCAGCTGGC GTAATAGCGA AGAGGCCCGC ACCGATCGCC CTTCCCAACA GTTGCGTAGC CTGAATGGCG AATGGCGCTT TGCCTGGTTT CCGGCACCAG AAGCGGTGCC GGAAAGCTGG
CGGTCGACCG CATTATCGCT TCTCCGGGCG TGGCTAGCGG GAAGGGTTGT CAACGCATCG GACTTACCGC TTACCGCGAA ACGGACCAAA GGCCGTGGTC TTCGCCACGG CCTTTCGACC

      6490       6500       6510       6520       6530       6540       6550       6560       6570       6580       6590       6600
CTGGAGTGCG ATCTTCCTGA GGCCGAYAC- GTCGTCGTCC CCTCAAACTG GCAGATGCAC GGTTACGATG CGCCCATCTA CACCAACGTA ACCTATCCCA TTACGGTCAA TCCGCCGTTT
GACCTCACGC TAGAAGGACT CCGGCTRTG- CAGCAGCAGG GGAGTTTGAC CGTCTACGTG CCAATGCTAC GCGGGTAGAT GTGGTTGCAT TGGATAGGGT AATGCCAGTT AGGCGGCAAA

      6610       6620       6630       6640       6650       6660       6670       6680       6690       6700       6710       6720
GTTCCCACGG AGAATCCGAC GGGTTGTTAC TCGCTCACAT TTAATGTTGA TGAAAGCTGG CTACAGGAAG GCCAGACGCG AATTATTTTT GATGGCGTTC CTATTGGTTA AAAAATGAGC
CAAGGGTGCC TCTTAGGCTG CCCAACAATG AGCGAGTGTA AATTACAACT ACTTTCGACC GATGTCCTTC CGGTCTGCGC TTAATAAAAA CTACCGCAAG GATAACCAAT TTTTTACTCG

      6730       6740       6750       6760       6770       6780       6790       6800       6810       6820       6830       6840
TGATTTAACA AAAATTTAAC GCGAATTTTA ACAAAATATT AACGTTTACA ATTTAAATAT TTGCTTATAC AATCTTCCTG TTTTTGGGGC TTTTCTGATT ATCAACCGGG GTACATATGA
ACTAAATTGT TTTTAAATTG CGCTTAAAAT TGTTTTATAA TTGCAAATGT TAAATTTATA AACGAATATG TTAGAAGGAC AAAAACCCCG AAAAGACTAA TAGTTGGCCC CATGTATACT

      6850       6860       6870       6880       6890       6900       6910       6920       6930       6940       6950       6960
TTGACATGCT AGTTTTACGA TTACCGTTCA TCGATTCTCT TGTTTGCTCC AGACTCTCAG GCAATGACCT GATAGCCTTT GTAGATCTCT CAAAAATAGC TACCCTCTCC GGCATGAATT
AACTGTACGA TCAAAATGCT AATGGCAAGT AGCTAAGAGA ACAAACGAGG TCTGAGAGTC CGTTACTGGA CTATCGGAAA CATCTAGAGA GTTTTTATCG ATGGGAGAGG CCGTACTTAA

      6970       6980       6990       7000       7010       7020       7030       7040       7050       7060       7070       7080
TATCAGCTAG AACGGTTGAA TATCATATTG ATGGTGATTT GACTGTCTCC GGCCTTTCTC ACCCTTTTGA ATCTTTACCT ACACATTACT CAGGCATTGC ATTTAAAATA TATGAGGGTT
ATAGTCGATC TTGCCAACTT ATAGTATAAC TACCACTAAA CTGACAGAGG CCGGAAAGAG TGGGAAAACT TAGAAATGGA TGTGTAATGA GTCCGTAACG TAAATTTTAT ATACTCCCAA

      7090       7100       7110       7120       7130       7140       7150       7160       7170       7180       7190       7200
CTAAAAATTT TTATCCTTGC GTTGAAATAA AGGCTTCTCC CGCAAAAGTA TTACAGGGTC ATAATGTTTT TGGTACAACC GATTTAGCTT TATGCTCTGA GGCTTTATTG CTTAATTTTG
GATTTTTAAA AATAGGAACG CAACTTTATT TCCGAAGAGG GCGTTTTCAT AATGTCCCAG TATTACAAAA ACCATGTTGG CTAAATCGAA ATACGAGACT CCGAAATAAC GAATTAAAAC

      7210       7220       7230
CTAATTCTTT GCCTTGCCTG TATGATTTAT TGGATGTT
GATTAAGAAA CGGAACGGAC ATACTAAATA ACCTACAA
```

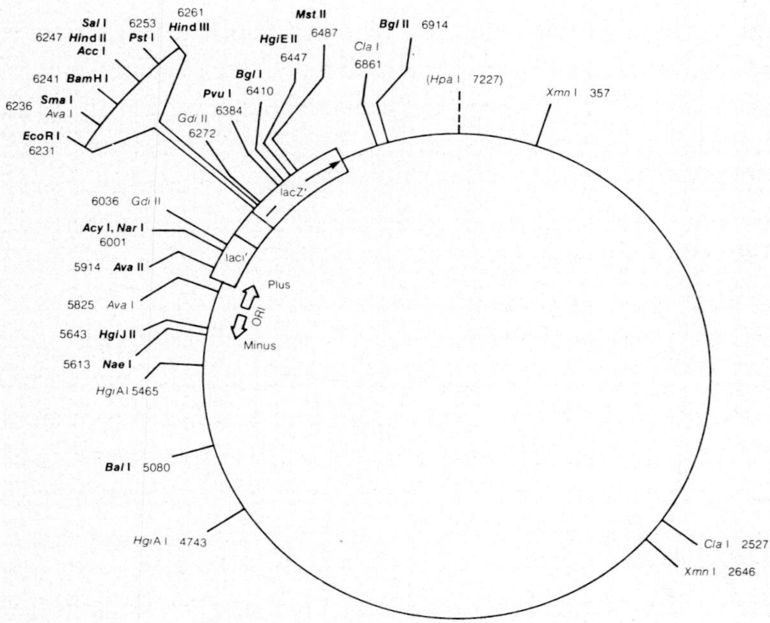

Appendix 8

Restriction map of M13mp8 (7229 base pairs)

Map of the restriction enzyme sites that occur once or twice in M13mp8. The position of the first base in each recognition sequence is given relative to the centre of the *Hpa* I recognition sequence present in wild type M13 DNA. (This sequence has been altered in M13mp8 and it is no longer recognized by *Hpa* I.) Enzymes that cleave M13mp8 only once are printed in bold type. The map depicts the viral (plus) single DNA strand, and the origins of plus and minus strand DNA synthesis. M13mp8 contains a 789 base pair fragment from the *E. coli lac* operon inserted clockwise of ORI after base number 5868. An additional 33 base pair synthetic DNA fragment that contains multiple unique restriction sites has been inserted into the amino terminus of the *lac Z* gene. Fragments cloned into these sites can be rapidly sequenced by the dideoxy chain termination method. For clarity, this section of the map has been expanded. Recognition sequences of the following enzymes are not present in M13mp8: *Afl* II, *Apa* I, *Asu* II, *Ava* III, *Avr* II, *Bcl* I, *Bss*H II, *Bst*E II, *Eco*R V, *Hpa* I, *Kpn* I, *Mlu* I, *Mst* I, *Nco* I, *Nru* I, *Sac* I, *Sac* II, *Sca* I, *Sna* I, *Sph* I, *Stu* I, *Xba* I, *Xho* I, and *Xma* III.

CHAPTER 6

Computer methods for DNA sequencers

(*Contributed by Rodger Staden, MRC Laboratory of Molecular Biology, Cambridge, U.K.*)

Introduction

The importance of the computer as a tool of the DNA sequencer has grown with the development of faster sequencing methods: the faster the method the more important the computer becomes. During the sequencing of the DNA of bacteriophage PhiX 174 [1] it became clear that it was necessary to use computers to store and analyse the data [2, 3, 4]. Later, while the very similar DNA sequence of bacteriophage G4 [5] was being determined the computer was used to compare and align the G4 sequence with that of PhiX 174. With the invention of the shotgun sequencing techniques (Chapter 4) the computer assumed a central role in the overall sequencing strategy. At the time of writing DNA sequencing is one of the fastest information gathering techniques of molecular biology and consequently is being used as a method to try to find answers to many diverse questions. Therefore it will become increasingly common that the DNA sequence will be one of the first things known about an organism: little will be known about the genes that the DNA contains prior to the determination of the sequence. For this reason the computer will become more important as an analytic tool. The computer sections of this book serve to introduce the use of computers in the field of DNA sequencing and to describe a particular set of programs for handling and analysing sequences.

The value of the computer as a tool of the DNA sequencer lies in its capacity to store and analyse data. Given suitable programs sequences can be stored in computers. Once a sequence is stored in a computer it can, (again given suitable programs) easily be edited, analysed and printed out in many forms. The following is a list of some of the things that can be done using simple programs: (1) store and edit the sequence; (2) produce copies of a sequence; for example single- or double-stranded printouts; (3) translate the sequence in all phases; (4) search the sequence for particular sequences such as restriction enzyme recognition sites; (5) compare one sequence with another to look for common subsequences; (6) search for simple secondary structures such as hairpin loops. Programs to perform all of these tasks will be described in this chapter. Computers have been used in this field for a number of years now and several programs have been made available through the literature but in this article I shall describe only those programs developed in Fred Sanger's laboratory in Cambridge and which are designed more for sequencers than sequence analysts. I shall give references to alternative programs at the end and shall point out specific weaknesses of the programs described.

The first programs that were developed here were for use with data produced by the 'plus and minus' method of sequencing. They performed simple tasks such as editing and storage, translation and searches for restriction enzyme recognition sites. These programs are still in use and are very simple to learn to use requiring no understanding of data storage and very little user input (i.e. the user of the program is required to supply very few commands to the program). Typical programs from this era are SEQEDT, SEARCH, SEQLST and TRANMT (see below). Our current method of sequence determination is the shotgun technique which accumulates data much more rapidly. Sequence data from this method require more complicated manipulations than those from earlier techniques and the speed of accumulation makes it necessary to design data handling systems that can give full account of the accuracy of the final sequence. Some of the functions available in these programs (for example DBFIX in DBUTIL) require the user to have a reasonable understanding of the way in which the programs operate on the data and how the data is stored on the disk. For these reasons it will be found that a large part of the computer section is devoted to describing the programs that deal with the shotgun data. Although these programs have many more options than others their use can be learnt fairly easily and in any sequencing project they are the ones most heavily used. It is these programs that produce the sequences that are analysed by the other programs described.

The analysis programs are divided into a number of classes: 1. General programs; 2. Base and codon total programs; 3. Search and comparison programs; 4. Translation programs; 5. Secondary structure programs; 6. Gene searching programs; 7. Protein program. Brief descriptions of the function of each of these programs will be given in this chapter and more detailed descriptions of how to use the programs (that is the actual commands used) will be given in the appendix.

The programs we have developed have been designed from a particular standpoint and I shall outline that here. Apart from the obvious criteria of faithfully performing the required tasks the programs developed here have been designed to be an integral part of the sequencing process and must be constantly available to the people who actually determine the sequences. This requires that the programs must run on fairly inexpensive computers so that sufficient of them can be bought, and that the programs are easy to use.

Making the programs easy to use is achieved in a number of ways:

(1) They are interactive which means that the user and the com-

puter communicate via the keyboard and that the programs should prompt for (i.e. request) all user input.
(2) All user input is checked for obvious errors. On finding an error the programs should either give the parameter some sensible value or reprompt with an error message. For example, if the user tries to print out a sequence of length 1000 bases from base 1 to base 1500 the program should recognise that the second parameter should be set to 1000 and do so automatically.
(3) User input is both minimised and standardised—standardisation making learning to use the programs easier.
(4) Where possible default values are allowed. A default value is one that is assigned by the program if the user does not specify one. For example if the user does not specify which part of a sequence he wishes to have printed out the program may use the default values of 'beginning' to 'end' and hence print out the whole sequence.
(5) Commonly used data such as restriction enzyme recognition sites can be stored on a disk file and selected by name.
(6) Output from the programs should be as descriptive as possible and where appropriate in a form suitable for publication.

I have divided the chapter into several sections: an introduction to computers to define some of the terminology; a method of coding for sequence data as read from gels; a description of a data storage and retrieval system for shotgun sequencing and a description of some simple analysis programs.

Introduction to computing

This section is to introduce some of the words used in computing and the components they refer to. Although the computer can mostly be treated as a black box it is often helpful to be able to have a general idea about what is happening and to know what some of the terms mean.

Hardware

The computer and its peripherals are often called 'HARDWARE'. The computer can be considered as being made up of 4 units:

1. The CENTRAL PROCESSING UNIT (CPU) decodes and executes the instructions of a computer program. It has circuits that can perform arithmetic and logical operations, e.g., add two numbers or compare them for equality.

2. The central processing unit has a MEMORY in which programs and data are stored during execution. This memory is made up of a number storage units called 'words' each of which can be accessed directly by the central processing unit using the 'address' of the word. Word sizes are measured in 'bits' or 'bytes' where one byte = 8 bits. One bit (short for binary digit) represents a single switch in the computers memory and can only take the values 0 or 1 i.e. 'on' or 'off'. For example the word size of the PDP 11 which we use is 2 bytes and so each 'word' comprises 16 switches. Memory sizes are measured in words or bytes and are typically of the order of many thousands of bytes. For example our PDP 11 has a memory of maximum size 64k bytes (note that in computer jargon k = 1024, which is 2 to the power 10). Each word can in general contain one instruction or data item.

3. The DISK is a magnetic device that is used to store data and programs in files. Files are organised collections of data such as programs or sequences and each is given a name (a file name). For example, the sequence of phix174 DNA is stored as a file in our computer with file name QXCS70. Both data and programs are stored on the disk under file names but are copied into the central processing unit's memory for processing. For example if we wished to analyse phix174 using a program we would have to tell the computer which program we wished to use by supplying its name and then tell the computer the name of the file containing the phix174 sequence (QXCS70). Disk capacities are much greater than that of memory and are often measured in megabytes.

4. The KEYBOARD is used for communication between the user of the computer and the program.

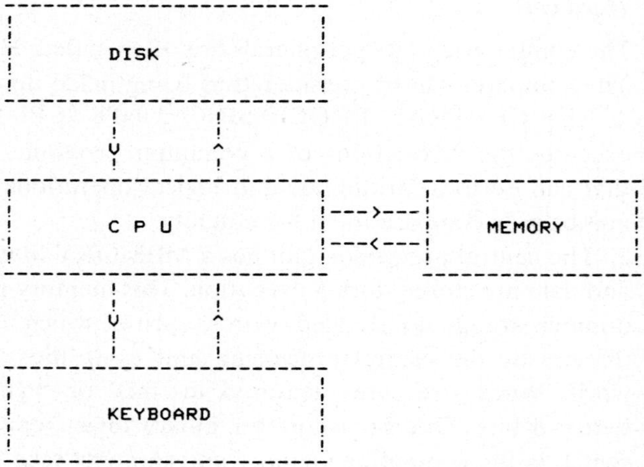

Programs

PROGRAMS are sets of instructions which can be used to solve particular problems. In solving these problems the programs perform operations on DATA. Programs are often referred to as SOFTWARE.

The operating system

The operating system is a program that supervises the operation of other programs and controls their input and output. Each type of computer has its own specific operating system that will generally be supplied with the machine when it is purchased. The operating system knows how the data and programs are stored on the disk and so acts as an interface between user-written programs (e.g. the sequence programs) and the hardware. It interprets messages from the keyboard and transfers data and programs between the central processing unit and the disk. When a user types a command on the keyboard to run a program it is the operating system that interprets that command and searches the disk for a program of the given name.

Users and strings

Throughout the computing sections the word USER refers to the person running the particular program being described. The word STRING means a string of characters. This string of characters may be a short sequence of the letters A, C, G, T that are used to define a DNA sequence (e.g. a restriction enzyme recognition sequence) or may be a sequence of characters that are commands to a program.

Quality of sequence data

One of the important requirements of any computer data storage system is to be able to show the accuracy of the data that it contains. The quality of a sequence can be assessed at two levels: firstly at the level of the individual nucleotide assignments as read from each gel; and secondly from the number of times each character in the final sequence has been given a particular assignment on different gels and whether or not it has been sequenced on both strands of the DNA. In order to record uncertainties when reading gels the codes shown below can be used. Use of these codes permits us to extract the maximum amount of data from each gel and yet record any doubts by choice of code. The storage and retrieval programs can deal with all of these codes and any other characters in a sequence are treated as dash (–) characters.

```
SYMBOLS FOR UNCERTAINTY IN GEL READINGS
       ---------------------------------------
   SYMBOL                 MEANING
   ------                 -------
     1           PROBABLY       C
     2              ''          T
     3              ''          A
     4              ''          G
     D              ''          C       POSSIBLY        CC
     U              ''          T          ''           TT
     B              ''          A          ''           AA
     H              ''          G          ''           GG
     K              ''          C          ''           C-
     L              ''          T          ''           T-
     M              ''          A          ''           A-
     N              ''          G          ''           G-
     R           A OR G
     Y           C OR T
     5           A OR C
     6           G OR T
     7           A OR T
     8           G OR C
     -           A OR G OR C OR T
     else = -
```

The storage and retrieval system described here records every gel reading that contributes to a sequencing project and can display them aligned correctly one above the other indicating their relative strandedness so that it is possible to assess the quality of the data at both levels.

A computer storage and retrieval method for DNA gel reading data:

The DB system

Our method of handling DNA sequence data is called 'the DB system' and is a system of files and computer programs [6]. The DB system has been designed to fulfil all of the storage and manipulation requirements for gel reading data produced by the shotgun method of DNA sequencing and is sufficiently general to cope with the data produced by any current sequencing strategy. I will describe its use for the shotgun method and so start with a brief outline of the shotgun strategy from the data handling point of view. This will lead into an outline of the requirements of such a system which should make an explanation of the actual system easier.

DNA fragments are selected randomly from a large pool and their sequences determined. The relationship between any pair of fragments is not known but is found by comparing their sequences. If the sequence of one is found to be wholly or partially contained within that of another for sufficient length to distinguish an overlap from a repeat then those two fragments can be joined. The process of select, sequence and compare is continued until the whole of the DNA to be sequenced is in one continuous well-determined piece. In order to make description easier we define the word CONTIG:

Definition of a CONTIG

A CONTIG is a set of gel readings that are related to one another by overlap of their sequences. All gel readings belong to one and only one contig and each contig contains at least one gel reading. The gel readings in a contig can be summed to produce a continuous consensus sequence and the length of this sequence is the length of the contig. The rules used to perform this summation are given under 'the consensus algorithm' and at any points in the sequence where there is not sufficient agreement between overlapping gel readings the algorithm places dash (–) characters. At any stage of a sequencing project the data will comprise a number of contigs each of which contains a set of gel readings. When a project is completed there will be only one contig and its consensus will be the finished sequence.

Requirements of the DB system

(1) Storage of original sequence for each fragment.
(2) Method to assemble and calculate a consensus for each contig either for comparison with new data or to produce the final sequence.
(3) Method to compare new gel readings with the consensus for all of the current contigs (The fragments may be cloned in either orientation and so the program compares not only the sequence of the gel but also its complementary sequence. We often refer to this as comparing both SENSES of the sequence).
(4) Method to add new gel readings to contigs (addition may add new data to left or right ends of contigs or simply confirm existing data by overlapping internally).
(5) Joining of two existing contigs (a new gel may overlap with two previously unrelated contigs which then need to be fused to form one contig).
(6) Complementing of any contig. (In order that two sections of data can be joined one of them may require to have its sense reversed so that both sections are in the same sense).
(7) Editing of individual gel readings.
(8) Editing of contigs.
(9) Examination by display of all gel readings covering any particular section of the sequence.

For computational reasons it is not practicable to store the gel readings as they would appear when aligned but it is better to store the data unassembled and to also record sufficient information for programs to align the gel readings during processing. The information used to assemble the gel readings is called relational information and it is constantly varying during a sequencing project.

Manipulations of types 4–8 described under *Requirements* will change the relationships of gel readings and manipulations 6–8 will change the actual sequences. In order to perform these manipulations the following types of information are required: (1) sequences from gel readings; (2) facts about individual gel readings and their relationship (if any) to others; (3) facts about individual contigs; (4) general facts about the number of gels and numbers of contigs. These then are the requirements of the system: we need some way of storing contigs, of assembling them during processing so that they can be examined, edited, added to, joined and formed into a consensus. Description of the DB system will be in two parts: first the data storage and then the programs.

Data storage

There are 6 types of file in the system:

1. The original sequence of each gel reading which we call the archive file because it is kept and never altered.
2. Temporary files of gel reading file names (see BATIN and DBCOMP).
3. Consensus sequences. These are calculated by program and their structure is described under the consensus calculation. The remaining files are part of the DATABASE in which the contigs are stored.
4. A working version of each gel reading. This is the version of the gel that is in the database and originally is an exact copy of the archive but it is edited and manipulated by program to align it with other gel readings.
5. The file of relationships. This file contains all of the information

that is required to assemble the working versions into contigs during processing; any manipulations on the data use this file and it is automatically updated at any time that the relationships are changed. The information in this file is as follows:

(1) Facts about each gel and its relationship to others (GEL DESCRIPTOR LINES):

(a) the number of the gel (each gel is given a number as it is entered into the database)

(b) the length of the sequence from this gel

(c) the position of the left end of this gel relative to the left end of the contig of which it is a member

(d) the number of the next gel to the left of this gel

(e) the number of the next gel to the right

(f) the relative strandedness of this gel, i.e. whether it is in the same sense or the complementary sense as its archive.

(2) Facts about each contig (CONTIG DESCRIPTOR LINES):

(a) the length of this contig

(b) the number of the leftmost gel of this contig

(c) the number of the rightmost gel of this contig.

(3) General facts:

(a) the number of gels in the database

(b) the number of contigs in the database.

6. The file of archive names. This is simply a list of the names of each of the archive files in the database.

Structure of the consensus sequence

The consensus sequence is calculated by program and is of a form suitable for input to any of the analysis programs described elsewhere in this article. This file comprises the consensuses of each of the contigs in the database and each contig consensus is separated from the one before by a title of 20 characters of the form ⟨-----LAMBDA.076---⟩ (where LAMBDA is the project name and gel 76 is the leftmost gel to contribute to this consensus sequence). The angle brackets ⟨ ⟩ are important as they are used by program DBCOMP to identify the ends of contigs.

Structure of the database files

1. The file of relationships

In order to use storage space efficiently this file has the following structure. The file is divided into 1000 lines of data; the general facts are stored on line 1000; facts about gels are stored from line 1 downwards; facts about contigs are stored from line 999 upwards.

As each new gel is added into the database a new line is added to the end of the list of gel lines. If this new gel does not overlap with any gels already in the database a new contig line is added to the top of the list of contig lines; if it overlaps with one contig then no new contig line need be added but if it overlaps with two contigs then these two contigs must be joined and the number of contig lines will be reduced by one and the list of contig lines compressed to leave the empty line at the top of the list. Initially the two types of line will move towards one another but eventually as contigs are joined the contig descriptor lines will move in the same direction as the gel lines. At the end of a project there will be only one contig line. The database is thus capable of handling a project of 998 gels.

Structure of the working versions file

The working versions of gel readings are stored in a file of 1000 lines each containing 512 characters. Gel 1 is stored on line 1, gel 2 on line 2 and so on.

Structure of the archive names file

This file has 1000 lines each 10 characters in length.

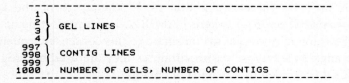

```
-------------------------------------------------
   1 )
   2 |
     }  GEL LINES
   3 |
   4 )
 997 )
 998 }  CONTIG LINES
 999 )
1000    NUMBER OF GELS, NUMBER OF CONTIGS
-------------------------------------------------
```

Programs in the DB system

1. BATIN
Short for 'BATCH INPUT'. Used for initial storage of gel readings on disk and for setting up a file of file names.

2. DBCOMP
Short for 'DATABASE COMPARISON'. Used to compare a batch of new gel readings with a consensus of the database and, if required, the cloning vector or some other sequence.

3. DBUTIL
Short for 'DATABASE UTILITY'. Performs all of the manipulations on data in the database, including editing, joining and calculating a consensus.

4. SCREENB
Short for 'SCREEN ON BASES'. One of the methods we use for determining whether or not a particular clone contains an unsequenced region is to perform a dideoxy reaction for only one of the nucleotides, and run out the fragments produced on a sequencing gel. This program is used to compare the resulting gel pattern with the consensus sequence.

BATIN

This program is used to store new gel reading data on the disk. Each time the program is used any number of separate gel readings may be entered to be stored in individual disk files. The program also writes another file which simply contains the names of all the gel reading files it has written during each run. This file of file names is used in the next stage of the processing by program DBCOMP. DBCOMP can access all of the gels in a batch simply by being supplied with the file of file names.

The user types each gel reading in turn and after each the program prompts to see if he wishes to type in another. If he does he is asked to supply a file name for the next gel and then the data.

When he has typed in all the gels he is given the option of having them all printed out.

DBCOMP

This program is used to search for overlaps between sequences. It uses the file of file names set up by program BATIN and compares each of the new gels (and their complementary sequences) with the consensus sequence and then each of the new gels with each of the other new gels. If it finds an overlap it reports the section of the database involved, and displays the match. It will also compare each of the new gels with any other sequences—for example, the sequence of the cloning vector.

When the gel readings are read in the program calculates their most likely version—i.e. it replaces all of the uncertainty codes by their most likely nucleotide assignments, so as to avoid missing possible overlaps. If an overlap is found the display is headed by the name of the contig involved.

The user input to the program consists only of the names of two files and a minimum number for the length of overlap (MIN). The first of these files contains the names of the sequences to be compared wih the gels—frequently only the name of the file containing the consensus sequence of the data gained so far during a project. This file of file names is not written by any of the programs described here but by an editing program supplied with the computer (often called a system editor). The second file of file names is the one written by program BATIN and contains the names of all of the gel reading files to be compared with the sequences from the first file. The minimum number for the length of the overlap is a minimum number of consecutive matching characters; only matches containing at least this number of consecutive identical characters will be reported by the program. There is no limitation on the number of sequence files or gel reading files that can be compared for each run of the program.

Below is an example showing the output from the program in which 3 gel readings have been compared, first in both senses with

```
LA341.36
STRAND      1
    1 25812
CONTIG NUMBER =    251
 25792      25802      25812      25822      25832      25842
       ATCGAATAAA ACTAATGACT TTTCGCCAAC GACATCTACT AATCTTGTGA TAGTAAATAA
       ********** ********** ********** ********** ********** **********
       ATCGAATAAA ACTAATGACT TTTCGCCAAC GACATCTACT AATCTTGTGA TAGTAAATAA
       1         11         21         31         41         51
 25852      25862      25872      25882      25892      25902
       AACAATTGCA TGTCCAGAGC TCATTCGAAG CAGATATTTC TGGATATTGT CATAAAACAA
       ********** ********** ********** ********** ********** **********
       AACAATTGCA TGTCCAGAGC TCATTCGAAG CAGATATTTC TGGATATTGT CATAAAACAA
      61         71         81         91        101        111
 25912      25922      25932      25942      25952      25962
       TTTAGTGAAT TTATCATCGT CCACTTGAAT CTGTGGTTCA TTACGTCTTA ACTCTTCATA
       ********** ********** ********** ********** ********** **********
       TTTAGTGAAT TTATCATCGT CCACTTGAAT CTGTGGTTCA TTACGTCTTA ACTCTTCATA
     121        131        141        151        161        171
 25972      25982      25992      26002      26012      26022
       TTTAGAAATG AGGCTGATGA GTTCCATATT TG
       ********** ********** ********** **
       TTTAGAAATG AGGCTGATGA GTTCCATATT TG
     181        191        201        211        221        231
STRAND      2
LA341.50
STRAND      1
    9 37207
CONTIG NUMBER =    251
 37179      37189      37199      37209      37219      37229
       CTCAATTGTA TCAGCTATGC GCCGACCAGA ACACCTTGCC GATCAGCCAA ACGTCTCTTC
          * *  ** ********** **** ****   ********* ********** **********
       TCAATTGTTA TCAGCTATGC GCCG3CCAG3 3CACCTTGCC GATCAGCCAA ACGTCTCTTC
       1         11         21         31         41         51
 37239      37249      37259      37269      37279      37289
       AGGCCACTGA CTAGCGATAA CTTTCCCCAC AACGGAACAA CTCTCATTGC ATGGGATCAT
       ********** ********** ********** ********** ********** **********

       AGGCCACTGA CTAGCGATAA CTTTCCCCAC AACGGAACAA CTCTCATTGC ATGGGATCAT
      61         71         81         91        101        111
 37299      37309      37319      37329      37339      37349
       TGGGTACTGT GGGTTTAGTG GTTGTAAAAA CACCTGACCG CTATCCCTGA TCAGTTTCTT
       ********** ********** ********** ********** ********** **********
       TGGGTACTGT GGGTTTAGTG GTTGTAAAAA CACCTGACCG CTATCCCTGA TCAGTTTCTT
     121        131        141        151        161        171
 37359      37369      37379      37389      37399      37409
       GAAGGTAAA
       *********
       GAAGGTAAA
     181        191        201        211        221        231
STRAND      2
LA301.10F
STRAND      1
STRAND      2
    1  2708
CONTIG NUMBER =    34
  2688       2698       2708       2718       2728       2738
       ACAGGTAAAC GGGTGGCCAA CAGTACAGAA AGACGGACGA AGGGTGGAGT TTACGGCCAC
       ********** ******  *          **     *    *  **  *   * *   * *
       ACAGGTAAAC GGGTGGDAAC AGTACAGAAA GACGGACGAA GGGTGGAGTT TACGGCCACT
       1         11         21         31         41         51
  2748       2758       2768       2778       2788       2798
       TTCCGTGTCT GACCTGAAAA AATATATTGC AGAGCTGGAA GTGCAGACCG GCATG
       * *          *   **** *     *          * *         * *
       TCCGTGTCTG ACCTGAAAAA ATATATTGCA GAGCTGGAAG TGCAGACCGG CATGA
      61         71         81         91        101        111
LARA3.F
STRAND      1
    1 12641
CONTIG NUMBER =    321
 12621      12631      12641      12651      12661      12671
       AGGCAATGGT GGGGATTGTC GGGAGTATCG GCAGCGCCAT TGGCGGGGCT GTTGGTGGCG
       ********** ********** ********** ********** ********** **********
       AGGCAATGGT GGGGATTGTC GGGAGTATCG GCAGCGCCAT TGGCGGGGCT GTTGGTGGCG
       1         11         21         31         41         51
 12681      12691      12701      12711      12721      12731
       GCGATCCGCG TCAGGCGGTA CAGCCATTCA GGCCGCTGCG GCG
       ********** ********** ********** ********** ***
```

```
     GCGATCCGCG TCAGGCGGTA CAGCCATTCA GGCCGCTGCG GCG
    61         71         81         91        101         111
STRAND      2
FIRST GEL  = LA341.36
 LA341.50
 NUMBER OF MATCHES =        0
 NUMBER OF MATCHES =        0
 LA301.10F
 NUMBER OF MATCHES =        0
 NUMBER OF MATCHES =        0
 LARA3.F
 NUMBER OF MATCHES =        0
 NUMBER OF MATCHES =        0
FIRST GEL  = LA341.50
 LA301.10F
 NUMBER OF MATCHES =        0
 NUMBER OF MATCHES =        0
 LARA3.F
 NUMBER OF MATCHES =        0
 NUMBER OF MATCHES =        0
FIRST GEL  = LA301.10F
 LARA3.F
 NUMBER OF MATCHES =        0
 NUMBER OF MATCHES =        0
```

a single consensus sequence, and then with one another, again in both senses. Any matches are displayed by placing the gel under the consensus and marking identical characters with asterisks. As can be seen the first gel matches perfectly but the others have insertions or deletions relative to the consensus. None of the gels matches with any of the others.

DBUTIL

This program performs all of the manipulations required on data gained by the shotgun method of DNA sequencing. It contains all of the necessary joining and editing functions and can calculate consensus sequences for contigs. All of the requirements outlined above (except 1 and 2 which are performed by BATIN and DBCOMP) are performed by this program. DBUTIL stores the data about the relationships between gel readings (i.e. which gels overlap with which) in the file of relationships and uses this information to assemble the working versions of the gel readings into the correct alignment during processing. Any changes in the relational information, such as change of neighbours or increase in the length of a gel reading, are achieved using DBUTIL which updates the database accordingly.

The functions offered by the program are:

1. PRINT will print the contents of the database in three ways—

(a) All contig descriptor lines followed by all gel descriptor lines.
(b) All contigs one after the other sorted, i.e. for each contig print, its contig descriptor line followed by all its gel descriptor lines sorted in order from left to right.
(c) Selected contigs: print the contig line and, in order, those gels that cover a user-defined region.

Below is an example showing the left of a contig from position 1 to 500. The total contig length is 48504, the left gel is 251 and the rightmost gel is 601. On each gel descriptor line is shown: the name of the archive version, the gel number, the position of the left end of the gel relative to the left end of the contig, the length of the gel (if this is negative it means that the gel is in the opposite orientation to its archive), the number of the gel to the left and the number of the gel to the right.

```
CONTIG LINES
               999     48504        0      251      601
GEL LINES
BLAT.001       251         1       70        0      772
LA313.38L      772        16      115      251      578
LAR14.REV      578        23      151      772      581
LA5S4          581        84     -257      578      561
LAHC17.S       561       208      256      581      565
LAHC17.L       565       356      254      561      410
LAY34          410       381     -234      565      418
LAYC13         418       385      227      410      434
LAYC36         434       389      225      418      209
```

2. DISPLAY will display on the printer or screen all of the gels covering any region of the sequence, lined up in register and showing their gel numbers, their strandedness and, underneath, their consensus.

Below is an example showing the left end of a contig from position 1 to 300. Overlapping this region are gels 251, 772, 578, 581 (in reverse orientation to its archive denoted by a minus sign) and gel 561. There are a few uncertainty codes and a few padding characters in the working versions, but the consensus (shown below each page width) has a definite assignment for every position.

```
               10        20        30        40        50        60
251    GGGCGGCGACCTCGCGGGTTTTCGCTATTTATGAAAATTTTCCGGTTTAAGGCGTTXCCG
772                    GGGTTTTCGCTATTTATGAAAATTTTCCGGTTTAAGGCGTTTCCG
578                          CGCTATTTATGAAAATTTTCC4GTTTAAGGCGTTTCCG
       GGGCGGCGACCTCGCGGGTTTTCGCTATTTATGAAAATTTTCCGGTTTAAGGCGTTTCCG

               70        80        90       100       110       120
251    TXCXTCTXCG
772    TTCTTCTTCGTCATAACTTAATGTTTTTATTTAAAATACCCTCTGAAAAGAAAGGAAACG
578    TTCTTCTTCGTCATAACTTAATGTTTTTATTTAAAATACCCT1TGAAAAGAAAGGAAACG
-581                          TTTTTATTTAAAATACCCTCTGAAAAGAAAGGAAACG
       TTCTTCTTCGTCATAACTTAATGTTTTTATTTAAAATACCCTCTGAAAAGAAAGGAAACG

              130       140       150       160       170       180
772    ACAGGTGCTG
578    ACAGGTGCTGAAAGCGAGGCTT2TTGGCCTCTGTCGTTTCCTTTCTCTGTTTT
-581   ACAGGTGCTGAAAGCGAGGCTTTTTGGCCTCTGTCGTTT1CTTTCTCTHTTTTTGTCCGT
       ACAGGTGCTGAAAGCGAGGCTTTTTGGCCTCTGTCGTTTCCTTTCTCTGTTTTTGTCCGT

              190       200       210       220       230       240
-581   GGAATGAACAATGGAAGTCAACAAAAAGCAGCTGGCTGADATT6TCGGTGCGAGTATCCG
561                               GCAGCTGGCTGACATTTTCGGTGCGAGTATCCG
       GGAATGAACAATGGAAGTCAACAAAAAGCAGCTGGCTGACATTTTCGGTGCGAGTATCCG

              250       260       270       280       290       300
-581   TACCATTCAGAACTGGCAGGAACAGGGAATGC1CGTTCTGCGAGGCGGTGGCAAGGGTAA
561    TACCATTCAGAACTGGCAGGAACAGGGAATGCCCGTTCTGCGAGGCGGTGGCAAGGGTAA
       TACCATTCAGAACTGGCAGGAACAGGGAATGCCCGTTCTGCGAGGCGGTGGCAAGGGTAA
```

3. EDIT will allow editing of any contig. It maintains alignments by making the same number of insertions or deletions in all of the gels covering the edit position.

4. COMPLEMENT will complement and reverse all of the gels in a contig, reorders left and right neighbours, recalculates relative positions and changes each strandedness.

5. JOIN will allow joining of any two contigs. The options available are:

(a) movement of join

(b) editing of either the left or the right contig

(c) display of the overlap between the two sections of sequence.

It is important that the gels are correctly aligned by using the edit functions on the separate contigs because once the join is completed the alignment is fixed.

6. ENTER allows entry of a new gel into the database. It is necessary to know beforehand if the new gel overlaps with any sequences already in the database (found by using DBCOMP). It automatically names and writes the working version of the new gel and then offers the following options:

(a) complementing and reversing of the new gel (to add it to existing data it may need its sense changed)

(b) adding to an existing contig either at a left or right end or internally

(c) movement of overlap

(d) display of overlap

(e) editing of the new gel. (It is only during entry into the database that gels can be edited individually. Once entry is completed all gels covering any position are edited together to maintain the alignments of the sequences).

(f) editing of the contig.

7. SEARCH allows a search to be made for a gel by its archive name so that its number and descriptor line can be examined.

8. FIX allows changes to individual data elements in the database and hence permits corrections to be made to mistakes. The options available are:

(a) changing of individual lines in the file of relationships

(b) movement of one part of a contig relative to another

(c) edit of a single gel independently of others

(d) deletion of a contig line

(e) renaming of a gel.

9. COPY copies the whole of the database, so giving a backup in case of problems.

10. CHECK performs a check on the logical consistency of the database.

11. SCAN gives a printout or summary of the quality of the data in a contig. The quality of the data depends on the number of times it has been sequenced and the particular uncertainty codes used in each gel reading. It divides the data into five categories and the quality codes are described below:

1. Well-determined on both strands and they agree. code = 0
2. Well-determined on the plus strand only. code = 1
3. Well-determined on the minus strand only. code = 2
4. Not well-determined on either strand. code = 3
5. Well-determined on both strands but they disagree. code = 4

The meaning of 'well-determined' is best explained by describing the algorithm that is used to examine the data for each strand.

Each of the possible characters in a sequence is given an individual value:

A, C, G, T, D, B, H, V, K, L, M, N = 1;
1, 2, 3, 4 = 0.75;
anything else = 0

For each position of each strand the program calculates five numbers: using the individual values for each code it calculates the sum for each of the 4 characters A, C, G, T (or their uncertainty codes) and the number of times the sequence has been determined for each position. The user of the program is asked to supply a percentage value before the scan is done and if the sum of any of the four bases at a position equals or exceeds this percentage then that position is called 'well-determined' for that strand. For example, a single code of 2 on one strand will be called 'well-determined' if the percentage is < or =75, but not if the percentage is >75. Or two A's aligned with one G will be called 'well-determined' if the percentage is <66.6.

The options available are:

(a) summary only for all or part of a contig
(b) printout of quality codes
(c) storage of quality codes on disk for processing by other programs.

12. CONSENSUS

This program calculates a consensus sequence for all or selected regions of the database using the rules outlined under 'consensus algorithm'. It writes this sequence to a disk file in a form suitable for input to other programs (e.g. DBCOMP). The user can:

1. calculate a consensus for the whole database and store it in one file;
2. calculate a consensus for selected parts of selected contigs and store it in one file;
3. calculate a consensus for selected parts of selected contigs and store them in separate files.

The consensus algorithm

Both the consensus function and the display routine use the rules outlined here to calculate a consensus from aligned gel readings. DISPLAY calculates a consensus for each page width it prints (it does not use the consensus sequence file calculated by the consensus function). It is worth noting that these rules differ from those used by the scan function: only defined uncertainty codes are counted by the consensus calculation whereas scan counts all characters. By ignoring undefined characters the consensus calculation is more likely to produce a definite assignment in the consensus, but scan, by counting all characters, can indicate places where padding has been used and hence show possible problems in the sequence.

Individual characters are assigned a value for the consensus calculation:

definite assignments, i.e. A, C, G, T, B, D, H, V, K, L, M, N = 1

probable assignments, i.e. 1, 2, 3, 4 = 0.75

any other character = 0

For each position in a contig we calculate 4 base totals—one each for A, C, G, T. These base totals are calculated by adding up all of the individual values for each of the characters that contribute to each base total at each position. When all of the gels covering a position have been added to the four base totals the programs calculate the sum of these four base totals for each position. The programs then look at each position to see if any of the four base totals is ⩾75% of the sum for that position. If a base total is sufficiently high its corresponding character is put into the consensus, otherwise a dash (–) is assigned.

SCREENB

A program to compare single A, C, G or T tracks with a consensus sequence. The user can compare any number of these single tracks with the consensus sequence by using a number of codes to define the gaps between the bands. The description below is for T tracks but any of the other 3 bases could be used. The T tracks are coded in the following way:

T = single T
X = single gap, i.e. single not T
Z = gap of any length
4 = gap of 4+ or −1
5 = gap of 5+ or −1
6 = gap of 6+ or −1
7 = gap of 7+ or −1
Y = gap of at least 7
all strings are finished by @ character.

For each match found the program displays the title of the contig involved, the match and the match position. Example of how to code T tracks:

TT5TXXTTTYTZTT@

This means search for 2 T's followed by a gap of 5+ or −1, 1 T, gap of 2, 3 T's, gap of at least 7, 1 T, gap of any length, and finally 2 T's. The program also searches for the complement of this string.

Safeguarding the system

It is advisible to copy regularly (using the copy function of DBUTIL) from say version 0 to version 1 in case of errors.

The give-up options in DBUTIL allow the user to change his mind about entering a new gel reading or joining two contigs without affecting the file of relationships. BUT if he has used the edit contig option from either of these two functions the edits will remain even though he has 'given up'. To leave the files completely unaffected the user could, if required, undo any edits before 'giving up'. There are various checks within the programs to protect the user from himself:

1. All user input is checked for errors—e.g. reference to non-existent gels or contigs, incorrect positions in the contig or gel. If an error is detected the programs usually reprompt for the input or ask the user if he wants to try again.

2. Before entering a gel reading the system checks to see if a gel of the same name has already been entered.
3. Join will not allow the circularising of a contig.
4. Both enter and join functions restrict the region that the user is allowed to edit (using edit contig) to the region of overlap.

Analysis programs

General
- SEQEDT
- SEQLST
- SQRVCM

Base and codon totals
- BASSUM
- CODSUM

Searching
- SEARCH
- SRCHMU
- CUTSIT

Homologies
- SEQFIT
- OVRLAP
- DIAGON

Translation
- TRANMT
- TRNTRP
- TRANDK

Secondary structure
- HAIRPN
- HAIRGU

Gene searching
- TRNA

Protein analysis
- MWCACL

General programs

SEQEDT

Introduction

A program for the storage and editing of sequence data. It can either be used to start a completely new file and store it on disk or to edit a file that is already held on the disk. The program reads the old file, if it exists, writes out the new file to disk and prints a copy of it on the keyboard. A new file is created on each run of the program which means that the old file is not lost and so can act as a backup. Positions in the file are defined by character numbers. Three edit commands are sufficient to perform any kind of change to the data. These are FIND, INSERT and DELETE, and they are described in more detail below. The program always starts off with the input array filled with dash characters which allows data to be inserted at any position leaving dashes at unedited points. If an input file of n characters is read in from disk it will overwrite the first n dashes but the rest will remain for expansion of the data if necessary. All edits are input from the keyboard together and must be in sequential order from beginning to end of the file.

General description of the editing process

The basic editing process can be divided into four sequential steps:
(1) Reading of input data into an array used by SEQEDT. (If there is no input file this array is completely filled with dash characters and editing can be thought of as being performed on an array of dashes.)
(2) Changing the data stored in the array on commands supplied by the user.
(3) Outputting the new or revised data to a new file (the old one remaining unaltered).
(4) Printing a copy of the new data on the keyboard.
There are three commands that function with respect to a movable reference point P. This pointer, P, is located between characters N − 1 and N where N is the number of the last character located using the FIND command. All numbers refer to the input file.

Commands

1. FIND F

The FIND command F/N/ positions the pointer P between characters N − 1 and N where N is some number, e.g. F/11/ puts P between characters 10 and 11 in the input file.

2. INSERT I

The INSERT command I/abc/ (where abc represent three characters to insert) adds the characters abc at position P, i.e.

between characters N − 1 and N. This does not alter the position of P relative to the input file. For example, suppose we have the character string lmnopqrst in the input file with pointer P positioned between q and r then the command I/abc/ will produce lmnopqabcrst in the output file. Any subsequent insertions will come after the position of character c.

3. DELETE D

The DELETE command D/M/ (where M is the number of characters to delete) skips the copying of characters N to N + M − 1 from the input file into the output file and so effectively erases them. This moves P on by M places to position N + M. For example, given the character string lmnopqrst in the input file with pointer P at position 3, the single command D/4/ results in lmrst in the output file. Changes are performed by using a combination of INSERT and DELETE commands. For example, input lmnopqrstu with pointer between q and r command D/3/I/abc will produce lmnopqabcu, i.e. rst has been replaced by abc.

SEQLST

A program to produce printed copies of sequence files. It can be used for both DNA and amino acid sequences although the double-stranded listing option (see below) is, of course, only applicable to the former. Although data is stored in a linear fashion in the computer with characters in sequential order the program can print the data as though it was stored in a circular manner, i.e. it can be used to list circular DNA sequences.

The user supplies the name of the data file and defines the region to be listed by character number. He is asked to select printing in either single- or double-stranded form. If he selects double-stranded printing the program creates the complementary strand of the input sequence. He is also asked to select either a 60 or 120 character line output. The program allows (for the 120 character line output) the user to define the number of blank lines he wishes to appear between the sequence lines.

Below is an example of the 60 character line output showing the left end of a consensus sequence with its title.

```
         10         20         30         40         50         60
<---LAMBDA .251-----> GGGCGGCGAC CTCGCGGGTT TTCGCTATTT ATGAAAATTT

         70         80         90        100        110        120
TCCGGTTTAA GGCGTTTCCG TTCTTCTTCG TCATAACTTA ATGTTTTTAT TTAAAATACC

        130        140        150        160        170        180
CTCTGAAAAG AAAGGAAACG ACAGGTGCTG AAAGCGAGGC TTTTTGGCCT CTGTCGTTTC

        190        200        210        220        230        240
CTTTCTCTGT TTTTGTCCGT GGAATGAACA ATGGAAGTCA ACAAAAAGCA GCTGGCTGAC

        250        260        270        280        290        300
ATTTTCGGTG CGAGTATCCG TACCATTCAG AACTGGCAGG AACAGGGAAT GCCCGTTCTG

        310        320        330        340        350        360
CGAGGCGGTG GCAAGGGTAA TGAGGTGCTT TATGACTCTG CCGCCGTCAT AAAATGGTAT
```

SQRVCM

This program reverses and complements a sequence file (i.e. it calculates the plus strand from the minus strand or vice versa). It reads in a sequence file from disk, calculates the complement of each character, reverses their order and writes out the resulting sequence to a new disk file.

Base and codon totals

BASSUM AND CODSUM

Programs to count base and codon totals over regions of a sequence. The codon totals are presented in the form of the genetic code box as is shown below.

```
CODON TOTALS OVER ALL GENES
 ===================================
 TTT  9.0 TCT  5.0 TAT  3.0 TGT  3.0
 TTC 12.0 TCC  4.0 TAC  8.0 TGC  4.0
 TTA 13.0 TCA  9.0 TAA  7.0 TGA  6.0
 TTG 11.0 TCG  8.0 TAG  0.0 TGG 11.0
 ===================================
 CTT  7.0 CCT  4.0 CAT  4.0 CGT  2.0
 CTC  4.0 CCC  1.0 CAC  0.0 CGC  6.0
 CTA  6.0 CCA  1.0 CAA  3.0 CGA  1.0
 CTG 11.0 CCG  7.0 CAG  2.0 CGG  5.0
 ===================================
 ATT  8.0 ACT  3.0 AAT  9.0 AGT  1.0
 ATC  2.0 ACC  3.0 AAC  2.0 AGC  5.0
 ATA  2.0 ACA  5.0 AAA 11.0 AGA  5.0
 ATG  6.0 ACG  7.0 AAG 10.0 AGG  9.0
 ===================================
 GTT  6.0 GCT  5.0 GAT  4.0 GGT  2.0
 GTC  8.0 GCC  2.0 GAC  6.0 GGC  1.0
 GTA  3.0 GCA  3.0 GAA  4.0 GGA  3.0
 GTG  4.0 GCG  9.0 GAG  7.0 GGG  1.0
 ===================================

TOTAL CODON COUNT =       333.
```

Search programs

These are used to search for short sequences such as restriction enzyme recognition sites. The main difference between them is in the form of their output.

SEARCH, SRCHMU

Programs to search for strings in a sequence file. Strings may be of any length and strings of different lengths may be searched for simultaneously. The output includes: a title, the found string, a section of the sequence around the matching site, the distance from the last match position, the position of the match, and is concluded by the number of matches.

Two types of search are possible: the user can either search for an individual string and output any matches found, or can search for several strings simultaneously before outputting the matches. The advantage of the latter is that the relative positions of the matches for the several strings are then shown. The program also permits a search for strings, of which not all the characters are known, by coding dashes in place of the unknown characters. Searches are performed over the whole of the sequence or restricted to any given region.

The program is often used in conjunction with an extra file containing strings and names of sets of strings. In the example given 'RENZYM.DAT' is the name of a file containing abbreviated names for restriction enzymes together with their respective cutting sites. This allows a search to be made for these cutting sites by the user just typing the abbreviation for the name of the restriction enzyme. Alternatively, by selecting option 'A' a search is made, in turn, for all of the strings in the file.

The usual way of supplying strings to the program is by typing them in from the keyboard. The format for typing in sets of strings or names is shown in examples. These three methods of string input are selected by the user whilst running the program. On output the first string of any set of typed-in strings will appear in the title. If the names file is used the name will appear in the title.

Below is an example of a run of search which shows the use of the strings option and then the names option. These are the actual commands and responses of a run of the program. The typing done by the user is double underlined (=) to make it easier to follow, but see the appendix for an explanation of the commands. This is typical of the interactive programs described in this chapter and demonstrates how easy it is to run them.

```
RUN SEARCH
==========
          PLEASE TYPE NAME OF FILE 1
QXCS70.SEQ
==========
          SELECT OPTION,TYPE A FOR ALL,N FOR NAMES,S FOR STRINGS
S
=
          IF REQUIRED,CHANGE SEARCH AREA
$ FIRST SEQ NO =
                  =
$ LAST SEQ NO =
               =
          TYPE STRINGS NOW
ATATA/GTGTG/GGCCT//TTTTTT/AAAAAA/CCC-CC//@
==========================================

           SEARCH FOR ATATA
 STRING          POSITION                                                DISTANCE
GGCCT                434      GTTTCCAGACCGCTTTGGCCTCTATTAAGCTCATTCAGGC      872
                                              -----
GGCCT                668      CCGTCAACATTCAAACGGCCTGTCTCATCATGGAAGGCGC      234
                                              -----
GGCCT               1172      TTCTCCATTGCGTCGTGGCCTTGCTATTGACTCTACTGTA      504
                                              -----
GTGTG               2296      CGTGATAAAAGATTGAGTGTGAGGTTATAACGCCGAAGCG     1124
                                              -----
GTGTG               2771      TATACCGTCAAGGACTGTGTGACTATTGACGTCCTTCCCC      475
                                              -----
GTGTG               3808      ACTGATGCTGCTTCTGGTGTGGTTGATATTTTTCATGGTA     1037
                                              -----
GGCCT               4487      GTTCAAGATTGCTGGAGGCCTCCACTATGAAATCGCGTAG      679
                                              -----
GTGTG               4732      TGCTATCAGTATTTTTGTGTGCCTGAGTATGGTACAGCTA      245
                                              -----
GGCCT               4876      AGCTTGCAAAATACGTGGCCTTATGGTTACAGTATGCCCA      144
                                              -----
GGCCT               4948      CGTTCTGGTTGGTTGTGGCCTGTTGATGCTAAAGGTGAGC       72
                                              -----

                      TOTAL OF MATCHES =    10
           SEARCH FOR TTTTTT
 STRING          POSITION                                                DISTANCE
AAAAAA               860      TGTAATGTCTAAAGGTAAAAAACGTTCTGGCGCTCGCCCT     2271
                                              ------
TTTTTT              2431      TCGCCATAATTCAAACTTTTTTCTGATAAGCTGGTTCTC      1571
                                              ------
TTTTTT              2432      CGCCATAATTCAAACTTTTTTCTGATAAGCTGGTTCTCA         1
                                              ------
TTTTTT              2653      TGATGCCGACCCTAAATTTTTTGCCTGTTTGGTTCGCTTT      221
                                              ------
CCCTCC              2699      TCTTCGGTTCCGACTACCCTCCCGACTGCCTATGATGTTT       46
                                              ------
CCCTCC              3933      ACCGTCAGGATTGACACCCTCCCAATTGTATGTTTTCATG     1234
                                              ------
TTTTTT              3975      TCCAAATCTTGGAGGCTTTTTTATGGTTCGTTCTTATTAC       42
```

```
                                                                    ------
                  TOTAL OF MATCHES =    7
N        SELECT OPTION,TYPE A FOR ALL,N FOR NAMES,S FOR STRINGS
=
         IF REQUIRED,CHANGE SEARCH AREA
$ FIRST SEQ NO =
                =
$ LAST SEQ NO =
               =
RENZYM.DAT PLEASE TYPE NAME OF FILE 2
==========
         TYPE R.ENZYME NAMES NOW
BGL1/HAE11//@
=============


         SEARCH FOR BGL1
 STRING        POSITION                                                      DISTANCE
GCCATAAGGC         4207     GATATGTATGTTGACGGCCATAAGGCTGCTTCTGACGTTC          5386
                                           ----------
                  TOTAL OF MATCHES =    1
         SEARCH FOR HAE11
 STRING        POSITION                                                      DISTANCE
GGCGCT              687     CTGTCTCATCATGGAAGGCGCTGAATTTACGGAAAACATT          2314
                                            ------
GGCGCT              872     AGGTAAAAAACGTTCTGGCGCTCGCCCTGGTCGTCCGCAG           185
                                            ------
GGCGCT              926     TAAAGGCAAGCGTAAAGGCGCTCGTCTTTGGTATGTAGGT            54
                                            ------
GGCGCC             1019     GTCTAATATTCAAACTGGCGCCGAGCGTATGCCGCATGAC            93
                                            ------
GGCGCT             1142     CGAGATGGACGCCGTTGGCGCTCTCCGTCTTTCTCCATTG           123
                                            ------
AGCGCC             1411     ATAACAACTATTTTAAAGCGCCGTGGATGCCTGACCGTAC           269
                                            ------
GGCGCC             2976     TTCTGCTCTTGCTGGTGGCGCCATGTCTAAATTGTTTGGA          1565
                                            ------
GGCGCT             3759     TGGCTCTTCTCATATTGGCGCTACTGCAAAGGATATTTCT           783
                                            ------
                  TOTAL OF MATCHES =    8
```

SRCHMU

This is an alternative to SEARCH that also gives output in map units and sorts fragments on length. Below is an example of the output.

```
         SEARCH FOR CCGGT
STRING           POSITION                       LENGTH                 ASCENDING LENGTH
              BASES  MAP UNITS            BASES  MAP UNITS

CCGGT           729  0.135                 3096  0.575              37  0.007
GTGTA           766  0.142                   37  0.007             337  0.063
CCGGT          1103  0.205                  337  0.063            1916  0.356
CCGGT          3019  0.560                 1916  0.356            3096  0.575

               TOTAL MATCHES =    4
```

CUTSIT

This program performs the 'all' search of SEARCH and SRCHMU and so produces a restriction map. Below is an example of the output which shows the position of each site in both base numbers and map units, the name of the enzyme and the recognition sequence.

19	0.003	HGA1	GACGC
28	0.005	HPA1	GTTAAC
28	0.005	HINC11	GTTAAC
53	0.010	HINF1	GAGTC
56	0.010	TAQ1	TCGA
110	0.020	TAQ1	TCGA
143	0.026	TAQ1	TCGA
155	0.029	MST1	TGCGCA
156	0.029	HHA1	GCGC
158	0.029	BBV1	GCAGC
160	0.030	ALU1	AGCT
162	0.030	AVA1	CTCGAG
162	0.030	XHO1	CTCGAG
163	0.030	TAQ1	TCGA
169	0.031	ALU1	AGCT
204	0.038	HINF1	GATTC
220	0.041	HGA1	GACGC
222	0.041	THA1	CGCG
232	0.043	MNL1	GAGG
286	0.053	HINF1	GAGTC
310	0.057	HINF1	GATTC
319	0.059	HINC11	GTTGAC
350	0.065	DDE1	CTGAG
352	0.065	HINF1	GAGTC
358	0.066	SFNA1	GATGC
392	0.073	HINF1	GAGTC
414	0.077	RSA1	GTAC
424	0.079	ECOP1	AGACC
433	0.080	HAE1	TGGCCT
434	0.080	HAE111	GGCC
436	0.081	MNL1	CCTC
445	0.082	ALU1	AGCT
480	0.089	MBO11	GAAGA
490	0.091	TAQ1	TCGA

Homology programs

SEQFIT

Program to look for similarities in sequences. A string of up to 200 characters is placed in every position alongside a region of up to 18,000 characters. In each position the total number of identical adjacent characters is counted. If this number is sufficiently high it is saved along with its position. When every position has been tried, the program prints out the total number of scoring positions and then sorts them into descending order on score. The program requests the user to supply the number of top scores he wants to see printed. These are written on the keyboard in a form similar to DBCOMP (see above), starting with the highest and working down.

The user is then asked to select, in turn, from four options to be used for the next search. The program will search for the complement of the current string, allow the input of a new string, allow the percentage of fit to be changed, or allow the search region to be changed. Once the options have been selected the search will start as before.

The percentage is defined as:—

$$\frac{\text{number of identical adjacent characters}}{\text{string length in characters}} \times 100$$

OVRLAP

This program is used for comparing sequences to look for regions of perfect homology; it can be used to find internal repeats or for finding identities between two sequences. A minimum length of overlap is specified by the user and any overlaps of at least this length are reported. The program also automatically calculates the complement of the second sequence and compares that as well and hence reports possible base-pairing regions.

To run the program the user is asked to supply the names of the two files containing the sequences to compare and the minimum overlap.

DIAGON

This program is used to find regions of homology. It will find internal repeats or common sequences between two sequences by displaying the two sequences in the form of a matrix and marking any regions of identity. One sequence is written along the x axis and the other along the y axis; any regions of identity will lie on diagonals and these are shown by the program. In the example below only diagonals of a least 3 consecutive bases are shown. The advantage of this method is that no homologies will be missed: both perfect matches and matches including insertions and deletions will be found.

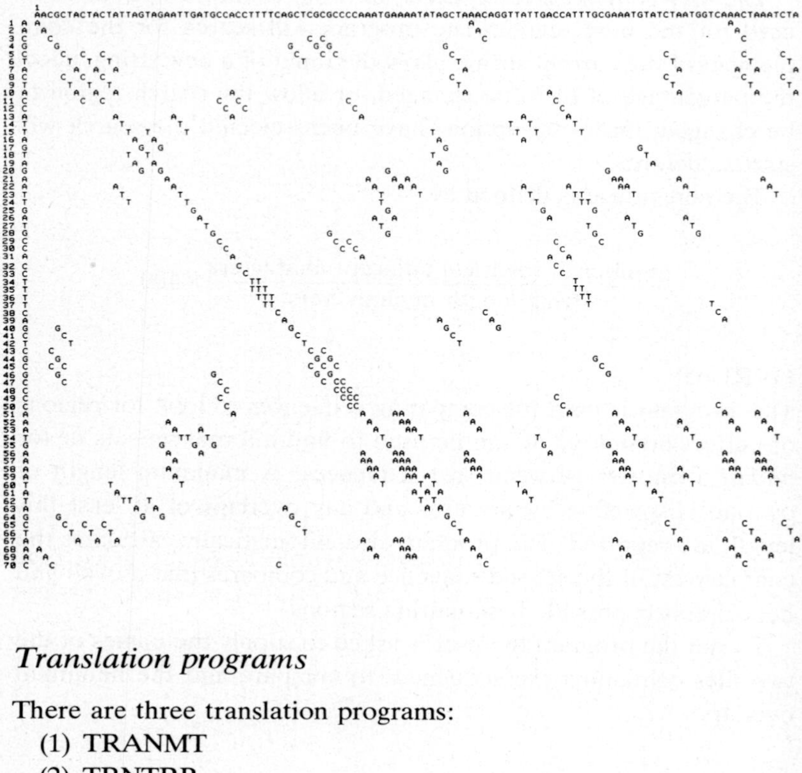

Translation programs

There are three translation programs:

(1) TRANMT
(2) TRNTRP
(3) TRANDK

TRANMT and TRNTRP write the translation and the DNA sequence on the printer in one and three letter amino acid codes. TRANDK writes the translation in one letter amino acid codes to a disk file for further processing. All three programs will treat sequences as circular so that any region may be translated. Genes may be overlapping. The programs offer the user a choice of genetic codes: standard, human mitochondrial (i.e. ATA = m, TGA = w, AGA, AGA = stop) or personal. If the user selects 'personal' he is asked to define deviations from the standard code by supplying the program with two numbers, one to identify the codon, the other the amino acid.

Below is an example of the output from TRANMT.

```
                                                   M  S  R  K  I  I  L  I  K  Q  E  L  L  L  L  V  Y  E  L  N  R  S  G  L
GAGTTTTATCGCTTCCATGACGCAGAAGTTAACACTTTCGGATATTTCTGATGAGTCGAAAAATTATCTTGATAAAGCAGGAATTACTACTGCTTGTTTACGAATTAAATCGAAGTGGAC
         10        20        30        40        50        60        70        80        90       100       110       120

           M  R  K  F  D  L  S  L  R  S  S  R  S  S  Y  F  A  T  F  R  H  Q  L  T  I  L  S  K  T  D  A  L  D  E  E  K
   L  A  E  N  E  K  I  R  P  I  L  A  Q  L  E  K  L  L  L  C  D  L  S  P  S  T  N  D  S  V  K  N  *
TGCTGGCGGAAAATGAGAAAATTCGACCTATCCTTGCGCAGCTCGAGAAGCTCTTACTTTGCGACCTTTCGCCATCAACTAACGATTCTGTCAAAAACTGACGCGTTGGATGAGGAGAAG
        130       140       150       160       170       180       190       200       210       220       230       240

 W  L  N  M  L  G  T  F  V  K  D  W  F  R  Y  E  S  H  F  V  H  G  R  D  S  L  V  D  I  L  K  E  R  G  L  L  S  E  S  D
TGGCTTAATATGCTTGGCACGTTCGTCAAGGACTGGTTTAGATATGAGTCACATTTTGTTCATGGTAGAGATTCTCTTGTTGACATTTTAAAAGAGCGTGGATTACTATCTGAGTCCGAT
        250       260       270       280       290       300       310       320       330       340       350       360

 A  V  Q  P  L  I  G  K  K  S  *
                               M  S  Q  V  T  E  Q  S  V  R  F  Q  T  A  L  A  S  I  K  L  I  Q  A  S  A  V  L  D  L  T  E
GCTGTTCAACCACTAATAGGTAAGAAATCATGAGTCAAGTTACTGAACAATCCGTACGTTTCCAGACCGCTTTGGCCTCTATTAAGCTCATTCAGGCTTCTGCCGTTTTGGATTTAACCG
        370       380       390       400       410       420       430       440       450       460       470       480

   D  D  F  D  F  L  T  S  N  K  V  W  I  A  T  D  R  S  R  A  R  R  C  V  E  A  C  V  Y  G  T  L  D  F  V  G  Y  P  R  F
AAGATGATTTCGATTTTCTGACGAGTAACAAAGTTTGGATTGCTACTGACCGCTCTCGTGCTCGTCGCTGCGTTGAGGCTTGCGTTTATGGTACGCTGGACTTTGTGGGATACCCTCGCT
        490       500       510       520       530       540       550       560       570       580       590       600

   P  A  P  V  E  F  I  A  A  V  I  A  Y  Y  V  H  P  V  N  I  Q  T  A  C  L  I  M  E  G  A  E  F  T  E  N  I  I  N  G  V
TTCCTGCTCCTGTTGAGTTTATTGCTGCCGTCATTGCTTATTATGTTCATCCCGTCAACATTCAAACGGCCTGTCTCATCATGGAAGGCGCTGAATTTACGGAAAACATTATTAATGGCG
        610       620       630       640       650       660       670       680       690       700       710       720

   E  R  P  V  K  A  A  E  L  F  A  F  T  L  R  V  R  A  G  N  T  D  V  L  T  D  A  E  E  N  V  R  Q  K  L  R  A  E  G  V
TCGAGCGTCCGGTTAAAGCCGCTGAATTGTTCGCGTTTACCTTGCGTGTACGCGCAGGAAACACTGACGTTCTTACTGACGCAGAAGAAAACGTGCGTCAAAAATTACGTGCGGAAGGAG

        730       740       750       760       770       780       790       800       810       820       830       840

    M  *
TGATGTAATG
        850       860       870       880       890       900       910       920       930       940       950       960
```

Secondary structure

HAIRPN, HAIRGU

These programs search for potential hairpin loops in nucleic acid sequences. HAIRPN searches for G–C and A–T basepairs only but HAIRGU includes a search for G–T basepairs. A hairpin loop can be defined by three integers, p, q and r. The position of the loop in the sequence is given by r, the length of the stem is given by p and the number of unpaired bases in the loop is q. Figure 1 gives an example where p = 5, q = 7 and r = 7.

When using the program the user types the name of the file containing the nucleic acid sequence and then defines values for p, q and r, as follows:

1. A minimum and maximum value for r to define the region of the sequence to search through.

```
      !--q----!
  !-p-!       !-p-!
TTGGCATTGCACACATGCCTT
      r

              A
            C   C
          G       A
           T     C
            T . A
            A . T
            C . G
            G . C
            G . C
            T   T
            T   T
```

Fig. 1.

2. A minimum and maximum value for q to define the range of numbers of bases to loop out.

3. A minimum value for p to define the minimum number of bases in the stem. Any shorter stems would be rejected.

Gene searching

TRNA

A program to search for tRNA genes [7]. The tRNAs that have been sequenced so far have two characteristics that can be used to locate their genes within long DNA sequences. Firstly, they have a common secondary structure—the cloverleaf—and secondly, particular bases almost always appear at certain positions in the cloverleaf. The cloverleaf is composed of four base-paired stems and four loops. Three of the stems are of fixed length but the fourth, the dhu-stem which usually has four base pairs, sometimes has only three. All of the loops can vary in size. The following relationships between the stems in the cloverleaf are assumed in the program: (a) there are no bases between one end of the aminoacyl-stem and the adjoining tuc-stem; (b) there are two bases between the aminoacyl-stem and the dhu-stem; (c) there is one base between the dhu-stem and the anticodon-stem; (d) there are at least three bases between the anticodon-stem and the tuc-stem.

The program looks first for cloverleaf structure and then, if required, for conserved bases. The sizes of the loops, the number of base pairs in the stems and the required conserved bases may all be specified by the user. The process of looking for the presence of conserved bases can reduce the number of potential structures found considerably and we refer to this process as 'filtering'. The user may also specify that an intron may be present in the anticodon loop. The level of the similarities of the tRNA's were assessed by studying the compilation of their sequences contained in reference [8].

Below is an example of the output from the program.

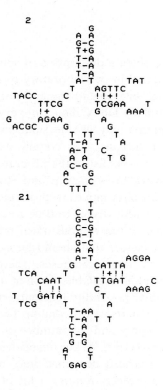

```
   2
              G
            A A
            G-C
            T+G
            T-A
            T-A
            T-A
            A-T     TAT
           T   AGTTC
 TACC     C    !!+!
     TTCG      TCGAA   T
     !+        G    AAA
G    AGAA       A
 ACGC     G      G
           T TT   T
           T-A T   A
           A-T  C
           A A   T G
           C-G
          A   G
          C   C
           TTT
  21
              T
            C T
            G-C
            C-G
            A-T
            G-C
            A A
            A-T     AGGA
           G   CATTA
  TCA     T     !+!!
     CAAT      TTGAT    C
     ! !!      C    AAAG
 T   GATA       T
  TCG     T      A
           T-AA
           T-A T T
           C A
           T-A
           G G
          A   C
          T   T
           GAG
```

Protein program

MWCALC

This is a program that calculates the molecular weight of a protein sequence, and counts the amino acids. Input to the program is a protein sequence in the one-letter amino acid codes (perhaps calculated by program TRANDK). Below is an example.

```
                        F     L     I     M     V     S     P     T     A     Y     *
NUMBER                 22.   34.   22.   11.   22.   26.   27.   31.   26.   22.    0.
WEIGHT               3234. 3842. 2486. 1441. 2178. 2262. 2619. 3131. 1846. 3586.    0.
PERC. BY NUMBER       5.29  8.17  5.29  2.64  5.29  6.25  6.49  7.45  6.25  5.29  0.00
PERC. BY WEIGHT       6.85  8.14  5.26  3.05  4.61  4.79  5.55  6.63  3.91  7.59  0.00

                        H     Q     N     K     D     E     C     W     R     G     -
NUMBER                 14.   22.   20.   15.   27.   15.    3.    6.   23.   28.    0.
WEIGHT               1918. 2816. 2280. 1920. 3105. 1935.  309. 1116. 3588. 1596.    0.
PERC. BY NUMBER       3.37  5.29  4.81  3.61  6.49  3.61  0.72  1.44  5.53  6.73  0.00
PERC. BY WEIGHT       4.06  5.96  4.83  4.07  6.57  4.10  0.65  2.36  7.60  3.38  0.00
  TOTAL MW =        47226.
```

Summary

In this chapter I have given a description of some of the programs that have been developed in our laboratory to aid DNA sequencing. The programs are very easy to use and provide a comprehensive method for handling DNA data for very large sequencing projects. These programs may not be sufficient for those people mainly interested in looking for complicated homologies or secondary structures. A good list of the programs available in 1980 is contained in an article by Gingeras and Roberts [9] to which readers should refer for alternatives to those described here.

The DNA sequencer who wishes to start to use computers faces many difficult decisions, often without much prior knowledge so I list here a number of factors that should be considered. In many places, particularly on large campuses, there will already be powerful central computers available, often at a cost for time used and data stored. Another possibility (currently mainly in the USA) is the use of computer networks or dial-up facilities. The cost of computers is falling rapidly and the number of different machines and languages is enormous. When considering these choices remember that the programs described here are interactive and that large sequencing projects require a lot of time spent at the computer terminal and a lot of disk storage. The position and number of terminals is important: it is best to have constant access to your machine and data. If buying a machine, bear in mind the following: why do you want the equipment and what is it going to do; what other uses might it have now (e.g. word processing) or in the future; will it do what you want (i.e., is software available for it in a language that it can use and does it have sufficient memory and storage)—remember that hardware is cheap and software expensive (this also means the costs of having programs written for another machine converted for your computer or operating system); who is going to do any conversions required or write new programs; who is going to help users and can this person do this and his own job?

The programs described in this book are written in FORTRAN and the original versions were run on PDP11 computers but recently we have converted to a VAX11/780, both of which are made by DEC (Digital Equipment Corporation). The VAX is a larger more powerful machine and so during the conversion some programs were combined and their options extended. The version of DBUTIL described here is the one for the VAX but the PDP11 version is very similar to use and contains all of the main options. All of our future programs will be written for the VAX but, where possible, will be written in such a way as to make it possible to convert them for use on smaller machines.

References

[1] Sanger, F., Air, G.M., Barrell, B.G., Brown, N.L., Coulson, A.R., Fiddes, J.C., Hutchinson, C.A. III, Slocombe, P.M. and Smith, M., 1978. *Nature 276*, 236–247.

[2] McCallum, D. and Smith, M., 1977. JMB *116*, 29.

[3] Staden, R., 1977. NAR *4*, 4037–4051.

[4] Staden, R., 1978. NAR *5*, 1013–1015.

[5] Godson, G.N., Barrell, B.G., Staden, R. and Fiddes, J.C., 1978. Nature *276*, 236–247.

[6] Staden, R., 1980. NAR *8*, 3673–3694.

[7] Staden, R., 1980. NAR *8*, 817–825.

[8] Gauss, D.H., Gruter, F. and Sprinzl, M., 1979. NAR *6*, r1–r19.
[9] Gingeras, T.R. and Roberts, R.J., 1980. Science *209*, 1322–1328.

Note added in proof

Since this chapter was written a number of new programs have been published, some of those described here improved, and a public library of nucleic acid sequences established. In January 1982 the journal Nucleic Acids Research devoted a whole issue to articles on the application of computers to research on nucleic acids [1]. This contains both alternatives to the programs described here as well as new analytic programs. The handling of 'shotgun' sequence data has largely been automated [2] and the DIAGON homology searching program greatly improved [3]. Copies of the nucleic acid sequence library are freely available to all [4].

Additional references

[1] The application of computers to research on nucleic acids, edited by D. Soll and R.J. Roberts. Nucleic Acids Research, volume *10*, number 1, January 1982.
[2] Staden, R., 1982, NAR *10*, 4731–4751.
[3] Staden, R., 1982, NAR *10*, 2951–2961.
[4] The European Molecular Biology Laboratory (EMBL) Nucleotide Sequence Data Library, European Molecular Biology Laboratory, Postfach 10 22 09, D-6900 Heidelberg, F.R.G.

Appendix to chapter 6

(*Contributed by Rodger Staden, M.R.C. Laboratory of Molecular Biology, Cambridge, U.K.*)

This appendix contains notes on how to start a database for the DB system and instructions on how to run some of the programs described in Chapter 6. When typing lines of commands for programs each line must be completed by a carriage return and in the descriptions below this is often referred to by cr.

Getting started with the DB system

DBSTART

This program is used to start a new database. It writes each of the files and sets the number of gels and contigs to zero. The files it creates are described below but note that the version number will be set to 0.

Running DBSTART

1. Start the program.
2. Type project name (only six characters for our file naming conventions).

Reserved file names

Each sequencing project is given a name for its files. Local file naming conventions will determine how names are set up. In our system file names are of the form ABCDEF.GHI

We give each project a 6-character name, for example, LAMBDA; DBSTART will then automatically name our file of relationships LAMBDA.RL0, the file containing the archived gel reading names LAMBDA.AR0 and the working versions of gel readings LAMBDA.SQ0.

Running the programs

BATIN

Running the program

1. Start the program.
2. Type name for file of file names—'FILE OF FILE NAMES' 'PLEASE TYPE NAME OF FILE 1'
3. Type name for next gel reading—'FILE NAME FOR NEXT GEL READING'
4. Type in data on lines of <80 characters. Finish data with an @ character. 'TYPE DATA IN NOW'
5. If you wish to enter another gel reading type Y or carriage return only. 'TO ENTER ANOTHER GEL READING TYPE Y' If you type Y control goes back to step 3.
6. When all the gels have been entered and step 5 is answered by carriage return only the program asks if you wish to have the gels printed. If you do, type Y and all the gels will be listed one after the other headed by their file names, or else type carriage return only and the program stops. 'TO GET GELS PRINTED TYPE Y'
7. The program stops.

DBCOMP

Running the program

1. Start the program.
2. Type name of file of consensus and/or vector sequence file names: 'FILE OF MASTER FILE NAMES =' This must be written beforehand using a system editor.
3. Type name of file of gel reading names: 'FILE OF GEL READING NAMES ='
This is written by BATIN.
4. Type a value for the minimum number of identical consecutive characters to define a match: 'MIN ='
5. Searching starts now.

DBUTIL

Running the program

1. Start the program.
2. Type name of database 'PROJECT NAME ='
3. Type version number 'VERSION NUMBER ='
4. Select 60 or 120 character lines for display: 'FOR 120 CHARACTER LINES TYPE Y'

(This is to allow the use of both wide and narrow keyboard output).

5. Select option by number: 'SELECT OPTION BY NUMBER' STOP = 0, ENTER = 1, PRINT = 2, DISPLAY = 3, JOIN = 4, COMPLEMENT = 5, EDIT = 6, SEARCH = 7, FIX = 8, COPY = 9, CHECK = 10, SCAN = 11, CONSENSUS = 12 'OPTION NUMBER ='

ENTER

ENTER allows entry of a new gel into the database system. It is necessary to know beforehand if the new gel overlaps with any sequences already in the database. (This is determined using DBCOMP.) It automatically numbers and writes the working version of the new gel and then offers the following options:

(a) Complementing and reversing of the new gel (to add it to existing data it may need its sense changed);

(b) Adding to an existing contig either at a left or right end or internally;

(c) Movement of overlap;

(d) Display of overlap;

(e) Editing of the new gel. (It is only during entry into the database that gels can be edited individually. Once entry is completed all gels covering any position are edited together to maintain the alignments of the sequences.)

(f) Editing of the contig. A basic requirement is that the data must be correctly aligned by the user when he enters it into the database. If the gels do not line up the user can use the editing functions to improve their alignment and this can always be achieved by making only insertions in the sequences. In this way no data need ever be deleted until sufficient agreement is found between the gels covering every sequence position. We use X or spaces as padding characters to achieve alignment so that problem areas stand out clearly in the lined-up sequences.

1. Type name of gel to enter 'NAME OF ARCHIVE ='.

If the gel has already been entered the program will type 'GEL ALREADY IN LINE X, ENTRY STOPPED' and return to DBUTIL. The program types the name of the working version of this gel.

2. Tell program if gel overlaps data already in the database: 'IF THIS GEL OVERLAPS TYPE Y' else type carriage return. If the gel does not overlap there is nothing else to do for entry and control passes back to DBUTIL.

3. Tell the program if the gel needs to be reversed and complemented to be in the same sense as the contig it overlaps (reported by DBCOMP) 'TO REVERSE AND COMPLEMENT THIS GEL TYPE Y' else carriage return only.

4. Tell program if the gel extends the contig leftwards: 'IF THIS NEW GEL EXTENDS THE CONTIG LEFTWARDS TYPE Y'.

5. Next prompt depends on step 4 and defines the position of the overlap.

(a) For a gel that extends the contig leftwards—'POSN IN NEW GEL OF LEFT CHARACTER IN OLD CONTIG ='

(b) For other overlaps—'POSN IN CONTIG OF LEFT END ='

If these numbers are out of range the program will reprompt.

6. The overlap is now displayed for its first 120 characters. The display consists of: on the top line the relative positions for the contig characters and then on succeeding lines the sequences from each of the gels that are in this part of the contig. At the left end of each gel is shown its gel number with the sign indicating its sense relative to its archive. On the next line down is shown the consensus sequence for the gels lined up above and then below that is the sequence of the new gel. Asterisks below this line show

mismatches between the new gel and the consensus and finally on the next line are the positions for the new gel.

7. The program then needs to know if the position of the left end of the overlap is correct: 'IF JOINT CORRECT TYPE Y'

If it is, type Y, if not, carriage return, and the program goes back to step 5.

8. The program now offers a number of options to allow the user to align the new gel correctly over its whole length with the data already in the contig. It is important that sufficient edits are made to the new gel or the gels in the contig at this stage to get the alignment correct because once entry is completed the alignment is fixed and cannot easily be changed (see DBFIX). Alignment can be achieved by making insertions or deletions but deletion of data requires the original gels to be checked. For this reason at entry we usually make only insertions to achieve alignment. We use X or spaces as padding characters to achieve alignment and so can distinguish padding characters from characters assigned from reading gels. Note: 'EDIT NEW GEL' will now allow the insertion of space characters so an X character must be used for this function. The options offered are:
'SELECT OPTION BY NUMBER', 'EDIT NEW GEL = 1, DISPLAY = 2, COMPLETE ENTRY = 3, GIVE UP = 4, EDIT CONTIG = 5', 'OPTION NUMBER ='

9. EDIT NEW GEL is identical to SEQEDT except that it is limited to 200 characters per edit command string. Note: if one deletes or inserts at the leftmost character in the gel the position of the gel relative to the contig changes and the whole gel will move left or right.

10. DISPLAY allows display of the region of overlap only. This is defined by the relative positions in the contig.

11. COMPLETE ENTRY adds all of the information about the new gel to the database and returns to DBUTIL.

12. GIVE UP returns to DBUTIL with no information about the new gel being added to the database.

N.B.: Both 'complete entry' and 'give up' options request the user to confirm these options are what he wants by typing 'if you are sure type Y'. If the user does not type Y the program goes back to step 8.

13. EDIT CONTIG allows editing of the contig, i.e., all of the gels covering any edit position are edited simultaneously so as to maintain the alignment of the gels. The program protects the user by allowing him to edit only within the region of overlap. See EDIT CONTIG described for the main options of DBUTIL.

PRINT

This will print the contents of the database in three ways:

(a) All the contig descriptor lines followed by all the gel descriptor lines in the order they appear in the file.

(b) All the contigs are printed one after the other with their gel descriptor lines in the order they have in the contig. Each contig is headed by its contig descriptor line.

(c) Selected contigs only. Each contig is headed by its contig descriptor line and followed by the gel descriptor lines that cover the region defined by the user.

Steps

1. If you want option (c) type Y, else carriage return: 'TO SELECT CONTIGS TYPE Y'

1(a). Define contig by left gel number: 'NUMBER OF LEFT GEL THIS CONTIG ='

1(b). Define region of contig to print over: 'RELATIVE POSITION OF LEFT END =', 'RELATIVE POSITION OF RIGHT END ='. Carriage return defaults to whole contig.

1(c). To select another contig type Y: 'TO SELECT ANOTHER CONTIG TYPE Y'.

2. For option (b) type Y, option (a) results if you type only carriage return 'FOR SORTED DATA TYPE Y'.

DISPLAY

DISPLAY will display on the printer or screen all the gels covering any region of the sequence, lined up in register showing their gel numbers, their strandedness and, underneath, their consensus.

(1) Define the contig you wish to display by the number of its left gel—'NUMBER OF LEFT GEL THIS CONTIG ='.
(2) Define the region of the contig you wish to have displayed. Carriage return only defaults to whole of contig. 'RELATIVE POSITION OF LEFT END =', 'RELATIVE POSITION OF RIGHT END ='.
Printing will then start and when finished control passes back to the main options.

JOIN
This function allows the joining of two contigs. It allows the user to correctly align the ends of the two contigs by editing each contig separately. It is important that the alignment achieved is correct because once the join is completed the alignment is fixed.
(1) Define the left contig by the number of its leftmost gel—'LEFT CONTIG', 'NUMBER OF LEFT GEL THIS CONTIG ='.
(2) Define the right contig by the number of its leftmost gel—'RIGHT CONTIG', 'NUMBER OF LEFT GEL THIS CONTIG ='.
(3) Define the position of overlap—'RELATIVE POSITION IN LEFT CONTIG OF LEFT CHAR OF RIGHT CONTIG ='.
The overlap must be of at least one character. If this criteria is not fulfilled the program prints 'ILLEGAL JOIN. TO RETURN TYPE −9' and then returns to step 3. If the user then types −9 the program returns to the main options in DBUTIL without making the join.
(4) The program then displays the join showing all the gels overlapping the join from the left contig, their consensus, all the gels from the right contig that overlap the join, their consensus and then asterisks to denote mismatches between the two consensuses.
(5) The program then offers the user options to enable him to align the two contigs. These options are: 'SELECT OPTION BY NUMBER', 'MOVE JOIN = 1, EDIT LEFT CONTIG = 2, EDIT RIGHT CONTIG = 3, DISPLAY = 4, COMPLETE JOIN = 5, GIVE UP = 6'.

a. Move join sends control back to step 3.
b. Edit left contig allows editing of the gels in the left contig (see EDIT CONTIG).
c. Edit right contig as above.

N.B.: The program protects the user by only allowing edits to be made in the region of overlap.

d. Display allows display of the overlapping region.
e. Complete join sets up all the values in the database and returns control to DBUTIL main options. After selecting this option the user must confirm it by typing Y when the program asks 'if you are sure type Y'. If the user does not confirm control passes back to step 5.
f. Give up returns control to the main options of DBUTIL without any changes being made to the database.

COMPLEMENT

This function will complement and reverse all of the gels in a contig. It automatically reverses and complements each gel sequence, reorders left and right neighbours, recalculates relative positions and changes each strandedness.

The only user input required is to define the contig to complement by the number of its leftmost gel. It will take a few moments for the program to complete this action. Control then passes back to the main options of DBUTIL.

EDIT CONTIG

EDIT CONTIG allows edits to be made to all of the gels covering any position in the contig. It maintains the alignments of the gels by always making the same number of insertions or deletions in all the gels. Note that these edits are immediately carried out and the 'give-up' options of ENTER and JOIN do not undo them. The user must undo them himself. Note that if this option has been entered from either 'enter' or 'join' options the program will restrict edits to the region of overlap. If the user tries to edit outside, the program reprompts for the position. To escape from

the insert or delete options the user can type a negative number for the position. To escape from the 'number of chars' stage the user can again type a negative number.
The options offered by EDIT CONTIG are:
'INSERT = 1, DELETE = 2, CHANGE = 3, RETURN = 4, OPTION NUMBER ='.

1. INSERT
(a) Type position to perform insertion: 'POSITION ='.
(b) Type number of characters to insert: 'NUMBER OF CHARS ='.
The maximum number allowed at any insertion point is 80.
(c) For each gel in turn the program then prompts for the characters to insert at the edit position: 'CHARS TO INSERT INTO GEL X ='.
If the user types only carriage return the program will automatically insert space characters. When all the gels have been done control passes back to the main options.

2. DELETE
(a) Type position to perform deletion: 'POSITION ='.
(b) Type number of characters to delete: 'NUMBER OF CHARS ='.
The program will then delete the given number of characters from all the gels covering the edit position and control will pass back to the main options.

3. CHANGE
This function allows characters in individual gels to be replaced.
(a) Type gel number to edit: 'GEL NUMBER ='.
The program responds with the relative position and length of the selected gel in case the user only knows the edit position relative to the gel. (The edit position must be relative to the contig.)
(b) Type position of first character to change: 'POSITION ='.
(c) Type the number of characters to change: 'NUMBER OF CHARS ='.
(d) Type of new characters: 'NEW CHARS ='.

The program will then replace the characters at the edit position and control passes back to the main options.

4. RETURN
This function simply passes control back to the part of the program from which the edit contig routine was entered.

SEARCH
This function allows a search to be made for a gel by its archive name. If the gel is found its gel descriptor line is printed, if the gel is not found the program prints 'NOT IN DB'.
1. Type name of archive: 'ARCHIVE NAME ='.
2. To search for another type Y: 'TO SEARCH FOR A GEL TYPE Y', else type only carriage return to get back to the main program.

DBFIX
This program exists to cope with any problems that occur with the database. Its use requires a good understanding of the file structure of the system. With imagination a skilled user can correct any mistakes and misalignments in his data.

It has the following options:
1. LINE CHANGE allows the user to change the contents of any line in the file of relationships. The line is selected by number, the program prints the current line and prompts for the new line. The user should type the new values, each followed by a comma. The program then asks the user to type 'Y' if he is sure he has made the changes he wanted. If the user types anything else the line is left unchanged.
2. SHIFT allows the user to change all the relative positions of a set of neighbouring gels by some fixed value, i.e., it will shift related gels either left or right. It can therefore be used to change the alignment of the gels in a contig. It prompts for the number of the first gel to shift and then for the distance to move them. It then moves rightwards (i.e., follows right neighbours) and shifts each gel, in turn, up to the end of the contig. (This means that only

those gels from, the first to shift, to the rightmost, are moved.) It then reminds the user to change the length of the contig accordingly (use line change).
3. EDIT allows the user to edit an individual gel independently of any others it may be related to. The commands are as for SEQEDT or EDIT NEW GEL in ENTER. The effect of this editing on the length of this gel or its relationship to others must be accounted for (if necessary) by use of the line change function. Maximum number of edit string characters is 200.
4. SQUASH is a function that deletes a contig line by moving down all the contig lines above by one position. It prompts only for the line to delete. It does not delete any of the gels or gel lines for the deleted contig but it does reduce the number of contigs on line 1000 by 1.
5. RENAME GEL is a function that is used to rename the archive names of gels in the database; it only changes the name in the .ARN file of the database: it does not change anything else.

COPY

COPY is a function that makes a copy of all of the data in all of the files in the database. It prompts for a version number for the new copy.

CHECK

There is no user input required once this option has been selected.

SCAN

The user is requested to select the contig and the section of it he wishes to scan. He is then asked to supply a percentage (see the scan algorithm). The program prints out a summary for the region in terms of the percentage number of positions that have been given each of the 5 codes and then asks the user if he wants to have the codes printed out in the form of a sequence. The user is then asked if he wants to store this sequence of codes in a disk file for further processing; if so he is asked to give a name for the file.

CONSENSUS

1. Type name for consensus file: 'NAME FOR CONSENSUS FILE ='.
2. If you wish to select contigs type Y or carriage return: 'TO SELECT CONTIGS TYPE Y'.
3. If selecting contigs supply number of left gel: 'NUMBER OF LEFT GEL THIS CONTIG ='.
4. If selecting contigs define region of current contig: 'RELATIVE POSITION OF LEFT END =', 'RELATIVE POSITION OF RIGHT END =', (zeroes give whole contig).
5. If selecting contigs the user can cycle round steps 3 and 4 by typing Y: 'TO SELECT ANOTHER CONTIG TYPE Y'.
6. To write another separate consensus file type Y: 'TO CALC ANOTHER CONSENSUS TYPE Y'.

If so, go back to step 3, or return.

SEQEDT

Running the program

Reference should be made to the examples especially when preparing the edit commands. These are best prepared before running the program and written out on paper so that they can be copied at the keyboard. It is easiest to use a keyboard listing of the file to be edited to write the edit commands on.

There are two modes of program operation.

A. Creation of a completely new file:

1. Start the program.
2. You do not wish to edit an old file, so type cr on receipt of the message: 'TO EDIT AN OLD FILE TYPE Y'.
3. Supply name of output file when asked: 'PLEASE TYPE NAME OF FILE 2'.
4. Type in edits.

A few seconds after you have completed the edits with an @ character, keyboard output will start giving a copy of the data just written to disk.

B. Editing an old file

1. Start the program.
2. You do wish to edit an old file, so type Y cr on receipt of the message: 'TO EDIT AN OLD FILE TYPE Y'.
3. Supply names of input (FILE 1) and output (FILE 2) files when asked.
4. Type in edit commands.

N.B.: The delete and insert commands may be used in reverse order and still produce the same result, e.g. D/3/I/abc/ is equivalent to I/abc/D/3/.

N.B.: To insert data at position 1 in a file do not find position 1 by command/F/1/I––, etc. The pointer is already at position 1 so just type /I/––.

SEQLST

Running the program

1. Start the program.
2. Supply name of the file containing sequence.
3. Define region to be listed (printed).
4. Select single- or double-stranded printing by typing 1 or 2. Typing cr only is equivalent to typing 1. If double-stranded output is selected there will be a short pause while the program creates the complementary strand.
5. Define number of gaps between lines (for 120 character line output).
6. Select 60 or 120 character line output.

When printing has finished the program requests the next start position for any more printing required. A negative start position will stop the program.

TRNTRP, TRANMT

Running the program

1. Start the program.
2. Supply the name of the file containing the sequence.

3. Select code.
4. Supply printing start and stop positions. These are sequence numbers to define which region of the sequence file is to be printed. The start position is always in phase 1 and determines the phase of all the genes. If zero values are supplied by the user typing only cr the whole file will be listed.
5. Supply first and last sequence numbers to define each gene or region to be translated. After each pair of numbers has been supplied there will be a slight pause while the program calculates the translation. Then the program requests the next pair of numbers. This will continue until the user supplies zero start and end positions by typing only cr. Printing will then start.
6. When printing has finished the program is ready for further instructions and user input commences at step 3. A negative printer start position will stop the program.

TRANDK

Running the program
1. Start the program.
2. Type name of input file containing DNA sequence.
3. Type name of output file to write the translation to.
4. Select code.
5. Type start and stop sequence numbers for each region you wish to translate.

A zero start position for a region will signal that you have finished defining regions and the program will write the translation to disk and stop.

SEARCH, SRCHMU

Layout of the names and strings file

This file must be laid out as shown in the example with names and strings separated by slashes. The end of any set of strings is terminated by an extra slash. The user may supply names in any order whilst running the program. Names such as HaeII and HaeIII must appear in the file in this order. Lines of input may be up to 80 characters in length.

Keyboard string layout
Individual strings are followed by a slash character and sets of strings are terminated by an extra slash. Lines of input may be up to 80 characters in length. Input is terminated by an @ character.

Keyboard name layout
Names are followed by a slash character. Lines of input may be up to 80 characters in length. Input is terminated by an @ character.

```
                Example
                -------
DDE1/CT-AG//
ALU1/AGCT//
HPA1/GTTAAC//
HPA11/CCGG//
RSA1/GTAC//
THA1/CGCG//
MBO1/GATC//
HHA1/GCGC//
HAE1/AGGCCT/AGGCCA/TGGCCT/TGGCCA//
HAE11/GGCGCC/GGCGCT/AGCGCT/AGCGCC//
HAE111/GGCC//
TAQ1/TCGA//
ECOP1/AGACC/GGTCT//
ECOR1/GAATTC//
ECOR11/CCAGG/CCTGG//
HGA1/GACGC/GCGTC//
HINF1/GA-TC//
MBO11/GAAGA/TCTTC//
```

Running the program
(1) Start the program.
(2) Supply the name of the file you wish to search through.
(3) Select option by typing A, N or S.
Option S is the one for which the user supplies strings from the keyboard as described in 'Keyboard String Layout'.
Options A and N require the operator to supply the name of the file containing names and strings. This file is arranged as shown in 'Layout of Names and Strings File'. All three options are described above.
(4) Change the search area if required by typing sequence character numbers to define the beginning and end of the region to be

searched. To leave the area unchanged simply type cr. If the region has not previously been defined and the user types only cr the region will be defined as the whole of the file.
(5) If option S is chosen supply the strings as described. If option N is chosen supply the names when the program prints the message 'TYPE R. ENZYME NAMES NOW'. The layout of this input is described in 'Layout of Keyboard Name Input'.
(6) When output is complete for all strings the program cycles round and requests the user to select his next option. If no option is selected, i.e., if the user types only cr, the program will stop.

SEQFIT

Running the program

Strings can be supplied to the program in one of two ways: they can either be typed in from the keyboard or extracted from a disk file. The program asks the user to select the string input mode at the beginning of the program run. This decision is binding for this run of the program, i.e., to change the string input mode requires the program to be restarted.
1. Start the program.
2. Define the string input mode. When the program types 'TO TYPE IN STRINGS TYPE Y', for string input from keyboard type Y, cr. For input from a disk file type only cr.
3. Supply name of disk file containing sequence.
(a). For string input from disk supply name of disk file containing strings.
4. Supply strings (a) by typing them in; (b) by defining a region of the second disk file. Strings input from keyboard are terminated by an @ character. Strings must be <200 characters in length.
5. Define region of file 1 to be searched. If only cr is typed the program will search the whole file.
6. Define quality of fit as a percentage. Once this has been typed the search starts. When this has been done the program prints out the total scoring positions. These are then sorted into

descending order and then the top ten scores and their corresponding positions are written out on the keyboard.

7. Define the number of these scoring positions to be printed out on the keyboard. These positions will then be printed out starting with the highest score and working down. When printing has finished the user is offered, in turn, four options. If he wants any of them he types Y cr, if not, only cr. The options are to:
 (a) search for the complement of the last string.
 (b) supply a new string.
 (c) change the region to be searched.
 (d) change the percentage.
 (Options a and b are mutually exclusive.) If no option is selected the program will stop.

8. When option selection is completed the program prompts the user to supply the relevant input for the options selected (as above) and then searching commences as before.

OVRLAP

Running the program

1. Start the program.
2. Type name of file 1.
3. Type name of file 2.
4. If required, restrict REGION of file 1 to search through. Carriage return for both will means the whole file is searched.
5. If required, restrict STRING from file 1 to compare with file 2. Carriage return will mean the whole file is compared.
6. Type MIN the minimum overlap. Comparison will now begin.

SQRVCM

Running the program

1. Start the program.
2. Type name of input file (FILE 1).
3. Type name of output file (FILE 2).

HAIRPN, HAIRGU

Running the program

1. Start the program.
2. Type name of sequence file.
3. Type start and stop positions to define region for program to search through.
4. Type MIN and MAX to define loop size, i.e., range of numbers of bases to loop out.
5. Type minimum number of pairings required for any loops, i.e., a minimum value for p. The search now starts and takes a few seconds. When it is complete the program asks the user to select for output sorting:
6. Type S for sorting on size of stem i.e., highest p first; or type C for sorting on centres of loop or type B for both forms of sorting printed one after the other. Output then starts.
7. When output is finished the program is ready for a new region definition. A negative start position will stop the program.

MWCALC

Running the program

1. Start the program.
2. Supply name of file containing protein sequence in one letter amino acid codes.
3. Define the region of the file to count over. Zero start and stop positions result in the whole sequence being counted, a negative start position stops the program.

TRNA

Running the program

1. Start the program.
2. Type name of file containing sequence: 'PLEASE TYPE NAME OF FILE 1'.

3. Define region to search: 'FIRST SEQUENCE NUMBER =', 'LAST SEQUENCE NUMBER =' (default is whole sequence, a negative first sequence number will stop the program).
4. Define a maximum length for the tRNA: 'MAX TRNA LENGTH =' (default = 92).
5. Define minimum scores for each of the stems: 'MIN SCORE AMINO ACYL STEM =', 'MIN SCORE TU STEM =', 'MIN SCORE ANTICODON STEM =', 'MIN SCORE D STEM ='.

Lowest base-pairing in loops from Gauss paper (reference 8 in Chapter 6)

	G–C, A–T	*G–T*	*Score*
Aminoacyl	5	1	11
TU	4	0	8
Anticodon	3	1	7
D-loop	1	1	3

6. Define a range of intron lengths: 'MIN INTRON LENGTH =', 'MAX INTRON LENGTH =' (defaults zero).
7. Define range of TU loop lengths: 'MIN LENGTH TU LOOP =' 'MAX LENGTH TU LOOP = (defaults 6, 9).
8. Select conserved bases filtering or not: 'TO FILTER ON CONSERVED BASES TYPE Y'.

If you choose to filter then you will be prompted to supply a score for each of the conserved bases.

Conserved bases used in filtering

Base number	*Base assignment*
88	T
10	G
11	C or T
14	A
15	A or G
21	A
32	C or T

Base number	*Base assignment*
33	T
37	A
48	T or C
53	G
54	T
55	T
56	C
57	A or G
58	A
60	C or T
61	C

9. Define minimum total score for conserved bases: 'MIN TOTAL SCORE ON CONSERVED BASES ='.

Searching then starts and any cloverleaf conforming to the above criteria will be displayed in both 1D and 2D.

References

ACHTMAN, M., WILLETS, N. and CLARK, A.J., 1971. J. Bacteriol. *106*, 529–538.

ANDERSON, S., GAIT, M.J., MAYOL, L. and YOUNG, J.G., 1980. Nucleic Acids Res. *8*, 1731–1743.

ATKINSON, M.R., DEUTSCHER, M.R., KORNBERG, A., RUSSEL, A.F. and MOFFAT, J.G., 1969. Biochemistry *8*, 4897–4904.

BARNES, W.M., 1978. Proc. Natl. Acad. Sci. U.S.A. *75*, 9281–83.

BARNES, W.M., 1978. J. Mol. Biol. *119*, 83–99.

BARNES, W.M., 1979. Gene *5*, 127–139.

BARRELL, B.G., 1978. In: 'MTP International Review of Science, Biochemistry' Vol. 17, 11 (Editor: B.F.C. Clark), University Park Press, Baltimore, Md.

BARRELL, B.G., BANKIER, A.T. and DROUIN, J., 1979. Nature *282*, 189–194.

BEACH, D.H., PIPER, M. and SHALL, S., 1980. Nature *284*, 185–189.

BEDBROOK, J.R., SMITH, S.M. and ELLIS, R.T., 1980. Nature *287*, 692–697.

BERG, P., FANCHER, H. and CHAMBERLIN, M., 1963. In: 'Symposium on Information and Macromolecules' (Editors: Vogel, H.J., Bryson, V. and Lampen, J.P.) pp. 467–483. Academic Press, London.

BERKNER, K.L. and FOLK, W.R., 1977. J. Biol. Chem. *252*, 3176–3184.

BOEKE, J.D., VOVIS, G.F. and ZINDER, W.D., 1979. Proc. Natl. Acad. Sci. U.S.A. *76*, 2699–2702.

BOLIVAR, F., RODRIGUEZ, R.L., BETLACH, M.C. and BOYER, H.W., 1977a. Gene *2*, 75.

BOLIVAR, F., RODRIGUEZ, R.L., GREENE, P.J., BETLACH, M.C., HEYNECKER, H.L., BOYER, H.W., CROSA, J.H. and FALKOW, S., 1977b. Gene *2*, 95.

BOLIVAR, F., 1978. Gene *4*, 121.

BREATHNACH, R., MANDEL, J.L. and CHAMBON, P., 1977. Nature *270*, 314–319.

BROWNLEE, G.G. and SANGER, F., 1969. Europ. J. Biochem. *11*, 395–399.

BROWNLEE, G.C., 1972. In: 'Laboratory Techniques in Biochemistry and Molecular Biology' (Editors: T.S. Work and E. Work) Vol. 3, pp. 2–265. North-Holland Biomedical Press, Amsterdam.

BROWN, N.L. and SMITH, M.J., 1977. J. Mol. Biol. *116*, 1–30.

BROWN, N.L., 1978. FEBS Lett. *93*, 10–15.

BRUTLAG, D., ATKINSON, M.R., SETLOW, P. and KORNBERG, A., 1969. Biochem. Biophys. Res. Commun. *37*, 982.

BUCHEL, D.E., GRONENBORN, B. and MÜLLER-HILL, B., 1980. Nature *283*, 541–545.

BURTON, K. and PETERSEN, G.B., 1960. Biochem. J. *75*, 17–27.

BURTON, K., LUNT, M.R., PETERSEN, G.B. and SIEBKE, J.C., 1963. Cold Spring Harbor Symp. *28*, 27.

CAMPBELL, V.W. and JACKSON, D.A., 1980. J. Biol. Chem. *255*, 3726–35.

CARDELL, B., BELL, G., TISCHER, E., DENOTO, F.M., ULLRICH, A., PICTET, R., RUTTER, W.J. and GOODMAN, H.M., 1979. Cell *18*, 533–543.

CHACONAS, G., VAN DER SANDE, J.H. and CHURCH, R.B., 1975. Biochem. Biophys. Res. Commun. *66*, 962–969.

CLEWELL, D.D. and HELINSKI, D.R., 1969. Proc. Natl. Acad. Sci. U.S.A. *62*, 1159–1166.

CRESTFIELD, A.M., MOORE, S. and STEIN, W.H., 1963. J. Biol. Chem. *238*, 622–627.

CZERNILOFSKY, A.D., DELORBE, W., SWANSTROM, B., VARMUS, H.E., BISHOP, J.M., TISCHER, E. and GOODMAN, H.M., 1980. Nucleic Acids Res. *8*, 2967–2984.

EDGELL, M.H., HUTCHINSON, C.H. and SCLAIR, M., 1972. J. Virol. *9*, 574–582.

ELLEMAN, T.C. and HINDLEY, J., 1980. Nucleic Acids Res. *8*, 4841–50.

ENGLAND, P.T., 1971. J. Biol. Chem. *246*, 3269–3276.

ENGLAND, P.T., 1972. J. Mol. Biol. *66*, 209–224.

ESHAGHPOUR, H. and CROTHERS, D.M., 1978. Nucleic Acids Res. *5*, 13–21.

FIDDES, J.C., 1976. J. Mol. Biol. *107*, 1–24.

FLINT, S.J., GALIMORE, D.H. and SHARP, P.A., 1975. J. Mol. Biol. *96*, 47–68.

FRIEDMANN, J. and BROWN, D.M., 1978. Nucleic Acids Res. *5*, 615–622.

FRISCHAUF, A.M., GAROFF, H. and LEHRACH, H., 1980. Nucleic Acids Res. *8*, 5541–5549.

GALIBERT, F., SEDAT, J. and ZIFF, E., 1974. J. Mol. Biol. *87*, 377–407.

GHOSH, P.K., REDDY, V.B., PIATAK, M., LEBOWITZ, P. and WEISSMAN, S.M., 1980. Methods Enzymol. *65*, Part 1, 580–595.

GRONENBORN, B. and MESSING, J., 1978. Nature (London) *272*, 375–377.

HAMLYN, P.H., BROWNLEE, G.G., CHENG, C.C., GAIT, M.J. and MILSTEIN, C., 1978. *15*, 1067–1075.

HAYWARD, G.S., 1972. Virology *49*, 342–344.

HEIDECKER, G., MESSING, J. and GRONENBORN, B., 1980. Gene *10*, 69–73.

HERRMAN, R., NEUGEBAUGER, K., PIRKL, E., ZENTGRAF, H. and SCHALLER, H., 1980. Mol. Gen. Genet. *177*, 231–242.

HINDLEY, J. and PHEAR, G.A., 1979. Nucleic Acids Res. *7*, 361–375.

HINES, J.C. and RAY, D.S., 1980. Gene *11*, 207–218.

HOUGHTON, M., STEWART, A.G., DOLL, S.M., EMTAGE, J.S., EATON, M.A.W., SMITH, J.C., PATEL, T.P., LEWIS, A.M., PORTER, A.G., BIRCH, J.R., CARTWRIGHT, T. and CAREY, N.A., 1980. Nucleic Acids Res. *8*, 1913–1931.

HSIUNG, H.M., BROSSEAU, R., MICHNIEWICZ, J. and NARANG, S.A., 1979. Nucleic Acids Res. *6*, 1371–1385.

JAY, E., BAMBARA, R., PADMANABHAN, R. and WU, R., 1974. Nucleic Acids Res. *1*, 331–353.

JEFFREYS, A.J. and FLAVELL, R.A., 1977. Cell *12*, 1097–1108.

JEPPESEN, P.G.N., 1980. Methods Enzymol. *65*, Part 1, 305–319.

KLENOW, H. and HENNINGSEN, I., 1970. Proc. Natl. Acad. Sci. U.S.A. *65*, 168.

LASKEY, R.A., 1980. Methods Enzymol. *65*, Part 1, 363–371.

LILLEHAUG, J.R., KLEPPE, R.K. and KLEPPE, K., 1976. Biochemistry *15*, 1858–1865.

LING, V., 1972. J. Mol. Biol. *64*, 87–102.

LIS, J.T., 1980. Methods Enzymol. *65*, Part 1, 347–353.

MAAT, J. and SMITH, A.J.H., 1978. Nucleic Acids Res. *5*, 4537–4546.

MANIATIS, T., JEFFREY, A. and KLEID, D.G., 1975. Proc. Natl. Acad. Sci. U.S.A. *72*, 1184–1188.

MARVIN, D.A. and HOHN, B., 1969. Bacteriol. Rev. *33*, 172–209.

MAXAM, A. and GILBERT, W., 1977. Proc. Natl. Acad. Sci. U.S.A. *74*, 560–564.

MAXAM, A.M., 1979. Unpublished data.

MAXAM, A.M. and GILBERT, W., 1980. Methods Enzymol. *65*, Part 1, 499–560.

MCDONELL, M.W., SIMON, M.N. and STUDIER, F.W., 1977. J. Mol. Biol. *110*, 119.

MESSING, J., GRONENBORN, B., MÜLLER-HILL, B. and HOFSCHNEIDER, P.H., 1977. Proc. Natl. Acad. Sci. U.S.A. *74*, 3642–3646.

MESSING, J., 1979. N.I.H. Recombinant DNA Tech. Bull. *2*, 43–48.

MESSING, J., CREA, R. and SEEBURG, P.H., 1981. Nucleic Acids Res. *9*, 309–321.

MILLER, J.H., 1972. In: 'Experiments in Molecular Genetics', C.S.H. Laboratory, Cold Spring Harbor, N.Y.

MUSHYMSKI, W.E. and SPENCER, J.H., 1970. J. Mol. Biol. *52*, 91.

NARANG, S.A., BROUSSEAU, R., HSIUNG, H.U. and MICHNIEWICZ, J.J., 1980. Methods Enzymol. *65*, Part 1, 510–620.

PAYVAR, F. and SCHIMKE, R.J., 1979. J. Biol. Chem. *254*, 7636–7642.

PORTER. A.G., BARKER, C., CAREY, N.H., HALLEWELL, R.A., THRELFALL, G. and EMTAGE, J.S., 1979. Nature (London) *282*, 475–477.

PROUDFOOT, N.J., 1977. Cell *10*, 559–570.

PERLMAN, D. and HUBERMANN, J.A., 1977. Anal. Biochem. *83*, 666–677.

RAY, D.S. and KOOK, K., 1978. Gene *4*, 109–119.

REDDY, V.B., GHOSH, P.K., LEBOWITZ, P., PIATAK, M. and WEISSMANN, S.M., 1979. J. Virol. *30*, 279–296.

RICHARDS, R.I., SHINE, J., ULLRICH, A., WELLS, J.R.E. and GOODMAN, H.M., 1979. Nucleic Acids Res. *7*, 1137–1146.

RICHARDSON, C.C., LEHMAN, I.R. and KORNBERG, A., 1964. J. Biol. Chem. *239*, 251–258.

ROTHSTEIN, R.J., LAU, L.F., BAHL, C.P., NARANG, S.A. and WU, R., 1979. Methods Enzymol. *68*, 98–109.

ROYCHOUDHURY, R. and WU, R., 1980. Methods Enzymol. *65*, 43–62.

SANGER, F., DONELSON, J.E., COULSON, A.R., KÖSSEL, H. and FISCHER, D., 1973. Proc. Natl. Acad. Sci. U.S.A. *70*, 1209–1213.

SANGER, F. and COULSON, A.R., 1975. J. Mol. Biol. *94*, 441–448.

SANGER, F., NICKLEN, S. and COULSON, A.R., 1977. Proc. Natl. Acad. Sci. U.S.A. *74*, 5463–5467.

SANGER, F., AIR, G.M., BARRELL, B.G., BROWN, N.L., COULSON, A.R., FIDDES, J.C., HUTCHISON, C.A. III, SLOCOMBE, P.M. and SMITH, M., 1977. Nature *265*, 687–695.

SANGER, F. and COULSON, A.R., 1978. FEBS Lett. *87*, 107–110.

SANGER, F., COULSON, A.R., BARRELL, B.G., SMITH, A.J.H. and ROE, B.A., 1980. J. Mol. Biol. *143*, 161–178.

SCHRIER, P.H. and CORTESE, R., 1979. J. Mol. Biol. *129*, 169–172.

SCHWARZ, E., SCHERER, G., HOBOM, G. and KÖSSEL, H., 1978. Nature *272*, 410–414.

SEIF, I., KHOURY, G. and DHAR, R., 1980. Nucleic Acids Res. *8*, 2225–2240.

SHAPIRO, D.J. and SCHIMKE, R.T., 1975. J. Biol. Chem. *250*, 1759–1764.

SHENK, J.E., RHODES, C., RIGBY, P.W.J. and BERG, P., 1975. Proc. Natl. Acad. Sci. U.S.A. *72*, 989–993.

SLEIGH, M.J., BOTH, G.W. and BROWNLEE, G.G., 1979. Nucleic Acids Res. *7*, 879–893.

SOEDA, E., KIMUNA, G. and MIURA, K., 1978. Proc. Natl. Acad. Sci. U.S.A. *75*, 162–166.

SMITH, A.J.H., 1979. Nucleic Acids Res. *6*, 831–848.

SMITH, A.J.H., 1980. PhD Thesis, University of Cambridge.

SMITH, A.J.H., 1980. Methods Enzymol. *61*, Part 1, 560–580.

SMITH, H.O. and BIRNSTEIL, M.L., 1976. Nucleic Acids Res. *3*, 2389–2398.

SMITH, H.O. and NATHANS, J., 1973. J. Mol. Biol. *81*, 419.

SMITH, H.O., 1980. Methods Enzymol. *65*, Part 1, 371–380.

STADEN, R., 1977. Nucleic Acids Res. *4*, 4037.

STADEN, R., 1978. Nucleic Acids Res. *5*, 1013.

STADEN, R., 1979. Nucleic Acids Res. *6*, 2601.

STADEN, R., 1980. Nucleic Acids Res. *8*, 817.

SUTCLIFFE, J.G., 1978. Nucleic Acids Res. *5*, 2721–2728.

SUTCLIFFE, J.G., 1979. C.S.H. Symp. Quant. Biol. *43*, 77–90.

SZALAY, A.A., GROHMANN, K. and SINSHEIMER, R.L., 1977. Nucleic Acids Res. *4*, 1569–1578.

SZEKELEY, M. and SANGER, F., 1969. J. Mol. Biol. *43*, 607.

TATE, W.P. and PETERSEN, G.B., 1975. Anal. Biochem. *67*, 263–267.

TU, C-P.D. and WU, R., 1980. Methods Enzymol. *65*, Part 1, 620–638.

ULLMANN, A., JACOB, F. and MOND, J., 1967. J. Mol. Biol. *24*, 339.

VAN DE SANDE, J.H., KLEPPE, K. and KHORANA, H.G., 1973. Biochemistry *12*, 5050–5055.

WELLS, R.D., HARDIER, S.C., HORN, G.T., KLEIN, B., LARSON, J.E., NEVENDORF, S.K., PANAYOTATOS, N., PATIENT, R.K. and SELSING, E., 1980. Methods Enzymol. *65*, Part 1, 327–347.

WINTER, G. and FIELDS, W., 1980. Nucleic Acids Res. *8*, 1965–1974.

WU, R., JAY, E. and ROYCHOUDHUNG, 1976. Methods Cancer Res. *12*, 87–176.

WU, R. and KAISER, A.D., 1968. J. Mol. Biol. *35*, 523–527.

YOUNG, R.C-A., LIS, J. and WU, R., 1980. Methods Enzymol. *68*, 176–182.

ZAIN, B.S. and ROBERTS, R.J., 1979. J. Mol. Biol. *131*, 341.

ZIFF, E.B., SEDAT, J.W. and GALIBERT, F., 1973. Nature (London) New Biol. *241*, 34–37.

Subject index

Created and Directed by Hans Höfer

INSIGHT GUIDES

MOSCOW

Project Editor: Wilhelm Klein
Introduced by Yevgeny Yevtushenko
Principal Photography: Fritz Dressler
Art Direction: Villibald Barl and Hans Höfer
Editorial Director: Brian Bell

HOUGHTON MIFFLIN COMPANY

APA PUBLICATIONS

MOSCOW

Second Edition (2nd Reprint)

Printed in Singapore by Höfer Press Pte. Ltd

Distributed in the United States by:
Houghton Mifflin Company
2 Park Street
Boston, Massachusetts 02108
ISBN: 0-395-66436-5

Distributed in Canada by:
Thomas Allen & Son
390 Steelcase Road East
Markham, Ontario L3R 1G2
ISBN: 0-395-66436-5

Distributed in the UK & Ireland by:
GeoCenter International UK Ltd
The Viables Center, Harrow Way
Basingstoke, Hampshire RG22 4BJ
ISBN: 9-62421-145-0

Worldwide distribution enquiries:
Höfer Communications Pte Ltd
38 Joo Koon Road
Singapore 2262
ISBN: 9-62421-145-0

ABOUT THIS BOOK

Imagine the difficulties of writing a guide to Peking in 1948 or to Saigon in 1975. Moscow's turmoil, when we began compiling this book, consisted mainly of political and economic warfare, but we felt the ground shifting constantly beneath our feet as we planned the book's contents and sought out writers and photographers. Glasnost and perestroika became much more than theoretical concepts, and the consequences of the collapse of the former USSR added unusual tension to the editing process.

Taking advantage of the new freedoms being eagerly devoured by Russian writers, Apa Publications opted for a unique approach. The uniqueness lay in the daring departure from the customary guidebook practice of sending writers into a country or city to observe, explore and report their findings. Instead, we sought out expert writers within Russia, pairing their insights with Apa's proven expertise in producing an internationally renowned series of guidebooks. What is on offer, therefore, is not Moscow as filtered through the sensibilities of foreign observers, but Moscow as seen through the eyes of native writers. Their views are refreshingly outspoken. Reading this book is the next best thing to staying with a family in the city, sharing their joys and frustrations about the place they know and love.

Together with its companion volume, *City Guide: St Petersburg*, this book represented a milestone in East-West co-operation – but also, at times, a trial in patience. As well as coping with a worsening food supply, the radical independence struggles of some of the republics and the corresponding relaxation of government control, we had also to integrate the ongoing micro-changes that are important for any city guide. Street names reverted to old designations, immigration and currency regulations were changing, and new tourist-related businesses sprung up all over the place. Life in Moscow changed more during the two years while this book was being written than in the previous 70. It seemed that the clock was being turned back, with impatience, to a pre-Bolshevik era.

To manage the project, **Hans Höfer**, Apa Publications' publisher, assigned **Wilhelm Klein**, whose previous Insight Guides range from *Burma* to *Austria*. It was while publishing the German edition of the *Monthly Review*, an American neo-Marxist magazine, back in the 1970s that Klein first gained an insight into the state of Soviet affairs. His ambition to get some of Russia's best writers to contribute to the present trio of books was realised when **Yevgeny Yevtushenko**, the country's most outspoken poet, agreed to participate. Yevtushenko, who lives in Moscow, knows its history and creative background as well as anyone, and his introductory essay provides a mouthwatering appetiser to the treasure trove of art and humanity contained in the city.

World-renowned writers

With Yevtushenko's help, many other contributors were lined up. **Anatoli Blinov**, who wrote the history section of this book, is a journalist by profession. He worked as an aide to Valentin Falin, a Secretary of the CPSU Central Committee, and is co-author of the book *The Soviet Economy: an Insider's View*.

The chapter about the undefeatable Muscovite was written by **Robert Tsfasman**, a

Klein

Yevtushenko

typical Muscovite. He graduated from the Moscow University's philological department and is editor-in-chief of the *Soviet Life Magazine*. He is author of numerous publications that appeared in Russia and abroad.

Igor Zakharov, who had already written for an Apa guide to the New Soviet Union, studied philosphy and wrote in this book about the "Private Life of Ivan the Terrible" and the "Brain Drain" in the Soviet Union.

The chapters about Moscow culture, museums and architecture were co-written by **Elvira Kim** and **Olga Kalinina**, art critics and geographers by profession. The theatre and cinema scene were described by **N. Zorkaya** and **Andrei Karaulov**. Zorkaya is a leading Soviet film critic and Karaulov is a drama critic who has written several books about art and politics.

Yevgeny Filiminov is an expert on Russian tourism. He already contributed to the New Soviet Union book, and works for Intourist. For this book, he wrote the chapter on the Golden Ring.

Vladislav Govorukhin, writer and editor at a well-known Moscow publishing house, spent more than a year assembling the details that form the backbone of this book: the Places section. He combed every important street and structure of Moscow and went time and again to research libraries to collect all those small details that bring to vivid life a city that, all too often, has been thought of as grey and uniform.

For this edition, the Travel Tips section has been updated by **Victor** and **Jennifer Louis**, who also contributed the chapter on life in Moscow today ("Capitalism and Chaos"). He is Russian, she is English, and they have worked as a Moscow-based journalistic team since the 1950s, acting as correspondents for a wide range of publications in the West and producing many guidebooks to the former Soviet republics.

Villibald Barl, Apa's Yugoslav-born art director, selected the photographs which bear the stamp of one of Germany's top photographers, **Fritz Dressler**. Dressler, who teaches photography at his native Bremen Academy of Art, made several trips to Moscow to shoot the rich material printed here. As with all Apa guides, his brief was not only to provide a pictorial extension to the written words but also to capture the essence of the living, new Moscow, to show what it is like to be a Muscovite in the 1990s.

The final touches

The final polishing of the book's text was carried out in Apa's London editorial office by **Christopher Catling** and **Marcus Brooke**, and proof-reading and indexing were completed by **Mary Morton**.

Welcome, then, to Moscow. We hope you will be as excited by the momentous changes taking place in this, one of the world's great capitals, as we were when we put together this book. And we hope that you will have an opportunity to follow up the insights offered by the Russian writers in this Insight City Guide by meeting and talking with the Russians themselves on their home territory. For, as that great traveller Karl Baedeker put it: "If ever a city expressed the character and peculiarities of its inhabitants, that city is Moscow."

Kim

Kalinina

Govorukhin

Karaulov

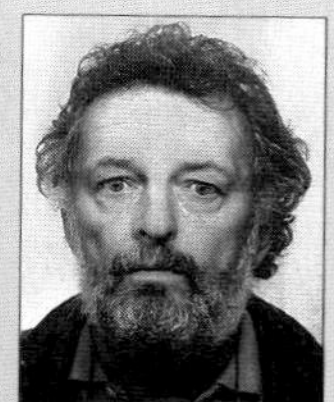

Dressler

History

Features

Places

—by Vladislav Govorukhin
(except where stated)

Maps

TRAVEL TIPS

For detailed information see page 257

ТРАНСПОРТ СССР
ВЫСТАВКА ДОСТИЖЕНИЙ НАРОДНОГО

СССР

ГОРУРАЗУ
МНУЮ·Ѿ
НЕАЖЕУСЕ
ЧЕСА·КАМЕ
НЬПРОЗЕЯ
ХТА·

MOSCOW, THE THIRD ROME?

***Yevgeny Yevtushenko** describes the changing moods of Russia's legendary capital.*

"Russia used to be run by St Petersburg and Moscow: now St Petersburg's role and the cultural period of the window cut through to Europe is over, now... but now the question is whether indeed St Petersburg and Moscow did run Russia... Was this really true? Certainly, not all Russia used to pour into St Petersburg and Moscow and gather there in crowds... and in actual fact, she ran herself, continually being rejuvenated by fresh influxes of new strength from her provinces and outlying regions..."

This is how Dostoyevsky accurately defined the significance of Russia's two main cities: the two main knots of her history which are made up of the tightly interwoven threads of human destinies, political intrigues, threads of blood and tears, ships' and gallows' ropes, the tie that the great Russian poet Yesenin hanged himself from, the barbed wire over the walls of the Butyrka and Lefortovo prisons, and the skipping ropes children have such fun with.

A village turned capital: Moscow, however, was the first small primordial knot to be tied when Russia was in its infancy. In relation to Moscow, St Petersburg appeared like a prodigal son, dressed in fancy foreign clothes and smoking tobacco, a son who was ashamed of his old-fashioned country bumpkin of a mother. Moscow is a village which has grown into a capital city, and even now, more than eight centuries after its foundation, it still retains its village ways.

In no other capital of the world will you encounter such an obsession for homemade jams and salted foods as in Moscow. Special jar lids for home preserves are among the countless items in short supply in Moscow's shops. Muscovites are superstitious, as country folk tend to be, and similarly fond of using home cures for common complaints. That's why on many window sills you will see spiky green aloe plants – supposedly the universal remedy for all ailments.

All the political news, all the rumours and all the gossip that has not been carried by the press spreads like wildfire through the city as though it were no bigger than a village. Muscovites tend to be rather rude in the streets but this comes from spending a quarter of their lives in shop queues. Long ago they virtually stopped using the verb "to buy" and, more aptly, started using the term "to get hold of."

When you visit these same Muscovites at home, however, they are as hospitable as country folk, and the entire contents of their refrigerators will be spread out on the table. Every family has at least one dish of its very own invention but the stomach of a Muscovite, no matter how radical, is always conservative. Even the élitist intelligentsia prefers the traditional country fare that was eaten in the days before Peter I. Most Muscovites, for instance, adore roast duck stuffed with apples, meat dumplings (*pelemeni*), grated beetroot with garlic, pickled cabbage soup with dried mushrooms, freshly salted cucumbers, marinated tomatoes and honey-soaked apples. All this is what Ivan the Terrible ate and his gastronomic tastes were no different from those of his serfs.

If Muscovites run out of alcohol at home, they never hesitate to go out on to the landing, ring their neighbour's doorbell and ask to borrow vodka. It is also customary to borrow anything else you need from neighbours, including tables and chairs, campbeds, cutlery, glasses, butter, sausage meat and frozen chickens. People borrow money from their neighbours and from colleagues at work: IOU notes are considered an insult.

There is also something rather countrified about the atmosphere in Moscow's leafy courtyards that nestle behind the box-like

Preceding pages: onion towers, typical of church architecture; monument to the worker and the woman collective farmer; guarding the mausoleum; the Exhibition of Economic Achievements; gold above the Kremlin; life's joyful moments. Left, the prophet Daniil, 15th-century icon.

concrete apartment tower blocks. *Babushki* (grannies) sit on benches, their knitting needles clicking slowly, while runny-nosed children play at their feet; nearby a local drunk snores peacefully on the shady grass, and old-age pensioners noisily bang down dominoes on a wooden table.

The heirs of the founder: The monument to Moscow's founder, Grand Prince Yuri Dolgorukiy, stands in the very heart of Moscow, and hiding coyly behind his steed's bronze tail is the Georgian *Aragvi* restaurant. In nearby streets suave Caucasians with narrow black mustaches, who slink like mountain leopards after their blue-eyed, blond-haired prey, will tell you jokingly: "That man Dolgorukiy was no fool – he built Moscow around a Georgian restaurant…"

That is practically all there is to joke about in the history of Moscow's foundation. Grand Prince Yuri was nicknamed Dolgorukiy (the Long-Armed) because of the greedy and single-minded manner in which he gathered lands, piece by piece, into his long arms. According to some legends, the lands of a boyar (as noblemen and landowners were then called) called Kuchka caught the Grand Prince's eagle eye. These lands included several villages, one of which, Vorobievo, was located on the site of the present-day Moscow University and on the same hilltop which brides and bridegrooms visit today to admire the view of the city. The boyar himself actually lived in a house very close to what is today one of Moscow's best theatres – the Sovremennik.

Kuchka had the reputation of being proud and outspoken. Dolgorukiy finally murdered him and for a long time afterwards the site of the murder was called Kuchkovo Field. Through this field ran Lubyanka Street, and here it was, in the infamous building also known as the *Lubyanka* and occupied by the secret police (known successively as the Cheka, NKVD, MGB and the KGB) that the long-armed heirs of Moscow's founder continued to torture and kill.

Yuri Dolgorukiy decided to outwit boyar Kuchka's ghost and win him over by marrying off his son, Andrei, to the murdered man's daughter. That, however, did not prevent Kuchka's sons from eventually killing the young man to avenge their father's death.

The following words of the monk Filotheus were cited in a 17th-century chronicle: "Two Romes have fallen, but the third stands, and there will be no fourth. Verily the town of Moscow is called the

Third Rome, for a sign hangs over it, just as it did over the First and Second. And different though it may be, it shares the same destiny – bloodshed" Prophetic words. While the Slavs in Kiev were being baptised in the Dnieper River, in Moscow they were being soaked in the blood of invasions, rebellions and executions.

For some time St Petersburg became the official capital, but Moscow went on being the national capital in the full sense of the word. The straight thoroughfares of St Petersburg had nothing in common with the tortuous, uneven Russian character with its numerous dark back streets and blind alleys. Moscow's celebrated side-streets, which, unfortunately, are fast disappearing, embodied the psychological geometry of the Russian soul. Quite understandably they were given names like *Krivokolenniy* (Crooked Knee) and *Krivoarbatskiy* (Crooked Arbat).

Moscow could not help but develop on the very site where it was first recorded in a chronicle of 1147 for, like a two-faced idol, it stood at the crossing of many ways with one face turned towards Asia and the other towards Europe. This was why, for many years, Moscow determined Russia's destiny, and why, even today, Westernisers and Slavophiles are still locked in their seemingly endless disputes. "Yes, we are Scythians, yes, we are Asians with greedy slanting eyes…" wrote the great poet Alexander Blok, even though he was a very highly educated European.

Foreign invaders: The Moscow princes received their royal titles from the Tatar khans and then, copying the khans' own wily tactics, gradually sewed together the fragmented patches of Russian lands. Over the next three centuries or so the Tatar overlords and conquerors turned into janitors, waiters and steam-bath attendants…

Every Russian was born in Moscow because that is where Russia herself was born. Moscow stood like a fortress, barring the Tatar hordes' way to Europe and receiving all the blows and humiliations Europe might otherwise have suffered. However, no sooner had Moscow cast off its foreign shackles than it was clamped in the shackles of czarism. Napoleon hoped that Moscow would greet him with jubilant peeling bells and the city fathers would present him with

Left, the wooden Lenin Mausoleum in the 1920s. Above, German prisoners-of-war being walked through Moscow.

the keys to Russia's capital on a silver platter; his hopes were soon dashed. Moscow rushed out of Moscow like water from a riverbed so that not a single drop would reach the invaders' parched lips. Napoleon found himself in the position of a seducer who had succeeded in getting into a woman's warm bed, only to see her running away, laughing at him derisively.

But the victors over Napoleon and Hitler's victors later on, were rewarded with the same humiliating ingratitude. So accustomed did Russia become to constant humiliations that she could no longer become accustomed to freedom. She was the last country in Europe to abolish serfdom and, after the October Revolution, exchanged one state of servitude for another, far more terrible than Czarism.

Then, after a short reprieve under Khrushchev, she exchanged the cult of Stalin's personality for that of impersonality. Yet it was in this officially stagnant but unofficially seething time that Moscow produced a young generation of poets who started gathering audiences in their thousands in public squares and palaces of sport. What's more, these audiences expected their contemporary idols to tell them the truth, which never appeared on the pages of *Pravda* (the Russian word for "the truth").

Eventually in 1962, Pravda published a poem against neo-Stalinism, the first of this theme to appear in our press. That poem was my own work *The heirs of Stalin*. The Moscow journal *Noviy Mir* was responsible for the earth-shattering publication of the first eye-witness account of Stalin's labour camps – *A Day in the Life of Ivan Denisovich* by Solzhenitsyn. And it was Moscow, the capital of the Party *nomenklatura* (the system whereby appointments to specified posts in government or the economic administration are made by organs of the Communist Party) that became the capital of nonconformism in 1968; it was then that a small group of young people came out onto Execution Place, in former times the site of so many executions, to protest against the Soviet tanks in Prague. From then on, the human rights movement headed by Andrei Sakharov began to gain momentum.

Economic issues: It was Moscow, too, that dealt the Party apparatus, its candidates and militant chauvinists a shattering blow by electing Professor Gavriil Popov as mayor. Popov is the author of the eloquent expression the administrative key-command *system*, editor of the journal *Economic Issues* and one of the leaders of the inter-regional group. He has the lumbering gait of a bear, shrewd Greek eyes behind a grey fringe of hair and a way of couching rebellious ideas in a gentle and slightly ironic style.

His life as a mayor is by no means easy because he has inherited a terrible legacy: a housing shortage, shop queues, ecological pollution, an increasing crime rate, speculation, prostitution and racketeering. The freeing of society has brought in its wake a dreadful unleashing of dark instincts and criminal elements.

There are now 9 million people living in Moscow. Every day Moscow has up to 3 million visitors, and up to 5 million on public holidays. The trouble is that Moscow's hotels can only accommodate 72,000 guests. Moscow has 1.4 million workers, 1.2 million scientific workers and 1.5 million students; 400,000 people are employed in its 14,000 cooperatives. There are more pensioners in Moscow than workers: 2.2 million of them, to be precise. In terms of average age, Moscow is the oldest capital in the world. Moscow's population is artificially controlled by the *propiska* (resident permit) system. Otherwise, it might easily double.

Both Gavriil Popov and his Leningrad counterpart, Anatoli Sobchak, resigned from the Communist Party shortly after their election. In their opinion, under the multiparty system, which has already been written into the Constitution and which is gradually being legalised, a mayor must serve the interests of the population as a whole and not those of any one party.

Several stations of the Moscow militia have announced that they too will no longer be subject to Communist Party control. Zaslavsky, the head of one of Moscow's district executive committees and a USSR people's deputy representing disabled people, has announced his intention to turn his district into a free enterprise zone. Popov and his colleagues in the Moscow City Council want to promote individual and collective initiatives and to do away with state monopoly. Will they succeed in turning a political democracy into an economic one? Will the opponents of the bureaucracy manage to oppose it in a constructive manner – in actions as well as words?

If democracy triumphs in Moscow and Leningrad, it will triumph everywhere else in the country. But what if it fails? Before the Revolution people used to whisper the following saying: "When nails are cut in Moscow, arms are cut off at the elbow in the provinces". So, if arms are cut off at the elbow in Moscow, what will happen in the provinces? The Third Rome? Why?

As a Russian and a Muscovite, I do not want Moscow to thrust its will upon the rest of the world or upon any part of it. Once Italy abandoned its imperial ways, it became a compact, enterprising country, and Rome, no longer considering itself to be the master of the world, would rather have the reputation of a charmer than a tyrant…

Messianic ideas and "Black Ravens": France should be grateful to Moscow for not letting the French Empire expand: Napoleon turned back from Moscow just in time. I grew up in Stalin's time when the propaganda machine did its level best to give Moscow the image of a Messianic city; and the capital city of the world upon which the despairing gazes of all the exploited peoples of the planet were fixed. At the same time cars nicknamed "Black Ravens" were speeding round Moscow, packed full of innocent people who had just been arrested. The clanking sound of a lift at night used to be terrifying for, you see, the secret police could burst into anyone's flat with an arrest warrant at any time. But at school we kept on singing with all our hearts:

Industrious,/ Illustrious,/ Forever victorious,/ Moscow of mine,/ The greatest/ The greatest love of mine.

I remember Moscow in 1941 when refugees drove their anxiously bellowing cows through Red Square and past the Bolshoi Theatre. I remember Moscow in 1944 when tens of thousands of German PoWs were led along the ring road. I remember seeing there a Russian woman who, submitting to the mysterious laws of female compassion, felt sorry for the people who had possibly killed her husband or brother, and feverishly began thrusting crusts of bread at them.

Working democracy: People's Deputies.

I remember that March day in 1953 when people crushed one another at Stalin's funeral and their breath was so dense you could see the shadows of the March branches quivering on it; all around lay piles of boots of every size and description which had been lost in the terrible scrum.

I remember a kiss under some tinkling spring icicles at Patriashiye Ponds (now Pioneer Ponds). I remember the day that Gagarin, the first man in space, flew into Moscow, and the public jubilation as though this heralded some unprecedented flight which still lay ahead of us all. And Pasternak, you know, also believed in the future of this great city when he wrote of it:

For the dreamer and the night-bird
Moscow is dearer than all else in the world.
He is at the hearth, the source
Of everything that the century will live for.

The century isn't over yet, don't forget… And the fact remains that Moscow, which once protected Europe from the Tatar hordes, has acted in the 20th century as a shield for all humanity by defending it from Fascism. Moscow, by inventing perestroika, has helped to destroy the Iron Curtain, the Berlin Wall and the psychological walls between various political systems. Perhaps Moscow, which today is still suffering from queues, will become the founder of perestroika not just of one country but of the whole human family? You see, there is really no such thing as a First, Second or Third World – all countries are merely different streets of one and the same large village.

Sakharov and the Ghosts of the Kremlin: It was the last day of Academician Sakharov's life, and it just so happened that I witnessed it. With hunched shoulders the academician came trudging out of the Kremlin's Palace of Congresses. It was the Second Congress of People's Deputies and, just as at the First, Sakharov had not been given the floor. He looked dreadfully tired; I would even say he looked worn out.

Sakharov's last words: On his return home he said to his wife: "Tomorrow there's going to be a big fight". These proved to be his last words. Watching him trudge wearily across the Kremlin's icy paving stones, unaware that the end was very close at hand, I found myself reflecting on what a tragic paradoxical situation it was – Sakharov in the Kremlin.

While he was the famous creator of the hydrogen bomb, on the secret list, thrice a

Hero of Socialist Labour, and an obedient brain of the State who, as he was later to admit, cried upon hearing of Stalin's death, Sakharov's appearance at the Kremlin to receive awards seemed quite natural and in keeping with the order of things. When pangs of conscience, the yeast on which all the Russian intelligentsia was leavened, finally joined his brain and heart together, making his brain disobey, the academician, once so submissive, became an inflexible heretic, and mounted the stake of his own free will. He was stripped of all his awards and professional duties, declared an "accomplice of world imperialism" and unlawfully exiled to Gorky. From then on, his name was only heard on foreign radio broadcasts and in the traditional liberal kitchens where the despairing intelligentsia used to pour their hearts out. In those days it was quite impossible to imagine Sakharov in the Kremlin.

Gorbachev's telephone call to Gorky brought this most gentle and reticent of rebels back to Moscow. He proved to be "ungrateful" as far as bureaucracy was concerned and went on fighting for freedom and democracy, arguing the case with Gorbachev himself when necessary. Sakharov proved to be well and truly a people's deputy for he had the courage to oppose the majority of deputies in the Kremlin Palace in the interests of the majority of the people. Having penetrated the Kremlin's walls by a rare miracle of history, he seemed like the ambassador of all the people who had been arrested and executed on the orders which had been signed in this same Kremlin.

As Sakharov set off from the Kremlin that day towards his home where death awaited him, gazing after him were the eyes of the great Russian architects of the Cathedral of Vassily the Blessed (St Basil's) in Red Square. (Those same eyes that had been put out on Ivan the Terrible's secret edict in case – heaven forbid – they might build a more beautiful church in some other country.) So were the insane eyes of Ivan the Terrible himself, squeezing the blood-stained staff with which he had killed his own son. Oh no, it was no mere chance that a deranged person once went rushing up to Repin's painting in the Tretiakov Gallery of Ivan the Terrible killing his son and slashed the czar's head with a knife, shouting, "Enough blood".

Left, Sakharov at a press conference. Above, Yeltsin and Popov at a rally for the People's Deputies.

Enough blood: Also watching Sakharov were the heads of the *streltsi*, the members of the military corps instituted by Ivan the Terrible, who had been beheaded on the orders of Peter I. Even now these heads are rolling invisibly in front of Lenin's mausoleum after bouncing off Peter I's boyish axe, which by then was already serving as both a murder weapon and a carpentry tool.

The young czar's record of 120 executions in one day of March 1697 pales in comparison with the one Stalin was to set in 1937 when he had approximately three million people arrested and annihilated. Sakharov was also being watched by the voluptuous, cannibalistic eyes of the hyena of the Kremlin jungle, Beria, and the yellow eyes of its Shere Khan, Stalin who, with the legitimate rights of a ghost, still pads around the Kremlin in his soft Caucasian boots with tiger-like stealth. "The Kremlin should not be lived in. The Reformer was right," warned Anna Akhmatova.

Yes, the atmosphere in the Kremlin is certainly rather daunting at times, and this is no illusion – it is an awareness of historical fact. The 15th-century Konstantin-Yelena Tower, for instance, was once popularly called the "Torture Tower" because it contained a rack – a wooden instrument of torture on which prisoners could be made to confess to any crime. They were stripped to the waist and stretched on this horizontal rack with their hands tied behind their backs in such a way that their arms were pulled out of their sockets. They were also tortured with a red-hot iron: three letters – vos – standing for "Disturber of the Peace" were branded on their foreheads, and their nostrils were torn. After being tortured to death, their corpses were flung onto the thoroughfare below to be carted away by relatives.

In 1606, the 6,000 rebels captured from Bolotnikov's army were executed in a conveyor-like manner: they were lined up in rows, struck over the head with a club, like cattle, and dumped under the ice of the River Yausa. When the so-called Copper Rebellion against the introduction of copper coins was put down in 1662, 19 people were hanged in Red Square and 12 had their arms, legs and tongues cut off.

A sophisticated method of torture was devised for Stepan Razin, the legendary peasant leader, who later became the hero of numerous folk songs. Razin withstood one

The *streltsi* on their way to execution.

hundred blows from a leather knout while on the rack; he rebuked his brother, Frola, on the execution block for groaning and not bearing the pain in silence. For three days icy water was dripped onto Razin's shaven head. On 6 June 1671 he and his brother were driven out into Execution Place (of which Mayakovsky was later to write somewhat flippantly: "Execution Place, so awful for heads to face").

Razin bore himself with extraordinary courage and, after crossing himself in the direction of the Cathedral of Vassily the Blessed, lay down on the block himself. He kept silent when the executioner cut off his right arm, and did not utter a sound when the executioner cut off his left leg to the knee. But when his brother shouted out, "I know the Sovereign's word" which meant that he wanted to stay alive by revealing information of importance to the czar, Razin, regardless of the blood gushing from his arm and leg, cried out to his brother, "Silence, cur!" At that moment the executioner cut off his head, stuck it onto a pole which, for a long time, stood in Red Square, the terrible rebellious power of its staring eyes striking horror in one and all.

Catherine II's accession to the throne in 1763 was marked by "lampoons unfit to be read" being burned in Red Square. During the Plague Riot of 1771 soldiers fired case-shot into the mob rushing towards the Kremlin with cudgels and demanding an end to the plague quarantines. Seventy-two of the rebels were beaten with leather knouts, had their nostrils torn, and were then dispatched to work on galleys.

Even the bell in the Alarm Tower, which had summoned the mob, had its tongue torn out, as though it was a living person, and then was exiled to Siberia. In 1775 the leader of the peasant revolt, Pugachev, after being betrayed by his fellows, was kept in a cage for two months to be jeered at and mocked by the people, and then executed and his head stuck onto a wheel spoke.

The forerunner of Russian democratic journalism, N.I. Novikov, whose apartment entrance looked straight over Red Square, was sentenced to 15 years of imprisonment for an article against serfdom. Another to live at one time in this blood-stained square was Alexander Radishchev who received the death sentence for his book *A Journey from St Petersburg to Moscow*, a sentence that was subsequently commuted to 10 years exile in Siberia. Radishchev's final words before committing suicide were: "Posterity will avenge me." His prediction came true as far as Czarism was concerned but then history turned out in such a way that independently minded liberals like himself became the target of vengeance.

Another independent minded person, Piotr Chadayev, suffered the tragic fate of being declared mentally sick because of his dissident views. The Czarist government had already developed – on a modest, amateurish scale – the tactics of using psychiatric hospitals to treat dissidents. These same tactics were to be so skilfully re-introduced in the 1960s by the KGB head, Yuri Andropov, who, for some incomprehensible reason, has recently earned the reputation of being something of a progressive.

A similar trick had obviously been prepared for Sakharov, but for some reason the past masters of this sort of game did not actually succeed in pulling it off. With regard to selflessness, courage and public spiritedness, Chadayev was the Sakharov of the 19th century. The Revolution of 1917 knocked the two-headed eagles of Czarism off the Kremlin's towers and replaced them with red stars. But have the hopes of the Russian intelligentsia been fulfilled with the merging of Russia's course with the civilised evolution of other peoples, described by Chadayev's heir, Sakharov, as a "convergence"? Whereas Shakespeare bitterly exclaimed that the links between times had fallen apart, the link between Czarism and socialism has proved to be the only convergence being secretly encouraged.

The formula of the former Chief Procurator of the Holy Synod, Pobedonostsev (who once pronounced the anathema on Tolstoy) was "Autocracy, Orthodoxy, National Identity". This was replaced at the First Congress of People's Deputies by the words of an Afghan war veteran and opponent of Sakharov. When he said: "Motherland, Power, Communism", most of the hall leapt to their feet, applauding, as though hypnotised.

Practically all the elements of the Czarist formula have remained intact except that Communism has replaced religion. The barbaric destruction of churches in Stalin's time has been explained as Communism's attempts to take the place of religion, as well as Communism's intolerance of such beautiful rivals as the stone and wooden churches.

Regardless of the fact that the entire Civil War had been fought under the slogan of "Beat the Gold Epaulets", Stalin reintroduced the generals' gold epaulets and even made himself a generalissimo. After being nurtured in the Russian intelligentsia's imagination, the Revolution began smothering its nurses, even while still in the cradle. No sooner had the Soviet government taken up office again inside the Kremlin than all the ghosts of the old Kremlin and the cobwebs of its intrigues and executions seemed to settle inside the Bolsheviks.

The revolutionary poet Demian Bely, who was living in the Kremlin at the time, gave Stalin some books from his library and one day incautiously reprimanded him for tearing open the uncut pages with his thick nicotine-stained finger. When Stalin came to power, he immediately evicted the over-scrupulous poet from the Kremlin. And he tore open the uncut pages of history in exactly the same way, leaving his greasy finger marks all over them.

Also worth noting is the following anecdote, which I heard as a young boy in Stalin's time and which might then have earned the person telling it 15 or so years in a Siberian prison camp:

Stalin summons Pushkin to his Kremlin office. "Good day, dear Comrade Pushkin, pride and glory of socialist realism."

"Good day, best friend of gymnasts and writers, Comrade Stalin!"

"How can I help you, O genius of mankind, I, a modest representative of the bureaucracy?"

"It's my editors, they've been giving me a hard time... They keep chucking out lines and rewriting others..."

Stalin picks up the receiver of his special government telephone. "Lavrentii Palich? (Beria, head of the secret police.) The pride and glory of socialist realism, Comrade Pushkin, is here with me, and his editors are pestering him. The whole business clearly smacks of a Japanese-American-Paraguayan plot against Soviet literature and the Soviet country as a whole. Wouldn't it be a good idea, as a precaution, to shoot 30 or 40 editors?"

Pushkin tugs at Stalin's sleeve in horror, shaking his head. Stalin shrugs his shoulders, "I'm sorry, Lavrentii Palich. Odd folk, these national geniuses. You really want to help them but they just don't let you."

Stalin hangs up and Pushkin gets to his feet, thanking him for being so sympathetic and obviously in a hurry to leave. "Well then, go and work with your goose quill to the glory of the doves of peace, Comrade Pushkin."

Pushkin goes out. Stalin picks the receiver up again. "Lavrentii Palich, sorry to trouble you so often. The pride and glory of socialist realism, Comrade Pushkin, has just left me and gone off towards the Borovitskiye Gates. Warn the loyal Leninist, D'Anthes, (the man who killed Pushkin in a duel) about this, will you..."

Sakharov's relationship with Khrushchev, who exposed Stalin's crimes and yet refused to listen to the great atomic scientist's warnings about the danger of nuclear experiments, and later his relationship with sentimental and kind-hearted Brezhnev who, on his Communist boyars' advice, moved his tanks across the Czech border and also got hopelessly involved in Afghanistan's civil strife – aren't they both like the relationship between Pushkin and Stalin in the anecdote?

Moscow cares: The old proverb "Moscow cares nothing for tears" may be put even more depressingly as "Moscow cares nothing for its geniuses". But would it be fair to say so? Wasn't Moscow among the many tens of thousands of people who joined Sakharov's funeral procession, weeping bitterly and with signs raised over their heads, saying "Forgive us". Doesn't Moscow visit Pasternak's grave and lay flowers there as a sign of gratitude and respect? After destroying its churches, isn't Moscow now collecting money little by little to restore them? Didn't Moscow, in February 1990, lead half a million Muscovites into its streets with slogans of "No to anti-Semitism" and "We're all for freedom but not for the freedom of Fascism" and "Party, give us a go at the wheel".

The echo of perestroika has breached the Berlin Wall. Reticent and awkward on the tribunes Sakharov proved a much shrewder politician than experienced ones such as Honneker and Ceaucescu and a much greater realist than Ligachev who claimed to be one. In the history of every city there are geographical and moral founders. Sakharov will remain in history as the founder of the new liberal Moscow. Berlin, Prague and eventually Moscow chose Sakharov's way.

The mania for Sovietising the names of towns and streets, so characteristic of the hungry, young and ambitious state brought into being in 1917, has been replaced by another mania for de-Sovietising these names which is being conducted by none other than the Soviets themselves.

For example, the street sign "Kalinin Prospekt" has been replaced with one for Vozdvizhenka Street and the monument to Kalinin, the first president of the USSR, has also been removed. Few tears were shed, since it had been a very long time since anyone put flowers on the monument's pedestal. Kalinin had not been brave enough to defend his wife who was thrown into one of Stalin's labour camps and who, according to witnesses, used to scratch the lice out of the seams of her prison clothes with a bit of glass from her broken spectacles. Now the city named after Kalinin has been renamed Tver.

When the Dostoyevsky scholar, Yuri Karyakin, in his speech at the First Congress of People's Deputies of the USSR in 1989, suggested removing Lenin's embalmed body from the mausoleum and laying it to rest in the ground, as Lenin had wished, not all present but certainly a great many were completely stunned as though he had said something blasphemous. Nowadays this issue is an ordinary discussion point.

In one way or another, the value of moral principles allied to politics is rising while that of mere politics, divorced from moral principles, is falling. The many monuments still marring Moscow's landscape are disappearing, but Sakharov's will remain. The streets still bearing names unworthy of Moscow will have their previous names restored but Sakharov Street will remain unchanged.

Crowds celebrate the failure of the 1991 coup.

CENTRE OF A MIGHTY EMPIRE

People who live in Moscow have every reason to identify their city with the origins of Russian statehood. The many centuries of Moscow's history are, as it were, inseparable from the troubles and hardships that became the lot of the Russian people. Anyone wishing to visit Moscow and its historical highlights can rest assured that the impression produced by the Eastern Slav citadel upon the inquisitive visitor is certain to be a memorable one, especially since it is generally conceded that an insight into a nation's past promotes understanding of its present life and mentality. Many things will thus be revealed in the character, aspirations and behaviour of the Russian people; and the veil over the "enigmatic Russian soul" will, perhaps, be lifted.

Historians say that the first settlements in the territory of present-day Moscow date to roughly the 3rd millennium BC. At the end of the 1st millennium AD, the Moscow Region was settled by Slavonic tribes called the *Vyatichi* and *Krivichi*. They did not remain independent for long: their lands became parts of principalities commanded by Chernigov and Suzdal overlords.

The year 882 brought the unification of the two largest cities in Rus, Kiev and Novgorod, to form the ancient state of the Eastern Slavs, the Kievan Rus, with its capital in Kiev.

The Prince of Suzdal: The 12th and the 13th centuries saw more and more Russian cities straying from centralised control. The ancient state was falling apart, a process which was accelerated by the Mongol invasion. Several independent principalities emerged, of which the largest were the Vladimir-Suzdal, the Galitzko-Volynskoye, and the Novgorod republic.

The first mention of Moscow appears in the chronicles of the year 1147, when the Prince of Suzdal, Yury Dolgorukiy ("The Long-Handed"), the son of Vladimir Monomakh, invited his ally, Prince Sviatoslav Olgovich of Novgorod, to Moscow.

The Kremlin of the 12th century, where the two chieftains met, was a small fort, with wooden walls and towers and it protected an area with a perimeter of about 1,200 metres (4,000 ft). It was surrounded by the huts of peasants and craftsmen, who hid behind its walls in times of danger.

Razed by the Mongols: Early in the 13th century Moscow became the capital of a small yet independent principality. Racked by frequent wars, the land passed from one ruler to another. The invasion of the Mongol hordes nipped the town's growth in the bud. Weakened by internecine strife, the Russian

princes were helpless before the invading armies. Batu Khan took Kazan and went on to Kolomna in 1237. After Kolomna fell, it was Moscow's turn. The town was razed, the population was reduced to one-third, and many people were taken away as slaves.

In 1263, Moscow got a new prince: Daniil, the son of Novgorod's Prince Alexander Nevsky (the famous conqueror of the Swedes on the Neva in 1242). Daniil was the first of the Moscow dynasty. He founded Danilov Monastery and Bogoyavlensky Monastery. By the year 1300, Moscow controlled Kolomna, Pereslavl and all the lands in the Moskva River basin.

Preceding pages: view of 18th-century Moscow; icon of Boris and Gleb, 1340. Left, the oldest mentioning of Moscow, 1147. Above, Mongols take Moscow in 1238.

Moscow consolidated its power in the struggle against its rivals during the reign of Ivan Kalita, Daniil's only surviving son. A shrewd tactician and no-nonsense politician, Kalita removed his political opponents with the help of the Mongols and added their lands to his realm. In Kalita's time, Moscow becomes the centre of the Russian church. The Assumption Cathedral, built there in 1326, was the first Russian church of stone and became the cathedral of metropolitans. Then came the church of Archangel Michael, the crypt of Moscow princes. After the old Kremlin was burned to the ground in 1331, a new one was built of sturdy oak. In 1326, Metropolitan Piotr moved his residence from Vladimir to Moscow.

A Kremlin of stone: The next ruler of Moscow, Dmitry Donskoy (1359–89) scored a number of victories over the Mongols. The heaviest blow that befell the invaders was delivered at Kulikovo Field in 1380. Prince Dmitry then built the first stone Kremlin (of white stone), together with additional fortifications and suburbs.

In the reign of Grand Prince Ivan III (1462–1505), Moscow conquered Tver, Novgorod, Pskov, and Ryazan. Politically, this completed the unification of the Russian lands around Moscow. The centralised Russian state was born. In 1480, it freed itself of the Golden Horde's yoke forever.

By the end of the 15th century, Moscow consolidated its international status: Ivan III married Zoe Paleolog, the niece of the last emperor of Byzantium. Russia was now regarded as the heir to the Empire of Byzantium. The two-headed eagle – the emblem of Byzantium – became the seal of Russia. Simultaneously, Russia established diplomatic relations with Western Europe – with Germany, Rome, Hungary and Poland.

THE PRIVATE LIFE OF IVAN THE TERRIBLE

Ivan IV, "the Terrible", (1533–84) inherited the throne of grand princes at the age of three. He considered himself the Deputy of Augustus, the Emperor of Rome. Perhaps it is for this reason that the regalia representing supreme power in Russia were similar, the famous *barma* (ceremonial shoulder-covers) and the *Monomakh* headdress. Many historians have tried, for several centuries now, to portray Ivan the Terrible as an outstanding statesman, a brilliant mind worthy of respect and sympathy, a predecessor of Peter the Great. (For this, they forgive him all his atrocities). Yet the majority view him as a bloodthirsty tyrant.

Nikolay Kostomarov, the prominent Russian historian, observed a striking similarity between Ivan the Terrible and Nero, despite the differences of circumstances and environment. Like Nero, Ivan was corrupted in his childhood years. Both Nero under the guidance of Seneca, and young Ivan under the guidance of Silvester the monk, accomplished many commendable things. Finally, when both got rid of their mentors, they proceeded to out-Herod Herod in depravity and sadism.

In their cruelty, both favoured the bizarre, the mannered, the theatrical. Nero started out by killing his mother; Ivan did not kill his mother – she died when he was still an infant – but made up for it by killing his son towards the end of his life. Nero set Rome on fire, and then tortured innocent Christians, trying to make them confess to arson in his "court of justice"; Ivan razed Novgorod to the ground and killed many more Russian Christians than Nero had Roman ones. He, too, accused his victims of heinous crimes which he, the stern yet just arbiter, set out to investigate.

Nero went to Greece to fool around with the arts and sciences, leaving Rome at the mercy of his underlings; Ivan fled to Alexandrovskaya Sloboda, where he played the monk while his *oprichniki* (elite guardsmen) plundered Rus. Both were greedy and self-interested. They devastated the provinces and harboured a particular hatred: Nero for heathen temples, Ivan for Christian monasteries. Nero boasted that he was the only Roman emperor who could reach the limits of arbitrariness; Ivan carried on about the enormity of his authority as a czar.

Nero was a coward and did not have it in him to kill himself when the moment had come; Ivan did not have to save himself from what he did to so many others, yet he exhibited cowardice and faint-heartedness several times in the course of his reign. Nero took pride in his talents as a poet, singer and artist; Ivan never missed an opportunity to air his gift as a rhetorician, theologian, historian – in a word, he loved to philosophise.

Ironically, the Czar of Muscovy was luckier in his respect than the Emperor of Rome. As far as we know, Nero's literary and artistic efforts were quickly forgotten. The tyrant of Moscow, on the other hand, is commended to this day for his "wit, humour, erudition, logic" and recognised as "the foremost writer of his time".

Other comparisons: Come to think of it, Ivan the Terrible can be compared to other monarchs, not just Nero. A curious picture awaits when we compare the Russian czar's private life with that of the notorious polygamist Henry VIII. Here, too, Ivan excels: Henry had only six wives, while Ivan... well, let's count!

Knowing no restraint since his childhood years, spoiled by the boyars (nobles) whom he so ruthlessly executed in his later life, Ivan remained a priest at the altar of dissipation till the end of his days. Historians know of only two weeks when he led a life that can be called decent; the two weeks after he first married.

In 1546 the 16-year-old czar married the youthful Anastasia Zakharina, who charmed him with her beauty and soft femininity. He had lost his virginity at 13; contemporaries say that he had several hundred lovers in the course of those first three years. And now, a

Ivan IV, after mortally wounding his son while in a rage.

week after his marriage, the boyars could not recognise their czar: gone were rough-and-tumble practical jokes with bears and jesters, the obscene songs, the whores who filled every room of the palace… Ivan was courteous, helpful towards the needy. He even released many prisoners from his dungeons. This change was believed to come from the influence of his young wife. Alas, things returned "to normal" in the third week of his honeymoon.

Be that as it may, his first marriage lasted for 13 years, in the course of which Anastasia, who lived the life of a recluse, bore six children. Disease and the never-ending insults of her husband wore the czarina out, and she died before the age of 30.

Orgies in the Kremlin: Ivan's second marriage was arranged in haste. Unsuccessful in Sweden and Poland, the Czar's emissaries brought him back a bride from the Northern Caucasus. Kuchenei was a daughter of the Czar of Kabarda. She accepted the Orthodox faith in Moscow, and was christened Maria Temriukovna; but, at first, she could not understand a word of Russian and could not make head or tail of what her husband said. This did not stop her from laying down the law in such a way that Ivan fulfilled every whim of his young savage. Accustomed to unrestrained and bloody pastimes in her own land, the czarina set the precedent of taking part in a four-hour mass public execution in Red Square. Drunken orgies in the Kremlin, in which the royal couple participated, were the talk of Moscow. Several causes are given for Maria's early death; many suggest that she was poisoned by her husband.

When Novgorod was taken and plundered, Ivan had 1,500 gentle girls brought to the city. The czar chose Marfa Saburova-Sobakina, and, even though she started to pine away after the engagement, Ivan proceeded with the wedding. Yet Marfa died of unknown causes before the marriage was consummated. When her tomb was opened this century, Marfa lay pale, untouched by putrefaction, as if she had been buried yesterday, not 360 years ago.

The Orthodox Church allows only three marriages. Yet there was no law for Ivan. He continued his search for a bride, and chose the "common-born" Anna Kolovskaya. Being wild and passionate by nature (not unlike Maria Temriukovna), Anna tried to influence her husband by participating in all his orgies and supplying him with a steady stream of lovers. Even this did not enable her to retain the throne. Less than a year passed before Ivan divorced Anna and forced her to take the veil. She spent the following 54 years in a monastery dungeon.

Ivan's fifth marriage was a replica of his fourth. Seventeen-year-old Anna Vasilchikove spent several months in the Kremlin, after which she died under dubious circumstances. (Some say she died later, after having been forcefully deported to a monastery). In 1573 Ivan married Maria Dolgorukaya. This marriage, his sixth, had been the most ill-fated of all: the bride, it turned out, had not been a virgin, and Ivan ordered her drowned. His next marriage was something of a surprise: he chose a deacon's widow, Vasilisa Melentievna, a comparatively elderly woman with two children. The marriage conformed to the previous pattern in that it was short-lived, ending after less than two years because of his wife's untimely death. (There is, however, a contemporary testimony which says that Vasilisa was buried alive for "exceeding whorishness".)

Three years before Ivan's death, A. Nagoi, the czar's favourite boyar, gave his niece, the beautiful Maria Nagaya, to the ageing czar. Even though the marriage was celebrated with great pomp, it ran contrary to all church rules, which made even many of the czar's favourites consider it illegal. At the same time, Ivan made his son Fyodor marry Irina (the sister of future Czar Boris Godunov) whom, shortly before, he had planned to marry. Ivan did not take his marriage seriously and willingly sacrificed the beautiful Maria for an opportunity to marry an English princess.

Left, the Kremlin in 1584. Above, Ivan the Terrible.

First, however, he tried his luck with Queen Elizabeth herself. When nothing came of this, Ivan sent his ambassadors to London, with the purpose of arranging his marriage with the queen's relative, Maria Hastings. Beside dynastic interests, this act had another goal: to lead Russia, plagued by a series of military failures, out of its total international isolation, and to serve as a prologue to a military union of Russia and England. He also hoped that, in the event of a palace coup (his lifetime pet fear) and forced exile to England, he would gain control of the Hastings principality, where he could live out his days with his sickly son. This plan, too, ended in failure: Queen Elizabeth rejected Ivan's proposals, citing "the bride's utterly shattered health."

The czar's desire to see an English woman – any English woman – as his wife was so great that Ivan threatened, as Ambassador Bows wrote from Moscow to London, to take his treasury, go to England and marry one of the queen's relatives if she refused to find a suitable bride for him. To prove that he was not kidding, he exiled Maria Nagaya (who had just borne him a son) from the palace. The next time she would see him would be when his body lay in state.

On the eve of his death, on 17 March 1584, the czar sent an envoy extraordinary to Sweden with an offer of marriage to a distant relative of the king, and an offer of unity to the king. The ambassador was informed the next day of the czar's death, which had caught up with him suddenly over a chessboard. Even Ivan the Terrible's mighty health could not endure more than 54 years of such orgy-filled life.

ПЛАНЪ
города
сочиненной
Ивана
Мичурина
въ 1739 году.
Москва рѣка
ЗЕМЛЯНОЙ ГОРОДЪ
БѢЛОЙ ГОРОДЪ
КИТАЙ ГОРОДЪ
Рѣка
Москва

After Ivan's death, the throne went to weak-willed Fyodor Ivanovich (1584–98). The actual power in the land belonged to Boris Godunov, the Czar's brother-in-law. Godunov was by all accounts a talented politician. After Fyodor died without leaving an heir to the throne, Godunov was elected czar (1598–1605). He relied on the church to keep the boyars (nobility) in line. When Russia got its Patriarch in 1598, the Russian Orthodox Church ceased to depend on Byzantium. The first Russian patriarch, Job, was appointed by Godunov.

Impostors in power: Plots were hatching among the boyars (the nobility was outraged at Godunov's policies). In 1591, Czar Fyodor's younger brother Dmitry, had died under suspicious circumstances. Some said Godunov had had him killed. These rumours were used by Sigizmund III, the King of Poland, who put "Czarevich Dmitry" (it was announced that he had miraculously survived) at the head of an incursion into Russia. The first incursion was followed by a second, with another impostor in command. It was False Dmitry II who managed, with the support of several traitorous boyars, to take Moscow.

These events fuelled strong popular sentiment. Merchant Kuzma Minin and Prince Dmitry Pozharsky (there is a monument to them in Red Square) stood at the head of a popular army. The army, composed largely of peasants, defeated the Poles and liberated the capital.

Russia would come under attack from Poland and Sweden on several other occasions. In 1617, Vladislav, the son of the King of Poland, invaded Russia, while Gustav Adolf, the King of Sweden, besieged Pskov. After peace with these countries was restored, Russia was in terrible shape. Moscow had been ravaged beyond all recognition.

The coming of the Romanovs: The new Czar Mikhail Romanov (1613–45), the son of Metropolitan Filaret, was far from sure of himself and relied heavily on his closest aides, particularly on his uncle, Boyar Morozov. In 1674 Moscow was shaken by an uprising which became known as the "salt mutiny" (because of the draconian salt tax imposed by the government). As if that was not enough, Czar Mikhail found himself with another uprising on his hands – the so-called "copper mutiny" was sparked by the ill-judged devaluation of the new copper rouble which sent prices rocketing.

Peter I (the Great) became Czar of Russia in 1672 and reigned until his death in 1725. He proved himself a great reformer and turned Russia into an international power. He also carried out innumerable domestic reforms. Peter introduced the system of recruits for regular army service and created the Russian fleet. In the last battle of the Northern War against Sweden, Russia won an outlet into the Baltic.

In 1702, Peter organised in Moscow the School of Navigation and the School of

Preceding pages: Poles in Moscow during the Times of Trouble. Left, Moscow map 1739. Above, Dmitry the Fake.

Artillery. That same year, he ordered the construction of the Comedy Chamber in Red Square – Moscow's first public theatre. A special "beard" tax was levied on Raskolniks and other stubborn men who would not shave. The tax was extracted from them at the Spasskiye Gates where stuffed figures, suspended from the gates for all to see, were dressed in examples of the type of clothes that all were recommended to wear.

A time of reforms: Peter also introduced, as of 1 January 1700, the new calendar – *ad dominum*, and not "from the creation of the world" as before. Peter's challenge to ancient Russian tradition and lifestyle was all-encompassing, touching upon street processions and masquerades, with their *Papa Princes* and *Messrs. Cardinals* (merrymakers and jesters), costumes, and even upon styles of speech and writing.

In 1712 Peter moved the capital from Moscow with its arrogant boyars to fledgling St Petersburg. Yet the czar never stopped developing Moscow, and its significance as the country's economic and cultural centre did not falter. Moscow remained a stronghold of merchants and gentlefolk.

"Peter I had no love for Moscow, where, with every step he took, he ran into remembrances of mutinies and executions, inveterate antiquity and the obstinate resistance of superstition and prejudice," as Alexander Pushkin wrote. "He left the Kremlin, which wasn't stifling, but which was too close for him, and departed for the distant shores of the Baltic Sea in search of leisure, space and freedom for his mighty and restless action... But Moscow, having lost its aristocratic glamour, blossomed in other ways: industry, with much support, livened and developed with extraordinary force. The merchants got rich and started to move into the palaces that the gentry vacated."

It was in Peter's day that the Kremlin became a fully-fledged administrative centre. The young monarch favoured the village of Preobrazhenskoye on the bank of the Yauza as his residence. He built a palace with a theatre, and a mock fortress (*Premburg*), used by the Preobrazhensky and Semionovsky Regiments during their war games. In and around Moscow, aristocrats started building what can only be described as architectural masterpieces: Count Sheremetiev's Palaces (in Ostankino and Kuskovo), Prince Yusupov's Estate (Arkhangelskoye), Merchant Pashkov's Mansion (near the Lenin Library).

In 1755, the first Russian university opened in Moscow as the result of the efforts of Mikhail Lomonosov, the famous scientist. Students were taught in Russian. By that time, foreigners lived in the territory of Moscow in large numbers. They came either by the invitation of the Russian government, or on their own initiative.

Foreigners in the city: Foreigners were approached with caution and, frequently, with distrust. The clergy tried to drive a wedge between its flock and the aliens, who were forbidden to hire people of the Orthodox faith as domestics. The general desire was to keep Catholics out of Russia, because Catholic "propaganda" was considered especially dangerous. Moscow merchants and craftsmen regarded foreigners as unwanted competitors. Foreign citizens were not allowed to buy palaces in Moscow and, as a result, special settlements for foreign subjects appeared in the city. The largest was the German settlement on the tributary of the Moskva River, the Yauza. It is interesting to note that all Europeans were called "German" because the Germans constituted the majority. The German Settlement was mainly inhabited by West Europeans; the other three settlements were occupied by Poles, Lithuanians and Greeks.

Understandably, diplomats were treated differently. A foreign ambassador was an extraordinary phenomenon back in 15th-century Moscow. It is noteworthy how totally un-Russian were the stiff, mechanical ceremonies of greeting in the court of the Russian czars. In Moscow, for example, a fine line was drawn between the three grades of ambassadorial rank: the highest honours went to the *great ambassador*; then came the *envoy* and then the *messenger*. Moscow received embassies with proper splendour and solemnity.

Left, Peter I. Above left, Sheremetiev, right, Lomonosov.

It may well be that these ceremonial receptions took root in the Russian character, producing the respect generally felt today in Russia towards any foreigner. The schism of Orthodoxy, which occurred in the 17th century (certain clerical parties opposed Patriarch Nikon in connection with an up-dated comparison of Biblical texts in Russian with the Greek originals), was destined to become an anti-feudal flag for many generations.

During the short reigns of the great reformer's successors – his grandson, Peter II and his niece Anna – Moscow again became the capital and a scene of palace coups. It was here that the mighty Menshikov (Peter the Great's crony) was defeated and exiled.

It was here that the conspiracy of the Golitsyns and the Dolgorukiys, who wanted to force the Empress to share her power with the aristocracy, was uncovered and punished. And it was here that Biron, the Duke of Kurland, Anna's favourite, started out on his reign of tyranny, which was to last almost a decade. Empress Anna built a new wooden palace, the Annenhof, in the Kremlin, and a summer palace in Peter's former retreat on the Yauza.

The "quiet" palace coup of 1762 heralded the dawning of the age of Catherine II, a native German princess. Her years on the throne (until 1776) are described by historians as "the era of enlightened absolutism". Keeping reforms down to a minimum, the Empress (it was Peter who had introduced the title of emperor in Russia) tried to reconcile serfdom, which continued to hold Russia in its vice-like grip, with budding bourgeois relations.

The plague: In 1771, the plague struck. Over 57,000 died. It was forbidden to bury the dead within the city limits. Hence the cemeteries beyond the Kamer-Kollezhsky Val (the *de facto* border of Moscow at the end of the 18th century – 37 km (23 miles) from the centre: the Vagankovskoye, Danilovskoye and Kalitnikovskoye Cemeteries. The territory of the Kremlin was cleared, and the old marketplace was burned to prevent the spread of disease.

In Catherine's time Moscow continued to grow. In 1775, the Empress divided the city into two parts; everything inside today's Boulevard Ring was to be considered Moscow proper, and everything beyond its limits was to be the suburbs. Numerous stone mansions went up; streets were paved and lit by oil lamps by night; a water-works was installed. Embankments appeared along the Kremlin section of the Moskva River, across which several bridges were built. It was even planned to rebuild the Kremlin. The population boomed, reaching 275,000 by 1811 – almost double the figure of 75 years earlier.

The golden age of the gentry: Free from the burdens of state service, the aristocrats mostly thought about their careers, built ubiquitous residences in and out of town, amused themselves lavishly, drank and fornicated. The Empress herself, as her contemporaries attest, was a great lover of amusements. Clubs became centres of the Moscow gentry. The English Club was founded in 1772, the Noble Assembly in 1783, the Merchant Assembly in 1786, and the German Schuster in 1819. "Moscow", a contemporary wrote, "is a remarkable haven for people with nothing to do but blow their wealth away, play cards, and pay endless visits...".

Napoleon in Moscow: The year 1812 brought the French intervention. Inspired by his victories in Europe, Napoleon's "Great Army" invaded Russia and pressed towards Moscow. "Once I take Moscow, I'll smite Russia through the heart," he believed. On 6 August, the French took Smolensk. Moscow plunged into panic, and many noble families hurriedly left the city. The Russian army retreated, avoiding a decisive battle. Then, on 26 August, battle took place near the village of Borodino. Napoleon's army, which was numerically stronger, received such a devastating blow that it never recovered. The military council of the Russian army meeting in the village of Fili near Moscow (Kutuzov's House there is now a museum) decided to surrender the city without fighting in order to preserve the army. People started to leave Moscow.

The fire of 1812: Meanwhile, Napoleon approached Moscow and waited for "the boyar deputation" to give him the keys to the city. But no one came – Moscow was empty. Soon after the French occupied it, the city perished in a terrible fire, the cause of which is still unclear to this day.

The fire lasted for six days. It was made all the stronger by a gale-force wind. "I myself remained in the Kremlin until the flames surrounded me," Napoleon remembered. "I then left for Emperor Alexander's country palace a mile or so from Moscow, and it may give you an idea of the force of the fire when I tell you that it was painful to put your palm against the walls or the windows which faced Moscow – to such an extent were they heated up. The sky and the clouds appeared to burn, it was a majestic and the most terrifying sight humanity had ever seen!"

Napoleon left Moscow, where no food remained. He ordered the Kremlin blown up. The vanguard of the Russian army stormed into the city, but managed to save only a few structures. The "Great Army" started running. Napoleon suffered defeat after defeat.

Moscow was rebuilt, of course. Recon-

Left, Napoleon and Alexander I meet in Tilsit in 1807. Above, the battle of Moscow, 1812.

struction was accompanied by the rapid growth of trade and industry. The leading industries were textiles, metal processing, tobacco and perfumery.

It is noteworthy how the merchants and the fledgling Russian capitalists imposed a ban on foreign imports, yet continued to export their own goods. Moscow merchants and trade companies developed ties with the Ukraine, Transcaucasia, and the Volga Region. Freight turnover grew after the first railroad to St Petersburg was opened in 1851. As the Moscow market expanded, the city saw the rise of its first families of businessmen – the Prokhorovs, the Novikovs, the Guchkovs.

Alexander I (1801–25) continued to suppress peasant movements. A liberal in word, Alexander was cruel and reactionary in deed. The atmosphere of general discontent spawned the revolutionary movement in Russia, which was mainly directed against monarchy and serfdom. Intellectuals from among the aristocracy founded several secret societies, aiming to overthrow the czar.

The secret societies: In 1817, one of the first such societies appeared in Moscow – "The Union of Prosperity", which united officers and civilians of noble ancestry. The moment of truth arrived in December 1825, when the revolutionaries came bearing arms to depose the czar – and were christened "Decembrists". Nicholas I was brutal to those who rebelled against him in St Petersburg's Senatskaya Square. Five officers were sentenced to death, and countless others condemned to forced-labour camps and exile in Siberia. The wives proudly accompanied their outcast husbands.

The second half of the 19th century saw the decay of serfdom. Hired labour, which was finding increasing use at factories and plants, proved to be several times more efficient than slave labour. And so, in February 1861, Alexander II issued a manifesto which declared serfdom null and void. The country entered the era of capitalism.

Moscow's industry was growing at a rapid pace. Capital was concentrated in the hands of several giant companies. They were owned either by merchant dynasties, or by peasants who got rich (as, for example, Savva Morozov, the famous patron of the arts). There was progress in transportation, as well. By the 1860s, Moscow was connected, by highways, to eight major cities. In 1899 the city got its first tram line and in 1902 its subway project.

The Moscow of that time is quite accurately described by Vissarion Belinsky, the famed literary critic: "There are many eating-places in Moscow, and they are always crawling with the kind of people who only drink tea there. These people drink up to fifteen samovars a day; they cannot live without tea, they drink it five times a day at home, and as many times on the town. Where but in Moscow can you work at the office, engage in trade, write novels, and publish journals for no other reason but your own amusement and recreation? Where, if not in Moscow, can you carry on so much about your labours, present and future, become famous as the world's most active person – and do absolutely nothing?"

An example for today: In 1906, Piotr Stolypin became head of the Russian government. Stolypin, who served as minister of internal affairs before his appointment, had all deputies from the social-democratic parties arrested and dissolved the Duma (parliament). Yet the Stolypin cabinet also carried out several progressive reforms. It gave greater elective rights to big business and pushed through agrarian reform, which essentially eliminated communal land ownership. Now it was the peasant who owned his land. Recently, progressive economists have been turning to the programme of the "reactionary minister" Stolypin in search of the rational grain that could save the present beleaguered economy.

As Russia plunged into World War I, its economic crisis got worse. The provinces, particularly Central Asia, were swept by a national liberation movement. The army seemed ready to turn on the government. Things got too hot to handle. The various political parties proposed different ways to deal with the crisis. Some demanded that the power of the czar be limited and that the scope of democratic freedoms should be broadened; others questioned peaceful methods and called for a *coup d'état*.

The uprising finally started in St Petersburg, and was supported in Moscow. The workers and soldiers from the Moscow garrison took over all government offices. Power passed to the "Committee of Social Organizations" and the Soviet (Council) of Workers' and Soldiers' Deputies.

Left, Nicholas II, the last czar with wife and officers. Above, the Czar's four daughters, Olga, Anastasia, Tatyana and Maria, who were murdered together with their parents.

In April 1917, Vladimir Lenin returned to Russia from Finland, where he had been hiding from the Provisional Government, and started his preparations for the socialist revolution.

The Bolsheviks in power: On 25 October 1917, another armed uprising racked Petrograd. There was to be no more Provisional Government: all power would belong to the people. That very day, the Second All-Russia Congress of Soviets declared Russia a Soviet Republic. It passed the Decree on Peace and called on all the warring sides to start negotiations. The first Soviet government – the Council of People's Commissars – was headed by Lenin.

The revolutionary forces required almost a week to gain victory in Moscow. It was only after reinforcements arrived from Petrograd, Vladimir and other towns, that the revolutionaries shelled the Kremlin and took power in Moscow. The old power structures – the Senate, the ministries, the Synod – were abolished. Industry and banks were nationalised. The state was separated from the church. Because the country remained at war with Germany, the old army could not be immediately disbanded. In January 1918, the Lenin government issued a decree on the creation of the worker-peasant Red Army and Fleet.

The government returns to Moscow: Peace talks started in Brest-Litovsk on 15 November. After peace was made in Brest, the Germans occupied the Baltic region and Finland and were now in a position to threaten Petrograd. Lenin therefore suggested that the government should move to Moscow. On 11 March, Moscow once again became the capital of the nation.

From 1918 until 1921, the republic fought for survival – not just at the fronts, but in the capital itself. In June 1918, the left-wing Socialist-Revolutionaries had assassinated the German Ambassador, Count Mirbach, in the hope of provoking another war with Germany. The assassination was followed by an armed uprising. To fight the counter-revolution, a special body was set up: the All-Russian Extraordinary Commission (the KGB's grandfather), headed by Lenin's comrade-in-arms, Felix Dzerzhinsky.

The young Lenin in 1897.

The execution of the Romanovs: Most leaders of the White Guard armies, which pressed towards Moscow and Petrograd, fought Soviet power under the banners of monarchy – they wanted to restore Czarism. Because of this, to do away with the apple of discord once and for all, the last of the Russian Emperors, Czar Nicholas II, was executed with all his family in the summer of 1918 in Yekaterinburg in the Urals (now Sverdlovsk), which was in danger of falling to White Czechs and Admiral Kolchak's troops. The execution, carried out without a court hearing, was recently been proved to have been done on the Kremlin's orders. Soon afterwards, the Czar's brother Mikhail was also assassinated. At the time it was said, however, that he made use of the rumours of his death, which the government disseminated, gave up his station in life, and wandered about Russia as a *starets* ("wise man") till the end of his days.

In December 1922, four republics – Russia, Transcaucasia, the Ukraine and Byelorussia – founded the Union of Soviet Socialist Republics.

The original plan for the formation of the new state, drawn up by Josef Stalin, envisaged autonomy status for the union republics; essentially, it was an infringement upon their rights. Lenin chose another form of unification – a voluntary alliance of independent republics with equal rights.

On 30 December 1922, the first congress of Soviets of the USSR adopted, on Lenin's principles, the Declaration and the Treaty on the Foundation of the USSR. In the summer of 1923, the first Soviet constitution was ratified. The constitution granted each republic the right to free secession from the union, and gave any socialist republic the right to join the union. Union authorities were placed in charge of foreign policy, economic planning, matters pertaining to borders, war and peace, the armed forces, and so on. Each union republic had a constitution of its own. Today, it is easy to see the result: nearly every republic wants out of the union. There is hope that the new union treaty will be more than just the piece of paper that the old turned out to be.

In the early 1920s, the cohort of Lenin's aides was increasingly dominated by Josef Stalin, who spearheaded the struggle against Trotskyists, right-wing opportunists and nationalists. He became General Secretary of the party's Central Committee and, once

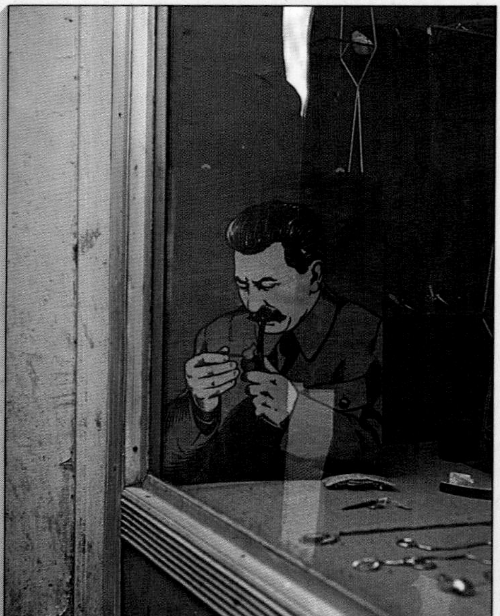

firmly established, proceeded to eliminate his political rivals.

Stalin unchallenged: Stalin gave *carte blanche* to the secret police, headed by the political adventurer, Lavrentiy Beria. Beria's reign of terror resulted in the suffering of thousands. The years of Stalin's regime crippled the development of self-consciousness, freedom of thought and the sense of dignity of the Soviet people. Yet it would be erroneous to blame Stalin and Beria for everything. The blame evidently rests with the entire nation, which passively tolerated every new insult the tyrants could think up.

In July 1938, Japan invaded Soviet territory in the area of Lake Khasan. The invasion continued into Mongolian territory with the aim of capturing Mongolia, cutting the Trans-Siberian railway and occupying the Soviet Far East. The attack was repelled, but the threat of aggression remained.

At about the same time, Stalin concluded a non-aggression pact with Hitler, which the foreign ministers of the two countries signed in August 1939. Today, it has been proved that there was an additional, secret protocol to the Molotov-Ribbentrop Pact, which would have divided continental Europe into Russian and German spheres of influence.

But the pact could not keep the Germans at bay for long, and on 22 June 1941, they invaded Soviet territory without declaring war. Stalin, caught after a recent "purge" in the army without the greater part of his commanding officer corps, refused to believe, despite numerous intelligence reports to the contrary, that Germany would attack so soon after signing the treaty.

The defence of Moscow: Moscow was the goal set before the group of German armies involved in the operations code-named "Centre". Powerful fortifications were erected on the approaches to Moscow to stop the Wehrmacht's drive. Blocking the most probable lines of attack were some 24,000 anti-tank "hedgehogs" and over 1,500 km (930 miles) of mined timber barricades.

The entire nation rose to protect Moscow. High Command HQ sent reinforcements from its strategic reserves – divisions from Siberia and units from Northwest and Southwest fronts. Army General Georgiy Zhukov, commander of the Western Front, was put in charge of Moscow's defence. The Germans were stopped on Moscow's threshold. For the first time in World War II, the Führer's armies suffered a major defeat. The Blitzkrieg was over, yet there were still four years before the triumphant spring of 1945.

The post-Stalin years: When Stalin died in 1953, Nikita Khrushchev was elected first secretary. He ushered in the *thaw*, as it is called today. Khrushchev's most commendable achievement was his elimination of the legacy left by Stalin. He presided over the improvement of living conditions and promoted international contacts. Being virtually uneducated, Khrushchev was incapable of devising a working strategy for political and economic development; he also committed numerous errors of a subjective nature during his term in office (1953–64).

Economic blunder: His failure to assess the political situation and errors in foreign policy finally bore bitter fruit: after a quiet palace coup, Leonid Brezhnev was installed in power in 1964. During his 18 years at the top Brezhnev proved strong-willed, but nevertheless managed to disappoint the aspirations of his patrons: " he changed his spots", concentrated enormous power in his own hands, and surrounded himself with a host of admirers. Economic blunders and the death throes of democracy were hidden behind self-congratulation, hypocrisy and triumphant rhetoric. The gap between word and deed, the arrogant irresponsibility and nepotism finally led to total corruption of the party apparatus and, ultimately, to economic crisis in the USSR.

Perestroika at last: The turning point in political development within Soviet society came with Mikhail Gorbachev's perestroika campaign of April 1985. Essentially, it was a turn away from the concept of the Soviet people's victorious march towards communism, which had been endorsed for decades.

The power in the land was returned to elective bodies – the Soviets of People's Deputies. The country also got a president, whose authority is beyond comparison even with that of the President of the United States. The planned centralised economy is to give way to a regulated market. Enterprises are getting new rights. A land act has been passed that provides for active land ownership by individual producers – first and foremost, the peasants themselves.

Ever since 1985, Moscow's capital status has kept it in the limelight of domestic politics. Gorbachev's perestroika started the crucial change. Its opponents attacked all radical measures directed towards economic improvement and the dismantling of the bureaucratic system. The Communist Party rapidly lost ground. One republic after another announced plans to secede from the Soviet Union. And while even the name of the country and many of its cities have changed, Moscow will remain Moscow, the living heart of this huge Eurasian country.

Left, still present in people's mind: Stalin. Above, "Boris, we are with you".

Pizza

Пицца
Хат
Express

Patriots are few and far between in Moscow. There is much cantankerous complaining and masochistic backbiting when Muscovites talk about their native town and themselves. Everyone seems to have a grudge against something. As with all generalisations, there is the risk of oversimplification: still, it seems worth making the effort and try and sum up the life of the typical Muscovite of today.

Traditionally, the native of this city is something of an intellectual – an engineer, doctor or librarian with an average income. He lives, as we often say, "on his salary alone", and often borrows money from colleagues or friends. Haunted by the housing problem throughout his life, he lives, at best, in a cramped co-op built with his scanty savings, or in a room in a "communal apartment" in some shaded sidestreet of old Moscow. "Communal apartments" (in which several families share the kitchen and the bathroom) still give shelter to something close to a million people, most of whom are true-blue Muscovites.

The dream of owning a car: The average Muscovite, even one whose monthly take-home pay is 300-400 roubles, clearly cannot afford an automobile, which costs between 7,000 and 10,000 roubles (between 20,000 and 30,000 on the black market). Yet, somehow, people manage. Many dream of having a car, but the honeymoon of the lucky owner is usually short-lived: garages are practically non-existent. This lends the owners of brand-new Ladas the sensitivity of a watchdog. They have been known to rise many times during the night to peer through the darkness – is someone out there fiddling with the doorlock of the compact?

Then there is the nightmare of spare parts. These are usually acquired through dubious characters at exorbitant prices (the state-run car industry is still in the Middle Ages). The list of woes has expanded recently and now includes fuel problems – the lines at petrol stations are, at times, several miles long. The only advantage Moscow has over a Western European city is that there are no parking difficulties. But problems are on the horizon: however slowly, the number of private cars is growing.

A broad outlook: Whatever the problem, people find ways to remedy the situation.

Preceding pages: the new joys of the Muscovites. Left, *Walking Down Tomorrow's Street*, painting by Jury Pimenov, 1957. Above, for the young, symbols have become decoration.

The vintage Muscovite is a person with a broad outlook and diverse interests. You won't catch him locked in an artificial little world, in the circle of everyday problems. He reads all the latest bestsellers, stands for hours to see avant-garde exhibitions and storms the ticket counters of international film festivals. In politics he is knowledgeable, radical and has become fervently anti-totalitarian.

Today, he is free to do what he likes: liberty has come to the land, democracy forges ahead, and he willingly takes part in meetings and demonstrations or even in the creation of new parties. This in a city where freedom of thought was traditionally punished with labour-camp sentences, exile to Siberia or, at best, banishment from the country – the past fate of the cream of Moscow's intelligentsia.

The typical Muscovite: Moscow is a town of celebrities. There are hundreds of people whose names are known at home and abroad: the ballerina Maya Plisetskaya and the poet Andrei Voznesensky, the cantatrice Yelena Obraztsova and the playwright Mikhail Roschin, the renowned economist and Presidential-Council member Shatalin and the historian Yury Afanasiev (now one of our most radical deputies) – a list that could be continued. There are also thousands of the less well known, who are, in their own way, no less interesting. They are the people who form society's "fertile layer". Most of them are rather eccentric; then again, as the old adage has it, "queer fish make life brighter".

An example is 63-year-old Muscovite Yury Mazurov. All his life, he has been collecting postcards dedicated to Moscow and he now has close to 13,000. It is by far the largest collection in town. You won't find some of Mazurov's prized cards even in the largest museum collections. "For me, postcards offer a glimpse of long-gone corners of the city, a remembrance of my early years. They show how Muscovites lived in past centuries, how they dressed, how they looked."

The gap between Moscow and the provinces has dramatically widened over the last few decades, encompassing food products and consumer goods, creature comforts, entertainment and spiritual interests. In the minds of millions of people – particularly the young – the myth that "you have to leave your hole and go to Moscow to see some real life" predominates.

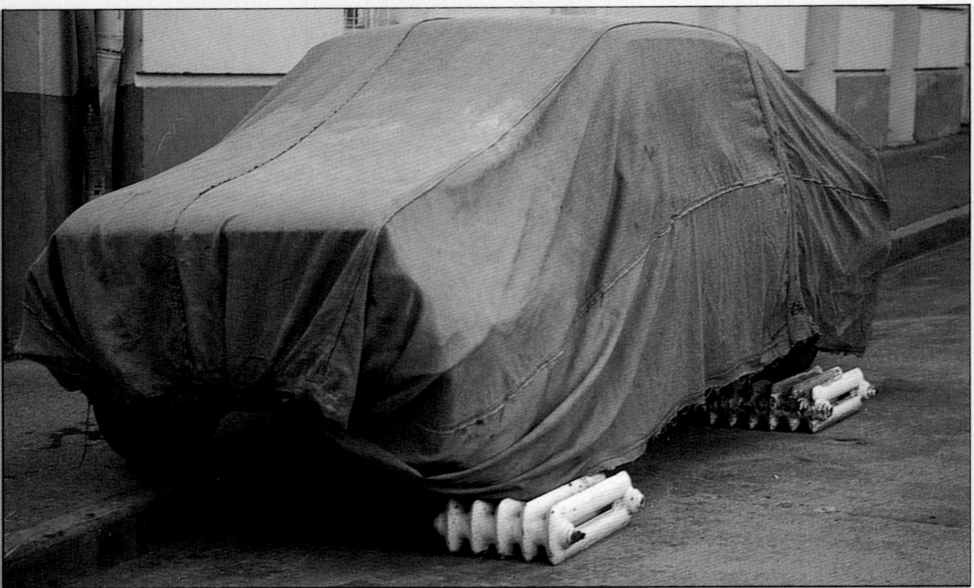

Migration from the provinces: In the meantime, Moscow is growing. Nine million live where there were 7 million 20 years ago. This has almost nothing to do with natural increment, because families are small (two children are a rarity, most often, there is a single child in the family, and sometimes even none at all). The main factor behind the population growth is the migration of out-of-town manpower. This, naturally, creates many problems, all of them formidable in their scope.

National sentiments: Like any megapolis, Moscow is a melting pot, a kaleidoscope of faces, languages, accents, religions. Russians form the traditional majority. There are many Ukrainians, Tatars, Armenians and Jews. Until recently, racial tensions were non-existent, despite a certain amount of officially approved discrimination. Jews, for example, found it difficult to attend colleges.

Today, when there is a broad awareness of mismanagement on the political level, social consciousness exhibits a paradoxical increase in nationalism, which sometimes acquires ugly forms. There are several reasons for this. Democratisation has fostered an awakening of national self-consciousness experienced even by Russia, the largest nation in the former Soviet Union, which, it transpires, has worse living conditions than most of the other republics.

This awakening of national pride is, in itself, laudable. Unfortunately, the majority of our people lead such a difficult life, and the crisis facing society is so formidable, that blind anger mounts and causes people to seek culprits. More often than not it is the people of other nationalities who are made scapegoats. Hence the black-shirted toughs from the notorious Russophile pseudo-patriotic society *Pamyat* ("Memory"), who shout anti-Semitic slogans and paste leaflets full of malicious slander against Jews in the streets. They enjoy little support, yet they certainly manage to poison the atmosphere.

<u>Left</u>, parking Moscow-style. <u>Above</u>, the young share the same dreams everywhere.

Foreigners in the capital: Foreigners occupy a special niche in cosmopolitan Moscow. Previously, foreign tourists, businessmen and scientists attracted by some international symposium or other, and members of endless delegations, stood out in the crowd – simply because they were differently dressed. Not so today: Muscovites, particularly the young generation, now wear the same kind of clothes as the foreigner (one cannot help wondering where all these chic items come from, considering the country's endless deficits). Yet foreigners are still easy to recognise by their untroubled, carefree expressions (the average Muscovite zooming through the rough-and-tumble of the street crowd usually has the resolute, tense expression of the Marathon runner).

Traditionally, foreigners have been treated with special deference, a quality which dates to olden times. When Peter the Great set out to fashion a European nation out of medieval Russia, he borrowed a great deal from the Germans, who shared an entire quarter of Moscow with other foreigners. Yielding to the czar's pressure, Moscow and St Petersburg aristocrats adopted the German way of life. Later, in the reign of Catherine the Great, Russian aristocrats started to emulate all things French.

Something of the sort is also going on today – with America as the standard this time. Indeed, the young generation is thoroughly Americanised – but only outwardly. Here and there, in the crowd, one sees T-shirts with logos of American universities (worn for the most part by people unable to make head or tail of these universities; they probably wouldn't know where to find Moscow University, either). American rock music blares from portable stereos and youngsters show off their American slang, distorted by hideous pronunciation. Many have a working knowledge of English, which is usually just good enough to direct the lost foreigner to the Cosmos Hotel. They may even accompany them there. People, worn out as they may be by everyday trivia, are still hospitable and, deep down, very amiable.

Hospitality: Moscow has always been known far and wide for its hospitality and willingness to welcome people from distant lands. Even in the days when it wasn't safe (to put it mildly) to associate with foreigners – until very recently, by the way – Muscovites ignored these restrictions. Today, when we are well on our way to becoming an open society, no one is surprised when schoolchildren from the US or France, who come to Moscow under educational exchange programmes, stay with the families of their Soviet peers rather than in hotels. There are also adult tourists and businessmen who, driven by a catastrophic lack of hotel rooms, rent apartments or rooms from Muscovites.

Every foreigner who has spent at least one evening in the company of a Moscow family will speak of the hospitality and amiability of the hosts – with a twinge of surprise at their uncanny ability to stack the dinner table with foodstuffs and exquisite bottles, considering the emptiness of the stores. But that secret must never be revealed to the guest.

A run-down service industry: It is a pity that a visitor's favourable impression of Moscow can be jeopardised by the various blackguards he encounters in our service industry – extortionist cabbies, slow-coach bellhops, untidy and somewhat rough-mannered waiters. Yet do not hasten to judge Muscovites by these people, because Muscovites suffer from them in the same way that you do.

During the long, long years of stagnation behind the iron curtain, official propaganda hammered home the idea that Muscovites lived in what was, for all practical purposes, the best city in the world – clean, green, safe and inexpensive. International observers painted intimidating pictures of New York ghettos, Bangkok's poisonous fumes, Paris-

ian prices, and lonely old men dying of hunger in unheated rooms somewhere in London. Moscow, meanwhile, it was implied, was just a breath away from becoming a "model Communist city".

Pollution: In the meantime, the former city fathers delivered speech after self-congratulatory speech – and destroyed the economy step by step, undermined the capital's infrastructure, and stole whatever they could lay their hands on (they were particularly partial to works of art from Moscow museums). Today, when many of yesterday's secrets have been revealed, it is clear that Moscow is not doing well at all. Half the population lives under what experts in the field call "ecological discomfort". In plain language, this amounts to air pollution (auto exhaust fumes and toxic industrial emissions), water pollution, excessive noise and vibration levels, and so forth. Members of the medical profession insist that at least every fifth Muscovite who falls ill is the victim of ecological hazards.

It also transpired that there are, in the capital, tens of thousands of old women and men who eke out an existence on a 900-rouble monthly pension with no one to care for them. Imagine what living on 900 roubles is like if the minimum standard of living is 800 roubles a month. This figure, by the way, came under fire from numerous economists, who consider it much too low on account of increasing inflation. While the minimum pension was raised in 1992, consumer goods prices have skyrocketed, making the situation even more difficult for the old.

Housing: We have already touched upon the housing problem. To get a better, larger and more comfortable apartment from the state, we have to wait for several years. Municipal authorities and trade unions usually regard families with many children as privileged and, for such families, the waiting is cut to a minimum. But for newlyweds whose children are still unborn, they have no choice but to crowd the apartment of one of their parents.

Demand is, as with many other things, way ahead of supply. "Looking-to-rent" ads abound, while "To let" signs are practically unknown. This naturally drives prices up. Ten years ago, a single-room apartment in a fashionable neighbourhood cost around 50 roubles a month; today, the price can be as high as 1,000 roubles (US$200).

There is a positive "but" to this situation. It may not be easy to get an apartment, true, but, once acquired, it is very inexpensive to maintain. Rent is low, especially by Western standards. A two-room apartment from the state costs around 60 roubles a month (gas and electricity included). You'll probably agree that it's not much, when the average monthly wage is around 1,900 roubles.

Most people would love to pay more in order to live more decently and it is not for nothing that people have a hard time finding a co-operative which will accept them. Co-

Left, the shops may be empty, the tables are full. Above, pollution has become a big problem.

operatives build houses and apartments with their members' money; a single-room apartment costs over 100,000 roubles, of which half can be paid on credit.

What the municipal authorities propose today is to sell state-owned housing into private hands. Many will find this change a welcome one: at last, my own apartment, to do with as I see fit! It can then be sold or bequeathed in a will. Until recently, when an elderly mother died, her son, registered in his wife's apartment, could not keep his late mother's place.

Social services: There has been much talk in the press about the fact that free social services, which have traditionally been hailed as a triumph of socialism, are not necessarily a blessing – there is a darker side to the matter. Free housing – after years of waiting. Free – but hopelessly retarded – education. Free – yet impotent – medicine. The doctors who work at outpatient clinics are overworked beyond the limits of human endurance. At times they don't even have the time to examine their patient properly. Hospital equipment is obsolete. There are shortages of the most elementary drugs. Naturally, the doctors do everything they can for the patient, but a surgeon's skills are not enough when it comes to post-operational nursing, and there is a hopeless shortage of hospital attendants since private nurses are unheard of. The picture is far from a pretty one, true, but the people of Moscow do not give in to the gloom. They are patient, since they have never been pampered.

Remembering the difficult post-war years, when decent going-out clothes were a rarity and everyone lived in overflowing "communal apartments", they remark: "before, we were poor, but life was happy and people were more friendly to each other." Muscovites are also saved by their sense of humour and briskness of character, a devil-may-care good-naturedness.

Daily chores: Let's try to follow the average Muscovite's day. Let it be a woman, because women have more responsibilities than men, and hence lead a more eventful life. On her way to work, she brings her youngest to the kindergarten, of which the child is sick and tired. But she thinks that the ingrate was lucky to have been admitted into a kindergarten of such favourable repute, with kind instructors and even a course in the basics of esthetics – and all for only 800 roubles a month.

Meanwhile her oldest is on the way to

school. Under his shirt, tied to a string around his neck, there is the key to the apartment – he will return home before his parents come back from work. Both parents are proud to see their son in a prestigious "special" school with "enhanced study of the English language". After he leaves – who knows? – he may be admitted to the still more prestigious Institute of International relations. Then it's plain sailing to the diplomatic Olympus. And who cares if the future Talleyrand loves to fiddle with all kinds of machines rather than reading Shakespeare in the original? In Moscow, as elsewhere, parents decide for their children.

The reading room Metro: If you look closely, you'll see that the entire underground carriage resembles a reading room on wheels: many people don't find other time to read – and so, the underground system comes to the rescue. The underground, by the way, is clearly a feather in Moscow's cap: it is clean, pretty, and efficient – trains arrive every minute in peak hours. It is also quite cheap – pay your 50 kopecks and go anywhere you want, with any number of transfers – stay there all day if you want to.

Buses, trolleys and trams also cost 50 kopeks regardless of the number of stops you travel. Surface transport, however, is much

With the youngest safely in kindergarten, our mother storms a bus (filled to overflowing), which takes her to the metro. There she has her first good luck of the day – some well-mannered young man surrenders his seat, giving her the opportunity to read. For the 20 minutes or so it takes to reach her destination, she might read a literary journal with a previously banned novel by one of the recent past's non-persons, now restored to his rights as a Soviet citizen.

Left, in a Moscow high-rise apartment. Above, the Moscow Metro.

less reliable than the subway. In winter passengers shake from the cold waiting for delayed buses at the stops. There are plans to double public transport fees, but that's hardly a problem – a rouble isn't money. The ancient adage about a penny saved is likely to produce a smile of condescension on the modern Muscovite's face.

Taxis aren't expensive, either – 6 roubles per kilometre – but difficult to flag down. It's easier to pay a private owner more to take you to your destination.

Arriving at the office, our heroine greets her colleagues, takes out her cosmetics bag and starts to work on her face. There is no time for that at home, but at work, you can always spare the required few minutes. The other women in the room are doing the same thing – an occurrence to which their male colleagues have long since become accustomed. In any case, the men don't have time for such trifles – they are on their way to the smoking room.

Shopping: During the lunch hour, the women hastily grab a bite and start to "distribute assignments". One is delegated to the nearest dairy store, another to the bakery, a third to a vegetable store. Each buys for three. This saves time, because the employees of neighbouring institutions aren't sitting idle – they are already standing in line.

After work, the first thing to do is to fetch her child from kindergarten. Once home, she thoroughly questions her elder son about his progress at school. Her husband has not been assigned any household chores because, after work, he and a friend stopped off for a drink on the way home. He is therefore unnaturally gallant. Trying not to breathe in the direction of his spouse, he sets the table with elegant movements. Over supper, between answering the children's interminable questions, the adults speak about almost everything, from troubles at the office to global problems, giving the highest echelons of power the dressing-down they seem to deserve. Before, parents tried to avoid discussing these things in front of children – God knows what they might tell at school, and then trouble would follow. Today, no one is afraid of anything any longer – a sure sign of society's recuperation.

Television: After supper, *Novosti* (a national news programme) is a must. The programme has become very interesting in the past few years, because it covers our life without colouring it pink – and life in the West without painting it black. After watching the news programme, some choose a movie, others a sports programme, and still others a televised discussion of the day's problems. It is interesting how Muscovites prefer TV programmes from Leningrad: more often than not, they are sharper, more scandalising and cleverer than what Moscow has to offer. As for the women, they rarely watch any programme to the end – there is cooking and laundry to be done. The men have it incomparably easier, even if they condescend to do the dishes.

There are, of course, elements of diversity: entertaining at home, going out, movies, theatres, and so on. But this gives rise to new problems. It isn't easy to get a ticket to a good theatre. A large proportion of the audience of the world-famous Bolshoi is made up of hard-currency paying foreigners.

Political life: The years of perestroika and glasnost have turned life in Moscow around, largely ridding it of its habitual patina of bleak indifference. Life has become politicised. Formerly off-limits to TV viewers, parliament sessions are now broadcast live. The man in the street is tired of being a cog in the soulless clockwork that is managed from the top. Everyone openly states his mind on each and every issue, and argues in favour of his platform.

During one of the last May Day parades in Red Square a group of demonstrators addressed the country's leaders from on top of the mausoleum and demanded their resignation. It was not a very tactful thing to do but it is significant that no one touched the rebellious loudmouths.

According to some estimates there are over 40 political parties and movements in Moscow alone. Previously, the word "party" could only mean the Communists. Today, it applies to the social-democrats, the greens, the liberal democrats, the anarcho-syndicalists, all of whom have their own platforms, programmes, ideologues, organisations and newspapers.

At one end of the city, thousands gather for a rally of perestroika's most radical supporters. Egged on by the approving hum of the crowd, speakers criticise Gorbachev for indecisiveness and softness towards conservatives. Meanwhile, others have gathered at the other end. They may have fewer supporters, but they are much more aggressive and passionate. "A market economy will lead to unemployment!" shouts a man who suspiciously resembles a typical *apparatchik*. "It will weaken the working class, it will help the shady wheeler-dealers. Who needs this return to capitalism?"

Left, for hard currency you can get everything. Above, the symbol is still valid.

In the centre of Moscow, in Pushkin Square, (where *Moscow News*, perestroika's most popular newspaper, has its offices) meetings and disputes go on round-the-clock. People form tight circles around those who engage in debate. When either runs out of wind he immediately gets help from the lookers-on. A grey-haired Stalinist faces a cocky student ("We fought with Stalin's name on our lips!" – "And he sent you to the labour-camps!"). An Azerbaijani and an Armenian argue over Nagorny Karabakh. A co-op worker proves to his state-employed opposite that he works more and harder than him and that there is nothing shameful about his comparatively larger income. "They are selling Russia out to the West!" comes the tortured scream from somewhere else.

Here and there, vendors of the "independent press" hustle their dubious leaflets printed on cheap paper. The small print is almost illegible, but the larger letters of the titles promise to uncover all the skeletons in the Kremlin's closets. Negotiating through

the crowd, not bothering to hide his disgust from these people who keep him from urgent business, an elderly sceptic mutters: "They've learned to read and write – when are they going to learn to work?" He is, in a sense, right – we have our Hyde Park, now we could really do with a City...

Revival of old values: We cannot work like the British or the other developed nations. We are more accustomed to taking everything apart than to building with patience and care. But as they say, "New is well-forgotten old." These "new old" things in Moscow include a revival of religion (even the word "renaissance" would not be too lofty here) that official ideology had helped everyone "forget". Churches are hopelessly crowded on religious holidays. Many young people in the churches look quite up-to-date and incongruous amid the ever-present old women. There is little, if any, reason to believe that it is a foppish desire to stand out that brings them here. Having lost the ideals that were hammered into their heads, our young grope for the truth, for a way towards self-improvement.

Moscow is a Russian city by its origins, and it is not surprising that the overwhelming majority of its churches belong to the Russian Orthodox Church. There is also a Catholic cathedral, a synagogue, a mosque and a Baptist prayer house. Few churches, however, are functional (they were all open before the revolution), yet with each passing year, more and more long-neglected churches which previously sheltered obscure offices or served as warehouses are being reinstated. The public has even come forth with a proposal to give St Basil's on Red Square back to the church, as the Assumption Cathedral in the Kremlin has been.

The church is a particularly active philanthropic institution. It does not limit its activities to generous donations to charity. Priests and parishioners visit hospitals and houses for the aged, they bring comfort, consolation and help to those who are unable to help themselves. Charity (aid to orphans, invalids, the poor) is also quite popular among atheists. It is another sign of "the new old", a revival of Moscow's golden tradition. In the days of yore, aristocratic ladies never disdained to lend a helping hand to homeless tramps. For us it would be a sin to neglect the less fortunate, if we really are to attain "humane socialism".

Since we mentioned the aristocracy, we should also say that it has been decided to revive Moscow's Assembly of the Gentry. The first meeting was attended by those descendants of noble families who had, by some miracle or other, survived the era of class struggle. The function of this "new old" association is not clear as yet, but our aristocrats now have a leader – Andrei Golitsyn ("Prince" Golitsyn), a well-known graphic artist who can trace his genealogy back to the 14th century.

New sights everywhere: Life really has become much more interesting. Before, could there have been, openly on sale, literature that was secretly photocopied and, at great risk, smuggled into the country? Could one have bought, with no trouble at all, tickets to a Western movie which our vigilante critics (who were allowed to attend the Cannes Festival) dismissed as talented yet hopelessly escapist and appealing only to man's base nature?

Even a simple walk can be fun. Take that vintage Moscow street, the Arbat. Picture galleries right on the pavement, concerts by street musicians, poetry, often bad, yet amusing and invariably imbued with great pathos by our homegrown authors – Krishna freaks, punks, and other picturesque groups of young people united by something you never knew existed. Fortunately, there are not many junkies among them.

It is easy to see signs of material stratification in our society – those very contrasts that our press used to criticise so much in the West. Above and over the surface of the Sadovoye Ring a suspicious-looking *nouveau riche* flies in his Mercedes; below, in the tunnel pedestrian crossing, an old woman stands, trembling hand outstretched in the beggar's pose. Beggars, by the way, have grown more numerous lately – they seem to feel that the cops won't hassle them any more and that the passers-by have become kinder.

The reverse side of the new times: Hotels are besieged by prostitutes (hard currency only). Our home-grown adventurers look at them with a twinge of sadness – they know that without dollars in your pocket, it is futile to hope for such exquisite female company. Moscow girls who dream of leading exciting easy lives regard these ladies of the night with malicious envy: an income such as theirs will forever remain a dream for a girl who keeps her virtue. The only way to make money and remain morally pure is to win, on a regular basis, beauty pageants and modeling contests, which have become very popular.

Growing crime has caught everyone unawares. Those papers that publish police chronicles report as many as one or two murders every day. If it is any consolation,

Left, an insider's tip: Moscow's nightlife. Above, the Moscow marathon is an international event.

most are committed in the traditional Russian manner – over a bottle. There is organised crime, too. The gangsters have automobiles, two-way radios and firearms, which are sometimes used in shoot-outs between rival mobs. It is then that the chance passer-by gets to see Chicago gangsters in action – and not on the screen.

Admittedly, the ordinary Muscovite has nothing to fear from the Mafia. The mobsters target the wealthy: co-op officials, collectors of paintings and antiques, certain joint-venture employees, and assorted dealers of the "shadow economy". Nevertheless, your chances of being mugged or beaten up are improving. Recently, private detective agencies have come to the aid of those who want personal safety for a moderate fee. If the militia have proved they cannot cope with criminals, then, say the cynics, the criminals themselves are all too eager to give you a hand.

The black market: Progressives are embarrassingly ready to admit that the country's life is poisoned – there is no other word for it, they say – by increasing numbers of speculators who make hefty profits thanks to shortages of the most basic goods. Cigarettes disappeared not long ago, for instance, and habitual smokers gladly pay six times the official price for a packet of cigarettes.

The profiteers sell whatever there is to sell – from vodka to furniture suites which cost many thousands of roubles. The country's only hope, it seems, lies with the market economy, which should take care, once and for all, of the shortages.

There is a variety of things good and bad in Moscow. Visitors from abroad usually experience the bright side of life; ordinary Muscovites often get the short end of the stick. But the winds of change are getting stronger. The power in the city now belongs to people of a new type – the progressive, independent thinkers and intellectuals.

Moscow's new mayor, the famous economist Gavriil Popov, and his young deputy, historian Sergei Stankevich – who has already proved himself a shrewd and principled politician – have set out, with the aid of their political allies and the most active sector of Moscow's population, to turn our town into a real "model city". It goes without saying that the people of Moscow are overwhelmingly on their side.

Above, most big hotels offer shows in their dining rooms. Right, Muscovites are sauna fans.

THE JOY OF BATH-HOUSES

In a small passageway at Okhotny Ryad, in a section of the former Marx Avenue, you will find Moscow's Central Bath-House, once known as the *Khludov Public Baths*, which are still the most popular in the capital, along with the Sandunov Public Baths, not far away, in first Neglinny Lane.

The Russians have always believed that taking a bath cures all ills. There is an old saying which runs as follows: the bath makes you sweat to get tough and to get slim.

The old tradition of taking a bath is an involved, multi-stage ritual. The bath-house itself consists, as a rule, of several premises, with the sweat-room coming first. It is not the cleanliness or exquisite workmanship of the interior, but the design – better or worse – of the sweat-room that generally prompts Russians to choose one particular bath-house in preference to any other.

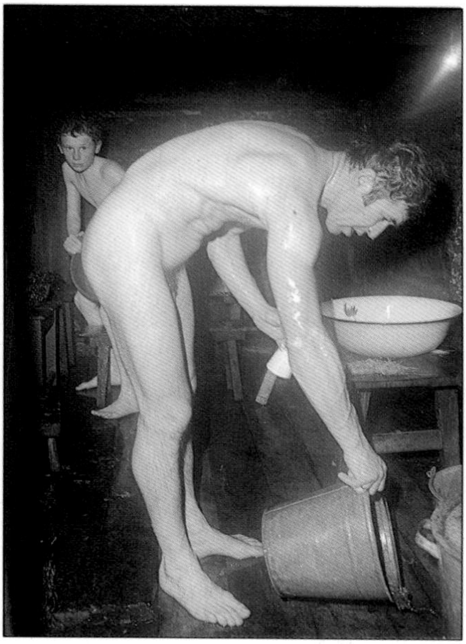

The temperature in it is not high; it is lower than that of Finnish saunas, and usually in the order of 60°–80° C (108°–144° F). But the humidity of the vapour makes you think it approaches 200°F. You enter the sweat-room, as a rule, with your head covered and long mittens on, equipped with a well-made and well-dried switch of oak or birch twigs. Before you enter the sweat-room, you must soak your switch in hot water to prevent the leaves from falling off when you strike with it.

The most respected man in the sweat-room is the one who knows how to handle the furnace: when to raise steam and how much, and what to feed the furnace with. Every sweat-bather has a formula of his own which he sticks to. Some like mint or eucalyptus added to the steam while others pour beer on the furnace. The heated-up and blazing bodies are properly whipped with switches.

You may occasionally shudder at the sight of a pair of husky fellows lashing their defenceless victim, seemingly more dead than alive. The latter, however, instead of pleading for mercy, smiles gratefully. Having got themselves to the point of half-fainting, with their legs about to give way, they then rush out to dip into icy water or, in winter, to dig into a snowdrift.

After a few spells in the sweat-room, you start the actual process of washing in a special wash-house where stone deck-benches and shower-baths are at your disposal. With a few simple gadgets, like a tub, wash-cloth and soap, you wash off what is still sticking to your body and soul.

Russian baths are big, as a rule, roomy enough for some 50 persons to wash at a time. Everybody willingly rubs anybody else's back and so hard, in fact, that the skin seems to be about to peel off. After washing most people prefer to relax in a lounge, a big hall with rows of benches and separate small compartments. You seldom go to a bath-house alone. It is common for a group of friends to do so in order to discuss all pressing and world problems in good humour and wash themselves at the same time.

It is also customary for bathers to drink all kinds of beverages to regain the moisture lost while sweating. Some prefer beer, others drink tea and still others treat themselves to various herbal potions. You may also enjoy a massage, a haircut and a pedicure.

In Russia you learn the tradition of going to public baths from childhood: fathers take their sons along and mothers take their daughters to initiate them into all the mysteries of the washing process. In short, you can hardly claim to be a Russian if you don't go to a bath-house.

SCIENCE PARKS AND THE BRAIN DRAIN

Arguably the worst problem of the USSR is the traditional inability to accomplish what is planned – in other words, a yawning gap between theory and practice. Many a grandiose plan has come crashing down because of this, starting from the lofty ideals of the October Revolution, the majestic five-year plans, food production programmes, the elimination of differences between manual and intellectual labour, and so on.

The result? Here is a humorous account by Mikhail Zadornov, a young satirist. "Developed countries laugh at the way we live. It is a vile thing for them to do. They should be grateful to us! The mission we chose for ourselves was all-important, and we accomplished it. We scared every Western country with the October Revolution and the way we live so much that they opted for another way of development and built socialism out of sheer terror!"

Past mistakes: Most Sovietologists agree that the USSR's level of inventiveness and theoretical science is reassuringly high but that the problem lies in putting these skills into practice. The main obstacles are the bureaucratic system, the absence of personal interest and initiative. In two or three spheres where the state spares no effort and creates ultra-favourable conditions – space exploration and certain types of munitions, for example – there are results. In all other spheres, almost none.

The situation, though, is beginning to change: not through "acceleration" of the scientific progress (as in the first year of perestroika), or through huge investments in machine-building or computerisation (as in 1986–87), but through the transformation of the economy. The programme adopted in the autumn of 1990 will once again make scientists, industrialists and peasants interested in the product of their labours.

It has finally dawned on Muscovites that empty shops, the low quality of their daily goods, inter-ethnic strife and political instability stem first and foremost from errors in the state's approach to education, science and culture. The state's economic mechanism retards the progress of science and technology, and drags down anything new. Only the market, progressives now believe, can reunite scientific and technological advance with the needs of the economy. The only solution, they say, is to integrate science, education and production. Leading scientists started talking about this a long time ago, but the adequately modern and appropriate form has only been found recently: the science park, the technopolis. Of all scientifically developed countries, the former USSR was the only one to ignore this movement for such a long time.

Finally there are results: Already, results have started to appear. The initiative, as usual, came from Moscow, from the new leadership of the State Committee for Education, which decided to use the thousands of first-rate scientists concentrated in the capital. The university and the technical university were chosen as the testing site. In 1990, a science park was established in Tomsk and an international seminar was held. Soon 10 further technopolises will be set up by leading higher schools in Cheliabinsk, Saratov, Leningrad, Dnepropetrovsk, Rostov and other cities.

The science park has no formal relationship to the higher schools. The university continues with its fundamental research, as all universities should. But when the time comes for the university to market the results of this research, the park begins to play its part, freeing the scientists from all the troubles associated with "implementation" by taking on all the duties of an intermediary. The technopolis will use a panel of experts to select only the most competitive projects, those that will find a use in industry and be marketable. In addition, the companies of the technopolis (in essence, it is an association of independent research and development, industrial and broker agencies) will provide employment to faculty members, post-graduates and students.

Intellectual property: There are huge problems still to be solved. The main one is the absence of legislation protecting intellectual property, which would give engineers, scientists and inventors the right to enjoy the results of their labour, to sell their ideas, to set up small and large companies, including high-risk ventures which would nevertheless have the opportunity of using bank credits, to develop successfully or to be driven bankrupt by the competition.

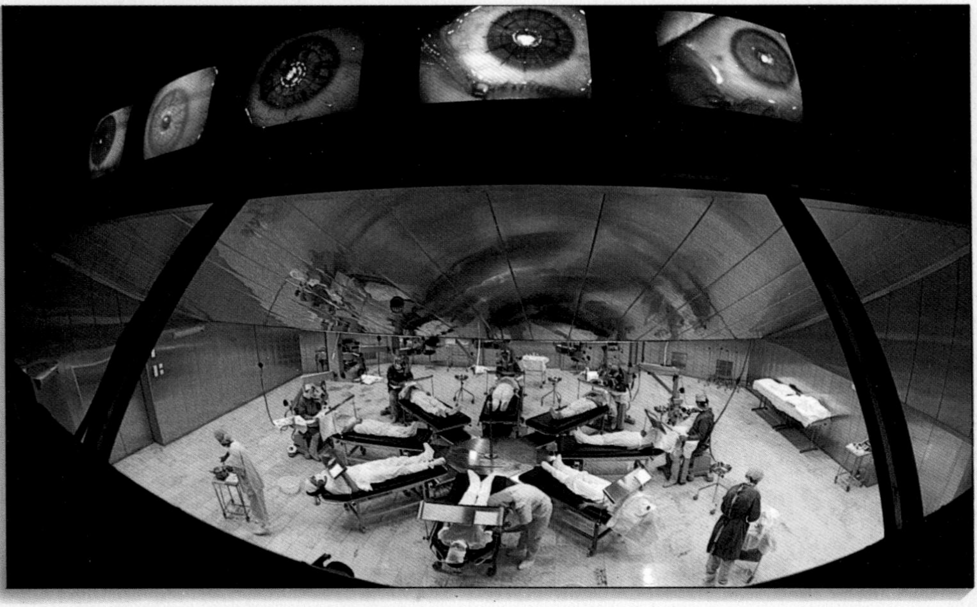

Finally, there is the problem caused by the "brain drain" from Russia reaching dangerous proportions and promising to accelerate when its citizens are allowed to travel freely. Russian scientists are arguing that the country's leaders must create the conditions which will enable them to capitalise on their creative potential, and to attract scientists and engineers from other countries. Technopolises are the tool on which many hopes are resting: they should, their advocates say, improve the conditions of research and life for scientists, the designers of high technology and new processes.

Left, the famous space monkey. Above, conveyor-belt eye operations.

CAPITALISM AND CHAOS

In the old days, before communism collapsed, no-one was even permitted to sell bunches of flowers at a street corner in Moscow. Suddenly, as the USSR mutated into the CIS, the pendulum swung to the other extreme and everybody could sell whatever they wanted, wherever they wanted. Stalls appeared, right on the steps of the best hotels, with underpants displayed alongside cases of champagne.

It was typical of Moscow that the approach to change was immoderate. Maria, a widow, was horrified to find that the spring bulbs she planted on her husband's grave, and more recently on her mother's, had been dug up and sold. Someone acquired a supply of fashionable shoes, and so potential customers balanced inelegantly on one leg on the pavement to try them on. Beggars held out a pleading hand, asking for a dollar or a mark. Who needed roubles any more?

The novelties that became available in previously puritan Moscow raised a few eyebrows. Whether it was the opening of a casino or a sex shop or the sale of flashy foreign cars, the first reaction was one of shock. The activity of street vendors became overwhelming. All manner of previously censored books appeared. Stalls were set up wherever there were most pedestrians. Street markets flourished, and it took a serious effort to remove the bazaar that had mushroomed in the very centre and relocate it in less obvious side streets.

But the centre of gravity shifted in a more crucial way. Once the Soviet Union split up, once the "unbreakable union of freeborn republics" (in the words of the national anthem) had been broken, Moscow was left not as the focus of a great empire but as simply the capital of the republic of Russia. In doing so, however, it became more cosmopolitan because the citizens of the other republics were now by definition foreigners. So, statistically, the number of visitors counting as foreigners vastly increased.

Some came to settle – like Anya, for instance, who arrived from the Ukraine. Her married daughter and two grandchildren already lived in Moscow, and, after retiring, she pulled up her roots and became one of the multitude of people sadly disorientated by the redrawing of the political map.

Attitudes to visitors changed, too. Once, foreign visitors were still special, respected, treated as honoured guests for whom only the best would do. All this is *passé*. Foreigners are no longer unusual, something from another world, another planet. Now they get cheated just the same as anyone else, in the restaurant and by the taxi drivers.

somebody else's room or flat is much cheaper for travellers than staying in a hotel.

It's difficult for first-time visitors to appreciate the scale of the changes. They won't glance twice, for example, at the Western advertisements they already know so well. But for Muscovites the images are something quite new. They had been brought up from earliest childhood to see red slogans with white lettering proclaiming "People and Party are United", "Forward to Communism", "Glory to the Communist Party" and so on, even shining at night in neon lights. Suddenly these all disappeared. No-one had paid any attention to them, of course, but the new advertisements are very different, significantly changing the city's appearance.

Local travel agencies have surfaced, although they are primitive, homemade affairs. They offer Russians all sorts of exotic trips to faraway countries – but at a price that is completely unrealistic for most people.

However, temporary home-swapping between Russian and foreign families, something once totally out of the question, has become increasingly popular. Even if the arrangement is not a direct swap, renting

Left, Yeltsin joins the line-up. Above, a broker gets to grips with the market in the Commodities and Raw Materials Exchange in Moscow.

The economic crisis hit the old age pensioners hard. Some regulars at the soup kitchens preferred not to drink their glass of tea there, but to take their tea leaves home for later or for breakfast, along with the teaspoon of sugar that went into each glass.

Take Klava, for instance: her wages as a cleaner are so low that she spends all her weekends and holidays earning a little extra by going out to cook or wash windows. Yura, on the other hand, is so sick of the whole situation that he prefers to go fishing instead of getting on with his interior decoration. Schoolboys equipped with cans of spray have been quick to learn the value of washing windscreens while the traffic lights are red – just as they do in London or New York.

Food parcels with a variety of basic groceries distributed in schools were welcomed by hard-pressed housewives. There was some black-marketeering, but sometimes it was better to sell a valuable tin of cocoa in order to buy expensive fresh vegetables.

tourists are now also shown the White House, an administrative building standing across the river from the Ukraine Hotel. It was here that, during the three days of the August 1991 coup, people gathered around Boris Yeltsin, defending the cause of democracy. Seventy years ago, when the Communists took office and Lenin was lending a hand to help clear the Kremlin, he joined a group of people who were carrying a log. In the years that followed more than 1,000 claimed to have lent a hand alongside him. In the same way now there are probably a million who talk about their loyalty to Yeltsin and Russia during the anxious and

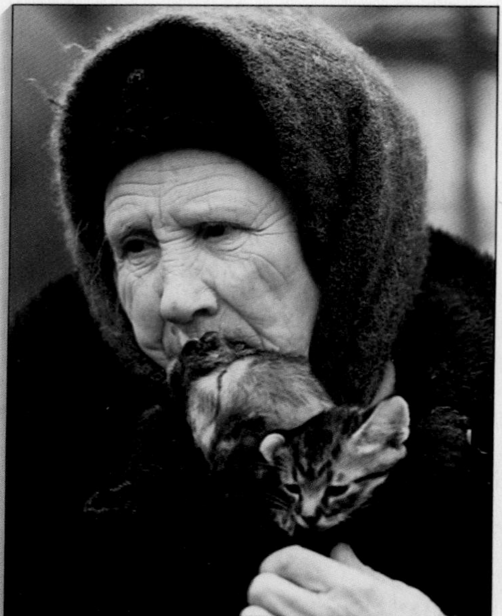

Further confusion has been brought about by the renaming of the city's streets, a process that is continuing. Don't be surprised if a passerby or even a taxi driver asks for the old name. Many Muscovites are newcomers and it will take years before everyone talks of Tverskaya instead of Gorky Street. The metro stations are causing the most trouble as they had no original names to return to. Sometimes they follow the name change of the street or square where they are located, but not always. Some stations have had their names changed three times already.

Apart from the city's historic buildings, decisive days that followed the coup.

Moscow lacks a Hyde Park Corner, and all attempts to keep demonstrations and meetings a bit out of the centre are futile. Different groups prefer Manezhnaya Square, next to the Kremlin, or Tverskaya Street. In 1991 the remaining scions of the old noble families gathered for a meeting in the October Hall of what used to be known as the Nobles' House. It has been used for fine concerts and dignified meetings, and is the building in which the bodies of Lenin, Stalin and other leaders lay in state before they were given a state funeral. In a similar way the merchants, or rather the descendants of prominent merchant families, have laid claim to the former Merchants' Club, which now houses the Communist Youth Theatre. And now members of the burgeoning *nouveaux riches* have organised a Millionaires' Club of their own.

Also under threat of eviction are the Lenin Museum, housed in the building designed for the City Council, and the Museum of the Revolution, using what was once the English Club. The Anglican Church of St Andrew's in the centre of Moscow, near the Conservatoire, now holds two services a month, while the rest of the time the Rossiya Recording Studio uses its fine acoustics to make gramophone records. The Catholics, not content to have their Church of St Louis back again, hold demonstrative open-air services in the street in front of another Catholic church near Gruzinsky Pereulok which is still closed. The Islamic Centre, which has opened in connection with the local mosque, provides a haven for visitors with both cultural and business interests, and at the synagogue they give information as to where kosher food may be purchased.

It is hard to list all the small but vitally important things which are happening in the capital. There are regular astrology programmes on television – horoscopes having long been an undercover science. Another novelty was the arrival of TV game shows with all kinds of prizes – among the most appreciated being packs of disposable nappies. Also new is the children's programme, not just designed for children but run by them. It promises to be a lively challenge to the extraordinary events now taking place in the grown-up world. Idealistic schoolchildren with their sharp eyes and fearless words have a marvellous platform to speak from.

A wide range of foreign newspapers is for sale, but where and how to find them is a different question. The *New York Times* has started appearing in Russian twice a week and other Western publishers are putting out feelers. There is a local, American-run FM radio programme and, late at night, live news telecasts from abroad receive simultaneous translation into Russian. More and more satellite dishes can be seen on the balconies and roofs, and there is a general feeling that life in Moscow is changing irreversibly.

After prices triple, an old woman tries to sell her puppy and kitten to buy food (far left) while would-be shoppers queue (left). Above, some people can still afford to dine at the Café Budapest.

Е. Гейтманъ

АЛЕКСАНДРЪ ПУШКИНЪ

THE MANY FACES OF MOSCOW CULTURE

You will have a hard time finding a Muscovite who doesn't consider him or herself to be at least a trifle more educated or the tiniest bit more cultured than "the rest of 'em" – that is, the unfortunate millions sentenced by fate to live outside the capital. Even janitors and cabbies look down, with heartfelt pride, upon their colleagues from the provinces. Not very modest, perhaps, yet you will find many capitals similarly afflicted. Some praise it as "enhanced awareness of personal predestination". Others dismiss it as "common bigotry". Both are probably wrong.

Since time immemorial, Moscow has been a Mecca for the enlightened and the industrious. The monasteries boasted impressive libraries and the first book was printed here in the mid-16th century.

During the Middle Ages the city became a haven for the fine arts. Rubbing elbows with Greek painters helped Russian artists to establish the foundation of what was later to become the Moscow school. The earliest records of musical life date to the 15th and 16th centuries (the Czar's Choristers); by Peter the Great's time, no festival or celebration could be held without music. Roughly in the same period, the first travelling theatre troupes of clowns and circus performers appeared in the land of the Russians. Russia, then, was far from "lost in ignorance and the darkness of cultural self-isolation".

Poets and writers: Moscow's books and magazines taught Russia "to speak and write in Russian," wrote poet Piotr Vyazemsky, a friend of Pushkin. In fact, it is difficult to think of even a lesser writer or poet who was not connected with Moscow in one way or another. Moscow was home for Pushkin, Lermontov and Dostoyevsky. Taganrog-born Chekhov experienced such an enormous creative upsurge here that he proclaimed himself "a Muscovite for eternity". Mikhail Lomonosov, the peasant son from the distant White Sea, came to Moscow and, a quarter of a century later, founded the university that still bears his name. "The oldest and the best university in Russia" gave the country's literature many first-rate talents: Nikolai Karamzin, Vasily Zhukovsky, Alexander Sumarokov.

As Alexander Herzen pointed out, when Peter the Great demoted the city from its rank of Czar's Capital, it nevertheless remained the capital of the Russian people.

Even though the cream of high society and the royal court resided in St Petersburg, Moscow managed to cling on to its cultural traditions. Away from rigid censorship and the stern eye of secret-police watchdogs, it became a breeding-ground for progressive thought and a haven for free-thinkers.

Pushkin, the great love of the literate: On 8 September 1826 Alexander Pushkin was conveyed from exile in Mikhailovskoye to the Kremlin for an audience with Czar Nicholas I. The poet was allowed to live in St Petersburg and in Moscow – under constant secret surveillance. The secret police did not

Preceding pages: tradition in the electronic age. Left, Pushkin as a child. Above, Kandinsky's *Improvisation*, 1911.

let the poet out of their sight. Pushkin's arrival in Moscow became a major cultural event celebrated by all. He brought *Boris Godunov*, written in exile, to Moscow. The first public reading of the play was on 12 September 1826. He was given a stern dressing-down from Benkendorf, the chief of the secret police, for reading the play without first submitting it to the censors.

Pushkin led a cheerful, if dissipated, life in Moscow. He was a frequent guest in the salon of "the queen of Muses and beauty", Zinaida Volkhonskaya. Here, he met Adam Mitskevich the Polish poet; here he first laid eyes on his "madonna" – Natalia Goncharova. He described his later life in Moscow as "married and merry". Memorials to Pushkin in Moscow consist of more than just the old houses and estates where the poet lived or which he visited. Pushkin's name was given to a street, a square, and an embankment, a museum and a theatre, a metro station and a library. He is still with us: he stands in bronze in the middle of Pushkin Square, high up on the pedestal, in a relaxed pose with a noble bearing, his face thoughtful and a little sad while, down below, crowds boil over at meetings as if excited by the poet's freedom-loving verses.

Mikhail Lermontov returned to Moscow from exile (imposed by Nicholas I for his poem *On Pushkin's Death*) in 1837. "Moscow is my native land, and will forever remain so for me: I was born there, I suffered there, and was too happy there!" he wrote.

On 25 February 1852, Moscow saw an extraordinary funeral procession as hundreds followed the hearse from the University to the Novodevichy Cemetery. It was in this way that Moscow paid its last respects to Nikolai Gogol. His popularity proved that he was loved and respected.

UNESCO pronounced 1981 Fyodor Dostoyevsky Year. The century of his memory and his 160th anniversary was celebrated throughout the world. Moscow is his native city. Here, in the courtyard of No. 12 Dostoyevsky Street he spent his childhood. Dostoyevsky travelled a lot, yet always returned to Moscow with joy.

Another literary Atlas, who was born in Yasnaya Polyana, also found acclaim in Moscow. "Any Russian looks at Moscow and feels that she is his mother," Leo Tolstoy wrote in *War and Peace*. Tolstoy's attitude to Moscow was contradictory, changing several times in his lifetime. He loved the city for the way everything Russian was concentrated and because it reflected so many centuries of history.

Anton Chekhov started his literary career in Moscow. In 1884 he graduated from the university's Department of Medicine, and found himself increasingly attracted to literature. Soon afterwards, the sign Doctor A. Chekhov disappeared from his front door, even though he joked that medicine was his "lawful wedded wife" and literature only his "mistress". In the last years of his life, Chekhov rarely visited Moscow, but he retained his ties with the Khudozhestvenny Theatre which staged his plays. *The Seagull* – that symbol of a Chekhov play – became

the theatre's emblem.

Vladimir Gilyarovsky was a contemporary of Chekhov and his bosom friend. A connoisseur of the epoch, he somehow did not fit into the mould of his contemporaries. Alexander Kuprin wrote that he could more easily imagine Moscow without the *Czar Bell* or the *Czar Cannon* than without Gilyarovsky. He was picturesque in everything – background, appearance, manner of conversing and, despite his childlike air, he had a multifaceted nature and brilliant talents. During the 19th century the city became a kind of a poet's workshop for all of these writers, a place where they polished their personal techniques.

The early 1960s saw the rebirth of an old tradition – "poetry nights". As fresh voices rang out, new problems were brought before the public eye. Yevgeny Yevtushenko, Andrei Voznesensky, Robert Rozhdestvensky, Bella Akhmadulina, representatives of the new generation, spoke about World War II, the future of the nation, and the role of the individual in the transformation of society. Yet the spiritual handcuffs and hy-

pocrisy of the epoch, now habitually referred to as the "stagnation", precluded the solution of those acute social and political problems which they raised in their poems.

Glasnost at last: Today, theatre posters once again carry the names of Gorky, Chekhov, Bulgakov, Mozhayev and Voinovich. People line up in front of bookstores and theatres and bring flowers to Pushkin's monument. There is now little doubt: a kind of a cultural turnabout is at hand. As perestroika brought dogmatic discipline crashing down, as the long-awaited Press Bill finally brought freedom of speech to the

Left, Lermontov and Dostoyevsky. Above, Chekhov and Tolstoy.

land, the reader was flooded by a torrent of literary publications churned out by all and sundry – state-owned, co-operative and even private companies. Independent information and telegraph agencies sprang up.

Desk-drawer writers and émigrés: Writers had gained their freedom at last, but an anticlimax soon followed. Ask any prominent publisher, and you're likely to hear that contemporary Soviet literature is rapidly losing moral and spiritual ground. Time will tell if this is true. Yet one thing is unmistakably clear: the 1990s are giving many grey-haired literary veterans a second youth. Writers such as V. Dudintsev, D. Granin, A.

Pristavkin, F. Iskander, V. Rasputin and Ch. Aitmatov – who for years wrote, as they say, "for the desk drawer" – have taken advantage of the changing times to make everything they wrote in the years of stagnation available to the reading public.

There are also double-edged publications on socioeconomics and history by the grandees of 20th-century Russian literature: Ivan Bunin, Vladimir Nabokov, Andrei Platonov, Mikhail Bulgakov, Varlam Shalamov, Vasily Grossman and Alexander Solzhenitsyn. In addition, there are the *émigrés* – G. Vladimirov, V. Voinovich, V. Aksyonov, I. Brodsky. The poor Russian reader, trained (or brainwashed) to take in mainly "trustworthy" and "permitted" facts, has suddenly been drenched by a Niagara of scandalous disclosures concerning the top leadership of the country, past and present. There are Utopian novels. There are even books about intimacy between the sexes! In a word, "non-party" literature is booming.

They say that the Muscovite has one undisputable advantage over the Westerner – the unique ability to read everything that turns up without a thought for more important things. But the pace of life has become so incredibly fast that there is no time nowadays even for that.

Painting: Moscow has been famous for its "masters of all arts" since the old days. The 15th century was blessed by the genius of Andrei Rublev. There's a museum dedicated to his work in the territory of the former Andronnikov Monastery, where the artist lived out his final years. In the 17th century, Simon Ushakov, an artist and engraver, organised a kind of an academy of arts in the Armory Chamber. The 18th century heralded a breakaway from medieval religious themes (mainly represented in icons) to secular subjects and, first and foremost, to the portrait. We have Peter the Great to thank for this – it was he who invited Western masters to Russia.

In the mid-1860s, Ivan Kramskoi became the leader of the 14 rebels – progressive artists who formed the Company of Travelling Artistic Exhibitions. At this time Ilya Repin was a frequent visitor to Moscow, where he painted several portraits. It was there that the artist befriended another titan of Russian culture: Leo Tolstoy.

The most poetic pieces of Russian national art came from Vasily Vasnetsov, who painted scenes from folklore, heroic epics and lyrical fairy-tales. Today there is a museum in the artist's fairy-tale house built according to his sketches.

A new phase in the development of Russian art was opened in the early 20th century by Konstantin Korovin, Vladimir Serov and Mikhail Vrubel. Korovin's work is "a feast for the eye". Attracted to the bright side of life, the artist neglects everything that interferes with harmony. Serov was a master portrait painter and an accomplished graphic artist. He was also one of the best-loved professors among the students of the Moscow Art School.

Passing by the freshly restored Metropol Building, you will no doubt notice the panels on the facade. Their author is Mikhail Vrubel, a master of monumental and easel painting, a graphic artist, sculptor and architect. In his imagination, he tried to enmesh the entire world of beauty, of all centuries, of all peoples and of all the arts, a concept totally in keeping with the times he lived in.

The new century evoked new images and brought new names – Kasimir Malevich, Vladimir Tatlin, Alexander Rodchenko, Vasily Kandinsky, all of whom tried to find the language of the "new era", the "revolutionary epoch". And the world of their art has opened for Muscovites in the last few years.

The first decade after the war brought, together with reconstruction of the war-ravaged economy, the worst period of Stalin's cult marked by the degeneration of lifestyle and, in the arts, a striving for the pseudo-heroic, the falsely pathetic, the parade-like, the thoughtless sense of "well-being". A great many works of this period astound with their triviality and uniformity.

The early 1960s brought large exhibitions. Most significant was the retrospective exhibition of Moscow artists which, unfortunately, was trampled underfoot by the party leadership and the old men from the USSR Academy of Arts. It was a lethal blow to progress, after which many artists had no choice but to flee the country.

A new brand of artists: Today's generation of artists knows nothing of war. Their work reflects a different era, marked by a protest against rigid unification of purpose. It is based on diversity of plots, the desire to cover new spheres of life and philosophical problems, such as art and outlook, the human being and the planet, the artist and the world, life and time. Many of these works still have a hard time getting to people, being exhibited, at best, in Arbat Street or in one of Moscow's parks. Some representatives of the new generation were more fortunate: their works were sold at Moscow's first Sotheby's auction. Some bureaucrats "responsible" for art are still trying to puzzle out how that happened.

Left, the writer, Aitmatov. Above, Repin's *Cossacks Writing a Letter to the Sultan.*

COMPOSERS AND CHOREOGRAPHERS

Russian music tantalises, attracts, enthralls. It evokes the olden days – and the modern era. Aeons ago, it sprang from the deft fingers of the *gusli* (psalters) players (who also sang and told tales), the throats of the merry *skomorokhi* (travelling circus actors), the deep chest of the peasant ploughing his land.

"Song is the soul of the people" is the adage that great Russian composers have endorsed on more than one occasion. In a way, Russian folk songs spawned Russian classical music, the music of Glinka, Dargomyzhsky, Musorgsky, Rimsky-Korsakov and Tchaikovsky. Moscow, naturally, was always a center of Russia's musical life. It gave the world Alexander Aliabyev (1787–1851), the author of the famous romance, *Nightingale*, Alexei Verstovsky (1799–1862), and Mikhail Glinka (1804–57). One of the fathers of the musical scene was Mikhail Rubinshtein (1835–81), an outstanding conductor and a brilliant pianist. He put together the first music classes in Moscow, which were later reorganised into the original Russian conservatories.

Tchaikovsky: On 1 September 1866, Arbat Street filled with carriages as the cream of Moscow's intelligentsia gathered for the opening of the conservatory, an event that coincided with the beginning of Pyotr Tchaikovsky's extraordinary career. Tchaikovsky led a singularly eventful life in Moscow – professorship at the conservatory, friendship with people like playwright Alexei Ostrovsky, actor Piotr Sadovsky (who lovingly called the composer *oriole*), and the charming, talented Italian primadonna Desiré Arto. Then, of course, he worked in Moscow on such greats as *Swan Lake*, the *Symphony No. 4*, and *Eugene Onegin*. Tchaikovsky would have been 150 years old in 1990. The jubilee coincided with the latest annual international contest bearing his name. And to this day, the pride of the repertoire of the young contestants is the great maestro's *First Piano Concerto*.

Left, Pyotr Tchaikovsky. Above, Mikhail Glinka.

Moscow is also the birthplace of Alexander Skriabin (1871–1915), the pianist and composer who was fated to write a new page in the history of piano music with his emotion-filled symphonies. The last thing the Skriabin Museum, near Arbat Street, resembles is a museum. In the museum-apartment where the author of *A Poem of Ecstasy* and *Prometheus* spent his last years, everything has been preserved intact. There are sheets of music, manuscripts, and Skriabin's philo-

sophical essays. There is also the apparatus Skriabin designed to enhance the effects of his music with flickering light.

The new times produced new names: Isaac Dunayevsky, Tikhon Khrennikov, Rodion Schedrin, and Dmitry Shostakovich. Isolated from the outside world in the 1950s and the 1960s, Soviet music went its own way, drawing on its roots in search of profound links between the past and the present. This direction, Soviet *avant-garde*, is associated with the names of A. Shnitke and S. Gubaidullina.

Jazz and Soviet rock: Musical life in the

1960s would have been nothing without jazz, which had to fight a long and uphill battle to come into its own in this country. The first international jazz festival of the USSR was organised in Moscow.

The early 1970s gave us our first rock and pop musicians. Both types were frowned upon by the powers that be – both the performers and the listeners were accused of following Western trends and leading an "alien" way of life. Contemporary music, thank heavens, combines every known type of folk, classical and pop music. Cultural-exchange programmes with Western countries are growing, and rock 'n' roll fans will

hall in the Scientists' House, the Beethoven Hall of the Bolshoi, the Olimpiisky on Prospect Mira, halls in Izmailovo and Luzhniki – it's simply impossible to name them all.

The concert hall named after Tchaikovsky has a perpetual exposition dedicated to the composer of *Swan Lake*, along with a collection of bow instruments by Russian and foreign masters. The pride of the collection are 14 Stradivarius violins, seven instruments by Amati and eight by Gvarneli. In accordance with tradition, these instruments are made available to outstanding performers on tour in Moscow.

probably recognise the names of such popular groups as Mashina Vremeni, Park Gorkogo, DDT, Chyorny Kofe and Rock Hotel, and individual performers such as Alla Pugacheva, Vladimir Kuzmin, Zhanna Aguzarova, and Valery Leontiev. There are also the young hopefuls – Alexander Malinin, Katia Semenova, Alexander Yegorov and Yelena Sysoyeva.

Abundant music halls: Moscow has many concert halls, with music to suit any taste: the Column Hall and the Oktiabrsky Hall in the House of Unions, the Major and the Minor in the Conservatory, the Tchaikovsky Hall, the

There's a place in Moscow which music lovers hold especially dear: the Gniesinykh Institute. Created as a music school in the late 19th century by Yelena Gniesina, the institute has educated several generations of Soviet musicians, composers and performers. There's always music in Gniesinykh House – the latest work of a modern composer, a simple folk ballad, finger-breaking jazz passages, pop songs, you name it.

Ballet: The magical world of dance! Who can remain indifferent to this eternally young art? Who does not know the joy ballet brings? Ballet came to Russia in the 17th century; the first troupes appeared in the 18th century. Over the centuries, Russian ballet-masters created veritable masterpieces. Today, the rating of Russian ballet is so high that there is hardly a ballet theatre in the world which does not stage ballets by Russian composers (mostly *Sleeping Beauty* and *Swan Lake*). Russian ballet is Anna Pavlova, Mikhail Fokin, Kasian Goleizovsky, and Galina Ulanova. Sergei Prokofiev said about Galina Ulanova, "She is the genius of Russian ballet, its intangible soul, its inspired poetry." Today, Ulanova is a teaching choreographer with the Bolshoi.

Maya Plisetskaya is another much-admired *prima ballerina* (unfortunately, Muscovites do not see her quite so often as foreigners do). Uncanny expression, passion, dynamism heralded the appearance of her individuality from her first steps on stage. Plisetskaya, it turns out, is no less gifted as a choreographer, which is easily seen from her productions of *Anna Karenina* and *Seagull* (music by Rodion Schedrin).

Where talents abound: The troupe of the Bolshoi is full of stars – such as Vladimir Vasiliev, Yekaterina Maksimova and Nina Timofeeva. The Bolshoi's school of choreography has also given the world many talents: Maya Plisetskaya, Igor Moiseev, Nadezhda Nadezhdina and many others. The school is considered to be among the best anywhere in the world.

New rhythms: There is more to dance than ballet, of course. New times bring new rhythms. Hence the steady popularity of Boris Sankin's Rhythms of the Planet Dance Ensemble, which, as the name implies, dance every dance imaginable. Be sure to see a performance of the plastic drama theatre headed by Gedriavichus. Lovers of contemporary ballroom dancing can both admire the intricate *pas* of classical waltz, rumba or fiery lambada and test themselves in dancing schools, or numerous community centres which welcome all to the enchanting world of dance.

In the evenings and on weekends, young people flock for modern rhythms to discos in Gorky Park, Sokolniki, the Palace of Youth, cafés and disco bars; elderly people find themselves more attracted to wind orchestras which, appearing more and more often in the streets and squares of the city, remind them of the days of their youth.

Left, in the Bolshoi. Above, Maya Plisetskaya in *The Humpbacked Horse*.

THEATRE AND CINEMA

The theatrical centre of Moscow is Teatralnaya Square, where three theatres form a semi-circle: the Bolshoi, still one of the world's best opera/ballet theatres, the Maly Drama Theatre (the oldest of its kind in Moscow – it is nearly 170 years old), and the Central Children's Theatre. Just around the corner is the Khudozhestvenny Theatre (founded in 1898) and, a little further up Pushkin Street, Moscow's second music theatre, named after Konstantin Stanislavsky and Vladimir Nemirovich-Danchenko.

Perestroika did not leave theatres untouched. The Khudozhestvenny Theatre split into two after a scandal in which the stars backed director Oleg Yefremov and the lesser-known members of this oversized troupe broke away to form their own Khudozhestvenny Theatre (named after Gorky, and not Yefremov's Chekhov).

To avoid misunderstanding, remember that the real Khudozhestvenny is on Kamergersky Pereulok. The new theatre, which is headed by Tatiana Doronina, is on Tverskoy Boulevard.

The new wave falters: Not far away, in Herzen Street, is the Mayakovsky Theatre, headed by director Andrei Goncharov, who has managed to maintain his reputation for two decades. As perestroika shuffled off all notions of leftist and official art, the playwrights of the new wave (as they were called in the 1970s) seem to have lost touch with reality. Meanwhile, the radicals – Mikhail Shatrov and Alexander Gelman – occupy orthodox positions. Gelman accepted the nomination to the CPSU Central Committee at the precise moment that thousands were leaving the party – people such as foreman Potapov, the character in the play *Protocol of One Meeting*.

Nowadays, there are no "angry" or "pro-Gorbachev" theatres: it is each for himself. Sovremennik, the popular theatre on Chisto Prudny, seems to prefer classical plays by Chekhov, Bulgakov and Olby, even though its latest productions are based on books that were banned before perestroika. Galina Volchek has come forth with a bold production of E. Ginzburg's *Hairspin Route*, about Stalin in 1937, while actor Igor Kvasha has turned to the prose of Voinovich.

Moscow's most popular theatre is the Leninsky Komsomol Theatre headed by Mark Zakharov. It is probably the only place where finding a spare ticket is always a problem. There used to be many such theatres in Moscow. Today, when everyone complains about the years of Brezhnev's stagnation, no one appears to be aware of a new period of stagnation resulting from perestroika, a stagnation more formidable than ever, particularly in the arts. Yet Mark Zakharov did not recognise that there was stagnation. Even in the 1970s he fought the "official patriotism of the epoch", and produced not what the minister of culture and his circle wished – but only what he himself wanted. For this reason, his plays from the 1970s are still in the repertoire: *Til and The Star of Joaquino Murietta*, for example.

A recent addition is Ostrovsky's *Wise Man*, and *Mourning Prayer*, a play based on a collection of stories by Sholom Aleikhem. It is a real masterpiece, the greatest theatrical event of the past few years, marked by Yevgeny Leonov's inspired acting. Leonov had just recovered from a severe heart attack. Maybe it's the return to this world that gives him such painful, heartfelt understanding of his character *Tevie the milkman*.

Ten minutes from Chekhov Street, in Triumfalnaya Square, is the Aquarium Garden, where director Pavel Khomsky staged *Jesus Christ Superstar* for the Mossoviet Theatre. Everything that Europe and America marvelled at 20 years ago is now in Moscow. Better late than never. The theatre also offers another wonderful play – Piotr Fomenko's production of Camus' *Kaligula*. Seemingly about the long-gone past, the play nevertheless discusses many of the moral issues of the Gorbachev epoch.

Then there's Merezhkovsky's old *Pavel I*,

Moscow's first erotic play.

the play which describes the last two days of the emperor. Not for nothing did Pavel Heifits decide to produce this for the Soviet Army Theatre (Kommuna Square) at this particular moment. The play shows a court unhappy with the emperor's reforms and his desire to make Russia a truly European country with the best army in the world.

When the once-powerful Yegor Ligachev came to see the play, the actors joked that they taught him to arrange palace coups. According to Heifits, Pavel I dies because his reforms are not needed by the people or by his closest supporters, who prefer to live as they'd always lived – merrily fornicating, thinking of nothing, and the devil take the country.

Some say there are 600 theatres in Moscow – the same number as in New York. There are over 50 state theatres, and the rest are small studios which eke out an existence in basements, garrets and even apartments. Unfortunately, almost all are professionally incompetent and more often imitate theatre than create something artistically valuable. Yet they include such world famous troupes as A.Vasiliev's School of Dramatic Art and V. Beliakovich's Theatre on Yugo-Zapad.

Avant-garde Russian cinematography: Can we speak about Moscow movies? Is there a breed we can call the Moscow moviemaker? Emphatically, yes! A few minutes spent in the conference rooms of the Moscow House of Cinema (Vasilievskaya Street) during a congress, plenary session or demonstration, will make you realise that Moscow moviemakers produce an atmosphere of their own with a unique tradition and charm. They can easily be spotted: they are good-natured, hospitable, slightly worried and always in a hurry. And with good reason – the movie locations map covers many different studios, movie theatres, clubs and concert halls. Moscow moviemakers produce for an appreciative audience: statistics say that the average Muscovite goes to the movies once a week, sometimes twice.

It's difficult to be everywhere. Yet the Moscow director is traditionally hyperactive, open to life and loath to be locked away on a set, in the narrow professional sphere. The giants of Soviet cinematography who worked in Moscow – including Sergei Eisenstein, Vsevolod Pudovkin, Mikhail Romm and Andrei Tarkovsky, and the lesser known but still living Elem Klimov, Andrei Smirnov, Rolan Bykov and Sergei Soloviev – have always reconciled professional interests with a broad artistic spectrum.

Avant-garde movies: In the century that has passed since the first movies were shown in Moscow (1896), the city became the cinema capital of the nation even before it became the capital of the USSR. The overwhelming majority of the 2,000 films released between 1908 and 1918 were produced in Moscow. After the revolution, Moscow headed cinematographic avant-gardism with such innovative masterpieces as *Battleship Potemkin*, *Mother* and *The Descendant of Ghengis Khan*.

Moscow is the seat of the highest cinematographic authority in the country – Goskino, which is simultaneously the buyer and the supplier of films produced in accordance with state plans. During the years of "stagnation", bureaucratic pressure mounted, and the movie producers of Moscow suffered the worst of it because they were so near. The conflict escalated and finally flared up at the 5th congress of the USSR Cinematographers' Union in May 1986. The democratic election of an alternative leadership for the Union heralded the beginning of a new period – "perestroika in action". The Union now supported an entirely different kind of committee and began taking new initiatives.

Over 250 previously banned films were released. Many new artistic associations were set up: ASK (American-Soviet Cine-initiative), the society of friends of cinematography, the Christian association, the association of women cinematographers, and so on. Newly formed guilds in all moviemaking professions proclaimed it their goal to protect the rights of union members from arbitrariness.

Hot-headed, excited people come together in the House of Cinematography near the Union building. They are the members of the capital's professional movie club. There, in the two viewing halls (Bolshoi and Bely) and in the adjacent cosy cafés, that are linked by a monumental fresco by Fernand Leger, new films are shown. The ritual of presentation, when the entire team comes out on stage, has developed into a cult.

Art and politics: The schedule of the House is crammed with events. Perestroika has brought frequent meetings where universal problems are discussed – the election of people's deputies from among moviemakers, dialogues with prominent leaders of state, scientists, assorted celebrities and with stars of the captivating and dramatic "political theatre" of our times.

When the New Year draws near, the House turns into a forest of green fir trees, under which merry groups of people sit down to celebrate. The *Kapustnik*, a stage-show dedicated to the issues of the day, is famous across the country and is, as a rule, always filmed.

The award of the Nika: In 1989, a new ceremony was conceptualised, organised and successfully presented as a yearly event – the award of *Nika*, the professional award of the Soviet Cinematographers' Union. Nika, the Greek winged goddess of victory, is the younger sister of America's Oscar and Europe's Felix. Each professional moviemaker votes on a lists of candidates, which are put together by the guilds of the Union (guilds unite directors, actors, scriptwriters, critics, cameramen, etc.). Presentation is a glamorous affair: beautiful girls bring out the fateful envelopes with the jury's decisions; the envelopes are opened, and the statuettes awarded by our most famous poets, singers, and ballerinas. It is a festival of talent and good luck.

The House is always under siege. Sometimes the militia has to lend a helping hand when there are too many people trying to crash the gates and see one of their superstar heroes or heroines. Recently, the House got its first rival – Kinotsentr, a young institution in an old Moscow district, not far from the mayor's office, the Zoo and the pretty church of John the Baptist. The monumental edifice made of rosy tufa incorporates an entire complex of large and small cinema halls, a museum, offices and co-ops, an exhibition hall and a restaurant. During the International Festival, which takes place every odd-numbered year in July, the entire square in front of the movie centre becomes a unique world of cinematography, decorated with banners and flags.

The Bolshoi ensemble tackles a Russian classic.

A City Of Many Styles

If one was to describe the architecture of Moscow in one word, one might call it multistyle. Is this necessarily a criticism, though? Some people argue that Moscow does not have a face of its own (as does St Petersburg, which was mainly built at the same time). Moscow, which is nearly nine centuries old, has a great deal of intermingled styles, which tend sometimes diminish the initial impact made on the visitor by a city that conveys a whole host of preconceptions. Yet, the more one looks, the more one can see a cunning plan behind it all, which reconciles many things (irreconcilable at first sight) into a general cohesive scheme.

The first wooden structures which remotely resembled a fortress appeared on Borovitsky Hill in the times of Yuri Dolgorukiy in the second half of the 12th century. In the 14th century, Ivan Kalita built a real town, which was called Kremlin. The main churches in Sobornaya Square also date to that time. Late in the 14th century, during the reign of Prince Dmitry Donskoy, the wooden walls of the Kremlin were destroyed by fire, and resurrected in white stone (limestone mined near Moscow).

Real work on the Kremlin started only under Ivan III, who hired Italian masters for the somewhat intricate job of combining the originality of ancient Russian architecture and the perfect finish of Renaissance cathedrals. By the beginning of the 16th century, the Kremlin ensemble and the merchant quarters to the northeast of the main walls were largely completed.

Gradually, Red Square was cleared of all houses (to prevent fires) and became the main marketplace of the city. Behind it, three streets fanned out: Nikolskaya, Ilyinka and Varvarka. In 1535–38 the space between them, Velikii Posad, was sealed off with another thick wall. The area was called *Kitai*, which meant "middle fort".

Preceding pages: onion-shaped towers crown Moscow's churches. Above, the entrance to the Economic Achievements Exhibition.

Medieval architecture: In those days, the place to settle down was at the junction of Moscow's three rivers – the Moskva, the Neglinka and the Yauza. Hence the historical settlements Zaneglimenye, Zayauzye and Zamoskvorechye. Within the limits of today's Boulevard Ring, there were wide streets: Volkhonka, Znamenka (*Kalinin Prospekt*), Tverskaya, Petrovka, Lubyanka, Solianka, Mjasnitskaya and Maroseika. In the late 15th century, all the residential quar-

ters within the limits of these streets were surrounded by a whitewashed brick wall and included into the so-called *White Town*.

Finally, the growing capital got another, this time wooden, circular wall in 1591, complete with an earthen wall and moat (it was 16 km/10 miles long). The town within these new limits was called *Skorod*. The new wall had gates (*zastava*), from which new roads ran. To the southeast and the southwest of Kaluzhskaya Zastava, there were two roads – Bolshaya Kaluzhskaya (Leninsky Prospekt) and Shabolovka; to the east of Semyonovskaya Zastava – Taganskaya; to the northeast of Vladimirskaya Zastava – today's Enthusiasts' Road. Beyond the gates, there were settlements, estates and monasteries, of which several stand to this day – Andronnikov, Donskoy, Simonov Monasteries and the Novodevichy Convent. In this way the city grew in accordance with the typical Russian radial-circular layout which is still adhered to in our day. The main characteristic of medieval architecture in Moscow is the tent-domed church (these had cupolas in the form of a multi-faceted pyramid with a cupola on top). Of those which still stand, the most characteristic is the Ascension Church in Kolomenskoye.

Moscow baroque: In the 17th century, Moscow churches were traditionally decorated with glazed ceramic tiles. A wonderful specimen survives in First Krutitsky Lane (No. 4) – the Krutitsky *teremok*, which is part of the partially surviving estate. In the last quarter of the 17th century, old Russian architecture was dominated by the style known today as *Moscow Baroque* or the *Naryshkin Style* (named after one of the wealthiest aristocratic families between the 16th and the 20th centuries). The best examples are the Pokrov Church in Fili (1693–94, 47, Novozavodskaya Street), and the Archangel Gavriil Church (also known as the Menshikov Tower, 15a, Telegrafny Lane), which combine the strict lines of a tiered tower and secular decor, making one think of European rococo.

Peter's reforms turned architectural development in Moscow full circle. The spotlight was claimed by public buildings. For instance, whole sectors of the city were given to rectangular layouts that appeared chopped up into squares – Nemetskaya Sloboda, Lefortovo on the Yauza. Wooden streets gave way to pavements, cemeteries were removed from the city limits, and several orchards and parks were founded.

The Classic architecture of the 18th century is connected with the names of two famous Russian architects: Vasily Bazhenov (1737–99) and Matvei Kazakov (1738–1812). Bazhenov's best creation in Moscow is, without doubt, the Pashkov House – the former Rumiantsev Museum and now a part of the Lenin Library. His second masterpiece is the Tsaritsyno Estate – an ensemble of the palace and several pavilions picturesquely scattered on the banks of a pond as befits a typical sample of romantic pseudogothics. Left unfinished owing to the wrath of Catherine II (she did not like the symbols of the Masons secretly introduced into the architecture) and partially redesigned by Kazakov, the ensemble stood in shambles for almost 200 years. It is being restored today.

Matvei Kazakov was by right celebrated as a great architect. Since he built so many houses, people even speak about "Kazakov's Moscow". He built the Senate in the Kremlin (the Supreme Soviet Building), the Petrovsky Palace (today the Zhukovsky Academy), the University, which incurred a lot of damage during the 1812 fire and was rebuilt by Gilardi, and the Golitsyn Hospital (today the First Municipal Hospital).

construction boom, there appeared, instead of the Zemlianoi Val, the Sadovoye Ring; the Neglinka River was channelled into a conduit, and Aleksandrovsky Garden was founded on the site of the river.

It was a time for new talents to appear, of whom the foremost was Osip Bove who redesigned Red Square and participated in the construction of the Bolshoi Theatre and the planning of Teatralnaya Square.

In the second half of the 19th century, Russian architecture was gripped by crisis. From the artistic point of view, the period was dominated by eclectics. New compositions and layouts were mechanically deco-

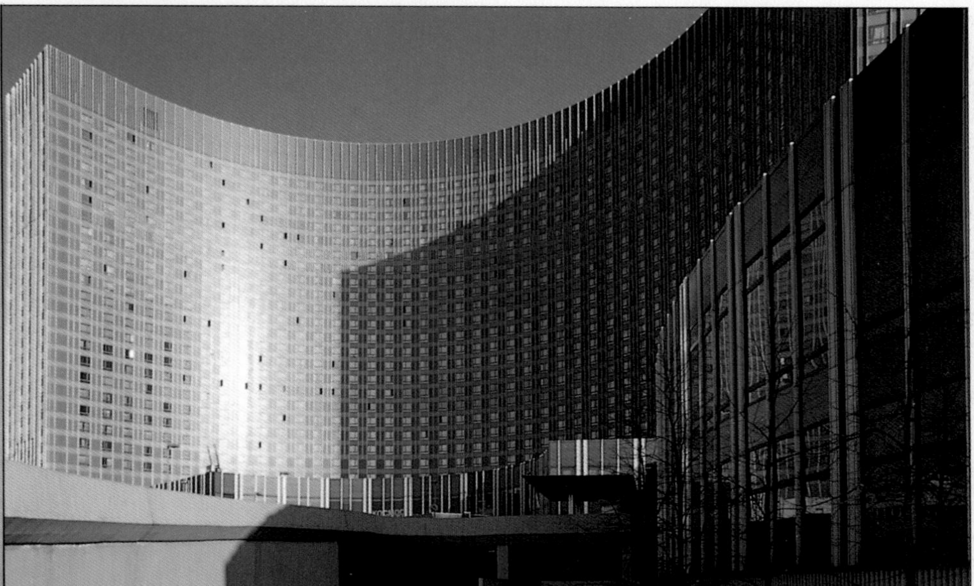

It should be mentioned that the two architects engaged in open competition which never seemed to end. Both were willing to rebuild the Kremlin as a fairy tale garden. Perhaps it is just as well that the Kremlin remained practically intact.

Reconstruction after the fire: The fire in 1812 put an end to the beauty of 18th-century Moscow architecture: 6,496 houses, 122 churches and over 8,000 trading facilities were burned to the ground. In the ensuing

Left, baroque facades and, above, the modern Cosmos Hotel.

rated with old forms and details borrowed from 17th-century Moscow churches. Witness the buildings of the Historical Museum, the Large Kremlin Palace and the Upper Market Rows (GUM).

The beginning of the 20th century was dominated by the Modern style. Its typical and most successful examples are the Riabushinsky Townhouse (Kachalov Street) – now the Maxim Gorky Museum, and the Yaroslavl Railway Station.

Budding capitalism brought utterly novel architecture: profit houses for different social groups, workers' and soldiers' barracks, large department stores, factories, banks, stock exchanges and railway stations.

After the October Revolution (the new broom sweeps clean), there came several plans for new "socialist" construction in the city, called "New Moscow" and "General Plan of Reconstruction in Moscow".

A time of "isms": The first decades of Soviet power turned out to be rather prolific architecturally, ranging from traditionalism (affinity for classical forms) and constructivism to avant-gardism and rationalism (laconic and economical). In the end, they all degenerated into gigantomania, self-congratulation in stone, and catering to mass

tastes. The buildings typical of that time can be found in Myasnitskaya Street (No. 39), Neglinnaya Street (No. 12), and in Red Square (Lenin Mausoleum). Another such project – the gigantic Palace of Soviets which was to have been topped by a huge Lenin statue – never materialised. This era also gave us the first metro stations, Kropotkinskaya and Mayakovskaya, which synthesise traditional architecture and the innovations of "proletarian culture".

In the 1950s and the 1960s, Moscow architecture was dominated by technologism – innumerable listless faceless boxes of concrete and glass (mainly of concrete), designed for the maximal possible number of consumers at minimal cost. These include the Palace of Congresses in the Kremlin, the Pioneer Palace on the Lenin Hills, Domodedovo Airport Terminal and Vozdrizhenka Street and Novy Arbat.

The 1970s saw little change, other than the fact that the boxes got a little higher. New residential areas designed for "self-sufficiency" appeared – microcities with populations sometimes exceeding 250,000.

The 1980s brought hope. It seems our architects are acquiring a taste for architecture. Some go in for complicated associations with ancient art, while others lean towards modernism.

There's a great deal of restoration going on, frequently by foreign contractors. Contemporary architectural experiments include Northern Chertanovo with its duplex apartments and underground garages, Krylatskoye and Strogino, and fledgling Solntsevo and Butovo. In a way, this also concerns the reborn Old Arbat, the redecorated Savoy and Metropol Hotels, and the newly built Slavianskaya Hotel.

The face of the city is made up of the houses in which people live and work. Moscow was historically a city of active silhouettes; its skyline is made up of church cupolas, spires of the skyscrapers built in the 1950s, and towering modern hotels.

The main orientation point is the Kremlin with the golden Ivan Velikii; it is easy to make out the Sadovoye Ring by the highrises in Smolenskaya Square, Kudrinskaya Square, Komsomolskaya and Lermontovskaya Squares. The Moskva meets the Yauza at the spot where there is another skyscraper (Kotelnicheskaya Embankment). Another highrise – Ukraina Hotel – stands at the bend of the Moskva River. There is a kind of creative paraphrase in the Moscow skyline, which combines Gothics with those Russian tent compositions that so amaze newcomers with their "sharp tops", colours and gold.

Above, Stalinist architecture: the foreign trade ministry. Right, one of the great 19th-century churches, the Temple of Christ Our Saviour.

THE TEMPLE OF CHRIST OUR SAVIOUR

Opposite Moscow's Fine Arts Museum are the extensive grounds of the Moskva Open-Air Swimming Pool. The site is a historically well-known place; sadly known, to tell the truth. Up to 1939 the Temple of Christ Our Saviour stood on this site. That giant structure had taken about 50 years to build.

The construction of this mammoth temple began on the site of the Alexei Monastery in 1839. The Cathedral was to immortalise the fame of Russian arms in the victory over Napoleon. The idea was that of Alexander I, but it was Nicholas II who succeeded in putting it into effect.

It was Konstantin Ton who designed the grandiose building in the best Empire-style tradition. The construction of the cathedral was finished in 1883. Pompous and bulky though it was, it did look fine.

For decades the cathedral stood as a kind of symbol of Moscow. But times changed, and so did tastes and mores. The newborn Soviet regime wanted to perpetuate its own existence. The idea of embodying the Republic of Soviets in an image of stone first cropped up, in fact, immediately after the signing of the Union Treaty in 1922. But those were trying times and the idea lay fallow until 1931 when it was brought up again.

It was then decided to build an enormous Congress of Soviets. Little by little more and more detail was added to the idea which, like Topsy, simply grew and grew. A special commission, under Molotov, Stalin's closest associate, was set up. An international contest involving many distinguished Western architects, such as Le Corbusier, Gropius and Mendelsohn, was held to produce the best design. Around 160 tenders were submitted. However, preference was given to a national group of architects, headed by Iofan. The design was for a four-stage cylindrically shaped building about 220 metres/725 ft high. But some clever heads announced that America had even taller skyscrapers. The project had to be revised as a matter of extreme urgency. The design work was personally monitored by Stalin.

Finally, in 1934, the blueprint was ready. The building was to be much higher than any American skyscraper. It was to be 420 metres/1,378 ft high, with a 70-metre (230-ft) high statue of Lenin on top. The main congress-hall would be able to seat 21,000. A frantic search got under way for a suitable place to site this giant. The one that was finally selected was the location of the Temple of Christ Our Saviour. It was hardly chosen at random since the former had, in a way, perpetuated the glory of the Russian Empire, while the latter was to do the same for the Soviet regime.

It was not enough to remove the temple alone for Stalin's monster to be put up. All adjacent structures had to be pulled down as well, because a vast square was to have been laid out in front of the Palace of Soviets for anniversary demonstrations and military parades. The Church, unwilling to accept the callous verdict of the authorities, resisted it to the end. Eye-witness accounts say that many monks taking shelter in the temple's underground cells never left that sanctuary. The temple was blown up and their bodies were buried in the rubble.

The job seemed to have been done and the time seemed to have come for construction work to start. Then, and only then, was it discovered that the floating bedrock could not support the pressure of such a giant mass of stone. The war, which broke out shortly afterwards, postponed this extremely "important" matter. After the war there seemed to be too many urgent things to do to distract manpower and building materials for bringing off such a pompous project, although the plans for building the palace weren't relinquished by the hot-heads in the Soviet leadership until 1959.

A giant circular pit was then excavated in the vacant ground on the site of the demolished temple. It was filled with water and a heated open-air public swimming pool, called the "Moskva", finally opened. Cleanliness, after all, is next to godliness.

In addition to the main pool, which is open all year, there are five shallower pools. The water, which never falls below 27°C (80°F), is changed three times a day. The pool's observation area provides panoramic views of the Moskva River.

THE PUSHKIN MUSEUM OF FINE ARTS

Moscow has many repositories of masterpieces such as the State Museum of Art of Oriental Peoples, the Museum of Decorative and Applied Arts, the Andrei Rublev Museum and others. But there is only one where the visitor can marvel at classical sculptures from ancient Greece and Rome and at renowned ancient and modern paintings.

The Pushkin Museum of Fine Arts, on Volkhonka Street, is relatively young. It was opened on 31 May 1912 when, in the presence of Czar Nicholas II, the doors swung open to the accompaniment of a cantata, written especially for the occasion, sung by a choir of 700. Roman Klein was the author of this architectural isle from Ancient Greece under the Moscow sky. Moscow owns this museum of world art that is second only to the Hermitage.

Intended to display a collection of copies, the museum gradually acquired a collection of originals. Antiques now occupy an area of 400 sq. metres (4,300 sq. ft). There are thousands of items, including the marble sarcophagus *Drunken Hercules*, a torso of Aphrodite, Corinthian vases, terracotta statuettes, beads, bracelets, rings.

The picture gallery began with a small collection of Italian paintings of the 18th and 19th centuries (a gift of the Russian diplomat M. Schekin). Muscovites will proudly tell you that the Museum now has six Rembrandts, five Poussins and Rubens, Jordans and Teniers canvasses. And the French collection! It is among the best in the world. Even the French will tell you that it's better to get acquainted with the Barbizon School in Moscow.

Gauguin's tropical pastorals represent the artist's vision of an integrated disposition, which fell victim to the civilisation of machines. *Bathing on the Seine* and *Nude on a Couch*, both world famous, reassert Renoir's reputation as "the painter of happiness". Some visitors will be burned by the red-hot disc of the sun in van Gogh's *Red Vineyards*, some will surrender to the colours and naivete of Henry Rousseau, the famous French primitive. Every collection, every name fuels admiration and pride.

Muscovites have grown accustomed, thanks to this Museum, to seeing on occasions, "all the flags" of art from different times, peoples and continents as "our guests". They have included Leonardo's inspired and

Left, the staircase of the Pushkin Museum of Fine Arts. Above, Karl Brullov's *Lady on Horseback*.

beautiful Mona Lisa, the treasures of the tomb of Tutankhamun, the redeemed glories of the Dresden Picture Gallery and the joys of the Prado.

The Museum never stops making Muscovites happy with new exhibits of painting and sculpture from throughout the world. Of course, like all other museums, there is the problem of funds for such basic necessities as the restoration of rooms and the acquisition of new works of art. The new city fathers are determined to remedy this situation, and there is reason to believe that they will succeed.

THE TRETIAKOV GALLERY

Everyone has heard about this world-famous museum of Russian and Soviet art. Everyone in Moscow knows where it is – in Lavrushinsky Lane, not far from the Kremlin beyond the Moskva River. The original facade of the gallery, which resembles either an ancient Russian *terem* (palace), or a portal of a church was designed by Victor Vasnetsov.

In 1856 Pavel Tretiakov, a young Russian merchant and industrialist, bought the works of Russian artists who were as young as he himself – Valery Yakobi, Mikhail Klodt. The collection grew and the owner decided to establish a public museum of national art, the first of its kind in Russia.

Initially, the paintings were kept in the Tretiakov residence, but as time passed, new halls were added, designed for the sole purpose of exhibiting the paintings. In 1873 the doors of these halls swung open to the public who could now view the collection that had by that time become famous in Moscow and was referred to, with increasing frequency, as the Tretiakov Gallery. The project took almost 40 years of Tretiakov's life. In 1892 he presented his collection of 3,500 paintings to the city. Today, there are over 50,000.

At first, Tretiakov bought the works of his contemporaries, preferably artists of democratic leanings – bright battle scenes by Vasily Vereschagin, the romantic *Princess Tarakanova* by Konstantin Flavitsky, *Bird-Catchers* by Vasily Perov. Then he filled a hall with spirited, inspired portraits by the old masters – Levitsky and Borovikovsky, the charming Rokotov, and several artists who were closer to Tretiakov's time – Tropinin, Venetsianov, Bryullov. Moscow had never seen such diversity!

Another major event for the gallery – and for the entire nation – was the purchase of such major works as *Ivan the Terrible and his Son Ivan* by Ilya Repin and *Morning of the Execution of the Streltsy* by Vasily Surikov. There, in that fairy-tale house in Lavrushinsky Lane, people got lost in the world of childhood tales, depicted by that kind sorcerer, Viktor Vasnetsov, in the epic forests of Ivan Shishkin, the delicate landscapes of Isaac Levitations, the portraits of soul-searcher and truth-lover Vasily Perov.

Sarayan's *A Date Palm, Egypt*, 1911.

A world of icons: Ilya Ostroukhov took an active part in the creation of the gallery. Himself an accomplished artist, Ostroukhov collected the icons which today form the bulk of Old Russian painting in the Tretiakov Gallery. There are ancient mosaics (10th–11th centuries), the *Vladimir Mother of God* – arguably the most famous icon in the world – a unique monument of Byzantine 12th-century art. It had a strange destiny. In the 12th century, Prince Andrei Bogolubsky brought it from Byzantium to Vladimir (hence the name), from where it made its way to a mightier overlord, the Prince of Kiev, and finally was installed for many centuries in the Assumption Cathedral of the Moscow Kremlin. The Vladimir Mother of God was the best-loved icon in Russia.

There are works by Andrei Rublev (the famous *Trinity*), Dionisius, and artists of the Stroganov School. Icons were painted on dry (kept for many years) planks of fir or limewood. Smaller icons were painted on a single board; larger ones required several boards joined by special splints. The functional side of the board was polished, after which a flax fabric was glued over the surface, to which a layer of chalk mixed with animal glue was applied. Only mineral paints were used. The icon was gilded with plate gold and covered with drying oil.

Another happy occasion for the gallery was the construction of the hall for Alexander Ivanov's *The Coming of Christ*, the largest painting in the Gallery. It provoked – and still does – rather heated debate.

In 1941 the treasures of the gallery were evacuated to Siberia, where they avoided the risks of war. In 1945, the hospitable doors of the gallery again opened to the public. Because the building is being restored at the moment, the bulk of the collection is on exhibit in Krymsky Val Street.

PLACES

A characteristic of Moscow, just as of many other old Russian cities, is its radial-circular layout. The **Kremlin**, or citadel, which had the form of an irregular ring, was surrounded, step by step, with new fortification walls, moats and ramparts. The gates in the wall were the starting points for a number of highways which radiated out, linking the city with other centres in Russia. As the rings were expanded, the defensive outposts on the approaches to Moscow were incorporated into the city, walls that hampered the growth of the city were pulled down, the moats were filled in and the ramparts were replaced by extensive blossom-filled boulevards and streets. Monasteries that doubled as fortresses were built on the most remote approaches to the city.

Five rings circle the city: The historical radial-circular layout of the city has survived to this day. The Kremlin wall still forms the nucleus of the city. Next comes the first ring, which is actually a semi-circle of Moscow streets, situated on the site of the former **Kitay-Gorod** walls. The second ring, also a semi-circle, is the **Boulevard Ring**, which is formed by wide boulevards that replaced the walls of the *Bely Gorod* (White City) in the 18th century. Next in turn is the **Sadovoye Ring** (Garden Ring), laid on the site of the former *Zemlyanoi Val* (Earthen Rampart). The name "Garden" is a complete misnomer and is based on tradition. This is the city's most important transport artery, along which traffic flows from early morning until late evening.

Further out is a circular railway built in place of the **Kamer-Kollezhsky Rampart** (named after the Kammer Kollegium, which was the name of the Ministry of Finance during the reign of Peter the Great), which was Moscow's customs border in the 18th century.

Finally, the fifth ring is formed by the **Moscow Circular Road** and marks the city's present boundary. Beyond this boundary stretches Moscow's "green belt" of forests and meadows, where many Muscovites like to spend their weekends. It covers an area of 1,800 sq. km (700 sq. miles).

Radiating roads and highways: A series of thoroughfares radiate from the centre, cutting across the rings on all sides of the city. **Bolshaya Lubyanka Street** (the former *Dzerzhinsky Street*) leads north to become **Mir Prospekt** and then the **Yaroslavl Highway**. Myasnitskaya Street (*Kirov Street*) leads northeast towards the **Shchelkovo Highway**. **Solyanka Street** goes east towards the **Entuziastov Highway** and **Ryazan Prospekt**. **Polyanka Street** leads south towards the **Kashira** and **Warsaw Highways**. Leading southwest, to the **Kiev Highway**, is **Bolshaya Yakimanka Street** and **Lenin Prospekt**.

Vozdvizhenka Street and the **New Arbat** (the former *Kalinin Prospekt*) lead west towards the **Minsk Highway**.

Preceding pages: Red Square; Moscow blondes; socialist art; view from the Hotel National. Left, the Kasan church in Kolmenskoye. Right, suddenly Lenin's statues are under threat.

Finally, **Tverskaya Street** (the former *Gorky Street*) runs northwest towards the **Leningrad** and **Volokolamsk Highways**.

Each historical period left its own inimitable imprint on the architecture of the city. Today, however, it would hardly be proper to speak about the Moscow of the days of Ivan the Terrible, Peter the Great or Catherine the Great. The past has merged with the present so closely that, with rare exceptions, it is difficult to single out any untouched architectural composition. Many structures have been pulled down or rebuilt, and every architect has proposed his own project for the development of the city.

Demolition and reconstruction: Much of the rebuilding occurred in the 1930s when the then still young Soviet government decided to make the city into a showpiece symbol of its triumphal advancement. The master plan for the reconstruction of the capital, adopted in 1935, envisaged the gradual redevelopment of the entire city on the basis of its historically formed radial-circular layout. As a result, the central avenues were expanded, the embankments were overhauled and the built-up areas round the Kremlin were cleared.

The architects, eager to please the leadership, did not give a moment's thought to the historic value of the structures they demolished. Thus, the majestic **Temple of Christ Our Saviour**, the **Sukhareva Tower**, the triumphal **Red Gate**, the **Monastery of the Ascension**, the **Monastery of St Michael's Miracle at Chonae**, the **Convent of the Passion of Our Lord** on whose site the Rossiya Cinema was built, dozens of churches, the fortress walls of the Kitay-Gorod, and quite a few other structures disappeared without trace.

Today, much is being said about the need to preserve and to restore old historical and cultural monuments. Practically the entire centre of Moscow has been proclaimed a protected zone and plans for restoring old churches and

Joyous ladies on 9 May.

monasteries are being brought forward. An "architectural legacy" programme has designated nearly 10,000 architectural monuments to be placed under state protection, and calls for the restoration of more than 1,500 buildings of historical or cultural value. In the future it is proposed that new housing projects should be built outside the city limits.

Orientation: Our tour through Moscow will follow the natural layout of the city. We will start with the **Kremlin**, the heart not only of Moscow but of the whole of Russia. From there we will visit **Red Square**, **Kitay-Gorod** and the inner city circle that surrounds the Kremlin walls. We will then visit four sections that surround the city centre.

Starting at the **Kropotkinskaya Embankment** we will survey the quarter where the Arbat is situated all the way to **Tverskaya Street** (*Gorky Street*), thereby following the radiating main routes that leave the city and its side-streets. The next section then covers the area between **Tverskaya** and **Myasnitskaya Street** (*Kirov Street*), followed by the quarter between Myasnitskaya Street and **Zayauze**. Finally we visit that part of the city that is on the other side of the Moskva River, **Zamoskvorechye**.

Street names: To find your way around Moscow today is not that easy. Leaving aside the Cyrillic lettering, many street, square and Metro names were changed in 1990 and 1991, giving back the historical names to squares and streets. This process of renaming is far from finished.

Metro stations, however, very often still carry the Soviet name, even though the streets or squares that they were named after have changed their name. Not all street signs carry the new names yet and while many Muscovites have to get used to them, others still use the names they grew up with. In this book we have printed the current names in bold and the former names (which may, in the future, again become the valid names) in italics.

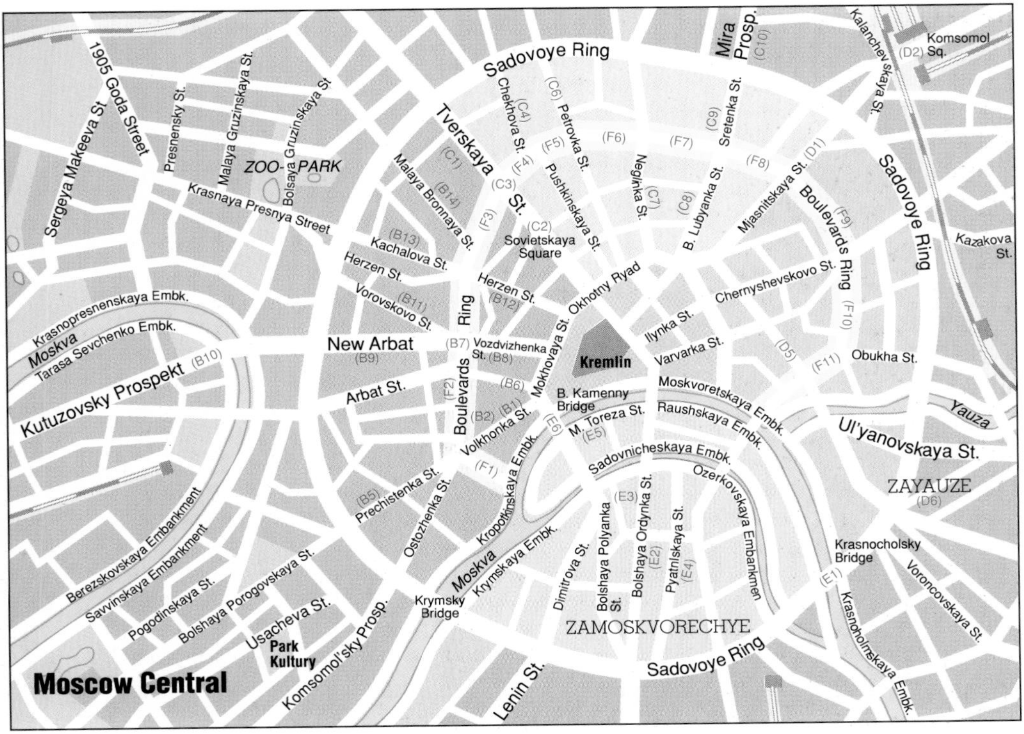

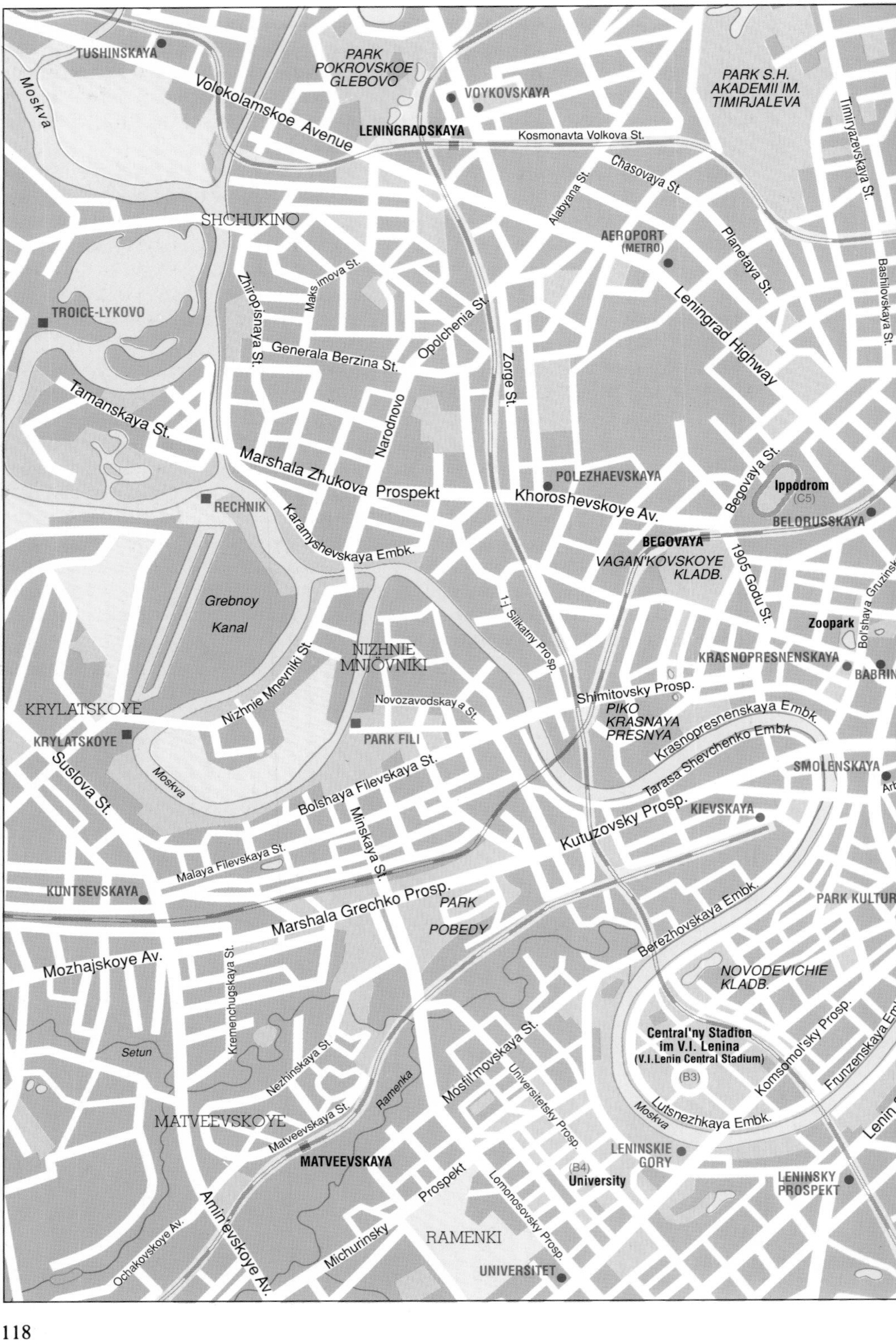

TUSHINSKAYA
Moskva
PARK POKROVSKOE GLEBOVO
Volokolamskoe Avenue
VOYKOVSKAYA
LENINGRADSKAYA
Kosmonavta Volkova St.
PARK S.H. AKADEMII IM. TIMIRJALEVA
Timiryazevskaya St.
Chasovaya St.
Alabyana St.
SHCHUKINO
AEROPORT (METRO)
Planetaya St.
Bashilovskaya St.
Maksimova St.
Zhiropisnaya St.
TROICE-LYKOVO
Opolchenia St.
Generala Berzina St.
Leningrad Highway
Zorge St.
Tamanskaya St.
Narodnovo
Marshala Zhukova Prospekt
POLEZHAEVSKAYA
Begovaya St.
Ippodrom
(C5)
RECHNIK
Khoroshevskoye Av.
BELORUSSKAYA
Karamyshevskaya Embk.
BEGOVAYA
VAGAN'KOVSKOYE KLADB.
1905 Godu St.
Grebnoy Kanal
1-j Silikatny Prosp.
Bol'shaya Gruzinskaya
Zoopark
NIZHNIE MNJÖVNIKI
Nizhnie Mnevniki St.
KRASNOPRESNENSKAYA
BARRIKADNAYA
KRYLATSKOYE
Novozavodskaya St.
Shmitovsky Prosp.
PIKO KRASNAYA PRESNYA
KRYLATSKOYE
PARK FILI
Krasnopresnenskaya Embk.
Tarasa Shevchenko Embk.
Suslova St.
Moskva
Bolshaya Filevskaya St.
SMOLENSKAYA
Arbat
Minskaya St.
Kutuzovsky Prosp.
KIEVSKAYA
Malaya Filevskaya St.
KUNTSEVSKAYA
Marshala Grechko Prosp.
PARK POBEDY
PARK KULTURY
Berezhkovskaya Embk.
Mozhajskoye Av.
Kremenchugskaya St.
NOVODEVICHIE KLADB.
Komsomol'sky Prosp.
Frunzenskaya Embk.
Central'ny Stadion im V.I. Lenina
(V.I. Lenin Central Stadium)
(B3)
Setun
Nezhinskaya St.
Mosfil'movskaya St.
Universitetsky Prosp.
Ramenka
Lutsnezhkaya Embk.
Moskva
MATVEEVSKOYE
Matveevskaya St.
Lenin St.
MATVEEVSKAYA
LENINSKIE GORY
(B4)
University
LENINSKY PROSPEKT
Prospekt
Lomonosovsky Prosp.
Amin'evskoye Av.
Ochakovskoye Av.
Michurinsky
RAMENKI
UNIVERSITET

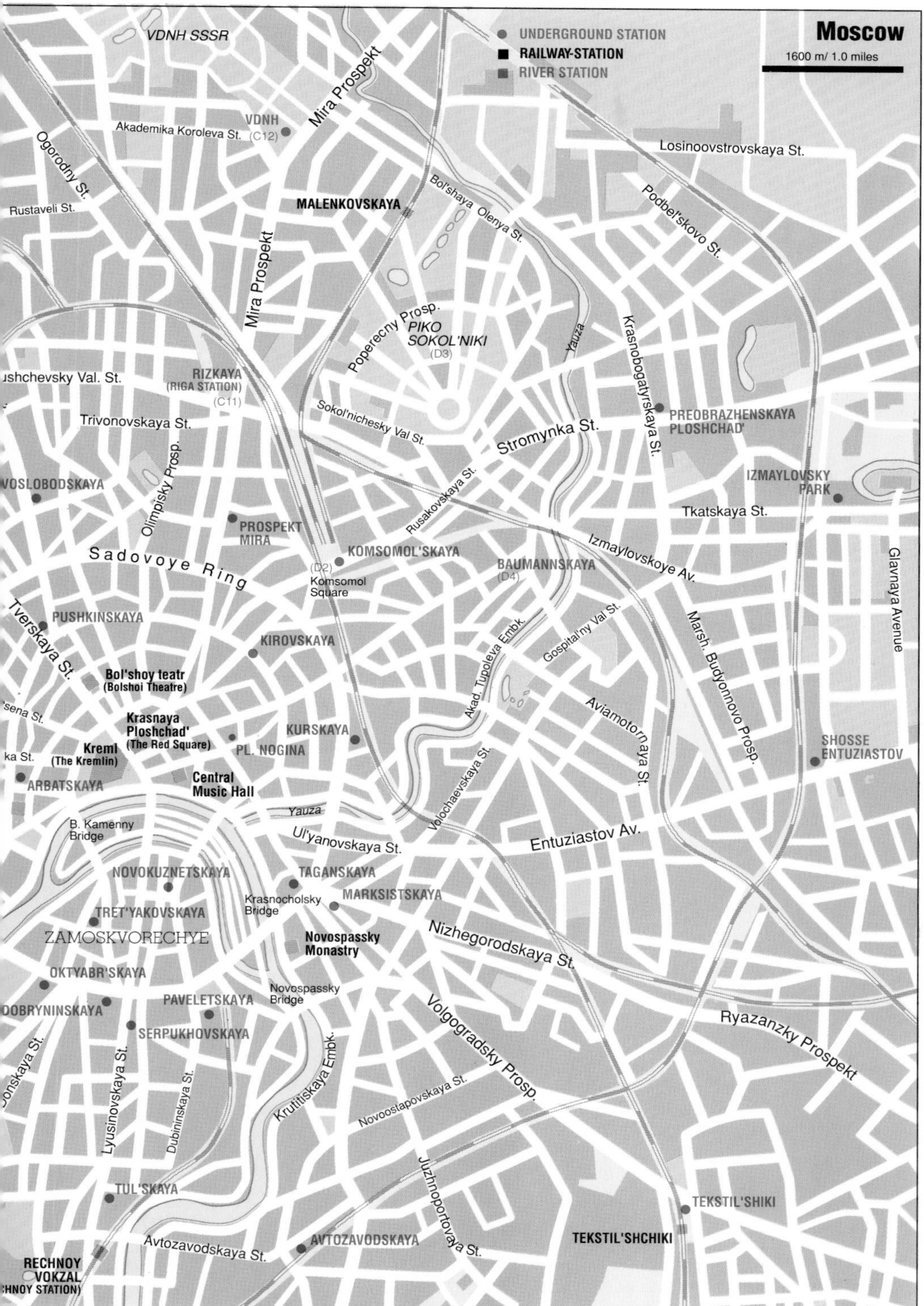
Moscow
1600 m/ 1.0 miles
UNDERGROUND STATION
RAILWAY-STATION
RIVER STATION
VDNH SSSR
VDNH
(C12)
Akademika Koroleva St.
Mira Prospekt
Ogorodny St.
Rustaveli St.
MALENKOVSKAYA
Bol'shaya Olenya St.
Losinoovstrovskaya St.
Podbel'skovo St.
Mira Prospekt
Poperecny Prosp.
PIKO
SOKOL'NIKI
(D3)
Yauza
Krasnobogatyrskaya St.
ushchevsky Val. St.
RIZKAYA
(RIGA STATION)
(C11)
Sokol'nichesky Val St.
PREOBRAZHENSKAYA
PLOSHCHAD'
Trivonovskaya St.
Stromynka St.
Olimpisky Prosp.
VOSLOBODSKAYA
Rusakovskaya St.
IZMAYLOVSKY
PARK
Tkatskaya St.
PROSPEKT
MIRA
Sadovoye Ring
KOMSOMOL'SKAYA
(D2)
Komsomol
Square
BAUMANNSKAYA
(D4)
Izmaylovskoye Av.
Glavnaya Avenue
Tverskaya St.
PUSHKINSKAYA
KIROVSKAYA
Akad. Tupoleva Embk.
Gospital'ny Val St.
Marsh. Budyonnovo Prosp.
Bol'shoy teatr
(Bolshoi Theatre)
Aviamotornaya St.
sena St.
Krasnaya
Ploshchad'
(The Red Square)
KURSKAYA
PL. NOGINA
Kreml
(The Kremlin)
ka St.
Volochaevskaya St.
SHOSSE
ENTUZIASTOV
ARBATSKAYA
Central
Music Hall
Yauza
B. Kamenny
Bridge
Ul'yanovskaya St.
Entuziastov Av.
NOVOKUZNETSKAYA
TAGANSKAYA
Krasnocholsky
Bridge
MARKSISTSKAYA
TRET'YAKOVSKAYA
ZAMOSKVORECHYE
Novospassky
Monastry
Nizhegorodskaya St.
OKTYABR'SKAYA
Novospassky
Bridge
DOBRYNINSKAYA
PAVELETSKAYA
Volgogradsky Prosp.
Ryazanzky Prospekt
SERPUKHOVSKAYA
onskaya St.
Lyusinovskaya St.
Dubininskaya St.
Krutitskaya Embk.
Novoostapovskaya St.
Juzhnoportovaya St.
TUL'SKAYA
TEKSTIL'SHIKI
Avtozavodskaya St.
AVTOZAVODSKAYA
TEKSTIL'SHCHIKI
RECHNOY
VOKZAL
HNOY STATION)

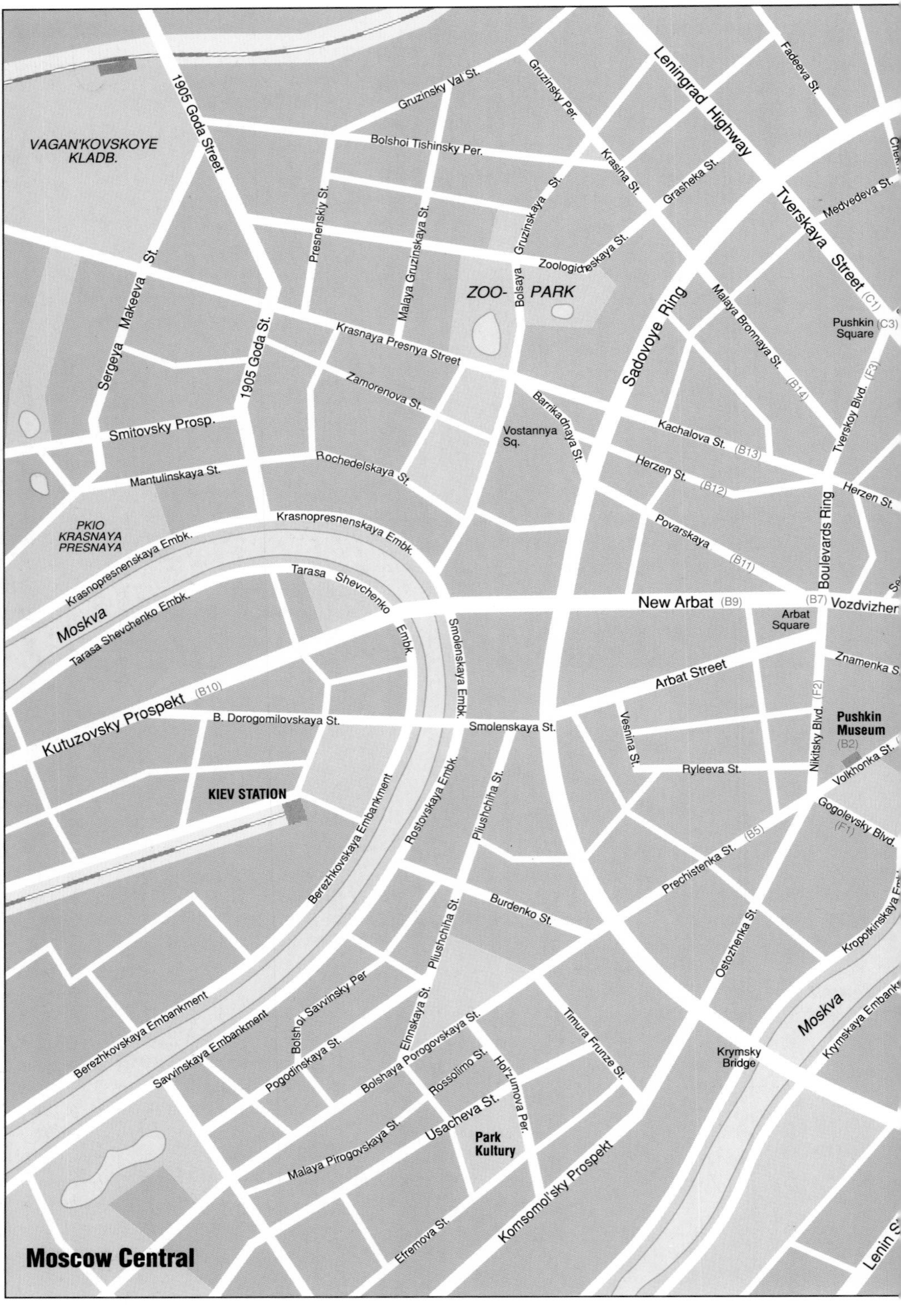

VAGAN'KOVSKOYE KLADB.
1905 Goda Street
Gruzinsky Val St.
Gruzinsky Per.
Leningrad Highway
Fadeeva St.
Bolshoi Tishinsky Per.
Krasina St.
Grasheka St.
Tverskaya Street (C1)
Medvedeva St.
Presnenskiy St.
Malaya Gruzinskaya St.
Bolsaya Gruzinskaya St.
Zoologicheskaya St.
ZOO- PARK
Sergeya Makeeva St.
1905 Goda St.
Sadovoye Ring
Malaya Bronnaya St. (B14)
Pushkin Square (C3)
Tverskoy Blvd. (F3)
Krasnaya Presnya Street
Zamorenova St.
Barrikadnaya St.
Vostannya Sq.
Kachalova St. (B13)
Smitovsky Prosp.
Rochedelskaya St.
Herzen St. (B12)
Herzen St.
Mantulinskaya St.
Boulevards Ring
Povarskaya (B11)
PKIO KRASNAYA PRESNAYA
Krasnopresnenskaya Embk.
Krasnopresnenskaya Embk.
Tarasa Shevchenko Embk.
Moskva
Tarasa Shevchenko Embk.
New Arbat (B9)
(B7) Vozdvizhen
Arbat Square
Smolenskaya Embk.
Znamenka St.
Arbat Street
Kutuzovsky Prospekt (B10)
B. Dorogomilovskaya St.
Smolenskaya St.
Vesnina St.
Nikitsky Blvd. (F2)
Pushkin Museum (B2)
Volkhonka St.
Ryleeva St.
KIEV STATION
Rostovskaya Embk.
Pliushchiha St.
Berezhkovskaya Embankment
Gogolevsky Blvd. (F1)
Prechistenka St. (B5)
Burdenko St.
Pliushchiha St.
Ostozhenka St.
Kropotkinskaya
Bolshoi Savvinsky Per
Timura Frunze St.
Moskva
Krymskaya Embank
Elnnskaya St.
Berezhkovskaya Embankment
Savvinskaya Embankment
Pogodinskaya St.
Bolshaya Porogovskaya St.
Rossolimo St.
Hol'zumova Per.
Krymsky Bridge
Usacheva St.
Park Kultury
Malaya Pirogovskaya St.
Komsomol'sky Prospekt
Efremova St.
Lenin
Moscow Central

LENINGRADSKY / SAROSLAVSKY STATION
Kalanchevskaya St.
Komsomol Sq.
(D2)
KAZANSKY STATION
Ol'hovskaya St.
Novokirovsky Prosp.
Novoryazanskaya St.
Novaya Basmannaya St.
Sadovoye Ring
Mira Prosp.
(C10)
(C9)
Sretenka St.
Petrovka St.
(C6)
Petrovsky Blvd.
(F5)
(F6)
Rozhdestvensky Blvd.
(F7)
Sretensky Blvd.
(F8)
Karla Marksa St.
Tokmakov Pereulok
Neglinnaya St.
(C7)
(C8)
B. Lubyanka St.
Mjasnitskaya St.
(D1)
Boulevards Ring
(F9)
Pushkinskaya St.
Kuznecky most St.
Sovietskaya Square
Bolshoi Theatre
Krivokolenny Pereulok
Kazakova St.
Okhotny Ryad
Pokrovka St.
Novaya Sq.
Petrovsky Blvd.
(F10)
Old University
Teatralnyi Proyesd
(A11)
The Arsenal
(A15)
Lenin Mausoleum
GUM
Ilynka St.
Staraya Sq.
Kitayshy Pereulok
Solyanka St.
(D5)
(A16)
St. Basil's Cathedral
Varvarka St.
Kremlin
Vorontsovo Polye St.
(A2)
(A5)
Grand Kremlin Palace
Cathedral of St. Michael the Archangel
Moskvoretskaya Embk.
Yauzky Blvd.
(F11)
Serebryanicheskaya Embk.
Kostomarovskaya Embk.
Yauza
B. Kamenny Bridge
(E6)
Sofiskaya Nab.- St.
(E5)
Raushskaya Embk.
Kotel'nicheskaya Embk.
Bernikovskaya Embk.
Nikoloyamskaya Embk.
Ul'yanovskaya St.
Sadovnicheskaya Embk.
Ozerkovskaya Embankment
M. Gor'kogo Embk.
Goncharnaya Embk.
(E3)
Tretyakov Gallery
ZAYAUZE
(D6)
Taganskaya St.
Marxistkaya St.
Voroncovskaya St.
Krasnokholmsky Bridge
(E1)
Krasnoholmskaya Embk.
B. Yakimanka St.
Bolshaya Polyanka St.
Bolshaya Ordynka St.
(E2)
Pyatnlskaya St.
(E4)
Novokuzneckaya St.
ZAMOSKVORECHYE
Sadovoye Ring
PAVELECKY STATION

A POET'S ADVICE TO VISITORS

***Yevgeny Yevtushenko** on how to come to terms with Moscow.*

I would not advise you to regard Muscovites scornfully as nothing more than a line of ants shuffling in shop queues. Muscovites are prepared to queue for other things besides Italian boots or vodka: they will also queue patiently for a subscription to the complete works of Dostoyevsky or tickets to a symphony concert at the Conservatory or a Salvador Dali exhibition.

I hope, too, that the poker-faced bureaucrats, the hard currency speculators and the prostitutes do not block your view of this city, which is actually inhabited by a great many decent and intelligent people. If some of them are still rather shy of foreigners, don't put this down to cowardice: for so many years our country was cut off from the rest of the world by the Iron Curtain and contact with foreigners was a punishable offence.

Learn a little Russian. Now people are gradually opening up. In the past we studied foreign languages purely as a formality, never imagining they might really come in useful. Now more and more young people are studying foreign languages not because they are forced to, but because they can't do without them. Do make the effort to learn a little Russian, otherwise you will certainly lose out. After living in a police state for so many years, Russians tend to be suspicious of interpreters and are reticent in their presence. Russians like confiding, however, and sometimes their confidences may suddenly come gushing out at you like Niagara Falls; in response, be as sincere as you can.

Experience public meetings. To understand the changes which have taken place over the past few years in Moscow, I would advise you to attend one of the large public meetings that are held

A pro-democracy demonstration.

from time to time. These meetings in Moscow today are unique as far as the intensity and sincerity of the discussions are concerned.

I would advise you not only to stroll along the old Arbat and past the art and handicraft stalls in Izmailovo Park but also to visit the small picture galleries which are springing up like mushrooms after a shower of rain. I would advise you to go not only to the Bolshoi Theatre or the Circus – traditional haunts of foreign tourists – but also to the drama theatres and studios, where you will understand and feel much even without a translation. The expressive visual language of the Taganka and Leninskiy Komsomol theatres is easily understood thanks to its tangible metaphors.

Don't miss Soviet films. Moscow has some top-class producers and actors, and this goes for its film industry as well. When people visit America from our country, they always go and see American films. When foreign tourists visit the USSR, they prefer to spend their time in foreign-currency cafés, and won't go near a Soviet cinema for fear of being bored silly by a film in the best traditions of socialist realism. Yet it is no accident that the latest Soviet films have won so many awards. Soviet film-makers, having coped with the Scylla of political censorship, now face the Charybdis of commercial censorship; nobody knows how they will fare.

Enjoying life.

Do visit the Moscow Metro. I'm recommending it not just because it is beautiful but also because you will see for yourselves how many people are reading – not just newspapers and journals but also books by Gabriel Garcia Marquez, Aba Kobo, Heinrich Böll, William Styron, and William Golding, and, of course, books of poetry. Moscow is a city of readers.

Kind souls at the cemeteries. Visit the grave of one of the Muscovites' favourite heroes, the poet and singer Vladimir Vysotsky, and you will see hundreds of fresh bouquets. Take a walk around Moscow's cemeteries where Muscovites come with a bottle of vodka on Sundays to converse with their deceased relatives and friends, and you will gain a better understanding of Moscow's suffering but kind soul. Don't forget to stand quietly in one of Moscow's churches when a service is in progress; the heavy drops of wax from the candles in front of the icons will slowly mark off the passing of time, and the barely audible music will link you with all of Moscow's living and dead.

According to one version, the Russian word for Moscow, *Moskva,* came from the Maris (a Finnish people) word *maska ava* meaning "mother bear". Numerous attempts have been made to kill this mother bear, and many cruel bullets still drift around her body. But she has still kept going and not become disillusioned with people. The expression "the hand of Moscow" used to inspire fear. Now this hand is held out to the rest of the human family, asking for support, not charity. Don't be afraid of it. Stretch your own out towards it.

GETTING YOUR BEARINGS

Making your acquaintance with a new city is always both exciting and difficult – even more so when that city is the centre of a multinational country and one that encloses an incredible wealth of history within its walls but which is also torn by contradictions.

The desire to learn about all the joys and delights of the unknown is tempered by the strangeness of the city and the absence of habitual associations. Some people prefer to plunge headlong into the sea of novel experiences. Others, by contrast, take it very gradually, step by step, prolonging the pleasure on the way.

At first sight everything in this capital of the first workers' and peasants' state will probably strike you as strange and even, perhaps, frustrating.

Splendour in the muddle: Alongside the architectural splendour which makes Moscow a hub for aficionados of baroque architecture and 19th-century town planning, there are untidy roads; hotels still lack the comforts you would expect for your money and there are not many cafés and restaurants offering the choice of dishes, the quality of food and the standard of service to which most foreigners are accustomed. Add to this a picture of residents who are perpetually worried about something and who rush single-mindedly about their various businesses.

And yet, you should not jump to conclusions. If you look at the city from a different angle, it will probably appear less unfamiliar, just as the apparent mess in your neighbours home does not necessarily mean that your neighbour regards it as any kind of disorder.

Preceding pages: ancient view of the Novodevichy Convent; Kitay-Gorod in olden days. **Left**, an Aeroflot hostess.

The reappearance of old values: For fairness' sake you have to see that sorting out today's chaos is no easy task for the leaders of this multinational country. One day the problems that had been piling up for many years suddenly made themselves felt and swept through every sphere of economic and political life. The numerous opinions and many recipes for salvation are as contradictory as they are varied. There is only one thing about which everyone agrees: namely, that this is no way to continue, not any longer.

Moscow, with its new democratically elected local government, is on the way to profound change. While concentrating on making life easier for its inhabitants the city authorities are also helping to unveil the hidden splendour that the city contains within its walls. The old street names are reappearing and churches are becoming churches again. Slowly, the wonderful atmosphere, that was so unforgettably narrated by our great poets and writers, and that once made Moscow not only a centre of power but also a centre of western culture, is re-emerging.

For the time being, however, problems are in store for the capital of this once-powerful empire which thought that it could leapfrog over a whole economic stage in its development. Having become aware of the repercussions of that leap, it is now facing what was regarded as already overcome: shortages, sometimes even of staple foodstuffs, a catastrophic devaluation of the rouble, increased unemployment, the loss of ideals and, as a consequence, a burgeoning black economy.

Today, you can often hear people saying that life in Moscow is not hard at all if you have enough money in your pocket. A foreigner coming to Moscow on a tourist visit will not be confronted with all the problems of this transitional period. They cannot, however, be avoided altogether, and you might face them right there where most foreigners arrive, in Sheremetyevo-II, Moscow's International Airport.

Arriving: What awaits the traveller during his or her first meeting with the capital of Russia? A squad of trigger-happy militia men with their sub-machine-guns? An ocean of charming

Moscow brides eager to vow love to every foreigner who comes along? It is neither. The story is much more prosaic. Having received your luggage and having experienced your first long queue at passport and customs control, you will at last get an opportunity to take a deep breath of the not very clear, yet very fascinating, Moscow air. They say that every city has its own smell. Apparently, this is also true about Moscow. Try it and you will find out for yourself what Moscow smells like.

When Sheremetyevo-II Airport was built back in 1979, it was thought to be a showcase for the country. Even those few Muscovites who had had the opportunity to see a good deal of the world regarded it as a wonder of modern architecture. Today, however, when it is not only the select few who fly abroad, it is evident that the airport is much too small and that it is already congested, although only 15,000 passengers pass through it each day.

Located not far from it is another airport, Sheremetyevo-I (Both these airports are named after the former estate of the Counts Sheremetev). It is a typical Soviet domestic airport. You will have to pass through it if you fly to Leningrad or to one of the cities in the Baltic republics.

The transfer from the airport to one of the Moscow hotels in an Intourist car or bus will take about 30 to 50 minutes.

Your first tour: The first thing that leaps to the eye when you enter the main artery, the **Leningrad Highway**, at its 23rd kilometer, is a memorial in the form of a group of enormous anti-tank "hedgehogs" on your right-hand side. It was here that the Soviet counter-offensive that repulsed the Nazis from the Moscow environs began in the winter of 1941. What was then open fields is now covered by apartment blocks.

Lying ahead is a small suburban town, **Khimki**, conspicuous for nothing in particular, that stretches as far as **Moscow's Circular Road**. This is the first, and the longest, of five rings cir-

German tanks reached this point.

cling the city centre and running for 109 km (68 miles) around the city. Built in 1962, it was a kind of city boundary until quite recently. Today, several new districts situated beyond the road have been incorporated into the city.

Leningrad Highway, as its name suggests, is the main thoroughfare linking Moscow with Leningrad. Built in 1713 and known as St Petersburg Highway until 1915, it was the straightest and shortest road connecting the two capitals of the Russian state.

A port of five seas: On the right-hand side you can see the **Khimki Reservoir** with its own rowing canal, where the water sports facilities of a number of sports clubs and societies are located. From a distance, a tall spire glittering in the sun and topped with a gilt star can be seen towering over the greenery of a public garden. It marks the **Northern River Port**. Before 1937 the star adorned the Spasskaya Tower of the Moscow Kremlin. The facade of the port building spans 350 metres/1150 feet. The central building resembles a three-decked ship with a captain's bridge, a mast, and anchors.

From here, modern ships and barges go to Astrakhan, Rostov-on-Don, St Petersburg, Volgograd, Nizhnij Novgorod (*Gorky*), Ufa, and many other Soviet cities. The building commands a magnificent view of a unique water development works, the **Moscow Canal**. Built in 1937, it links the city with the waters of the Volga River, and since 1953, with the Don, thus making Moscow a port for five seas – the White Sea, the Baltic, the Caspian, the Sea of Azov, and the Black Sea.

When entering the flyover near the Voikovskaya Metro Station, you will observe a rather curious sight. On either side of the flyover are hundreds of trolleybuses – one of the principal means of public transport in Moscow – in an open-air trolleybus yard. In Moscow there are about 70 trolleybus routes whose total length is in excess of 1,000 km/630 miles.

The Red Star is still omnipresent.

Next on the left-hand side of the Leningrad Highway are the buildings of the Moscow Aviation Institute where the various types of Soviet aircraft are designed. However strange this may seem, this highly popular Soviet college has also turned out quite a few talented writers, journalists, artists, actors and even diplomats.

The building with a colonnade houses an industrial art school, training students in various genres from folk to applied art. Ahead is a square where the road forks: it is the starting point of the Volokolamsk Highway. In the centre of the square is the 27-storey tower of the **Gidroproyekt Institute**, responsible for all water development works in the country. It is here that Promethean programmes like the one that was to make some of the Siberian rivers flow in the opposite direction, towards the arid areas of Central Asia (a plan which was fortunately dismissed) are worked out.

A village in the city: The high-rise buildings on your right-hand side, just after the road fork, hide from view a rather unusual residential district consisting of one and two-storey houses. They were built to look like traditional Russian peasant log cabins, each having a large plot of land attached to it but provided with every modern amenity. These are the houses of the **Sokol experimental housing co-operative**, built in the 1930s. The very idea of renouncing megalomania which was so characteristic of the period and providing people with individual houses, or rather small private estates, in the era of collectivisation is simply amazing.

It is not difficult to guess that it was by no means ordinary peasants who had the luck to live in this experimental "village district". The streets in the settlement are named after celebrated artists such as Surikov, Polenov, Kiprensky, Shishkin, and Venetsianov. From the late 1940s to the early 1950s, the entire area round the Sokol Metro Station was a kind of proving ground for testing new construction methods.

Spring's first sun on Red Square.

On the same side of the thoroughfare, at 71 Leningrad Prospekt, is the highly popular **Moscow Chamber Musical Theater** headed by Boris Pokrovsky.

The building at 37 Leningrad Prospekt, also to your right, is the **Moscow City Air Terminal**, built in 1960. From here passengers are transported by express coaches to the various airports round Moscow – to Vnukovo, Sheremetyevo I and II, Domodedovo, and Bykovo. Many of the streets in this area, which was formerly known as Khodynskoye Field, are named after noted airmen and cosmonauts. On 18 May 1896, celebrations were organised here on the occasion of the coronation of Nicholas II. A great crowd gathered to enjoy themselves but during the distribution of royal gifts some 1,400 people were crushed to death in the frantic throng.

On the other side of the avenue stands the building of the **Zhukovsky Military Aviation Engineering Academy** with a semi-circular red brick wall and graceful turrets. The Academy occupies the former Petrovsky (Peter the Great's) Coaching Palace built by the architect Matvei Kazakov in 1775–82. Here, the imperial family would rest before entering Moscow on their way there from St Petersburg.

The Sports Avenue: It is no exaggeration to say that Leningrad Prospekt is the capital's major sports avenue. Located along it are three large stadiums and numerous sports complexes. To your right you can see the sports complex of the **Soviet Army Central Sports Club** and then, on the opposite side near the Dynamo Metro Station, the **Dynamo sports complex** built in 1928. The Dynamo Sports Society, the oldest in this country, was established in 1923 on the initiative of Felix Dzerzhinsky, Lenin's close associate. Farther on along Leningrad Prospekt on your right is still another sports complex, the **Krylya Sovetov (Soviet's Wings) Sports Palace**.

On the same side of Leningrad Prospekt, at its crossing with Begovaya Street, is the **Young Pioneers' Stadium**, opened in 1920. It is the country's biggest training ground for young athletes.

Beyond Begovaya Street, in Khodynskoye Field, is the **Hippodrome**. It is adorned with bronze figures of horses by the sculptor Pyotr Klodt and was built at the end of the 19th century with funds provided by the Racing Society. The Hippodrome is a venue for both national and international horse and trotting races. On weekends both Muscovites and visitors who are fond of equestrian sports and those not averse to some betting come here to root for their favourites.

At the intersection of Leningrad Prospekt with Raskova Street stands what was one of the most popular turn-of-the-century Moscow restaurants, the **Yar**. Famous for its Gypsy choir, it was a favourite haunt of the city's gilded youth at the time.

The building, completed in 1910, was subsequently rebuilt on more than one occasion and finally, in 1950, became part of the **Sovetskaya Hotel**. In the 1960s, the Sovremennik (Contemporary) Studio Theatre gave performances in the hotel's concert hall. Today, it houses the **Romen (Romany) Gypsy Theatre**, founded in 1931. Thus, the Sovetskaya Hotel is no less popular today than the restaurant itself used to be during its heyday.

Where the "truth" is printed: A little further to your left you will see Pravda (Truth) Street branching off from the thoroughfare. It is named after the **Pravda Publishing House** and Printing Works owned by the CPSU Central Committee. The national dailies *Pravda* and *Komsomolskaya Pravda*, the magazines *Ogonyok* and *Rabotnitsa*, and quite a number of other periodicals are printed here.

After Byelorussian Square, the Leningrad Highway becomes Tverskaya Street (the former *Gorky Street*) and continues to the city centre.

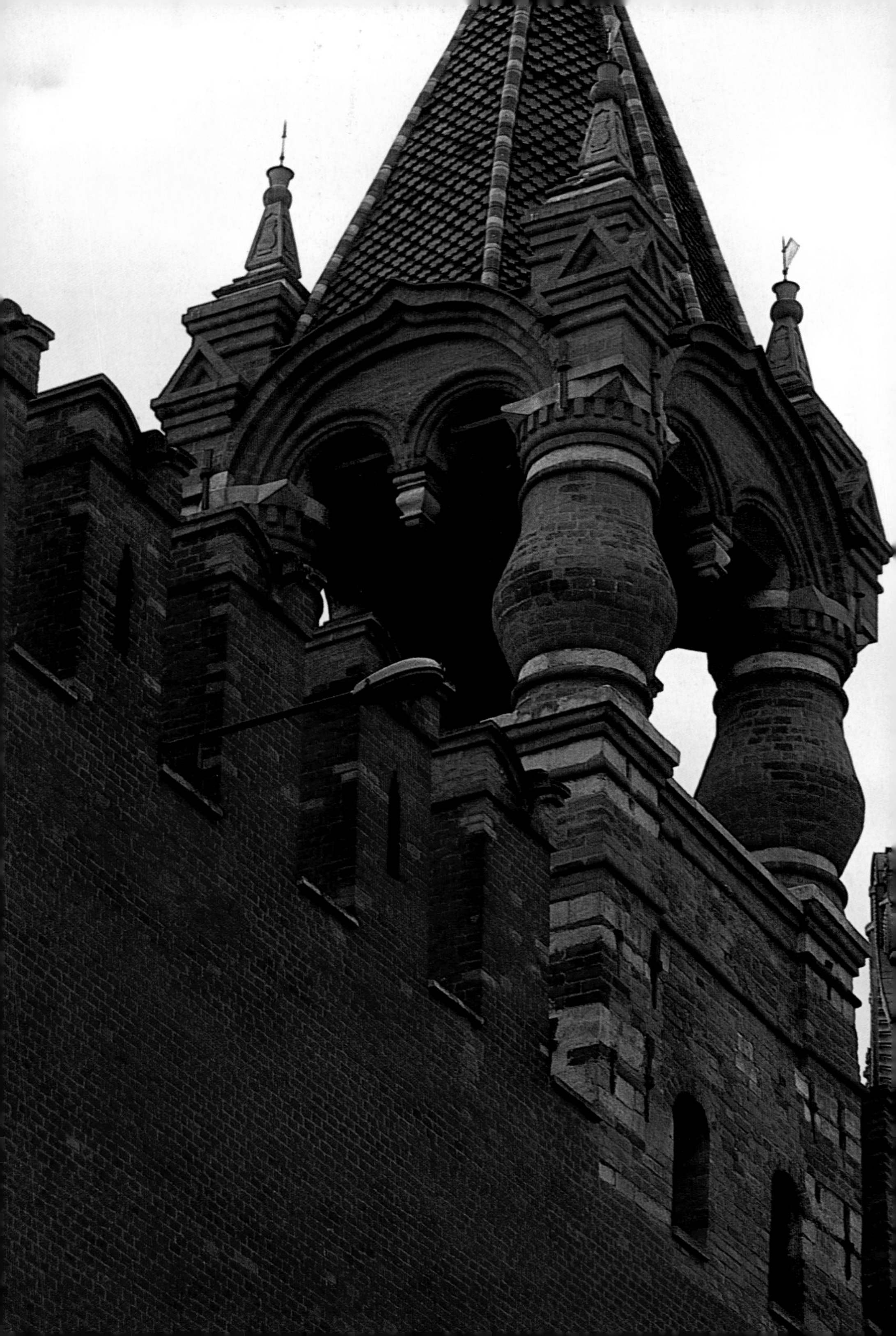

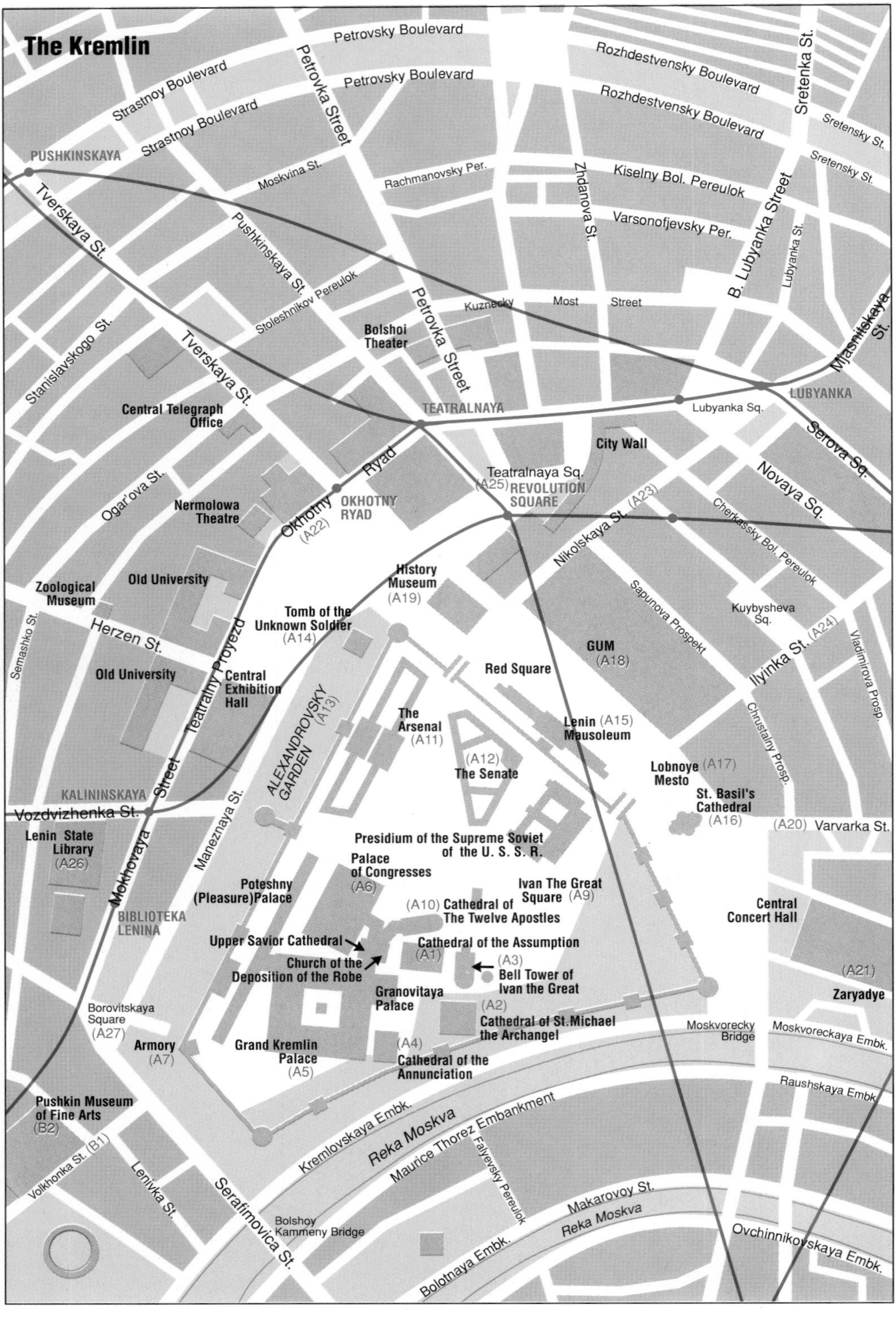

The Kremlin
Petrovsky Boulevard
Petrovsky Boulevard
Strastnoy Boulevard
Strastnoy Boulevard
Petrovka Street
Rozhdestvensky Boulevard
Rozhdestvensky Boulevard
Sretenka St.
Sretensky St.
Sretensky St.
PUSHKINSKAYA
Moskvina St.
Rachmanovsky Per.
Zhdanova St.
Kiselny Bol. Pereulok
Varsonofjevsky Per.
B. Lubyanka Street
Lubyanka St.
Tverskaya St.
Pushkinskaya St.
Stoleshnikov Pereulok
Kuznecky Most Street
Petrovka Street
Bolshoi Theater
Mjasnitskaya St.
Stanislavskogo St.
Tverskaya St.
Central Telegraph Office
TEATRALNAYA
Lubyanka Sq.
LUBYANKA
Serova Sq.
City Wall
Novaya Sq.
Ogar'ova St.
Okhotny Ryad
Teatralnaya Sq.
(A25) REVOLUTION SQUARE
OKHOTNY RYAD
Nermolowa Theatre
Okhotny (A22)
Nikolskaya St. (A23)
Cherkassky Bol. Pereulok
Zoological Museum
Old University
History Museum (A19)
Sapunova Prospekt
Kuybysheva Sq.
Tomb of the Unknown Soldier (A14)
Semashko St.
Herzen St.
Teatralny Proyezd
GUM (A18)
Ilyinka St. (A24)
Vladimirova Prosp.
Old University
Central Exhibition Hall
Red Square
ALEXANDROVSKY GARDEN (A13)
The Arsenal (A11)
Lenin (A15) Mausoleum
Chrustalny Prosp.
(A12) The Senate
Lobnoye (A17) Mesto
KALININSKAYA
Street
St. Basil's Cathedral (A16)
Vozdvizhenka St.
Maneznaya St.
(A20) Varvarka St.
Lenin State Library (A26)
Presidium of the Supreme Soviet of the U. S. S. R.
Palace of Congresses (A6)
Poteshny (Pleasure)Palace
Ivan The Great Square (A9)
Mokhovaya
(A10) Cathedral of The Twelve Apostles
Central Concert Hall
BIBLIOTEKA LENINA
Upper Savior Cathedral
Cathedral of the Assumption (A1)
Church of the Deposition of the Robe
(A3) Bell Tower of Ivan the Great
(A21) Zaryadye
Granovitaya Palace
(A2) Cathedral of St.Michael the Archangel
Borovitskaya Square (A27)
Moskvorecky Bridge
Moskvoreckaya Embk.
Armory (A7)
Grand Kremlin Palace (A5)
(A4) Cathedral of the Annunciation
Raushskaya Embk.
Pushkin Museum of Fine Arts (B2)
Kremlovskaya Embk.
Reka Moskva
Maurice Thorez Embankment
Falyevsky Pereulok
Volkhonka St. (B1)
Lenivka St.
Serafimovica St.
Makarovoy St.
Reka Moskva
Bolshoy Kammeny Bridge
Bolotnaya Embk.
Ovchinnikovskaya Embk.

THE KREMLIN

"The Kremlin sits on a high hill like the crown of sovereignty on the brow of an awesome ruler", wrote Russia's great poet Lermontov back in 1833. His impression is as valid today as it was then.

The Kremlin rises 25 metres (82 ft) on **Borovitsky Hill** and is traditionally regarded as the inviolate nucleus of Moscow. Muscovites sometimes dub their city the Third Rome, for Moscow, just as the great city of Rome, was also built on seven hills, though the grand dukes of Moscow didn't think of Rome – they considered themselves to be the inheritors of Byzantium's glory.

The ensemble of the Moscow Kremlin developed over many centuries. In the days of Dimitri Donskoy, in the 14th century, the original oak walls and towers were replaced by white-stone. In the 15th century, during the reign of Ivan III, the present walls of red brick were built under the supervision of the Italian architects Marco Ruffo and Pietro Antonio Solari.

The Kremlin wall, which forms an irregular triangle when seen from above, is 2,235 metres (7,300 ft) long, from 3.5 metres (11.5 ft) to 6.5 metres (21 ft) thick and from 5 metres (16.5 ft) to 19 metres (62 ft) high. Above the crenelated walls with bifurcated merlons are 19 elegant towers – 18 battle towers and the graceful tent-shaped **Tsarskaya Bashnya (Czar's Tower)** – four of which have entrances. Opposite the **Troitskaya (Trinity) Tower** is another, smaller one, the **Kutafya Tower**, built in the early 1500s as a bridgehead watch-tower. Connected with the Troitskaya Tower by the **Troitsky Bridge**, it has long been one of the official entrances to the Kremlin, used by Moscow czars and by Napoleon.

Kremlin, meaning "citadel" in Old Russian, supposedly derives from an Old Slavonic word denoting a fortified place. The principal stone structures, such as the Palace of Facets, the central cathedrals and a number of service buildings, were erected over a brief period from the late 15th to the early 16th centuries.

Preceding pages: the Saviour's tower.

Stars replaced the eagles: Until 1935, five of the Kremlin towers were topped with double-headed eagles – symbols of czarist autocracy. Eventually they were taken down and gilt stars were put up in their place. In 1937 they were replaced by stars faced in three-layer ruby glass. They are mounted on swivels so as to turn sideways into the wind.

The round corner tower on the far left (looking from across the Moskva River) is called the **Vodovzvodnaya (Water) Tower**. The name dates back to 1633, when devices for raising water were installed and the first pressure water supply in Russia was used to carry water to the royal palaces and gardens.

Standing in the middle of the south wall is the **Taynitskaya Tower (Tower of Secrets)**. The name of the tower comes from the secret well that was hidden inside it. At one time the Tower of Secrets had great strategic significance and a supplementary fortification was attached to it with entrance gates and an underground passage to the river.

Immediately beyond the **Moscow River Tower**, which stands at the south-east corner, is the **Konstantino-Yeleninskaya (St Constantine and Helena) Tower**. It was popularly known in the days of old as the **Torture Tower**, because of the torture chamber inside.

Not far from it is the **Tsarskaya Tower**, the youngest in the Kremlin. To be precise, it is not really a tower but a stone turret placed on the wall. In olden times a wooden tower stood here, from which, as legend has it, the young Ivan the Terrible watched executions in Red Square; hence its name, the Little Czar's Tower.

The corner **Borovitskaya Tower** stands at the foot of Borovitsky Hill, where the Kremlin arose. The name comes from a pinewood (*bor*), which once covered the whole hill upon which

Yuri Dolgorukiy founded his fortress. This tower differs from the others by its gradational form.

Cathedral Square: In the days of old, the central streets of Moscow such as Borovitskaya, Troitskaya and Nikolskaya streets converged in Cathedral Square. It is here that the oldest Kremlin structures are located.

The imposing building in the center of the square {*map reference* **A1**} is the white-stone **Cathedral of the Assumption**, (*Uspensky Sobor*, also called Cathedral of the Dormition). It was built between 1475 and 1479 on the site of a 14th-century church of that name and was the first cathedral to be built in the Kremlin. In 1475, the unfinished walls and vaults of the cathedral collapsed, apparently because the weakness of the mortar was aggravated by an earthquake.

The Italian architect Aristotile Fioravanti was commissioned to build a new cathedral. This was no easy task; it involved the skilful combination of the traditions of Old Russian architecture with Italian building techniques. Before starting work, Fioravanti visited Vladimir and Novgorod and thoroughly studied the architecture of Old Russian cathedrals. He decided to build the vaults and drums of the cathedral from brick made to his own formula, which he believed to be stronger than stone. Five years of careful and steady effort resulted in the emergence of a majestic cathedral crowned with five gilt domes.

The throne of Monomakh: The cathedral contains many fine examples of Old Russian applied art. Among them is the southern door, known as the **Korsun Door**, which was made in 1405 for the Dormition Cathedral in Suzdal. This oak door is covered with copper sheets on which you can see 20 Biblical scenes and inscriptions in Old Slavonic made in gold on black lacquer.

Next to it is the **Throne of Monomakh**. Twelve carvings depict the Grand Prince Vladimir Manomakh (1113–25) receiving the "Crown of Monomakh" from the Byzantine Emperor Con-

The centre of power.

stantine IX Manomachus. This throne, made for Ivan IV (the Terrible) in 1551, supports the claim that Moscow is the heir to Byzantium.

The amazingly spacious interior is decorated with beautiful frescoes. Most of these were executed by the famous Russian artist Dionysius in 1514 but some date from a later period. There is also a collection of unique icons including the **Vladimir Icon of the Mother of God** (11th/12th centuries) and the **Trinity** (14th century).

The Cathedral of St Michael the Archangel {A2}: Near the southern wall of the Kremlin, to the east of the Assumption Cathedral, is the Cathedral of St Michael the Archangel (*Arkhangelsky Sobor*), which was built between 1505 and 1508 under the direction of another Italian architect, Alevisio Novi. This cathedral was the burial place of Moscow grand dukes and czars. Here again you can see the result of combining the techniques characteristic of Old Russian buildings with elements of the architecture of the Italian Renaissance. Of exceptional interest is the interior of the cathedral with its 18th-century carved gilt iconostasis and its frescoes painted by masters of the Armoury under the supervision of Simon Ushakov in the mid-17th century. The wall paintings include over 60 so-called *parsunas*, stylised portraits of historical personalities buried in the cathedral.

Under the cathedral vaults are 46 tombs, including those of grand dukes Ivan Kalita, Dimitri Donskoy and Ivan III, and of the Czars Ivan the Terrible and Mikhail and Alexei Romanov. Among the portraits painted on the columns are those of grand dukes Yaroslav Vsevolodovich and Alexander Nevsky.

The Bell Tower of Ivan the Great {A3}: On the eastern side of Cathedral Square is an architectural complex which includes the Bell Tower of Ivan the Great, a campanile and the Filaret Annex. In 1329, Ivan Kalita built the Church of St John Climacus on the site of the present

In the Palace of Facets.

bell tower. Two centuries later when the church fell into disrepair and was pulled down, a new two-storey church, which formed the basis of today's bell tower, was erected in its place. In the mid-16th century the Italian architect Petroch the Younger built a five-storey campanile for the large bells on its northern side.

In the late 16th century Czar Boris Godunov decided to build a new great church in Moscow. Still another tier crowned with a gilt dome was added to the bell tower. A quarter of a century later the master builder Bazhen Ogurtsov erected a four-storey belfry with a hipped roof, the **Filaret Annex**, on the northern side of the campanile. In the 19th century still another addition was made to the ensemble: the **Dormition Bell**, weighing more than 70 tons, was hung in the central opening of the campanile.

When completed, the bell tower reached a height of 81 metres/266 ft. The enormous structure rests on wooden piles, each no longer than 1.5 metres (5 ft) and 25 cm (10 in) in diameter. Despite their seeming fragility, they have been excellently coping with their task to this day.

According to historians, Boris Godunov prohibited, under pain of severe punishment, the building of any bell tower, higher than his creation in the Kremlin.

The Cathedral of the Annunciation {A4}: On the western side of Cathedral Square stands the **Cathedral of the Annunciation**, built of brick by master builders from Moscow and Pskov in 1484 to 1489. The cathedral served as the domestic chapel of Moscow grand dukes and czars. The floor of the cathedral is faced in jasper.

Its iconostasis, with icons painted by Theophanes the Greek, Prokhor of Gorodets and Andrei Rublev as early as 1405, is of exceptional artistic and historical value. The frescoes on its walls were executed by Feodosi, the son of Dionysius. In the 16th century, a high white-stone porch was added to the cathedral specially for Ivan the Terrible.

Left, construction of the Bell Tower. Right, Iconostasis in the Cathedral of St Michael the Archangel.

Initially the cathedral had three domes. During its reconstruction in the mid-17th century, another six domes were added.

The Grand Kremlin Palace {A5}: Immediately to the west of the Cathedral of the Annunciation stretches a great palace with a high wrought-iron fence along the southern wall of the Kremlin. This three- storey building with carved white-stone platbands round the windows is the Grand Kremlin Palace. It was built between 1838 and 1849 in neo-Russian style by the architects Konstantin Thon and Richter on the site of an older palace damaged by the great fire of 1812. This palace was the Moscow residence of the Russian emperors. The ensemble of the palace includes 15th to 16th-century chambers, royal private suites and a winter garden.

The largest hall in the palace has a seating capacity of approximately 2,500 and is used for the congresses of the Peoples Deputies of the Russian Federation. It was built in the 1930s by combining the former **Alexandrovsky (St Alexander)** and **Andreyevsky (St Andrew) Halls**. The first congresses of the Comintern were held here.

The second largest hall in the Kremlin is the ceremonial **Georgievsky Hall**, named after St George the Victorious. The hall is richly ornamented with stucco moulding and 18 convoluted zinc columns, each supporting a statue of Victory crowned with a laurel wreath, sculpted by Giovanni Vitali.

In the tall niches along the walls are marble slabs engraved in gold with the names of units that distinguished themselves in battle, and of officers and men awarded the Order of St George. Today the Georgievsky Hall is used for state receptions and official ceremonies.

During the revolution and the Civil War, a priceless collection of paintings evacuated from the Hermitage was kept in the cellars of the Grand Kremlin Palace. In the 1920s the collection was returned to Leningrad. The octagonal **Vladimirsky (St Vladimir) Hall** con-

Left, the Czar's cannon. Right, the Czar's bell.

nects the Grand Kremlin Palace with a building dating to the 16th and 17th centuries.

The Palace of Facets: Next to the Cathedral of the Annunciation and adjoining the Grand Kremlin Palace is the Palace of Facets, one of the oldest buildings of the Moscow Kremlin. Its facade, finished in faceted white stone, overlooks Cathedral Square. The palace was built by the Italian architects Marco Ruffo and Pietro Antonio Solari between 1487 and 1491.

In days of old, the Palace of Facets was the place where the *Boyar Duma*, the main legislative body of Russia, held its sessions. It was also here that foreign ambassadors were received and victories celebrated. Today the palace is occasionally used for sessions of the country's supreme bodies of state power and, as before, for state receptions.

The interior of the palace is decorated in traditional Russian style. Its spacious square hall is 9 metres (30 ft) high, has an area of some 500 sq. metres (5,389 sq. ft) and is covered by four cruciform vaults supported by a central pillar. In the 1890s, the Belousov brothers, masters from Palekh (a village renowned for its distinctive paintings), reproduced on the walls and the vault subjects taken from the 16th-century murals that originally covered the walls of the palace. They followed an exact description left by the famous Russian icon painter Simon Ushakov .

The Terem Palace {A8}: Adjoining the Palace of Facets is the Terem Palace. This multi-storey structure incorporates part of a 16th-century palace. The palatial ensemble includes the **Church of the Nativity of the Virgin-over-the Vestibule** whose side chapel is the **Church of the Resurrection of Lazarus** (late 14th century), the oldest building to have survived in the Kremlin. Very close to it is the **Upper Cathedral of the Saviour**; its frieze and the drums of its 11 domes are decorated with multicoloured tiles. In 1680 they were all united by a single roof.

The 13th-century Monomakh crown in the Armoury.

Adjoining the Upper Cathedral of the Saviour is the 16th-century **Lesser Golden Chamber**, thus named from the decor of its interior, featuring ornaments covered with gold leaf.

Standing next to the Terem Palace is the single domed **Church of the Deposition of the Robe**, built in the late 15th century by master builders from Pskov as a domestic chapel for Moscow metropolitans. The frescoes in this church were executed by Sidor Osipov and Ivan Borisov.

The Kremlin Palace of Congresses {A6}: Wedged in between the Patriarch's Chambers and the Armoury is a monolithic block of concrete and glass, the Kremlin Palace of Congresses. It was built in 1961 by a group of architects headed by Mikhail Posokhin. A feature of the Palace of Congresses is that it is sunk some 15 metres (50 ft) into the ground. It has about 800 rooms, including a huge auditorium that seats 6,000 people. Before 1960 there were service buildings on its site and earlier still, as was successfully established during archaeological excavations, the white-stone chambers of Natalia Naryshkina, the mother of Peter the Great, and, even before that, in the 14th century, the palace of Sophia Paleologos, the wife of Ivan III.

The Armoury {A7}: Standing along the western wall of the Kremlin is the Armoury, the oldest museum in Moscow. The emergence of the Armoury as a depository for Moscow grand dukes and czars dates back to the 16th century. The Armoury owes its name to the fact that it was once the Kremlin workshops which manufactured, purchased and stored weapons, jewellery and articles used in the palace.

The Armoury was transformed into a public museum at the start of the 19th century. Today's Armoury was constructed in 1844–51 by Konstantin Thon and Nikolai Chichagov.

The collection in the Armoury includes battle and ceremonial sidearms made by Russian, West European and

The eternal flame.

Oriental masters, Russian crowns, sceptres and orbs, the insignia of the orders of the Russian Empire, precious tableware, church plate, precious fabrics and royal ceremonial attire and ceremonial carriages, each of which is a work of art in itself. The oldest of the relics is the **Kievan Seal** found during archaeological excavations in the Kremlin.

Also housed in the Armoury is a remarkable exhibition: the USSR Diamond Fund, which contains some of the finest diamonds in the world.

The USSR Diamond Fund Exhibition: Seven legendary gems are the pride of the USSR Diamond Fund, which was brought to Moscow from St Petersburg soon after the outbreak of World War I.

The largest of the seven, the **Orlov** diamond, was brought to Russia from India. It was found over 350 years ago on the site of the ancient Indian city of Golconda. At one time it was known as the **Great Mogul** and adorned the throne of Shah Nadir of Persia who conquered Delhi. Subsequently the British recut the diamond and sent it to the Bank of Amsterdam. Here it was bought by the Armenian merchant Lazarev, who then sold it to Count Orlov for 400,000 roubles in gold. Count Orlov presented it to Catherine the Great on her name day in 1773 and, to show her gratitude, the empress named it after her favourite. In 1914 the gem fell out of the mounting on Nicholas II's sceptre. This accident made it possible to measure the weight of the diamond: it was 189.62 carats.

The second largest diamond, **Maria**, weighing 106 carats, was given the name of the woman geologist who found it in 1967 in a diamond mine in Siberia.

Another legendary gem, the **Shah**, also comes from India, where it was found over four centuries ago. On its facets are three microscopic inscriptions detailing its history. In 1591 it belonged to the ruler of the city of Ahmednagar and in 1641 its owner was one of the Great Moguls. In 1825, the Persian Shah sent it to St Petersburg to make amends for the assassination of the Russian ambassador to Persia, Alexander Griboyedov, the Russian poet.

The fourth of the famous gems is a flat diamond with an area of just 7.5 sq. cm (1.16 sq. in). This gem is also from India; its previous history, however, remains unknown. One thing is known for sure about it: there is no other flat diamond like this anywhere in the world.

The great 412.25 carat red spinel (similar to a ruby), which adorns the Russian imperial crown made by the jeweller Pozier in the 18th-century, is the collection's fifth best-known gem.

Few visitors will spare a second glance for a flat, bottle-green stone whose size is 7 x 4.7 cm (2.75 in x 1.85 in). Yet this is the mysterious chrysolite, a precious trophy brought by the German crusaders from Palestine. This gem was long time one of the most valuable items in Cologne Cathedral. Then, no one in Europe knew where chrysolites could be found. It was only in the early 20th century that long-abandoned deposits of this stone were found on one of the islands in the Red Sea.

Finally there is a flat emerald (136 carats) which made a long journey from Colombia and across Europe before arriving in Russia.

Ivan the Great's Square (A9): To the east of Cathedral Square is Ivan the Great's Square. In days of old it was a place where czar's *ukases* (orders) were announced and it was always crowded with people. Because of the incredible noise the criers often had to shout at the top of their voice. This gave rise to the Russian phrase "to cry for all of Ivan Square to hear".

A gun that never fired, a bell that never rang: The famous **Czar Bell**, cast of bronze in 1733–35 by Ivan Matorin and his son Mikhail here in the Kremlin, is displayed in the square. The bell was intended for the Bell Tower of Ivan the Great. During the Great Fire of 1737 the bell was still lying in its casting pit and attempts to extinguish the fire led to water falling on the red-hot bronze. The result was that an 11-ton chunk fell off the bell. In 1836 the bell was raised from the pit and put on a special pedestal. It weighs over 200 tons.

Not far from the Czar Bell, also on a special pedestal, stands the **Czar Cannon**, the largest artillery piece of the period, which is no less famous than the Czar Bell. It was cast by Andrei Chokov in 1586. This formidable gun has a bore of 890 mm (35 in) and weighs 40 tons. The gun was never fired, just as the bell was never rung.

Behind the cannon is a five-domed building, the **Cathedral of the Twelve Apostles {A10}** (*Sobor Dvenadtsati Apostolov*). It forms a long structure together with the **Patriarch's Palace** (*Patriarshiye Palaty*) and houses the **Museum of 17th-Century Life and Applied Art**.

Newly-weds must visit the Tomb of the Unknown Soldier.

To the southeast of Ivan the Great's Square at the foot of the hill stretches the **Tainitsky Garden**. Situated here were the Monasteries of St Michael's Miracle at Chonae and the Monastery of the Resurrection. They were badly damaged during the artillery bombardment of the Kremlin by revolutionary troops in October 1917. Later on, they were pulled down.

The Arsenal {A11}: The Arsenal occupies the northern corner of the Kremlin. It was built on the order of Peter the Great in 1701 for the storage of weapons and as a museum of military trophies. After the fire of 1737 the Arsenal had to be rebuilt. In 1812, Napoleon's retreating troops blew the building up and it was only in 1828 that it was restored. Playing a significant role in its restoration was the celebrated architect Osip Bove. Along the southern facade of the Arsenal are cannons captured by the Russian troops from Napoleon's Grand Army.

The Senate {A12}: South of the Arsenal is the former Senate, built in 1776–87 by Matvei Kazakov. It was intended to house two departments of the Senate, the Judicial Department and the Department of the Nobility, which was trans-

ferred from St Petersburg to Moscow. After the Soviet government headed by Lenin moved to Moscow in March 1918 the government and Bolshevik Party leaders took up residence in this building. In the spring of 1819 the red flag of the Soviet Republic was raised above its huge dome. Lenin was given a flat in the Senate building and it was here that he lived and worked from March 1918 to May 1923. **Lenin's flat-museum** was opened here in 1953.

Standing next to this building, close to the Kremlin wall, is the **Amusements Palace** built for the Boyar Miloslavsky in 1651. It was named thus because entertainments for the court members were staged in it.

The Presidential Residence: Near the Senate is another building, constructed in the 1930s on the site of former monasteries. Initially this housed a military school and, after 1958, the Kremlin Theatre. Today it is the residence of the President of the USSR. It is also here that the standing committees of the Supreme Soviet of the USSR meets. The building has a hall seating 1,000 in which the two Houses of the Supreme Soviet of the USSR hold their meetings.

The Alexandrovsky Gardens {A13}: On leaving the Kremlin by way of the Troitskiye Gate, you will find yourself in the Alexandrovsky Gardens laid out in the early 1820s above the covered channel of the Neglinka River. In the early 18th century Peter the Great ordered earthen bastions to be put up around the Kremlin. The Neglinka River was thus channelled into a special moat. Later, during the reign of Catherine the Great, the scheme was extended and practically the whole of the Neglinka River was piped underground.

Near the **Middle Arsenal Tower** at the foot of the Kremlin wall is a somewhat unusual structure reminiscent of the ruins of an ancient Greek temple. This is a grotto named **Ruins**, a typical example of 19th-century park and garden architecture, built by Osip Bove in 1821.

Opposite the grotto, beside the central alley, stands a light grey obelisk. Originally set up in 1913 to commemorate the 300th anniversary of the House of Romanov, the obelisk was crowned with a double-headed eagle. After the revolution the eagle was removed and the names of the czars were erased from the obelisk and replaced with the names of Marx, Engels, Plekhanov and other Communist theoreticians. This, the first monument of revolutionary Russia, was unveiled in 1918.

Behind the fence, at the foot of the Kremlin wall, is a low structure of red granite and black marble with an eternal flame burning in its centre. This is the **Tomb of the Unknown Soldier {A14}**. In 1966 the remains of an unknown soldier, who gave his life for the motherland during the battle with the Nazis on the approaches to Moscow, were brought here from the common grave at the 41st kilometre on the Leningrad Highway and ceremonially buried by the Kremlin wall.

Left, Lenin: he brought the Bolsheviks into the Kremlin. Right, Ivan IV Bell Tower and the Cathedral of the Annunciation.

RED SQUARE

Having passed the gently sloping cobblestone **Istorichesky Passage**, you will reach the main square of the capital, Red Square (*Krasnaya Ploshchad*). It did not always bear this name. Before the 15th century, it was a market centre called *Veliky Torg* (Great Marketplace). After the fire of 1403, which destroyed more than half the wooden structures in the square, it was called *Pozhar* (Fire Site).

Only in the mid-17th century, when official ceremonies began to be held here and the *ukases* (decrees) of the czar began to be announced in its southern part, was it given its present name.

At that time, the square was adorned with several new stone structures and tall hipped roofs were added to the Kremlin towers. People began to call the square *Krasnaya*, which then meant beautiful but now means red.

Facing the square is one of the most beautiful Kremlin towers, the **Spasskaya (Saviour) Tower**. Its gate is the main one in the Kremlin. The tower was built in 1491 to the design of Pietro Antonio Solari. The first clock was installed in it in the late 15th and early 16th centuries. The clock that adorns the tower today was made in the mid-19th century by the Vutenop brothers. During the October 1917 uprising the clock was damaged by a shell and, on Lenin's instructions, was repaired by the clockmaker N. Behrens. Since that time the carillon has played the *International* instead of *God Save the Czar*. In days of old the Spasskaya Tower was regarded as holy.

Where Napoleon lost his hat: According to the chronicles, in 1812 Napoleon, elated by his victory, decided to enter the Kremlin on his white horse through the Spasskaya Gate. He did not bother to take off his general's hat. As he was riding under the arch of the gate his horse, suddenly startled, reared; the hat fell of the head of the military leader, who himself very nearly fell of his horse. After that the belief that the tower was actually holy became even more firmly established in peoples' minds.

Along the Kremlin wall to the right and left of the **Senatskaya (Senate) Tower** is the burial place of the most prominent statesmen and public figures of the Soviet period. The proletarian writer **Maxim Gorky**, Lenin's wife **Nadezhda Krupskaya**, the first Soviet cosmonaut **Yuri Gagarin**, and the inventor of the Soviet atomic bomb **Igor Kurchatov** lie buried here. A number of noted military commanders and the "father of all peoples", **Josef Stalin**, are also buried here. Stalin's embalmed body was transferred here from the Mausoleum, having lain for several years alongside Lenin's body.

Lenin's embalmed body: The body of **Lenin** was placed in the **Mausoleum** {*map reference* **A15**} upon his death in 1924. The original wooden mausoleum was later replaced by the present granite structure built in the style of a pre-Columbian temple. Inside the **Funeral Hall** (*Traurny Zal*) constant air temperature and humidity are maintained. Since August 1924 Lenin's body has been exposed to view in a sarcophagus around which visitors walk. Leaving the mausoleum you come to a kind of Soviet pantheon where the most honoured Soviet citizens are buried.

During World War II Lenin's body was moved to the Urals but was returned in April 1945. After Stalin died, in 1953, his body was placed next to Lenin's. The mausoleum was then known as the Lenin-Stalin Mausoleum. In 1961 Stalin's remains were, however, removed and buried under the Kremlin wall.

Changing of the guard: Above the Mausoleum there are stands from which, for several decades now, the leaders of the Soviet state greet Muscovites during parades on national holidays. The entrance to the Mausoleum is called the **Sentry Post No. 1**. The guards manning it are changed every hour. The sentries, together with a corporal of the guard,

Preceding pages: the Kuzma Minin and Dmitry Pozharsky Monument in front of St Basil's. Left, the Cathedral of St Basil the Blessed.

come marching from the Saviour Gate and take exactly 2 minutes 45 seconds to reach the Mausoleum. A huge crowd of people always gathers to watch the changing of the guard.

St Basil's, the epitome of Russia: Nearby, on the southern side of the square stands the festive-looking **Cathedral of the Protecting Veil**. It is also known as the **Cathedral of St Vassily the Blessed** (St Basil's) **{A16}** after God's fool, who was well-known in Moscow at the time of its construction and whose grave is next to the cathedral. For many people around the world, the picture of this cathedral has become a potent symbol of Russia.

The cathedral was built between 1555 and 1561 on the order of Ivan the Terrible to commemorate the victory over the Kazan Khanate. It is believed to have been erected by a Russian master builder by the name of Postnik Barma. Having created this complex composition of nine tower-like churches built on a common foundation, the great master builder lost his sight. Tradition has it that the cruel czar ordered the master to be deliberately blinded so that he could not create anything else that would be as beautiful.

Incidentally, contrary to Russian tradition the cathedral was built in a market square, apparently still another manifestation of the wilful czar's capricious character. Each of the nine domes of the cathedral forms the top of a separate church, each dedicated to a saint on whose feast day the Russian army achieved a victory. The inner walls are decorated with patterns featuring Russian folk motifs. After decades of use as a branch of the History Museum, St Basil's was finally returned to the Orthodox Church for the Easter festivities of 1991.

In front of the cathedral is a monument to two popular heroes, **Citizen Minin** and **Prince Pozharsky**, who saved the city from an invasion by Polish troops led by the impostor False Dmitri II in 1612. The Monument, executed by the

The Lenin Mausoleum on Red Square.

famous sculptor Ivan Martos, was erected by public subscription.

The dreadful white-stone platform: Near the cathedral is a small stone mound, the so-called **Lobnoye Mesto {A17}**, or Place of Execution. This round platform of white stone was built in the 16th century in front of the main gate of the Kremlin on the crest of the hill where it starts sloping downward. The czar's decrees were proclaimed here and public executions were carried out.

Red Square has witnessed many historical dramas. In 1698 it saw the cruel execution of the *streltsi*, royal musketeers, who, incited by Sophia, Peter the Great's sister, rose in revolt against the czar. In 1671 Stepan Razin, the leader of a peasant uprising, was cruelly executed in Red Square and a century later it saw the end of still another rebel, Yemelyan Pugachev.

Once the shoppers' dream: Along the entire length of the square stretches the building of the former Upper Trading Arcade, now the State Department Store, better known as **GUM {A18}**, built in 1890–93. It contains three passages running throughout the length of the building. It is covered with a glass roof and so the interior is always relatively light no matter what the weather is like outside.

Before the revolution this huge building contained several hundred small private shops where you could buy just about anything you could possibly wish for. Today the situation is different in that the store is overcrowded with people even though it is conspicuously lacking in goods. Hopefully, this situation will not last for very much longer.

On the northern side of the square stands the building of the **History Museum {A19}**, established in 1872. The architect Vladimir Sherwood tried to combine in this building the styles of different architectural periods.

On display in the museum's 57 halls are over 300,000 exhibits, including primitive ancient tools, articles of everyday life, historical chronicles – in a word, everything that might conceivably be regarded as a historical relic.

Next to the History Museum is still another building of red brick, the former City Duma (City Hall) of Moscow, built in 1840. In the 17th century the small structure in its courtyard housed the **Mint** where silver and gold coins were made, and one of the annexes was used as a prison known as the **Yama** (Pit). In 1775, Yemelyan Pugachev, the leader of a peasant war, was kept here before his execution in Red Square.

Today the former City Hall building houses the **Central Lenin Museum** whose displays document the life of the proletarian leader, practically minute by minute.

Until the 1930s, the **Voskresenskiye Gate** of Kitay-Gorod, which contained a chapel built in 1680 that held the **Iberian Icon of the Mother of God**, stood in Kitaisky Passage. However, like the other sections of the Kitay-Gorod wall, the gate was pulled down on the orders of Stalin.

The long hall of the GUM department store.

KITAY-GOROD

Beyond Red Square, bordered by the **Moskva River** to the south and by **Okhotny Ryad Street** (a part of the former *Marx Prospekt*) on the north, is the oldest residential district of Moscow, Kitay-Gorod. The name of the area, however, has nothing to do with *Kitay*, which is the Russian name for China. It is derived from the Old Slavonic *kita*, which meant a bundle of stakes, for these bundles formed the basis of the earthworks which surrounded the original area.

The *posad* (trading quarter) emerged outside the Kremlin wall back in the hoary past. It was connected with the Kremlin by way of Velikaya (Great) Street. In the 14th century it was known as *Veliky Posad* (Great Trading Quarter). At that time it was surrounded by a rampart to which a moat and wooden fortifications were added in the early 16th century.

In the mid-16th century a stone wall in the form of a horseshoe was built round the *posad*. The wall began near the Beklemishev Tower, passed along the bank of the Moskva River and then went round the *posad* on the east and north to join the Kremlin Wall near the Arsenal Tower. The wall had 14 towers altogether, including six with gates. This was Moscow's second ring. Today only two small fragments of the fortress wall, one not far from the northern entrance of the Rossiya Hotel and the other near the Revolution Square Metro Station, have survived.

A 16th-century ground plan: At the end of the 16th century the *posad* area began to be built over with boyars' and clergymen's mansions. In the 16th and 17th centuries monasteries, foreign embassies and trade missions began to emerge here. In the late 19th and the early 20th centuries Kitay-Gorod became the financial and commercial district of Moscow, where banks, insurance companies and wholesale warehouses were concentrated.

Preceding pages: Russian women are dress-conscious. **Left**, staircase in the GUM.

The basic layout of Kitay-Gorod had taken shape by the mid-16th century. Three streets radiated from the Kremlin gate towers. In time, they were crossed by streets and lanes. (To reach Kitay-Gorod take the Metro to Revolution Square, Dzerzhinskaya or Ploshchad Nogina station.)

Varvarka Street {A20}: Vasilyevsky Slope, which begins immediately behind St Basil's Cathedral, affords a splendid view of the district on the opposite side of the Moskva River, known as the Zamoskvorechye. Walking down the slope you will reach one of the busiest streets in Moscow, **Varvarka Street** known until a short time ago as *Razin Street*. Varvarka was the oldest road from the Kremlin to Ryazan and Kolomna. In the 14th century Prince Dimitri Donskoy returned by this road after victory at the Battle of Kulikovo. The street was named after the Church of St Barbara (*St Vavara*) built at the start of the street by the architect Alevisio in the early 16th century.

The first thing that leaps to the eye in Varvarka Street is the incredible mixture of periods and shapes: here, old churches stand next to monumental neoclassical structures which, in turn, contrast with the modern building of the huge Rossiya Hotel.

On the right side of the street is the already mentioned **Church of St Barbara** built to the elegant design of the architect Matvei Kazakov in the 18th century on the site of Alevisio's original 16th-century church. Today the church houses a branch of the All-Russia Society for the Protection of Monuments of History and Culture.

Further along is a striking structure with narrow asymmetrical window apertures and a steep wooden roof which dates to the 15th and early 16th centuries. This is the **Old English Residence**, which was set up after the establishment of diplomatic and trade relations between England and Russia. It is called the Old Residence because in 1636 the English were offered a new building in Moscow outside the Ilyinskiye Gate and this older one was sold to Boyar Ivan Miloslavsky.

The building was rebuilt more than once and by the turn of the century it was no longer considered of any value as an architectural monument. In the 1950s it was even planned to pull it down. Restorers, however, have succeeded in reconstructing its original facades and interiors.

Neighbouring upon the Old English Residence is the austere single-domed **Church of St Maxim the Blessed**, built in 1699 with a bell tower added two centuries later. Today it houses an exhibition hall of the **All-Russia Society for the Protection of Nature**. Also located on this side of the street is the former **Monastery of the Icon of the Mother of God, "The Sign"**, including the five-domed Cathedral of the Icon of the Mother of God. "The Sign" was built by stonemasons Fyodor Grigoryev and Grigory Anisimov in 1679–84. After the fire of 1737 the facades and interiors of the monastery were overhauled. The monastery itself arose in the second half of the 17th century on the grounds of the Old State Court. Its Brethren's Building with monastic cells and a two-tier cathedral have survived to this day. The cathedral, resting upon oak piles, was built in 1864 and then rebuilt so often that, by the beginning of the 20th century, it had virtually lost its original appearance. Concerts of chamber music and Russian choral singing are performed in its hall.

No. 10 Varvarka Street was the mansion of the boyar family Romanov and dates back to the 16th and 17th centuries. The old building was rebuilt several times and, as a result, its appearance was changed considerably. In the 19th century the architect Richter carefully restored the mansion in old Russian style. Today, it houses a museum devoted to everyday life and applied art in the old Moscow.

The view from Red Square towards the Kremlin.

The left side of the street was mostly built up in the 18th to early 20th centuries. The structures seen here include a former shopping arcade, followed by the low vast building of the **Old Merchant's Court**. Further on are the early 20th-century buildings of commercial and industrial companies.

The last side street running off to the left of Varvarka Street, **Ipatyevsky Lane** (*Pereulok*), contains a complex of 17th to 18th century monuments. Here stands a mid-17th century mansion known as the **Borovskoye Podvorye**, which housed the workshop of Simon Ushakov, the noted Russian icon-painter who painted quite a number of churches throughout the country.

In neighbouring **Nikitnikov Lane** is the **Church of the Trinity in Nikitniki**, built in the 1630s. Its oldest part is the side chapel of **St Nikita the Martyr**, the burial place of the merchants Nikitnikov with whose money the church was built. The murals in it were executed by Yakov Kazanets and Simon Ushakov, masters from the Armoury. Today the church is a branch of the State History Museum.

Zaryadye: The district on the right side of Varvarka Street {*map reference* **A21**} was formerly known as the **Zaryadye** (Beyond the Stalls). In the 15th and 16th centuries a main street linking the Kremlin with wharves on the Moskva River passed through here. At that time the district was inhabited by merchants and artisans and, later on, by boyars and clergymen. By the turn of the century the structures in this district had fallen into utter disrepair and it became a hotbed of disease and crime.

In 1960 most of the houses were pulled down and the construction of the **Rossiya Hotel**, the biggest hotel in the USSR, was begun. It took seven years to build the hotel. Its four 12-storey wings form an enormous cube with an inner courtyard. The central part of its northern wing is surmounted by a 23-storey tower in the shape of a ship's bridge. The hotel can accommode 5,300. Its facilities include several restaurants, numerous bars,

Homeless protester on the Red Square in 1990.

the 2,600-seat **State Concert Hall**, the **Zaryadye Cinema** and a Beriozka shop that sells goods for hard currency. To this may be added poor service standards and excessive rates: one has to pay for its "central location". But it need not remain this way; the new attitude that can be felt all around Moscow could infect even the Rossiya.

To the southeast of the hotel is one of the oldest structures in the Zaryadye, the late 15th century **Church of the Conception of St Anne in the Corner**, called thus because at one time the church stood in the corner formed by the eastern and southern walls of the Kitay-Gorod that was pulled down in the 1930s.

The Kitaysky Passage (Metro Ploshchad Nogina): Going up from the church is Kitaysky Passage, named after the Kitay-Gorod wall on whose site it was laid in the 1930s. Facing the passage on its right side is a building, erected in the mid-18th century, which today houses a **military academy**. Before the revolution it was the Royal Home for Orphans and Illegitimate Children. It also housed the Fine Arts Classes, the predecessor of today's Choreographic School of the Bolshoi Theatre.

At the spot where Kitaysky Passage meets Varvarka Street the Varvarskiye Gate of Kitay-Gorod once stood. Here begins former *Varvarskaya Square*, which was renamed **Nogin Square** after Victor Nogin, an eminent statesman and party leader of the period of the revolution and the early years of Soviet rule.

In Soviet times the square was almost entirely rebuilt. Of its old structures only the 16th/17th-century **Church of All Saints in Kulishki** (Kulishki was the area's ancient name) has survived.

On the right is the massive grey building of the former Business House with a porch in the form of a classical portico, built in 1913. Today the building is occupied by a number of ministries.

From the centre of the square the **Ilyinsky Gardens**, which were laid out in 1882 and named after the former *Ilyinskiye Gate* of Kitay-Gorod, run off. In the upper part of the gardens stands a **Monument to the Russian Grenadiers** who fell in the battle of Plevna in 1878. The monument was erected by the architect and sculptor Vladimir Sherwood from subscriptions collected by the men and officers of the Grenadier Corps. Inscribed on the monument are the names of those grenadiers who gave their lives for the liberation of Bulgaria from the Ottoman yoke.

On the left side of the gardens is **Starya (Old) Square**, which emerged back in the 17th and 18th centuries. At one time it was one of the busiest market squares in Moscow where, according to historians, one could buy practically anything. At the turn of the century tenement houses were built here in place of the market. In 1923 the Central Committee of the CPSU occupied the building at No. 4 and the building next to it. It now houses the Moscow City and Moscow Regional Committees of the CPSU. On the right side of the Ilyinsky

Sta. Varvara gave the name to the quarter.

Gardens is **Serov Passage** (*Proyezd*), named after a noted test pilot. Until 1939 it was known as *Lubyansky Passage*. It should be noted that the historical names of quite a few Moscow streets have been restored and this process of "rehabilitation" is still going on. Very often it is the old pre-Soviet names that are now being used again.

In the centre of Ilyinsky Square, at the former Ilyinskiye Gate, the building of the **Polytechnical Museum** begins. It was built in different phases over a period of 30 years and was completed in 1907. The central part of the building (1877) and its right wing (1896) were built in traditional 17th-century Russian style while its left wing, added in 1903–07, is an example of art nouveau.

The museum was opened in 1872 to popularise natural science. Today its displays, occupying about 80 halls, explain recent thinking on various branches of the economy and recent achievements in science. The museum is also noted for its wide range of historical displays, from antique mining lamps to models of spacecraft. The lower floors of the building house the **All- Union *Znaniye* (Knowledge) Society**, the principal non-governmental body engaged in promoting knowledge of science, technology and politics.

On the left side of **Novaya (New) Square**, at No. 12, is the building of the former **Church of St John the Divine**, built between 1825 and 1837. Since 1934 the building has housed the **Museum of the History and Reconstruction of Moscow**. The museum was founded in 1896. At that time, it occupied the Sukhareva Tower, which has not survived.

Lubyanka (*Dzerzhinsky*) Square: The facade of the left wing of the Polytechnical Museum overlooks a circular precinct which carried, until recently, the name of Felix Dzerzhinsky, one of Lenin's closest associates and the first head of the Cheka, the All-Russia Extraordinary Commission for Combating Counter-revolution, Sabotage and

Kuybishev Street leading toward the Kremlin wall.

Speculation, the predecessor of the present KGB. Today the square has been given back its historical name, Lubyanka Square, originally conferred by the inhabitants of Novgorod, who settled here in the 15th century, in memory of the Lubyanitsa Street in their native city.

It was in this square in August 1991, after the failed coup against Gorbachev, that cranes hauled down the monument to Dzerzhinsky, erected in 1958 in place of a fountain. Opposite the Polytechnical Museum is the huge building of the **Detsky Mir (Children's Paradise) Department Store**, built in 1957 on the site of the former Lubyansky Shopping Arcade. In the 15th to 18th centuries this was the Gun Foundry. Master founder Andrei Chokhov, who cast the Czar Cannon in the Kremlin, worked here.

Frightful cellars: The massive yellow building to the right, at the spot where **Myasnitskaya** (*Kirov*) and **Bolshaya Lubyanskaya** (*Dzerzhinsky*) **Streets** radiate from the square, has been occupied by the KGB since 1947. This building, which was formerly a tenement house of the Rossiya Insurance Agency, was erected in 1899 on the site of the former Royal Secret Dispatch Office. Its deep cellars were once a dreadful prison. The dispatch office was liquidated on the orders of Emperor Paul I after the death of Catherine the Great. The cellars remained and, in the days of Stalin's regime, they were once again used for their terrible purpose.

Okhotny Ryad Street {A22} (*Marx Avenue*): Descending from Lubyanka Square is the former Marx Prospekt (*Prospekt Marksa*), again divided, as in olden times, into three sections: **Okhotny Ryad (Hunters' Row)**, **Theatre Thoroughfare** and **Mokhovaya (Moss-Grown) Street**.

The solid grey buildings on the right side were constructed in the late 19th century. At present they house various ministries and departments. In a small passageway between some of these buildings you will find **Moscow's Central Bath-House**, once known as the *Khludov Public Baths*, which are still the most popular in the capital, along with the **Sandunov Public Baths**, not far away, in the first Neglinny Lane.

There is a small public garden immediately opposite the bath-house entrance. You will see, standing on a pedestal, the monument to Russia's printing pioneer, **Ivan Fedorov**, who printed the first Russian book in 1563. Next to it stands a tower with a decorative roof in neo-Russian style. Through an arched gateway, you pass into the **Tretyakov Passage** linking Okhotny Ryad Street (*Hunters' Row*) with **Nikolskaya Street {A23}** (*25 October Street*).

The tower and the gateway, designed by architect Kaminsky, were erected in 1870, funded by the Tretiakov brothers, well-known collectors and patrons of art. Adjacent to the gateway is a surviving fragment of the ancient **Kitay-Gorod wall** and the **Metropol Hotel**.

The Metropol is one of Moscow's finest and best-known hotels. It was built between 1899 and 1905, a typical specimen of Russian modern style. In recent times the building has been almost entirely reconstructed by Finnish and Swedish construction companies. The only piece of exterior decoration reminding you of its former appearance is the inlaid panels made after drawings by Vrubel. The Metropol is not only magnificent, it is also one of the most expensive hotels in Moscow.

Returning to the heart of Kitay-Gorod we will now walk along **Nikolskaya Street,** which we reach by walking through a narrow passageway between the remnants of the Kitay-Gorod wall that still remain behind the public garden. Nikolskaya Street (the former *October 25 Street*) is named after the Nikolskaya (St Nicholas) Gatetower of the Kremlin. The latter, incidentally, named after the 14th-century Greek St Nicholas Monastery, had been reconstructed again and again until it lost its original appearance completely.

If you turn right, you will see the

lateral facade of GUM, the State Department Store, mentioned earlier on. By walking up to **Sapunov Thoroughfare** behind the store, you will find an outline of the foundations of the former Kazan Cathedral built here in the 17th century with the money of Prince Dmitry Pozharsky. The cathedral, like many other architectural monuments, was dismantled during Stalin's time, but there are plans to rebuild it.

In the courtyard of 9 Sapunov Passage is the surviving ensemble of the **Icon of Our Saviour Monastery**, founded back in 1600. The many-tiered cathedral was reconstructed in the early 18th century, using decorative Baroque elements. Next to it, on the left-hand side of Nikolskaya Street are the surviving remnants of the former **Slav-Greek-Latin-Academy** – the first institution of higher learning to have appeared in Moscow. The Academy was founded by Silvestre Medvedev in 1687, converted from a former monastery school. Mikhailo Lomonosov, the great Russian scholar, studied here. In the early 18th century it had over 600 students from all over Russia. In 1814 the Academy was reorganised into a theological academy and transferred to the Trinity St Sergius Monastery in the town of Sergiyev Posad (*Zagorsk*). Today, the building houses the **Moscow State Institute of Historical Records**.

Another building, with two high spires, stands out among the structures in this street: the former **Synodal press** was built in 1810–14. Over its entrance is a sundial topped by a lion and a unicorn – the coat-of-arms of the Royal Print Yard of 1553. It was here that Ivan Fedorov produced the first printed Russian book, *The Apostle*. In the courtyard of the former printing plant is one of Moscow's oldest civilian structures – the **Chambers**. Ivan Fedorov lived here in his day.

You will find one of Moscow's best-known and most popular restaurants, the **Slav Bazaar**, at No. 17 Nikolskaya Street, on your left coming from Red

The Polytechnical Museum at Ilyinsky Square.

Square. The building was reconstructed late in the 19th century to house a hotel and a restaurant. In 1930 it was used to accommodate a puppet theatre then, in 1966, the restaurant was reopened. The rich decor of the interior, the distinctive layout and a wide choice of Russian dishes have earned it the love of many residents and visitors.

The forgotten Russian cuisine: You should beware that the finer points of Russian cuisine have been all but forgotten, to be supplanted by more diversified West European food. Until recently it was difficult to name any typically Russian drink or dish except *vodka*, *kvass* (Russian rye-beer), *borscht* (highly seasoned Russian soup of various ingredients including beetroot), and, perhaps, *pelmeni* (meat dumplings).

Today, this modest menu has been appreciably diversified by the efforts of historians and cookery experts. They have rediscovered what seemed to have been irretrievably lost – recipes for preparing *rasstegais* (open-topped pastry), chitterlings, Poltava cutlets, Moscow pancakes and many other dishes.

In the next house, at No. 21 Nikolskaya Street, there is still one of Moscow's first chemist's, once known as the **Ferein Drugstore**, after its owner's name. The right side of the street was occupied by the so-called **Sheremetev Coaching Inn**. Today it houses all kinds of offices, stores and workshops. But, to the delight of Muscovites, the new city authorities have already drafted an ordinance requiring most of the present occupants of Moscow's historical downtown area to leave their dominions. Under the latest Moscow reconstruction plan, the entire area within the limits of the former Kitay-Gorod will become a pedestrian zone. New shops and numerous cafés and restaurants will open there.

Ilyinka Street {A24}: This runs parallel to Nikolskaya Street, until a short while ago called *Kuibyshev Street* after Valerian Kuibyshev, a Party leader and statesmen. This street links Red Square with the **Ilyinsky Public Garden**. Prior

Dzerzhinsky's monument as it used to stand outside the dreaded KGB building...

to the Revolution, this was the central street of the city of Moscow. Most of the buildings date from the late 19th century and the early 20th century, designed in what was considered to be a modern style, typical of those times. They housed the central stock exchange, banks and trade offices.

A small square, **Ilyinsky Sacrum**, stands at the intersection of this street with a passage of the same name. On the opposite side of the street is a building with a semicircular facade built in 1873–75. At one time it housed the Merchant Stock Exchange. At present it is home to officials of the Chamber of Commerce and Industry.

In **Ilyinka Passage**, branching off to the left, is the almost block-long ensemble of the former **Monastery of the Epiphany**, founded by Prince Daniil of Moscow back in the 13th century. The monastery has been reconstructed more than once. At present it incorporates the Cathedral of the Monastery of the Epiphany, monks cells, built in 1693–96, and the belfry that was built next to them in the 18th century.

Restoration work in the monastery that took place in 1986 led to the discovery of the base of the walls of a white-stone church, the **Epiphany of Stone**, that stood here in the mid-14th century. It is the oldest stone structure in Moscow's trading quarter.

On the left of Ilyinsky Square is the solid building of the **Ministry of Finance**, amalgamating the premises of the former Petersburg and Azov-Don Banks.

Almost the entire right side of the street is built up with lavish premises of the former banks of the late 19th century and the early 20th century. An entire block between Ilyinka and Varvarka Street is occupied by the mammoth building of the former **Arcade (Gostiny Dvor)**, built in 1791–1805 by the famous architect Giacomo Quarenghi. You can easily recognise it by the massive Corinthian columns along the facade. The building was restored after the 1812

...and after it was toppled in 1991.

fire. Nowadays it houses various organisations, stores and a car park.

The building of the former **Church of Ilya the Prophet**, constructed in the 16th century, has survived at the very end (or rather, the beginning) of the street, nearly abutting onto Red Square.

Crossing the Square and passing the History Museum we leave Kitay-Gorod proper to arrive at **Revolution Square** with the **Monument to Karl Marx** in its centre. From here we will continue along the former Marx Prospekt to the Borovitskaya Gate on the opposite side of the Kremlin wall. Until 1961 Danton's "chopped off" head reposed here on a solid granite base in Revolution Square – something like an object-lesson for all revolutionaries. Then, apparently considering that the implication was much too obvious, the head of one unfortunate revolutionary was replaced by that of a more fortunate one.

Moscow's centre of the arts: The square between the two lateral facades of the Metropol and Moskva Hotels is called **Teatralnaya (Theatre) Square** – *Sverdlov Square* {**A25**} until a short while ago). Theatre Square got its name in the late 18th century – after the Royal Peter Theatre which stood here at the time. It burnt down in 1805, and the new building, designed by Osip Bove, appeared as late as 1825. It housed the **Grand (Bolshoi) Imperial Theatre** founded back in 1776.

The Bolshoi Theatre building burned down again and was reconstructed by architect Albert Kavos in 1856. The reopening of the Bolshoi Theatre was rushed to coincide with the coronation of Alexander II. In consequence, a huge crack appeared in the theatre's wall in 1890. The Neglinka River, flowing virtually beneath the building, had eroded the foundation and the building subsided. You can see the crack when you have a side view of the Bolshoi Theatre's portico.

The base was underpinned at the end of the 19th century. During a matinee in 1906, however, the wall of the audito-

The ladies' dress department at the GUM.

rium curved down unexpectedly. There was a panic in the hall. Many exit doors jammed and horror-stricken spectators had to try to push their way through. Only in 1921 was the nearly destroyed wall underpinned again. Incidentally, the theatre never closed during these repairs.

The Bolshoi Theatre has often been the venue for important social and political gatherings, such as the Congresses of Soviets and sessions of the Comintern. Lenin frequently spoke there. Recently the stage was placed, for the first time, at the disposal of the Orthodox Church for its celebration of the *Millennium of the Baptism of Rus*.

When the Bolshoi Theatre company is on tour, the stage is taken over by foreign troupes such as the French Grand Opera, the Swedish Royal Theatre and Britain's Royal Shakespeare Company.

From the Bolshoi you can see the **State Academic Maly Theatre** on your right. The Maly Theatre Company performed here before the building became its permanent home. In 1938–40, after it became the company's property, the building was substantially reconstructed. The Maly Theatre has been for many decades Russia's leading classical drama theatre.

Immediately opposite the Maly is the **Central Academic Children's Theatre**, built in 1821. The Children's Theatre was formed in 1921 and has performed in this building since 1936.

Manege Square: Walking along **Teatralny Passage** (as *Marx Prospekt* is now called from here on) you reach Manege Square. The square is dominated by two huge buildings – the **Moskva Hotel**, built in the 1930s on the site of the former Hunters' Row (as the local market was called), and the premises of the Royal Manege built in 1817 by the architects Montferrand and Bove. It was used for parades of the Moscow military garrison and for cavalry unit exercises.

Since 1831 this building has served as host to various exhibitions, including an international exhibition of automobiles and bicycles. It has also been used to stage musical recitals. After reconstruction in 1951, it housed the **Central Exhibition Hall** hosting major national and republican art displays.

The building is of a unique design – it is 166 metres (545 ft) long and 45 metres (148 ft) wide and is covered with a roof resting on wooden beams that lack any central supports.

The white-columned, bright-green building opposite the Moskva Hotel, near the Theatre Thoroughfare, is the **House of Trade Unions**. It was built in 1770 for Prince Dolgorukiy of the Crimea. Later the building was bought by Moscow nobility as a convenient home for their club – the **Assembly of Nobility** – which was frequented by all of Moscow's high society, nobility and intellectuals.

The **Hall of Columns** and the **October Hall** of the House of Trade Unions are often used to host festivities and social functions. In 1931 the playwright

Monument to Karl Marx on Revolution Square.

George Bernard Shaw was honoured here on the occasion of his 75th birthday. The building was also the scene, in the 1930s, of the notorious repressive frame-up trials. It has also been used as the place where the bodies of former statesmen lie in state: Leonid Brezhnev and Yuri Andropov were laid here in recent years, just as Lenin and Stalin had been in earlier times.

Next to the House of Trade Unions is the 11-storey building of the **State Planning Committee of the USSR**, built in 1932–35. This mammoth building of granite embodies the heavy burden of decision-making that the nation's top planning agency has to bear.

Moscow's main street starts immediately opposite the History Museum. Until quite recently it was called *Gorky Street*, but, following a decision of the Moscow City Soviet of Peoples Deputies, its original name, **Tverskaya**, has been restored.

At the very corner of Tverskaya Street is the six-storey **National Hotel**, built in 1902 in a rather unusual eclectic style. In the years of the Revolution Vladimir Lenin lived here in Room 107. The National Hotel – one of the capital's most popular and fashionable hotels – has been reconstructed and reopened in all its former splendour.

The house nearby is the headquarters of **Intourist** – the National Tourist Organisation. The United States Military Mission had its headquarters here during the years of World War II.

A little further on along the **Theatre Thoroughfare** (*Marx Prospekt*) you will find the old buildings of the Lomonosov Moscow State University, founded in 1755. Originally, the university was housed in a chemist's shop that stood on what is today the site of the **History Museum**. Today's University buildings were erected in 1786–93 and were designed by the architect Kazakov. The building caught fire and was restored in 1817–19. At present it houses an exhibition of rare books and manuscripts.

Bolshoi Theatre at night.

The University's second building, close by, was taken over in 1833–36 and was reconstructed in 1904. Adjacent to the lecture halls is the building of what was once the University's **St Tatyana's Church** and which was a theatre before that. In 1953 nearly all the faculties moved to a new complex on the Lenin Hills. Only the Department of Journalism, the Institute of the Countries of Asia and Africa, the editorial offices of several magazines and the University Museum remained here.

A library of 36 million books: A long block on the right-hand side of **Mokhovaya Street** (as this last stretch of former *Marx Prospekt* is now called) houses the premises of the **State Public Library {A26}**.

The present building was erected in 1928–40. The facades are adorned with medallions and sculptural portraits of many famous writers and scientists. Adjacent to the building is a nine-storey repository containing close to 36 million books, representing practically all of the world's languages. When the Metro was being constructed, the building subsided and a huge crack appeared across its wall.

It must be said that the books in this repository, especially those in the basement premises, are kept in appalling conditions, and many inestimable manuscripts have been fatally impaired. But the Library lacks the means for major repairs, while the city authorities have not until now been able to offer financial assistance.

Opposite the library building is the former mansion of Prince Shakhovsky, built in 1821. Since 1950 it has been used as a memorial museum to Mikhail Kalinin, who was President of the Presidium of the Supreme Soviet of the USSR throughout the Stalin era.

Mokhovaya Street ends at **Borovitskaya Square {A27}** onto which the Kremlin gate of the same name opens. Several old streets – Volkhonka, Znamenka and Bolshaya Polyanka – converge here.

The Lenin State Public Library.

THE ARBAT AND BEYOND

This chapter covers the quarter of Moscow between the Kropotinskaya Embankment of the Moskva River and Tverskaya Street, an area once called **the noblemen's Moscow.**

Volkhonka Street {*map reference* **B1**} starts at Borovitskaya Square outside the Kremlin. This is another of Moscow's old streets and one that has been uncommonly lucky, having been neither renamed nor reconstructed. The entire right side of the street is made up of small mansions, typical of the noblemen's Moscow of the 18th and 19th centuries. In the courtyard of No. 8 Volkhonka Street are the surviving 17th-century chambers of the Volkonsky princes. This street, which once led to Smolensk, may well have been named in their honour.

Moscow's finest museum: Beyond the facade of the Volkonsky chambers stands a building of dark-grey stone, with a massive colonnade and a glass roof: the **Pushkin Fine Arts Museum** {**B2**}. The Museum was opened in 1887 upon the initiative of Professor Tsvetayev of Moscow University. Originally, the museum was conceived as an extensive visual aid for art students. It had a collection of casts and copies of works of art from the ancient world and Western Europe. The modern building was constructed between 1898 and 1912.

As mentioned earlier, Volkhonka was the noblemen's suburb of Moscow. The seat of the Golitsin princes was at No. 14 Volkhonka Street. The neighbouring house, which was the seat of the Dolgoruky princes, is now home to the Institute of Philosophy of the Academy of Sciences of the USSR.

Beyond **Kropotkin Square** (Metro Kropotinskaya), Volkhonka Street splits into **Ostozhenka** (*Metrostroyevskaya*) and **Prechistenka Streets** (*Kropotinskaya*). The former took its name, in the 17th century, from the local *ostozhie*, or meadow. Both sides of the street are lined by former guesthouses. Today the **Ostozhenka Co-operative Café** occupies the ground floor of the corner-house on the left: a welcome retreat in the bustling city. On the right are some 17th-century chambers which once belonged to the princely family of the Golitsyns.

The **Conception (Zachatievsky) Monastery**, in Zachatievsky Lane, is considered to be the oldest structure in the whole district. It was founded in 1584. Unfortunately, little has survived – only a few fragments of the rampart and the overgate church can still be seen. Despite restoration work, daily services are still being held.

No. 38 Ostozhenka – a large house with a massive portico – belonged to Pyotr Yeropkin, Governor-General of Moscow. Alexander Pushkin, while a boy, attended balls here. From the early 19th century it housed a commercial college and, on the 30th anniversary of the Soviet regime, the house was turned over to the Moscow Institute of Foreign Languages.

Yet another green-coloured building on the left side is the former **Kadkovsky Lycée**. In the first few years after the Revolution it housed the offices of the People's Commissariat for Education and has now become the Diplomatic Academy.

Beyond the Garden Ring: Ostozhenka continues as **Komsomolsky Prospekt** (Avenue) which was laid out in the late 1950s. It leads to the Luzhniki Central Stadium and, beyond the Bridge, to Moscow University.

The avenue cuts through what was the 17th-century Khamovniki weavers' settlement. Almost at its very beginning, on the right side is the five-domed brightly coloured **Church of St Nicholas in Khamovniki**, dating to the late 17th century. The church was fortunate in having been permitted to function as a church throughout almost all the years of the Soviet system.

Turn right behind the church to arrive

Preceding pages: portraitists on Arbat Street. **Left**, western decadence under the Kremlin walls.

in **Leo Tolstoy Street**, where the great writer's urban estate still stands. Today, it is his memorial museum. The left side of Komsomolsky Avenue is occupied principally by blocks of flats and a variety of shops. Prominent against a generally monotonous background is the former **Patronage House**, an early 19th-century building with a tall four-columned portico. Now the house is occupied by the Board of the Union of Writers of Russia.

Across the street are three early 19-century buildings of the Khamovniki Barracks. Here, Alexander Griboyedov, the Russian poet who so pointedly ridiculed Moscow high society, joined the Hussar Regiment. Today, the former barracks house the Faculty of Military Conductors of the Moscow Conservatory of Music.

Near the Frunze Metro Station is the massive cube-shaped building of the **Youth Palace**. This is said to have been built with the money earned by teams of young workers on *subbotniks* – unpaid working days, named after Subbota (Saturday). Possibly that is why it took nearly 20 years before young people were able to gain admission. Now it is used for exhibitions and festivals.

In a small public garden, awash with greenery, not far from the Frunze Embankment is the **Moscow Choreographic College**, the main training ground for Soviet ballet dancers. Founded as far back as 1773 it boasts such famous graduates as Maya Plisetskaya, Marius Liepa and Yekaterina Maximova. The General Staff of the Soviet Armed Forces has its headquarters nearby on Frunze Embankment.

Home of the 1980 Olympics: Komsomolsky Avenue ends at the Central Lenin Stadium in Luzhninki. The **Stadium** {**B3**} (Metro Sportivnaya) was built in the 1950s on what was the site of Luzhnikovo Village and a small sports stadium. More than 140 different sports-installations were built on an area of 180 hectares (445 acres), comprising a swimming pool, tennis court, various sports grounds, a covered Palace of Sports and the Grand Sports Arena with a football pitch and stands to seat almost 100,000. This was the venue of the 1980 Moscow Olympics.

The sports complex incorporates a splendid public garden with numerous cafés and restaurants. Not far from the Grand Arena, on the Moskva River Embankment, is one of the most popular co-operative restaurants – **Olympus**.

Cross the river by the two-tier Metro Bridge, which was built in the late 1950s and which was then considered to be just about the peak of engineering art, and you will be in **Vernadsky Avenue**. At this point, the Moskva River curves to form a kind of peninsula.

On the right side of the avenue are the former **Vorobiev Hills** (Sparrow Hills), as they were known from as early as the 15th century. Here Peter the Great and Catherine II both had country palaces in the 17th and 18th centuries. In the 19th century there were many *dachas* (cottages) here.

A city within the city: In the late 1940s, Stalin ruled that the new **Moscow University** {**B4**} (Metro Universitet) be built on the heights of the Vorobiev Hills. A 35-storey building was erected in the space of four years. The central section stands about 240 metres (790 ft) high. The top floor houses the university museum. The central building has 18-storey towers adjoining it on four sides. These contain student dormitories and are adjoined by four more nine-storey wings providing dormitories for post-graduate students. Finally there are four 12-storey buildings with apartments for the teaching staff.

The entire university complex incorporates about 60 academic faculties providing instruction for a total of over 30,000 students from all over Russia and from various foreign countries. The university has its own research institutes, observatories, libraries, botanical gardens, sports complexes and camps.

Spread around the university area is a small student township complete with shopping centres, laundries, cafés and cinemas. Moscow University is like a small city, with the same unwieldy administrative machinery as any typical Russian city. The university recently became independent from the Ministry of Education and is now sole proprietor of all its property.

At the fork of Vernadsky and Lomonosov Avenues are the new premises of the **Moscow Circus**, built in 1971. Nearby is a children's music theatre. Vernadsky Avenue ends at the junction with Lenin Avenue in Troparevo, a district known for the late 17th-century **Archangel Michael's Church**.

The road to the convent: Returning to Kropotkin Square (by Metro to Kropotkinskaya) we will now walk along **Prechistenka Street** {**B5**}. This street (the former *Kropotkinskaya*) was named after the icon of the Novodevichy (New Maiden) Convent. In the 16th century it served as the main road from the Kremlin to this convent. It is easy to see, by observing the architectural development of the street, how the tastes of the Moscow nobility changed over the years. Many of the mansions still standing belonged to the families of famous Decembrists.

Peace-Rock Festival in Luzhniki Park.

The **Leo Tolstoy Memorial Museum**, at No. 2 Prechistenka Street, once belonged to Lopukhin, the poet. The Tolstoy Museum, which was founded before the Revolution, is one of the few museums in Moscow whose basic function remained the same after the Revolution.

The building across the street has been known as the **House of Scholars and Scientists** since 1922. At one time the house belonged to the Military Governor of Moscow, Arkharov, whose name has passed into common parlance as a symbol of vulgarity and cruelty.

From 1921 to 1923, No. 20 Prechistenka Street was the home of the ballet school. The famous American ballet dancer Isadora Duncan (the wife of the great Russian poet Sergei Yesenin) lived in one of the apartments.

The large two-storey house with a six-column portico and a huge loggia belonged to Prince Dolgorukiy at the end of the 18th century. The neighbouring house, No. 21, which belonged to the millionaire and patron of art, Savva Morozov, in the early 20th century, now accommodates the offices of the Presidium of the Academy of Artists of the USSR.

The first Moscow co-operative café opened several years ago at No. 36 Prechistenka Street. It is known as **Fedorov's Café** after its organiser. It has a pleasing interior and background chamber music. In a small beer bar in the semi-basement you can enjoy good Austrian draft beer.

Moscow's Latin Quarter: We will start our walk through the Arbat quarter at **Znamenka Street** {**B6**} (former *Frunze Street*), a small street running from Borovitskaya Square, just outside the Kremlin, to the Arbat. As early as the 14th century, this was the road to Novgorod the Great. At the beginning of the street is a small house with caryatids, built by the architect Tyurin for his own occupation in the early 19th century.

Further along is No.12, which was Count Vorontsov's house in the mid-18th century and which accommodated Prince Urusov's opera-house at the end of that century. This was the forerunner of the Bolshoi Theatre.

Across the street, at No. 19 Znamenka, is **Apraxin's palace**, a late 18th-century structure designed by the Italian architect Camporesi. Performances in the central hall of the palace were staged by the Moscow Imperial Theatre which subsequently split into the Bolshoi and the Maly Theatres. Late in the 19th century it was the seat of the Alexander Military College. Today, this building, along with most of the surrounding monumental structures, belongs to the Ministry of Defence of the USSR.

In Marx-Engels Street, which crosses Znamenka, are the chambers of the 17th-century **Royal Pharmacy Yard**. It once belonged to German chemists who laid out a medicinal herb garden on Vagankovo Hill.

A Bohemian nest (Metro Kropotinskaya/Arbatskaya): **Sivtsev Vrazhek Street** is a reminder of the days when a gully (that's what *vrazhek* means) was cut through here, with the River Sivets running at its bottom. After the river was "imprisoned" in a conduit, the street above became the Bohemian nest of writers, poets, artists and actors. At No. 9, Vuchetich, the famous sculptor, lived. Alexander Herzen preferred No. 27, a house built in Moscow's classical style. Today, it houses a **literary museum**.

On the right side of this side-street is a mansion behind an iron grille. Aksakov, the writer, lived here in the late 1840s. Opposite, at No. 31, were the homes of celebrated World War II marshals – Bagramyan and Bagritsky. Mikhail Sholokhov, the writer, lived at No. 33.

Changing times? Sivtsev Vrazhek Street is also famous for a massive edifice of red-black granite. This relic of Sta-

Want to shake hands with Gorby?

lin's day was, for many years, an outpatients' clinic for top party and government bureaucrats. In theory, such bastions of privilege no longer exist, thanks to democratic forces. In fact, however, they still function – all that has changed are the signs above the doors.

In the past few years many co-ops have moved into the Old Arbat area. Co-op cafés, restaurants, photographic studios and small shops continue to spring up like mushrooms after rain.

In **Plotnikov Pereulok**, which forms a cross-roads with Sivtsev Vrazhek Street, is the old **Oktiabrskaya Hotel** – the first Party hotel in Moscow (No. 12). Nearby are several multi-storeyed brick houses, which are architecturally out of place. They are inhabited by the once-almighty pillars of party and state. In **Maly Vlasievsky Pereulok** is the recently restored 17th-century **Vlasii Church**, with its ancient frescoes. It is now a concert hall.

Running parallel to Sivtsev Vrazhek Street from Plotnikov Pereulok to Gogol Boulevard is **Ryleeva Street**, which has a large Empire-style mansion (built in the 1830s) where Alexander Pushkin stayed when in Moscow. Today, it is occupied by the Council of the All-Russia Society for Protection of Historical and Cultural Monuments. No. 15 with the wooden mezzanine is where Ivan Turgenev, the writer, lived in the 1830s.

No. 35 **Starokoniushenny Pereulok** was the former residence of millionaire Maecenas Schiukin and the home of his famous collection of old European paintings. When Schiukin gave them to the city in 1918 the paintings were moved to the Pushkin Museum of Fine Arts. At the very end of the street is a wooden tower decorated with lace-like carvings, which was built in 1871 for demonstration at international exhibitions.

Cosy Moscow: The entire block to the right of Prechistenka and all the way to Arbat is criss-crossed by very narrow intertwining streets, always quiet and serene. All are bracketed under one

Trees protected from the cold at the Moscow University.

name: *Story Arbat* (Old Arbat) {**B7**}.

This has long been considered one of Moscow's most privileged districts. The tenants may have changed since the 1917 Revolution, but not the principle.

In this quarter, **Vesnin Street** (the former *Denezhnyi-Money Lane*), named after a famous architect, branches off from Shchukin Street towards Arbat. Here, at No. 5 Vesnin Street, where the German Embassy was located in the first few years after the revolution, the terrorist, Blumkin, assassinated the German Ambassador, Mirbach, on July 6, 1918.

Today many houses in the Old Arbat are occupied by foreign embassies and trade missions. There are several monuments of early 19th-century wooden architecture surviving in Vesnin Street, such as No. 9 and No. 11 – wooden houses which belonged to officers who participated in the 1812 War against Napoleon. Vesnin Street ends at the high-rise building of the Ministry of Foreign Affairs of the USSR.

The church on Devil's creek: "A kopeck candle in Arbat burned down Moscow", says a 15th-century chronicle. The fire of 1365 was the worst in the history of Moscow. A small candle was left burning in the wooden "All-Saints Church" that stood on a stream with the peculiar name of Chertory (*chyort yego ryl* means "the Devil dug it"). The resultant fire, which started where Volkhonka Street meets Prechistenka Street, razed the city to the ground.

Muscovites claim that the place was no site for a church: it was damp, boggy, cursed. The Devil's Creek meandered (or, as they said in the days of yore, "hunchbacked") throughout the length of Arbat Street. In consequence, the Arbat wall with the drive-through gates was anything but arrow-straight – a tricky place to negotiate in a coach.

The Alekseevsky Convent was built on the site of the ill-starred church, but did not survive for long. Nicholas I, wishing to glorify Russia's military might, built the enormous Temple of

At the Moscow Circus.

Christ Our Saviour in its place. Everyone knows what happened to that cathedral under Stalin.

A name with many meanings: There is still no consensus as to the origin of the word "Arbat". Some believe that it comes from Arabia, from the ancient settlements of Arab merchants who found their way into Rus in the 7th century; others say it derives from the Mongol word *arba* – a "sack for collecting tribute". Still others claim it stems from the Latin *arbutum* – "cherry", because there were once cherry orchards on the high Vagankovsky Hill, planted by German apothecaries. Most probably, however, the name comes from the Russian *gorbaty*, meaning "hunchbacked place", which, owing to the Moscow tradition of omitting the initial consonant and pronouncing "o" as "a", reduced "gorbat" to "arbat".

One way or another, Arbat has always been one of the best-known streets in Moscow. It is still the only pedestrian street, though it will soon be joined by others. The facades of its historical mansions were restored several years ago and the street was paved with cobblestones. After pretty lamp posts and ornate flowerbeds were added the street assumed the old Moscow look – or so its restorers claim. But judge for yourself.

The Arbat is a short street, only a kilometre (half a mile) long, yet it is famous worldwide. In the 16th and 17th centuries this was a district of artisans. Hence the names of the side-streets – **Plotnikov (Carpenter's), Serebryany (Silversmith's), Kolashny (Baker's), Starokoniushenny (Equerry's)**.

Early in the 18th century, Arbat became the aristocratic district of Moscow. In the 19th century it was populated by lawyers, artists and top-level state officials. The early 20th century added the homes of industrial magnates and bankers.

Arbat Street starts in **Arbat Square** {**B7**} (Metro Arbatskaya) where Gogol Boulevard meets Vozdvizhenka Street. Here, on the boulevard, is the pleasant

Moscow's nightlife has many faces.

and inexpensive **Praga Restaurant**, a late 18th-century coaching inn that was revamped in 1955 by a team of architects from Czechoslovakia.

Arbat is full of fun. There are amateur street-bands and young poets out to conquer the world; there are artists and artisans out to air their creations and there is Moscow's first video-rental shops which opened a few years ago and caused a small revolution in culture and technology. One immediately feels the unique atmosphere of the area. You can have your portrait sketched here in a few minutes, listen to a favourite tune, buy a painting or a souvenir. Chekhov's words still ring true: "the Arbat and its adjoining side-streets may be the most pleasant on earth."

Arbat has always attracted celebrities. The **Literary Café** at No. 7, was a haunt for Mayakovsky and Yesenin in the 1920s. Leo Tolstoy was a frequent guest at No. 9, where his niece, Obolenskaya, lived. On both sides of the street are houses built by speculators, rebuilt in the 1930s. Their ground floors are occupied by small shops and cafés.

The New Arbat, once called Kalinin Prospekt.

No. 35, which dates to the turn of the century, is a huge house whose facade is decorated with statues of knights. It belongs to the **Ministry of Culture**. The first floor is occupied by the **Samotsvety (Precious stones) Shop**. During the early days of perestroika, the shop's prices seemed outrageous (some items sold for more than 100,000 roubles at a time when the average monthly income was 250 roubles). But the goods did not remain on the shelves: some were sold, and the rest were probably hidden so as not to traumatise society's "less equal" have-nots.

If the jewels were to reappear today there would probably be a mile-long line in front of the shop. The shady wheeler-dealers of the underworld, who made their fortunes in the years of perestroika, would gladly invest that black market money in something that could be sold for convertible currency.

On the opposite side of the street is another large building – the **Academic Vakhtangov Theatre**. It was founded in 1921 by Vakhtangov, the outstanding director. The theatre has been losing points with its patrons, probably because of its lacklustre productions. The street bearing to the right also bears Vakhtangov's name. Here, at No. 11, the composer Scriabin lived and worked in the early 20th century. Today there is a museum in the house. No. 12 across the street houses the **Schiukin Higher School of Theatre** and the **Opera Studio of the Moscow Conservatoire**. Among the graduates of the school are numerous stars of Soviet stage and screen.

A block ahead is **Spasopeskovsky Pereulok**, which runs parallel to **Vakhtangova Street**. It owes its name to the former church of **Spas-na-peskah (Savior-on-Sand)**, which was built in the 17th century. A concert hall is soon to be opened in the former church. This side-street ends in a spacious square with a garden. Ftorov, the banker, chose the square as the site for his townhouse which was built in neo-Empire style (1913–14). In 1933 it was turned over to the US Embassy and was promptly renamed **Spaso House**.

There are also some modern structures in Arbat Street. No. 42, built in 1987 by a team of Georgian architects, houses the **Georgian Centre of Commerce and Culture** – *Mziuri*. On the ground floor you can buy souvenirs from Georgia and in the basement you can try national Georgian dishes.

Pushkin's happiest moments: Before we say goodbye to Arbat Street, let us dwell on No. 53, which is the pride of the street. The small, azure two-storey mansion with an open-work balcony and metal grille is the only apartment Pushkin ever owned in Moscow. On 18 February 1831, the young poet brought his wife here after their wedding in the Church of the Great Ascension. And here he spent the happiest three months of his life before moving to St Petersburg. It was from here that he went with wife and friends to the Assembly of the Gentry; it was from here that he rode to Neskuchny Garden. In 1875, Pyotr Tchaikovsky lived in the house. After the place was restored in 1986, it reopened as the **Pushkin Museum**. Arbat Street ends in Smolenskaya Square (Metro Smolenskaya). On the right is the central **Smolensky Supermarket**.

Vozdvizhenka Street {**B8**} (the former *Kalinin Prospekt*) owes its name to the 16th-century Krestovovozdvizhensky (Exaltation of the Cross) Monastery that unfortunately no longer exists. The region beyond Arbat Square is now called New Arbat. It became a fashionable avenue only when aristocrats started to build their residences here in the 19th century.

Facing the Manege Building is the former **residence of Prince Gagarin** (early 19th century). The executive committee of the Communist International worked here in the 1930s.

How to get a ticket: Today there is a

The German President at the Pushkin Museum on Arbat Street.

theatre-ticket office on its ground floor. It isn't easy for Muscovites to buy tickets for performances in Moscow theatres, for roubles at least To get a ticket, you must go through many hours of waiting in an endless line (there is another way – to avoid standing, you register your name and then come back at regular intervals – sometimes the ticket vendors choose the most "convenient" hours for this purpose, say, 2 or 3 o'clock in the morning). Needless to say, this makes theatre-ticket vendors quite popular.

The corner building on the right of Vozdvizhenka Street is the office of the Presidium of the Supreme Soviet. When built in the late 19th century, it was the **Peterhof Hotel**.

The adjacent building was owned by Count Razumovsky in the late 18th century, and then sold to Prince Sheremetev. Late in the 19th century, the **Moscow Duma** (municipal council) used the place until it moved to new quarters on Kremlevsky Proyezd. Today, it belongs to a department of the privileged "Kremlin" Clinic.

Granovsky Street, which was named after the celebrated historian, goes off to the right. It was the home of many famous Soviet scientists and statesmen and dates to the late 19th century. After the revolution, it was invaded by military top brass and party bureaucrats. Frunze, Kosygin, Voroshilov, and, for a time, even Khrushchev, lived here.

The opposite side is occupied by the imposing grey-black building of the **State Lenin Library** (Metro Biblioteka Lenina), which was built after the end of World War II. The old building is the former Tolyzin estate (late 18th-century). Today, you will find the **Schiusev State Museum of Architecture** here. No. 9 once belonged to Leo Tolstoy's grandfather, Count Volkonsky, and now belongs to the Russian Committee for UNESCO Affairs.

The former home of respected Anarchists: On the opposite side, beyond the small garden to the right, is a rather peculiar

The Arbat is Moscow's St Germain des Prés.

mansion, which was built in 1899 in picturesque pseudo-Moorish style. The building belonged to the **Union of Anarchists** in the first years after the revolution. Sergei Yesenin lived here for a short time in 1918 (he wrote the script for the film *Beckoning Dawns*, dedicated to the October Revolution).

Before World War II, the building housed the Proletkult (Proletarian Culture). Since 1959, it has belonged to the Society of Friendship with Foreign Countries.

At the crossroads with Kalashny Pereulok, which leads to the right just before the Boulevards Ring, is an eight-storey building with a turret, built in the 1920s in the constructivist style. The house was the first Soviet skyscraper. Here, in **Arbat Square**, the old part of the avenue comes to an end.

New Arbat {B9}: New Arbat was designed by architect Mikhail Posokhin, who served for a long time as Moscow's chief architect and city planner. The high-rises on both sides of the avenue create the impression of airiness and distance, even though the street is barely a kilometre long. Until recently, the avenue bore the name of Mikhail Kalinin. But there was nothing about this former head of state to merit such a great honour, and the name did not stick.

New Arbat Street is the main western thoroughfare of the capital. It connects the Trinity Gates of the Kremlin with the Kalininsky Bridge over the Moskva River and consists of two parts: the first, from the Kremlin to Arbat Square (now called Vozdvizhenka Street), is lined with houses built mostly in the 18th and 19th centuries; the second, from Arbat Square to Kalininsky Bridge, was built in the early 1960s.

The left-hand side of the New Arbat is lined with 26-storey twin towers, which resemble open books. The upper floors are occupied by assorted ministries and departments. The ground floor is a huge supermarket and department store, with numerous service agencies. A colour polyscreen was installed on one of the

A crowd on Arbat Street.

facades in 1985. It is used for screening television programmes and international advertisements.

The wide pavements make New Arbat a favourite promenade. But the throbbing life here is different from that on Tverskaya or on Old Arbat. The tempo is more subdued and controlled.

New Arbat ends near Kalininsky Bridge, which was built in 1957. A wonderful panorama of the river bend and the high-rise **Ukraina Hotel** can be enjoyed from this spot. In the distance is the concrete-and-glass **International Trade Center**, consisting of an administrative building with offices of foreign firms and banks, a hotel building and several restaurants.

Nearby, on the quay, is another cosy hotel and fashionable floating restaurant – the Russian-American **Alexander Blok**. The hospitable owners of the ship serve good food and good coffee. Close to the quay is the mayor's office, completed in the late 1960s.

On the right, high on the bank of the Moskva River is a white building with a golden clock – the **Parliament of the Russian Federation**.

Beyond the New Arbat: Crossing Kalininsky Bridge, the continuation of New Arbat is called **Kutuzovsky Prospekt** {**B10**}, (Metro Kutuzovskaya) named after Mikhail Kutuzov, the Russian general. The avenue was built in the late 1930s in place of the Old Dorogomilov Quarter. At the very beginning of the avenue, on the right, is one of Moscow's seven skyscrapers – the **Hotel Ukraina**, built in 1957. The 170-metre (560-ft) tall hotel can accommodate more than 1,600 people.

Most buildings on Kutuzovsky Prospekt date to the mid-1950s. From the very start it was intended to be a privileged area. Foreign diplomats, famous scientists, actors and artists, political and state figures live here. Leonid Brezhnev occupied two floors at No. 26. Another former party boss, Yury Andropov, also lived in the right-hand wing of that house.

A floating restaurant on the Moskva.

At the junction with **Bolshaya Dorogomilovskaya Street** (at the site of the former Dorogomilovskaya Zastava Gates) stands an obelisk erected in honour of the Hero City Moscow. Behind the stele, in a glass building, is the popular **Khrustalny (Crystal) Restaurant** and, on the ground floor, a branch of Pizza Hut.

Even though the avenue is by far the widest in Moscow traffic jams are not uncommon, because the normal traffic flow is often blocked to make way for some VIP escort. Kutuzovsky Prospekt is the road that leads to the *dachas* of the *nomenklatura*.

Stalin's orderly: It all goes back to Stalin. The "father of all peoples" built an out-of-town residence beyond **Poklonnaya Hill** in Volynskoye where he lived from the beginning of the war until his death. From here his body was brought to the House of Unions for a final farewell. Given to assorted phobias and manias, Stalin tried to spend as much time alone as he could. There was only one person he trusted: his manservant. Yet even he suffered the fate of hundreds of thousands of slandered "enemies of the people".

Stalin's orderly, who had been with him since the days of the revolution, showed his zeal by stalking the house at night in his coarse woollen socks to collect the scraps of paper that his "teacher" discarded. After examining every suspicious paragraph, he tore these papers into confetti for fear that even a passing thought of the Great Leader should fall into the hands of "the uninitiated". The orderly's activities were detected by equally zealous servants and reported to Stalin. Without even bothering to hear the man out, Stalin gave the order to get rid of him.

Legends about this terrible man abound. Some say that Stalin's *dacha* was connected to the Kremlin by a tunnel with a special train; it is, however, too early to tell whether this is true.

The avenue ends near the **Triumphal Arch**, built in 1829 to 1834 by Osip

They all love Mickey Mouse.

Bove in commemoration of Russia's victory in the 1812 war against Napoleon. The arch was initially situated near Tverskaya Zastava. During the reconstruction of the 1930s it was removed and later rebuilt in its present location. Over 150 sculptures and 12-metre (40-ft) high iron columns were cast in the process. The arch is decorated with the coats of arms of Russia's 48 provinces.

To the right of the arch is the **Borodino Panorama**. This museum was built in 1962 on the 150th anniversary of the battle at Borodino. Inside is a 150-metre (500-ft) long canvas depicting a key moment in the battle. The huge painting is the work of Rubo (1912).

The left side of the avenue embraces the remains of Poklonnaya Hill, which is named after the events of 14 September 1812. Here, Napoleon, enthused by his easy victory, awaited (in vain, of course) for the boyars to bow and present him with the keys of the city. In the 1980s the hill was cut down to create space for a gigantic monument commemorating the heroism of the Soviet people in World War II. The decision unleashed a wave of indignation and protests against the thoughtless destruction of historically valuable terrain. Construction was stopped and a contest for a better project was announced. Today, construction is being carried out in accordance with this new and, hopefully, better project.

The village of Fili: Beyond Poklonnaya Hill, Kutuzovsky Prospekt changes its name to **Mozhaiskoye Highway** and then to **Minskoye Highway –** the main road to the West (Minsk, Brest and Western Europe). The highway passes through the former villages of Kuntsevo, Troekurovo, Fili and Mazilovo, which were engulfed by Moscow in the 1950s. In the village of Fili, at No. 6 **Novozavodskaya Street**, is one of the best specimens of 17th-century Naryshkin baroque – the **Pokrov Church**, which is included in the list of outstanding architectural monuments drawn up by the UNESCO Committee for the Protection of Cultural Heritage.

Handrail in the Gorky House.

Return to Arbat Square (by Metro to Arbatskaya) and continue our walk inside the Garden Ring on **Povarskaya Street** {**B11**}. Lost among the modern giants is a miniature church, which generates a change of mood and makes one want to escape the imposing heaviness of the concrete structures. It is the only surviving church in the area – the 17th-century **Church of Simeon Stolpnik**. Here, in days of yore, Count Sheremetev secretly married the serf actress Parasha Kovaleva-Zhemchugova.

Nikolai Gogol was a member of the church's community. Today, the church belongs to the **Society of Nature Protection**. Povarskaya Street's name stems from Povarskaya, or Cook's, Village. Together with Arbat Street it was once one of the most fashionable streets in Moscow. In the final years of the 19th century, the street was lined with lime trees and grandiose mansions, many of which are now occupied by foreign embassies.

The huge building on the left houses the **Supreme Court**, the country's highest judicial body. The court examines most important cases connected with crimes against the state. Beria was tried here and, in the final years of the USSR, the court building saw a host of Brezhnev's corrupt cronies.

No. 25 Povarskaya Street belonged to Prince Gagarin in the 19th century and in 1937 was turned over to the **Gorky Institute of World Literature**. This institute, which numbers many prominent Soviet writers among its graduates, is famous throughout the land.

The next block is almost entirely taken up by the **State Musical Institute** named after the Gniesyn sisters. The institute was opened in 1944 on the basis of a network of music schools founded by the Gniesyn sisters in 1895.

A club of former political prisoners: No. 33 Povarskaya Street is the **Studio Theatre of Cinema Actors**, a kind of a stage school for movie actors. The house was built in the 1930s in the constructivist style as a club for "former political prisoners and exiles".

On the opposite side of the street is the sprawling townhouse of the Dolgorukiy family, which dates to the late 18th century. In 1920, the **House of Arts** was opened here. Blok, Yesenin and Pasternak read their poems here; in 1930, Vladimir Mayakovsky arranged his exhibition entitled "*Twenty Years of Work*". Today, the building belongs to the Writers' Union. Next door, in the former Alsufiev residence, is the **Literary Union**. Writers come here for professional conferences, and to meet with their readers and foreign guests.

One hundred metres from **Sadovoye Ring**, is **Trubnikov Pereulok**, an interesting high-rise building decorated with eagles and lion heads. The basement of this house is where Count Golitsyn, the famous winemaker, stored his collection of wines. After restoration, the cellars were turned over to the Museum of Industrial Architecture. Povarskaya Street ends near the Sadovoye Ring in **Kudrinskaya Square**.

Herzen Street {B12}: This short and narrow street (*Ulitsa Gertsena*) appeared in place of the former highway to Novgorod, which existed in the 15th and 16th centuries. Its old name, *Bolshaya Nikitskaya*, comes from the Nikitskiye Gates of the White City. The street is cut in two parts by the Boulevards Ring. The first part is lined with 18th to 19th century mansions. The quarter with the outlying Kachalova, Granatny and Spiridonovka streets is replete with the garish, late 19th-century townhouses of industrial magnates and bankers.

At the very beginning of the street, on the Okhotny Ryad side (Metro Okhotny Ryad former *Prospekt Marksa*) are the old University buildings. They form a kind of gate, situated as they are on either side of the street. Practically the entire block beyond is filled with buildings which are connected with the university in one way or another.

A delegation arriving at the foreign trade ministry.

Geniuses of music: At the crossing with Belinskogo Street (which leads to the right) is the Zoological Museum of the University. No. 13 in Herzen Street belongs to the foremost school of music in the country – the **Moscow Conservatoire**. The house was built late in the 18th century and belonged to Princess Dashkova, who was President of the Russian Academy. In the 19th century it was purchased by Count Vorontsoff and then, in 1871, by the Moscow Conservatoire, founded by professor and composer Rubinstein. In 1901, the gala hall of the conservatoire opened its doors. The house has seen Tchaikovsky, Rachmaninoff, Skriabin and many other famous composers. Since 1958, international contests for violinists, cellists, pianists and vocalists have taken place here every four years.

Several houses on Herzen Street are connected with the names of prominent composers, actors, directors and writers. Shostakovich, Kabalevsky and Khachaturyan lived on the corner in No. 8. Today, it houses the board of the Composers' Union.

On the corner with Sobinovsky Pereulok is a three-storey brick house dating to the mid-18th century. This was the Paradise Theatre which produced Chekhov's first plays. After the revolution the theatre was renamed Theatre of Revolutionary Satire. Today its stage belongs to the **Moscow Mayakovsky Theatre**. Next-door, at No. 23 is the **Folk-Studio-Theatre Near Nikitskiye Gates**. The **Films rerun movie theatre** has been here since 1939. People come here to see old favourites. Nearby is a pleasant co-op café, **Near Nikitskiye Gates**, which specialises in Russian and European cuisine.

The Gate to the White City: The crossroads of Herzen Street and Suvorovsky Boulevard form the **Square of Nikitskiye Gates**, the former gates of the White City. On the right is the central entrance to the TASS building. TASS is Russia's largest telegraph agency, with correspondents in more than 300 coun-

Painting of Shostakovich composing.

tries. Until recently, it was practically the only supplier of information from within the Soviet Union for Soviet and foreign mass media. It now faces considerable competition from several new independent news agencies such as Postfactum and Interfax. For a long time, TASS did not seem to notice its competitors, first trying to ignore them and then to buy them out. These efforts failed and we are now witnessing a fierce struggle. The young news agencies have braved all the difficulties and found clients. Of course, TASS does not stand idle. It is in the process of creating new departments for youth and video programmes. TASS also sponsors annual Russian hit parades.

Kachalova Street {B13}: Herzen Street divides near Nikitskiye Gates. The right branch is Kachalova Street, which is named after the famous actor of that name. In the centre of the square, where the two streets meet, stands the **Church of the Great Ascension** (late 18th to mid-19th century). Two famous Russian architects – Matvei Kazakov and Osip Bove – worked on this church. In 1831, before the church was complete, Alexander Pushkin and Natalia Goncharova were married here. Behind the church is another – the 17th-century **Fyodor Studit Church**.

At the very start of Kachalova Street, opposite the Church of the Great Ascension is one of Moscow's most curious mansions, built in modern style in 1901. The facade is decorated with ornate stained-glass windows and wood carvings and the building has a glass roof. Maxim Gorky lived here between 1931 and 1936. Like most houses where our geniuses lived (even if it was only for a short while) it is now a museum.

Spiridonovka Street used to be named after the writer Alexei Tolstoy, who lived on the adjacent Spiridonievskaya Street at No. 2. Here he wrote his novels *Peter the Great*, *Ivan the Terrible*, and many other works. It is also a museum.

On the opposite side of the street, at No. 17, was the **Bukhara House of**

Muscovites love colourful shows.

Enlightenment – a boarding school for children from Bukhara. That was in 1923. Now, in its place, is the Reception House of the Ministry of Foreign Affairs. The Africa Institute of the Academy of Sciences occupies No. 30.

The surrounding area was occupied in former times by artisan settlements – *Bronnaya* (Armorers), *Kozia* (Wool-Spinners), and, in the 16th and 17th centuries, *Patriarshaya* (the Patriarch's fishponds and grazing pastures).

Granatny Pereulok used to be named after Schusev, a famous Soviet architect. This street runs to the left of Spiridonovka Street. Here, at No. 3, is where the architect lived. The house is now occupied by the Board of the Architects' Union – the men at the helm of Russian architecture.

The Patriarch's Ponds: If you turn right at the end of Granatny Pereulok, walk up Zholtovskogo Street and enter a small square formed by the intersection of three streets – Zholtovskogo, Mickiewicza and Malaya Bronnaya – you will reach the place known as Goat's Marsh in the 16th century. Three ponds were dug here to supply the Patriarch's household with fish. Only one of them survived – Patriarch's Pond. Closer to the centre of the square is a **monument to Krylov** the fable-writer, surrounded by bronze characters from his fables. Muscovites associate the place with another writer: Mikhail Bulgakov. It was the site chosen by the author of *Master and Margarita* for his novel, to which the co-op café *At Margarita's* now attests.

Malaya Bronnaya Street {B14}: This street owes its origins to the ancient road from the Kremlin to Tver. Most of its surviving buildings date to the late 19th and early 20th centuries. It starts at the **Moscow Theatre** on Malaya Bronnaya. Before World War II, there was a Jewish theatre here. Its star actor, Mikhoels, was killed by Stalin's henchmen. Before the revolution the entire area was Moscow's Latin Quarter. The group of houses on the left side offered lodgings for students at moderate prices.

Training of a future driver.

MOSCOW PROPER

Our tour around the centre of Moscow now brings us into the quarter between **Tverskaya Street** {*map reference* **C1**} and **Myasnitskaya Street**. These two main thoroughfares, which limit this old Moscow quarter, have had their original names restored recently (from *Gorky Street* and *Kirov Street* respectively). We will start by walking along Tverskaya Street from Okhotny Ryad towards the Garden Ring.

The old and the new: Tverskaya Street is special not only because it is Moscow's main street but also because it blends different epochs, different styles, different moods and different impressions. Architecturally it tells the story of Moscow.

Like many other main streets in the world, it lacks cohesiveness, being a conglomeration of separate microcosms, each quite different in terms of rhythm and atmosphere. In a subtle way, they convey the mood of the past, the present and even the future.

Its history starts in the 14th century, when a road to Tver was built here. In the 17th century, the *Tverskaya-Yamskaya Sloboda* (Quarter) appeared outside the earthen city ramparts.

Early in the 18th century Tverskaya assumed a festive air – it was by that time the parade entrance to the city, and the wealthiest and best-known Moscow aristocrats lived there. By the end of the 19th century the street was filled with shops. In 1901 it was illuminated with electric lights – the only one in Moscow to be so privileged.

The original street was not as long as it is now – it ended near Sadovoye Ring, where the Triumphal Arch once stood.

The street starts with houses built in this century (Metro Okhotny Ryad, the former *Prospekt Marksa*). It is the widest part of the otherwise narrow Tverskaya. On the left, are two hotels: the **National**, which dates to the turn of the century, and the **Intourist**. Each has its fans and its character. The doorman standing guard at the National will never condescend to visit his colleague next door, while the currency dealer near the Intourist entrance will never invade the National's sphere of influence.

Preceding pages: at the Hotel National. **Left**, Moscow is no tame city.

Nearby, is a small house with an arched entrance – the **Yermolova Theatre**. At the end of the 19th century, a hotel and a shop stood here and, after the revolution, Meyerkhold's Theatre was located at this spot.

At the crossing with Ogareva Street is the huge building of the central telegraph authority and the Ministry of Communications. It was built in 1927 as the central studio of Soviet radio. Today, Moscow telegraph offers myriad services, including personal telex and telefax numbers for those who don't own their own machines.

At the crossing with Kamergersky Pereulok is the popular **Rossiiskiye Vina (Russian Wines) Shop**, one of the best-loved shops in town. Formerly the choice was larger and the shop stocked the best wines from all over Russia.

The passage owes its name to the theatre that Stanislavsky and Nemirovich-Danchenko founded in 1898. Its emblem – the seagull – floats over the entrance. Today, **Khudozhestvenny Theatre** has two companies and two stages – the original one here and a new one in Moskvina Street. In the small house next door is a second-hand bookstore, **Pushkinskaya lavka**.

At the corner of Tverskaya Street and Ogareva Street is a monumental building with a ground-floor gift shop. Its outer wall is lined with granite captured by the Russian troops during the battle of Moscow in 1941. The granite was brought by the Nazis from Finland. They planned to use it for a monument commemorating the victory of Germany over the Soviet Union.

Stoleshnikov Pereulok is a side-street which branches off to the right of Tverskaya Street. It is one of the busiest places in Moscow and designated to become Moscow's second pedestrian area. The name comes from the village of Stoleshniki (tablecloth weavers), which existed here in the 16th and 17th centuries.

In the second half of the 19th century the area became the shopping centre of Moscow. No. 9 was the home of probably the best-known reporter in Moscow: Vladimir Giliarovsky, the author of *Moscow and Muscovites*.

The other attraction of the area is the **Literary Café Stoleshniki**. The basement of this old building, which dates to the 17th century, has recently seen the ceremonial opening of two halls – **Reporter's Hall** and **Moscow and Muscovites**, where Giliarovsky Medals are awarded every year to the writers of the best articles about Moscow.

Opposite, at No. 7, is one of the largest liquor stores in the capital. If you wish to buy a bottle of vodka, you must bring an empty one along to the store: there is enough vodka around – but a severe shortage of bottles.

The dark red building with white columns in **Sovietskaya Square** {**C2**} is the headquarters of the **Moscow City Soviet** – the city's legislature. Built in 1782 by Matvei Kazakov, the house belonged to the Governor General of Moscow, Count Chernyshov, the general whose troops took Berlin in 1760. When the street was rearranged in 1939, the building was moved back 14 metres (46 ft) and two floors were added. Until recently it was traditional in the Moscow Soviet to register Vladimir Lenin as deputy No.1 before the first session of the newly-elected deputy corps. But times change, and so do traditions.

The square in front of the building was once the site of the ceremonial changing of the guard. Today Yury Dolgorukiy, the founder of Moscow, sits high above the square on his mighty steed on a pedestal. The building on the right side of the monument was formerly the Dresden Hotel (favoured by such literary lights as Turgenev, Nekrasov and Chekhov). Composer Schuman also

The world's most successful McDonald's branch.

stayed here. Today the ground floor is occupied by the Georgian **Aragvi Restaurant**. Representatives of this fun-loving nation laud the foresight of Prince Yury, who went out of his way to build a city around the Georgian restaurant.

No. 10 is the **Tsentralnaya Hotel** (built in 1911 as the Liuks Hotel). In the first years after the revolution Sergei Yesenin lived here; between the 1920s and the 1940s most of the people who stayed here worked for the Communist International.

Moscow's shopping paradise: No. 14 is well-known by Moscow shoppers as **Yeliseev's**. The house was built in the late 18th century. In the 1820s, it belonged to Zinaida Volkonskaya, the "queen of Muses and beauty", whose celebrated salon was attended by Pushkin, Vyazemsky and Baratynsky. At the end of the 19th century the ground floor became a shop where the rich merchant from St Petersburg, Yeliseev, decided to open a branch of his Petersburg grocery store.

As you have probably surmised by now, the entire right-hand side of the street is filled with shops: each caters to its own distinct clientele. It is, after all, difficult to mistake the sybaritic idleness which attracts buyers to the **Gift Shop** for the intimate breath of **Estée Lauder** cosmetics or to the Spartan scarcity of the **Dieta foodstore** from the hustle and bustle amid the glittering bronze and crystal at **Yeliseev's**. For those who seek food for the soul, there are two bookstores, **Akademkniga** and **Druzhba**, which stock books in many foreign languages.

Pushkin Square {C3} is one of the liveliest places in Moscow. Bordered by a large movie theatre, the editorial offices of newspapers and magazines and, best of all for many locals, by Russia's first **McDonald's** restaurant, the plaza attracts tens of thousands of young people every day. In the last few years it has also been the site of political rallies and angry demonstrations.

The square first appeared in the late

All dresssed up – but not, perhaps, for McDonald's.

18th century and, until 1931, was known as *Strastnaya* because of the Strastnoy Monastery, which was built near the Tverskiye Gates of the White City in 1640 and which was torn down in the reconstruction of the 1930s. Its place was taken by the **Izvestia Building** and the **Rossiya Movie Theatre**.

In front of the Rossiya, a small garden with fountains was built around the **Monument to Alexander Pushkin**, erected in 1880 with money voluntarily donated by the public.

Under the square are three metro stations – Tverskaya, Pushkinskaya and Chekhovskaya – which are connected by long tunnels. Here the first part of Tverskaya comes to an end. The next section stretches to Mayakovsky Square.

On the left side of the street, in the red house behind the antique grille, are the mighty lions described by Pushkin in his *Eugene Onegin*. After its reconstruction by Gilardi in 1831 the house was purchased by the English Club – one of the cultural nests of Moscow's aristocracy. Leo Tolstoy and Alexander Pushkin attended its famous balls. The house was turned over to the Museum of the Revolution (now called **Museum of Russian Revolutions**) in 1924. The exposition reflects the history of the Russian revolutions from 1905 to 1917. Next door, at No. 23, is the **Stanislavsky Drama Theatre**, founded in 1935.

Chekhova Street {**C4**} starts at Pushkin Square and is a continuation of Pushkinskaya. The street commemorates Pushkin, who lived here in the 1890s. On the right-hand side of the street is the **Church of the Birth of the Virgin in Putinki** (*Putinki* means crossroads), which dates to the 17th century. The delicate miniature church is one of the best tent-roofed structures in town. It still has its 17th-century frescoes – a souvenir of when the Polish House stood behind the church where foreign ambassadors were received.

The next house was built in neo-classical style early this century. In the first years after the revolution it was home to

At the Mayakovskaya Metro Station.

a Communist University and Lenin gave a lecture there. In 1933, the house was turned over to the Theatre of Working Youth, later renamed **Lenin Komsomol Theatre**.

Beyond Sadovoye Ring: The junction of Tverskaya and Sadovoye Ring forms **Triumfalnaya Square** (Metro Mayakovskaya). It was known as the *Square of the Triumphal Arch* before 1935. In the 15th century, ceremonial greetings were extended here to foreign ambassadors; in the 18th century the place was the entrance to Moscow from the new capital of St Petersburg. Today, the **Monument to Mayakovsky**, erected in 1958, stands in the centre of the square.

The character of the square is determined by several key buildings. On the left side, down Sadovoye Ring, is the **Tchaikovsky Concert Hall** with its massive colonnade. The house was built for Meyerkhold's Theatre in the 1930s. Opposite is a white edifice with picturesque turrets, which was built in 1946 for the **Pekin (Beijing) Hotel**. The hotel has a recently restored restaurant, with a wide selection of Chinese dishes prepared by chefs from China.

Near the Tchaikovsky Hall is the modern building of the **Satire Theatre**, one of the most popular in Moscow. On the right, the square is completed by the **Sofia Restaurant**. Below the square is the Mayakovskaya Metro station. The city block to the left of Tverskaya, formed by 1st and 2nd Brestskaya Streets, holds the office of Moscow's chief architect and the **Museum of Moscow Reconstruction**.

The **Central Cinema House** at the corner of 2nd Brestskaya Street and Vasilievskaya Street shows new films. The board of the Union of Cinematographers, which meets in this house, organises international film festivals and meetings with film makers.

The gateway to Western Europe: Tverskaya Street ends in **Tverskaya Zastava Square** {**C5**}, which was the site of the gates in the Kamer-Kollezhsky Wall on the road to Tver. In 1870 a

Some of Moscow's newspaper vendors are quite famous.

railway station was built near here. Known first as Smolensky and then as Aleksandrovsky Station, it was rebuilt in 1909 and was renamed **Byelorussia Station** after the revolution. It is Moscow's main gateway to Western Europe and is where most foreign visitors who do not come by plane first set foot on Moscow soil.

Most squares on Tverskaya Street are named after a famous figure: "Yury The Long-Armed" in Sovietskaya, Alexander Pushkin in Pushkinskaya. The last square belongs to Maxim Gorky (who "owned" the entire street until recently). Each personage adds something to the overall atmosphere. Proud and unbending is the great prince as, with his outstretched arm, he orders the construction of a great city. Pushkin is pensive, proudly alone, oblivious of the hustling crowd below. The gigantic Mayakovsky on Triumfalnaya Square is full of dynamism and energy. Gorky, on the other hand, looks tired, as if just off the train. The street reflects this pattern, changing from arrogant pride to pensive melancholy to chaotic rambling.

Petrovka Street (C6): (Metro Ploshchad Teatralnaya, former *Sverdlova*) This practically parallels Pushkinskaya Street. In the old days a road here along the Neglinka River led to the village of Vysokoye, which stood on a hill in the White City. Petrovka Street starts in **Teatralnaya Square** with the old building of the Muir and Merrilees Department Store. Today, the building belongs to **TsUM** (Central Department Store).

In the old days, Petrovka Street, Kuznetsky Most (Blacksmith Bridge) Street and Neglinka Street were Moscow's shopping streets. In the 18th century they were lined with small shops and stores owned by various trade companies. These offered the latest in Paris fashions, cheeses from Holland and Switzerland, guns from Germany and America – in a word, anything that a person could wish.

Early this century the street was filled with high-rise administrative and residential houses. Opposite TsUM was the Deprais Wine Store. The next building is Petrovsky Passage, once one of the largest shops in town. The house, built in 1906, resembles the GUM building in miniature. The next monumental white edifice with the colonnade is the former seat of the Raevsky family. Today, the building houses the *Russkiye Uzory* (Russian Lace) shop and a salon of art.

A short distance from Boulevards Ring is the tall spire of the **Vysokopetrovsky Monastery**. This institution was founded in the 14th century on the bank of the Neglinnaya River in honour of the victory over the Mongols at Kulikovo Field. In 1612, the monastery was an invaluable stronghold in the war against Poland. The monastery has a collection of monuments, most of which were built by the Naryshkin family in the late 17th and 18th centuries.

In the middle of the area is the main church standing above the Naryshkin family crypt. **Sergiy of Radonezh**

Muscovites are attracted by western luxury articles.

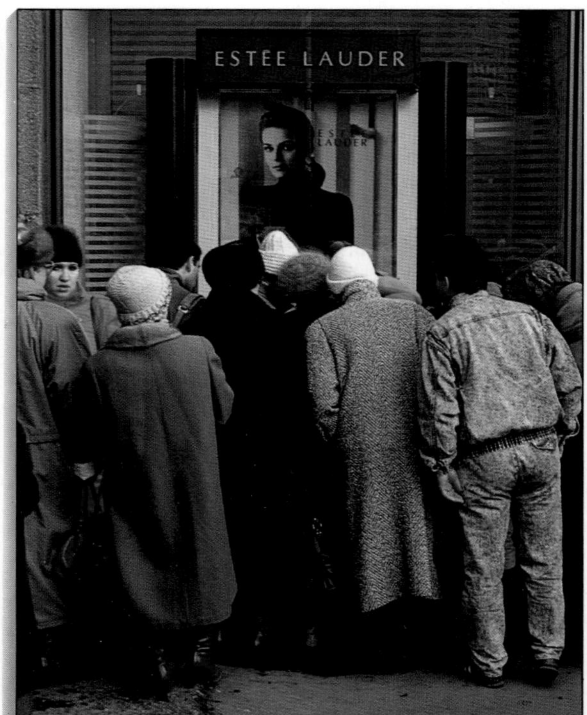

Church and **Peter the Metropolitan Church** are next to it.

The last section of the street is called **Karietny Ryad** (Carriage Row). The name refers to the time when, in the late 18th century, carriages were sold here. Some of the shops which sold the carriages have survived.

On the right side is a structure with high arches. The vast area behind this was rented by Schiukin, the industrialist, in 1894. He founded the Hermitage Garden here. Today the park is home to the **Theatre of Miniatures**, the **Summer Mirror Theatre**, the **Maly Concert Hall** and an outdoor concert stage.

The street above the water: The next street that leads from the city centre towards Garden Ring is **Neglinka Street** {**C7**}, which runs parallel to Petrovka Street and is one of old Moscow's youngest streets. The name comes from the Neglinnaya River, channeled through an underground conduit at the order of Catherine the Great. The conduit, however, was too narrow and caused flooding, particularly during the spring thaw. In 1975 a larger pipe was installed. Opposite the Kremlin's Water-Raising Tower the Neglinnaya falls into the Moskva.

Contemporaries say that the Neglinnaya River was the sewer of Moscow. Although dumping waste into the river was prohibited by law, the rich aristocrats who lived in the street built drainage systems running directly into it. The "services" of the underground river were also used by criminals who used it to dispose of bodies and "hot" loot. The bodies and corpses of dead animals frequently jammed the pipe and caused flooding, which affected even the Bolshoi Theatre.

A club, a casino and an old Moscow bathhouse: No. 9 **Pushechnaya Street** housed the German Schuster Club – an aristocratic establishment for Moscow gentry. Today, there is another club here – the **Club of Art Workers**.

On the corner of Pushechnaya Street and Rozhdestvenka Street is one of the most expensive and fashionable hotels in Moscow – the **Savoy** – built in 1913 by the architect **Velichkin**. After the war it was renamed the Berlin and neglected. Recently, the hotel opened its doors after restoration under its original name. For hard currency it offers all the amenities of hotels in the west, including a casino.

Behind the sheet-music shop on the right of the street is the old building of Sandunov's Baths, dating to 1808. The owner of the best bath-house of the time was actor Sandunov. The baths have been restored, and seem to be regaining their former glory. On the left side is **Uzbekistan**, an Uzbek restaurant.

Now comes the last radial road in this chapter, **Bolshaya Lubyanka** {**C8**} (Metro Lubyanka), the former *Dzerzhinsky Street* that was named after the chairman of the Cheka, the forerunner of the KGB. It leads into Sretenka Street, where it becomes Prospekt Mir. Nos. 11 and 13 housed the Cheka and the Cheka Club where Lenin spoke during the first anniversary of the Revolution in 1918.

A little further on, the palatial mansion at No. 14 was built by Rastrelli and figures in Tolstoy's *War and Peace* in a scene during the fire of Moscow. At the end of the street is the **Cathedral of the Sretensky Monastery** that goes back to 1395 and contains recently restored frescoes from the 18th century.

Where the icon was met: Beyond the Boulevards Ring Bolshaya Lubyanka Street becomes **Sretenka Street** {**C9**}. The name derived apparently from the word *Vstreteniye*, meaning "meeting". In 1395 the Vladimir Icon of the Mother of God was brought from Vladimir to Moscow to be placed in the Cathedral of the Dormition and was solemnly met at the gate of the White City. In the 17th century all the surrounding area was occupied by artisans' quarters, where printers, founders and gunsmiths had their shops. In the 19th and early 20th centuries the street was built over with tenement houses and shops.

At the very beginning of the street, on

the left side at the junction with the Boulevards Ring, is the small **Church of the Dormition at Pechatniki**, built in the late 17th century. Most of the houses lining the street are now being restored and its historical appearance reconstructed. The architectural make-up of the street at its further end, near the Kolkhoznaya Metro Station, was spoiled in the 1930s after the Sukhareva Tower was pulled down and the square was cleared for the construction of the Metro. At its junction with Sadovo-Sukharevskaya Street, stands the **Church of the Trinity at Listy**, which was built in the 17th century.

On the other side of the Sadovoye Ring, Sretenka Street becomes **Mir (Peace) Prospekt** {**C10**} (Metro Prospekt Mira), which was formerly the *First Meshchanskaya Street*. The street emerged on the site of an old road that led to the Trinity – St Sergius Lavra, Yaroslavl, and then on to the north of Russia.

In the 17th century the *Meshchanskaya Sloboda* (Commoners' Quarter) inhabited by immigrants from Byelorussian and Ukrainian towns sprang up here. The street was named after this quarter. During the 19th century bankers' and industrialists' mansions and tenement houses began to be built here. In the Soviet period, in the 1930s and 1940s, most of the houses were rebuilt and the street was widened and extended.

A wedding palace: On the left side of Mir Prospekt is No. 5, the house of the Perlovs, well-known tea merchants. It was built in the late 19th century in renaissance style. On the opposite side of the avenue is the early 18th-century mansion in which Jacob Bruce, Peter the Great's comrade-in-arms, lived. No. 18 on the same side of the avenue, built by the architect Bazhenov in the late 18th century, has been converted into a **Wedding Palace** where young couples have their marriage ceremony.

Moscow fashion and Russian rock: Practically opposite the palace is the centre of **Vyacheslav Zaitsev**, the leading Moscow fashion designer. His designs have been quite successful in West Europe and in Japan.

On the right side of the avenue, near No. 26, are the grounds of the country's oldest **Botanical Gardens**. Founded in 1706 by Peter the Great for growing medicinal herbs, it was known as the "apothecary's garden". For more than 250 years the gardens have been managed by Moscow University.

In the late 1970s the **Olimpiisky (Olympic) Sports Complex** was built opposite the Prospekt Mira Metro Station. Soviet and foreign rock and pop music concerts are periodically held in the complex's central arena.

In front of the monolithic sports palace is the miniature **Church of St Filipp the Metropolitan of Moscow**. It was built in the late 18th century and, like so many other buildings of that time, was designed by Matvei Kazakov.

Further along, on the same side, is **Riga Railway Station Square**. It

The Vostok rocket catapulted the USSR into the space age.

emerged on the site of the Krestovskaya Gate of the Kamer-Kollezhsky Rampart in the early 19th century. On its left is the building of the **Riga** (formerly, *Vindava*) **Railway Station** {**C11**}, built at the end of the 19th century. From here trains run to the Baltic republics.

One casualty of the "second Russian revolution" in this area has been the disappearance of the famous Rizhsky Market and the even more famous flea and thieves' market. This used to be the only place in Moscow where trade in all types of home-made goods produced in the old USSR was officially permitted. Here you could buy a fur hat, a national costume, a military uniform and many other items. Not infrequently, stolen goods were sold here.

Immediately ahead is the Krestovsky Flyover under which the Oktyabrskaya Railway passes. Branching off from Mir Prospekt near the VDNKh Metro Station is **Academician Korolyov Street** named after the chief designer of Soviet spacecraft. The street leads to yet another historical district of Moscow, **Ostankino**. In fair weather the tall spire of the **Ostankino TV Tower** can be seen practically from anywhere in Moscow. The 540-metre (1,770-ft) TV tower was built in 1967.

Gagarin, the first man in space.

On the right side of the avenue opposite the VDNKh Metro Station is the colossal **Cosmos Hotel**, built in the late 1970s by a joint team of Soviet, French and Yugoslavian builders.

Facing the hotel is the entrance to the **Exhibition Complex** {**C12**}, the biggest permanent exhibition in Russia. The tradition of arranging exhibitions of this type, demonstrating Soviet achievements, originated in 1939. At that time each of the 15 Union Republics had its own pavilion built by Moscow architects in pseudo-national styles. Everything was designed to demonstrate the indestructibility of the multinational union and the might of the world's first socialist state.

In the course of time, the pavilions were shuffled and then divided by industries. This was a kind of first warning bell, indicating that the republics, being completely dependent on the central government, had nothing to demonstrate individually. Now, more than 70 pavilions situated on an area of 234 hectares (578 acres) offer a vivid picture of what the seemingly incurable socialist economy could produce.

In front of the central entrance to the exhibition stands the sculptural composition, *Worker and Woman Collective Farmer* by the sculptor **Vera Mukhina** and the well known architect **Iofan**. The original idea was to place a frontier guard with a dog on the pedestal, thus personifying the inviolability of the country's borders. However, since the idea of "inviolability" had also the connotation, "no one will be let out", the frontier guard was replaced by the neutral figure of a worker.

Beyond the Severyanin Railway Station, Mir Prospekt becomes **Yaroslavl Highway** and leads to Archangelsk far in the north of Russia.

СССР

IN THE FORMER FOREIGN QUARTER

In this chapter we will follow Myasnitskaya Street (the former *Kirov Street*) to Nemetskaya Sloboda, the ancient German settlement, and will then cross the Yauza River to Zayauze.

Myasnitskaya Street {*map reference* **D1**} emerged in the late 15th century on the site of a road leading to the Moscow grand dukes' estates in the environs of Moscow. Its name, meaning Butchers' Street, goes back to the 17th and 18th centuries when the quarter was full of butchers' shops. In the late 18th century aristocratic mansions began to be built here. Subsequently, it became one of the fashionable business streets in Moscow.

The first part of the street up to the Boulevard Ring is occupied by monumental pre-revolutionary structures, which formerly belonged to banks and various trading firms, and by apartment houses. At the crossing with Komsomol Lane stands one of the most interesting buildings in art nouveau style built at the end of the 19th century by the famous architect Shekhtel. At one time it was the headquarters of the trading house of Kuznetsov, the noted manufacturer of porcelain and the owner of the Dulyovo Porcelain Factory.

Home of Armenians and Poles: In nearby **Armyansky (Armenian) Lane**, named after the *Armyanskaya Sloboda* (Armenian Quarter) which was here in the 17th and 18th centuries, stands a mansion with stone lions guarding the entrance gate. At one time it housed an Armenian school, set up in 1814 with funds provided by the merchants Lazarev. Later, it was converted into an Institute of Oriental Languages, also known as the Lazarev Institute. Today, the building houses the **Armenian Embassy**.

Branching off from Myasnitskaya Street to the left is **Marchlewski Street**, which was called *Milyutinsky Lane* before the revolution. At the beginning of the street, near the former Polish Quarter, is the Roman Catholic Church of **St Peter and Paul**, built in the mid-19th century. It survived the anti-religious attitudes of the last decades.

One unusual house in Myasnitskaya Street, designed in a Chinese style, was built in the late 19th century by Perlov to promote the selling of Chinese tea which his firm imported. Today the shop on its ground floor still sells tea which is, however, far below its former standard.

The next building, No. 21, was erected in the late 18th century by the architect Vassili Bazhenov for a rich nobleman, Yushkov. Later, it was turned into a school of painting, sculpture and architecture. Such noted artists as Shishkin, Levitan, Serov and Vasnetsov taught here. In 1872 an exhibition of the Association for Mobile Exhibitions, the famous *Peredvizhniki*, was held at the school. In the early post-revolutionary years such painters as Petrov-Vodkin, Saryan and Malevich studied and taught here. Before the revolution the poet Vladimir Mayakovsky appeared here before the public more than once. Subsequently the school became the **Surikov Moscow Arts Institute**.

The last building on the right near the Myasnitskaya (former *Kirovskaya*) Metro Station is the **Moscow General Post Office**. It was also here, at No. 40 Myasnitskaya Street, that the first post office was opened in Moscow in 1700. The present building of the GPO, reminiscent of some romantic medieval edifice, was built in 1912.

Stalin's fallout shelter: Beyond **Turgenev Square** Myasnitskaya Street runs parallel with the wide Sakharova Prospekt. At No. 37 is a house built in the early 18th-century by the architect Osip Bove. It belonged to Soldatenkov, a merchant and patron of the arts, and was a favourite meeting place for Moscow writers. The house was also famous for its excellent library and splendid picture gallery. In the years of World War II, the Soviet Supreme Commander-in-Chief,

Preceding pages: a friendly face. **Left**, logo on a train at Jaroslavl station.

Josef Stalin, took a fancy to this house. His choice was not entirely accidental, for the house stood near the Kirovskaya Metro Station, one of the deepest in Moscow, which was connected with the house by a special underground passage. In the event of an air attack he could take shelter deep underground in no time at all.

Further along, at No. 39, is a building of glass and concrete, the work of the French architect **Le Corbusier**, built in the early 1930s. Today it houses the USSR State Committee for Statistics.

Where the trains converge: Myasnitskaya Street comes to an end at the Sadovoye Ring not far from **Komsomol Square** {**D2**} – "Three Stations Square" (Metro Komsomolskaya). On the southeastern side of the square you can see the central facade of the **Hotel Leningrad**, one of the famous Moscow wedding-cake skyscrapers. Russia's first casino opened here at the end of the 1980s.

This place was known since the 17th century as *Kalanchevskoye Field*. At one time a royal coaching palace with a high watchtower (*kalancha*) stood here, and the czar would stay in this place overnight on his way from the Kremlin to the village of Krasnoye.

The **Leningrad** (formerly *Nikolayevsky*) **Railway Station**, the oldest in town, was built in 1849 by the architect Konstantin Thon. Characteristic features of St Petersburg architecture can clearly be seen in its design. The railway, which linked the two capital cities, was one of the first to be built in Russia. It was laid practically as the crow flies across forests and marshes; its total length is 650 km (400 miles).

Starting point of the Trans-Siberian: The neighbouring **Yaroslavl Railway Station** was built in the early 20th century by the architect Shekhtel. Its design was strongly influenced by art nouveau. Under the arch over the entrance to the station building are the ancient coats of arms of Moscow, Yaroslavl and Arkhangelsk. The sturdy pillars and fanciful turrets are reminiscent of the

On Three Stations Square.

traditional image of northern Russia. The Yaroslavl Station is the starting point of the world's longest railroad, the **Trans-Siberian Railway**, which runs to the Pacific shore of Russia. Its total length is 9,300 km (5,770 miles); travelling time is 143 hours.

On the opposite side of the square is the biggest and the busiest of all the Moscow railway stations, the **Kazan Station**. This was built between 1913 and 1926 by the architect Shchusev. From here trains run to the Caucasus, Central Asia, Western Siberia and the Altai. Every day up to 150 long-distance trains arrive at this station.

Two million passengers daily: In all, the daily flow of passengers through the three stations exceeds 2 million. This does not mean that all these people have Moscow as their destination. Many people are travelling to other places. The structure of the Russian railway network is such that all the railways intersect in Moscow and you have to travel to the capital even if you wish to reach some other place. Tickets are often in short supply, since it is impossible to forecast the number of passengers travelling in any one direction. Because of this, many people are compelled to spend the night at the station, for, as is well known, hotels in Moscow are always overbooked.

To brighten up the cheerless stay of the unbidden visitors to the capital, the **Moskovsky Department Store**, the biggest in Russia, has been built on the right side of the square. For some time, however, the purchase of certain goods is only possible if you have a Moscow registration card.

Beyond Komsomolskaya Square the road goes northeast by way of the **Sokolniki Park** {**D3**} to the wooded tract of *Losiny Ostrov* – area 11,000 hectares (27,000 acres) – then to **Bauman Street** and a district that was at one time known as *Nemetskaya Sloboda*, the German Quarter {**D4**}.

In the former Foreign Quarter: The German or, rather, the Foreign Quarter

Waiting for the Trans-Siberian Railway.

– for all foreigners at that time were called Germans – emerged in the 16th to 17th centuries. It was inhabited by immigrants from various European countries who came to Russia to serve the Russian government. The German Quarter was distinguished by a clear cut layout, numerous Protestant churches and small taverns and by its inhabitants, who wore outlandish garments and spoke strange languages.

During the reign of Peter the Great the **German Quarter** and the neighbouring village of **Preobrazhenskoye** (Metro Preobrazhenskaya) became the czar's residence until the transfer of the capital to St Petersburg. From that time the names of a number of lanes such as Aptekarsky Lane, Poslannikov Lane, Starokirochny Lane, and Lefortovsky Lane (called after the Swiss-born general Franz Lefort, Peter the Great's comrade-in-arms) have survived.

The mansion in today's **2nd Bauman Street**, presented to Lefort by Peter the Great for the entertainment of foreign guests, has been preserved to this day. In the 18th century it was bought and rebuilt by Prince Alexander Menshikov. Today, it houses the **Central State Archives**. Alongside it stands a palace which belonged first to Bestuzhev-Ryumin, then to Count Orlov and finally to General Field Marshal Kutuzov. Upon its reconstruction by Gilardi and Grigoryev, the palace was turned over to the Moscow Higher Technical School, now the **Moscow State Technical University** (until recently the Bauman).

Catherine's former Palace, rebuilt from the Petrovsky (Peter the Great's) Palace by a group of architects headed by Kazakov and Camporesi in 1774–96, has survived. Today, this huge building with a colonnade of 16 Corinthian columns houses the **Malinovsky Academy of Armoured Troops**. Adjoining the palace is Petrovsky Park, which now belongs to the headquarters of the Moscow Military District.

In this district is the old **Vedenskoye Cemetery**, also known as the **German**

Green Moscow.

Cemetery, where Lefort and French pilots from the Normandy-Niemen Division who died fighting the Nazis during World War II lie buried.

The **Cathedral of the Epiphany at Yelokhovo**, built between 1837 and 1845, is by right regarded as one of the most beautiful structures in the district. The first church on its present site was laid down by Peter the Great in commemoration of the annexation by Russia of the Caspian lands in the northern Caucasus. Later, the church was rebuilt and consecrated in honour of the victory over Napoleon. Not long ago a record of the birth of Alexander Pushkin, who was born on May 26, 1799, in the house of Titular Counselor Skvortsov in Malaya Pochtovaya Street, was found in the parish register. Behind the cathedral is a small and cosy co-operative restaurant bearing a historical name, **Razgulyai**, that serves Russian cuisine.

In nearby **Preobrazhensky Rampart Street** the late 18th and early 19th century ensembles of the **Preobrazhenskaya (Transfiguration) Old Believers' Community** and the **Old Believers' Monastery of St Nicholas** have partially survived. From 1918 this has been the centre of the *Bespopovtsi* (Priestless) branch of the Russian Old Believers. The Churches of the Exaltation of the Holy Cross and of St Nicholas, built with the participation of Vassili Bazhenov, have also survived here.

Return now to the city centre (by Metro to Ploshchad Nogina) and to a quarter of Moscow on the opposite side of the Yauza River. But before crossing the river, we will walk along **Solyanka Street** {**D5**}. Its name is one of the oldest in Moscow. It led from the Varvarskiye Gate of Kitay-Gorod to the city of Vladimir and dates back to the 17th and 18th centuries. At the very beginning of the street, where it meets Pokolokolny Lane, stands the **Church of the Nativity of the Most Holy Virgin at the Spit** (1800).

The Jewish centre of Moscow: In Arkhipov Street, branching off to the

Taganskaya Square.

right, and uphill towards Maroseika Street, is the building of the **Moscow Choral Synagogue**, built in the late 19th century. It is the spiritual centre of Moscow's Jewish community.

In neighbouring Starosadsky (Old Gardens) Lane, named after the grand duke's gardens that used to be here, is the building of the **Historical Library**, the biggest library of historical literature in Russia; the stock, based generally on private collections, runs into more than 3 million books in various languages.

The next structure, the **Church of St Vladimir in the Old Gardens**, dates to the early 16th century, It was built by the architect Alevisio Novi, the same architect who built the Kremlin. Today the church is a book depository.

At the corner with Zabelin Street stands a grim building with blind walls and towers. This is the former St John Convent founded in the 16th century. Here the mysterious Nun Dosifeya, better known as the Princess Tarakanova, the natural daughter of the Empress Elizabeth Petrovna, was held in confinement on the orders of Catherine the Great. In the 18th century the convent was rebuilt and after the revolution its buildings were turned over to the Cheka, the predecessor of today's KGB.

Across the Yauza: We will now cross over the Yauza by way of the Astachov bridge to **Zayauze {D6}**. At the spot where the Yauza flows into the Moskva River rises one of the seven Moscow skyscrapers. This 176-metre (586-ft) high building was built in 1952. One of the wings of the skyscraper houses the **Illusion Movie Theatre**, well-known and popular for showing masterpieces of world and Soviet cinematic art. It is the only movie theatre of its kind in Moscow and getting a ticket for a show is no easy matter. During the construction of this tall building the remains of an old riverside quay, dating back to the 15th century were found. At the time the Yauza was still navigable.

The church on "lousy" hill: Taganskaya hill begins beyond this monolithic building. It was once known as **Shvivaya Hill**, which, perhaps, is a corruption of the word *vshivaya*, meaning lousy. On its slope lies the former *Shvivaya Street*, now called **Goncharnaya** or Potter Street. In this street is the shining white **Church of St Nikita the Martyr Beyond the Yauza**, built in the early 16th century on the very crest of the hill.

At the very end of Goncharnaya Street is the **Church of the Dormition of the Mother of God in Gonchary**, whose name is indicative of the existence of a *Goncharnaya Sloboda* (Potters' Quarter) in this part of the city. The five-domed church is of a traditional shape and was built in the mid-17th century; it is decorated with coloured tiles showing full-size human figures. Not far from it, near the entrance to the Taganskaya Metro Station in **Upper Radishchevskaya Street**, is the late 17th-century **Church of St Nicholas at the Taganskiye Gate in Bolvanovka**.

The street ends at Taganskaya Square which emerged at the Taganskiye Gate back in the 17th century. From the square, roads led to the Monastery of St Simeon the Stylite, the New Monastery of the Savior, and to the Alexeyevskaya, Semyonovskaya and Vorontsovskaya Slobodas (Quarters).

A library of 132 languages: Another main thoroughfare in Zayauzye is **Ulyanovskaya Street**. This emerged back in the late 17th century on the site of a coachmen's settlement. Known before the revolution as *Nikolo-Yamskaya Street*, it was the ancient road to Vladimir. At the beginning of the street, on its left side, stands the modern-looking multi-storey **Library of Foreign Literature**, erected in 1967 by the architect Chechulin. Today the library, the biggest of its kind in the country, has a stock of some 5 million books and periodicals in 132 languages.

At the end of the street, on the right, is the **Church of St Simeon the Stylites**, built at the end of the 18th century by the architect Matvei Kazakov.

A suburban street scene.

In The Bend Of The River

We continue our way through Moscow by crossing the Moskva River by way of the **Krasnokholmsky Bridge** {*map reference* **E1**}, which leads from Zayauze to **Zamoskvorechye**, another historical district. In the north it is confined by the Moskva River and in the south, it is surrounded by the semi-circle of the Sadovoye Ring.

The first settlements in this area are believed to have emerged in the 13th and 14th centuries. It is known for certain from historical chronicles that Ivan IV, the Terrible, settled the *streltsi* (royal musketeers) in *Zarechye*, which is an older name for this area.

The artisans' district: The main thoroughfare in the district, crossing it from north to south, **Ordynka Street** {**E2**}, emerged on the site of a historical road leading to the Golden Horde. During the 16th century, artisans quarters known as *Kadashevskaya*, *Khamovnaya*, *Ovchinnaya*, *Kozhevennaya* and *Kazachya Slobodas* came into being on both sides of the road. Today their names can still be traced in the names of some streets.

The emergence of many different artisans' quarters resulted in the construction of separate churches in practically every district, because they are an integral part of every Russian settlement. One of the most beautiful 17th-century Moscow churches is the **Church of the Resurrection at Kadashi**, located in 2nd Kadashevsky Lane.

Most of the houses in **Bolshaya Ordynka Street** that still stand date from the 19th and early 20th centuries. The most prominent example of neoclassicism is the former house of the merchant Dolgov, built at the end of the 18th century to the design of the architect Bazhenov. Later on, after a fire, the house was rebuilt by Bove. It now houses the **Institute of Latin America** of the Russian Academy of Sciences. Immediately opposite is the **Church of the Icon of the Mother of God, "Joy of All the Afflicted"**. Like the house, the church was erected by Vassili Bazhenov and then rebuilt by Bove. The **Church of St Nicholas** at Pyzhi, built in the mid-17th century with money donated by the *streltsi* regiment under the command of Colonel Pyzhov, is another adornment of this street.

Further up, on the right side, beyond a wrought-iron screen with gates, is the ensemble of the former **Cloister of St Martha and Mary**, built in the early 20th century as a charitable institution for war invalids. The ensemble includes a **Church of the Protecting Veil** and hospital buildings.

At the very end of the street is still another church, the late 18th-century **Church of St Catherine**. In **Shchetininsky Lane** the museum of the artist Tropinin displays a collection of paintings by a number of early 19th-century Moscow artists.

The Tretiakov Gallery: In **Lavrushinsky Lane**, named after Lavrushina, a former landlady, are several attractions. The principal one, located nearer to the embankment, is the building of the universally famous **Tretiakov Gallery** {**E3**}, the outstanding museum of Russian and Soviet art.

The facade of the building, erected in the early 20th century, was designed by the artist Viktor Vasnetsov. Next to it is the house in which Tretiakov, the celebrated founder of the gallery, lived. Today the museum has 20 buildings, including the State Picture Gallery in Krymsky Rampart Street where there is a permanent display of paintings from the Tretiakov Gallery.

Branching off to the right from Dobryninskaya Square and leading into the heart of Zamoskvorechye is still another thoroughfare of the district, **Pyatnitskaya Street** {**E4**}, named after the **Church of St Parasceve** which has, however, not survived.

The entire block at the very beginning of the street, on its right side, is occupied by the 1st Model Printing House, built

Preceding pages: Muscovites are a smiling people. **Left**, Tretiakov, founder of the eponymous gallery.

in 1876. Located in Chernigovsky Lane, which branches off to the right, is the 16th-century **Church of St John the Baptist at the Pinewood** and, next to it, the late 17th-century **Church of St Mikhail and Fyodor of Chernigov**.

In **Klimentovsky Lane**, across the street, are several old historical monuments. The building at No. 7, erected between 1762 and 1770, was once the Church of St Clement the Pope, after whom the lane is named. This huge building, whose walls and towers are decorated with Corinthian columns, is about to be converted into a concert hall.

Pyatnitskaya Street ends at the **Chugunny (Iron) Bridge** across the Bypass Canal. Having walked over the bridge to the other side of the canal, you will arrive at the two buildings of the former Bucharest Hotel (now called **Balchug Kempinski**), built in 1898. Both of them stand in **Balchug Street**, whose name means "mud" in Tatar. It is not accidental that the street has a Tatar name: there were Tatar settlements here in the days of the Tatar-Mongol yoke.

Along the **Sofiyskaya Embankment** {**E5**} are many old mansions. No. 34 was a hotel in which the American delegation stayed that was sent to Russia by US President Lincoln to express the American people's gratitude for Russia's support in preventing an Anglo-French intervention in Northern America in 1865. The next building, formerly Kharitonenko's Mansion, houses the **British Embassy**, the first embassy of any Western power to be opened in the Soviet Union following the revolution.

Proceed towards **Bolshoi Kamenny Most** {**E6**} and you will reach **Repin Square**, once known as *Bolotnaya* (Boggy) *Square*. This was the location before the revolution of the well-known Bolotny Market. It was here that Yemelyan Pugachev, the leader of the insurgents in the Peasant War, and his comrades-in-arms were executed in 1775. After World War II, at the time when the 800th anniversary of Moscow was celebrated, a public garden with fountains was laid in the square.

Beyond Repin Square, stretching from the Bolshoi Kamenny Bridge, is the shortest street in town, **Serafimovich Street** (known as *Vsekhsvyatskaya Street* before 1933). Its length is 363 metres (1,200 ft).

The horseshoe-shaped island, washed on the one side by the Moskva River and on the other, by the Bypass Canal, terminates at the spit. It was covered with granite in 1935 when the new Maly Kamenny, Maly Moskvoretsky and Maly Chugunny Bridges were built.

When the **Kamenny Bridge**, one of the first bridges over the Moskva River, was built at the end of the 17th century under the supervision of the *Starets Filaret* ("starets" is a title that was given to people viewed as spiritual leaders in Russia), it was regarded as the "seventh wonder of the world". The dry wooden arches of the bridge served as an excellent shelter for homeless people and thieves. The present bridge was built in its place in 1938.

In season, fruit vendors from the vicinity come to the city.

Branching off to the right from Bolshaya Polyanka Street is **Bolshaya Yakimanka Street**. Here a floating restaurant, the **Yakor**, is moored alongside the embankment of the Bypass Canal. At the bend of Bolshaya Polyanka Street your attention will be attracted by the festive-looking **Church of St Gregory Thaumaturgus**, built between 1667 and 1669. Near this pillarless five-domed church, the gate of Kadashevskaya Sloboda once stood.

The entire right side of Bolshaya Yakimanka Street is occupied by three modern buildings of unusual architecture – at least, unusual in Moscow. Built in 1982–85, they form a single cohesive group. The first one houses a shop for newlyweds.

The building in the centre, with huge and strange-looking cubes on top, is the **President Hotel**, a deluxe hotel owned by the former CPSU Central Committee. Formerly, guests were received here free of charge – or rather, at the expense of the state. Now that the more insistent aspects of democracy are making their impact, accommodation and services provided by the hotel have to be paid for, preferably in hard currency.

On the ground floor of the next building is the **House of Toys**, the biggest toy shop in Moscow which has become a rival of the famous Detsky Mir (Children's World) Department Store in Lubyanka Square. The decor of the facades of the three buildings was designed to be in harmony with the nearby **Church of St Ioann the Warrior**, an example of early 18th-century Baroque.

The street ends at the spacious **Kaluzhskaya Square**, formed by the crossing of the Sadovoye Ring, passing in a tunnel under the square, and Lenin Prospekt, leading into the southwestern part of the capital.

Lenin Prospekt is the scientific centre of Moscow, with the capital's major research, medical and educational institutions being concentrated in this area. It ends in the new residential districts of Tyoply Stan and Troparyovo.

A familiar scene from many Muscovites' apartments.

THE BOULEVARDS RING

We will continue our acquaintance with the city by going along the third city ring, formed by different boulevards. The **Boulevards Ring**, which is a chain of 10 boulevards merging into one another, emerged in the late 18th and the first half of the 19th century. The walls and towers of the White City, erected back in the 16th century, originally stood on this site. Late in the 16th century, Czar Fyodor Ioannovich had over 7,000 masons from every part of Russia brought to Moscow to build the czar's White City.

The construction work, supervised by the master builder Fyodor Kon from Zvenigorod, resulted in the emergence of an immense semicircular brick wall, about 10 km (6¼ miles) long, up to 20 metres (63 ft) high and up to 6 metres (20 ft) thick. The wall had 27 towers, including 10 with gates.

In compliance with the master plan for the development of Moscow, adopted in 1775, the walls were pulled down and boulevards were laid in their place, while the gates were replaced by squares. Some of the squares still bear the names of the old gate towers. Mansions were built on both sides of the boulevards, and later, in the early 20th century, tenement houses. The foundations of the old walls, however, have survived under the grass covered central reservation of the present-day boulevards.

The Boulevards Ring is not really a ring at all, but a horseshoe which begins on the embankment of the Moskva River at the **Kropotkinskaya Metro Station** and ends where the Yauza flows into the Moskva River. We will start our tour at the Kropotkinskaya Metro Station.

It should be noted that the entire area adjoining the Boulevards Ring is regarded as a protected zone, which implies careful preservation of historical and cultural monuments.

Preceding pages: sometimes you'll need a dictionary to communicate. **Left**, little girls love dressing up.

The first interesting edifice on the boulevard has an elaborate roof and a pointed pediment (No. 29) and was designed in 1901 (by Vasnetsov) as a picture gallery for Tsvetkov, the collector, whose paintings are now in the Tretiakov Gallery. During World War II, the building housed the Free French Mission. Today, this is the residence of the French military attaché.

The first boulevard starting in **Prechistenskiye Vorota Square** is today known as **Gogolevsky** {*map reference* **F1**}; it was *Prechistensky* until 1924. Its right side is lined with mansions dating from the 18th and the 19th centuries. Its left side is younger, going back to the turn of the century. No. 6 used to belong to Tretiakov, the collector, and then to the millionaire Riabushinsky. In the late 1980s the building was turned over to the Soviet Culture Fund. In the adjacent No. 10, built by Matvei Kazakov, the Naryshkins, also a wealthy family, had their town estate in the 19th century.

A little further on we come to **Arbatskaya Square** (Metro Arbatskaya) where one of Moscow's most famous streets, **Old Arbat**, starts. On the right side is the Khudozhestvenny Movie Theatre, one of Moscow's oldest.

Leading off to the right and ending at the Kremlin's **Borovitskaya Gate** is **Znamenka**, a small but very old street (13th century), which takes its name from the Church of the Icon of the Mother of God, "The Sign". It is crossed by Marx-Engels Street where the czar's apothecary lived in the 17th century. Here, on Vagankovsky Hill, German apothecaries planted medicinal herbs. Not far away is the 18th-century Dormition Convent.

Nikitsky Boulevard {F2}: Beyond Arbat Square runs **Nikitsky Boulevard**. On the left side, at No. 7a, is the former townhouse of General Talyzin, where Gogol lived between 1848 and the time of his death. Today, fittingly, it houses a museum.

Opposite, No. 8a, the **House of the Press** (opened in 1920) saw such literary lights as Vladimir Mayakovsky and

Ilya Ehrenburg. Today, the house belongs to the Union of Journalists.

The most architecturally attractive edifice on the boulevard is the former **Lunins's House** at No. 12, built in 1820 by the architect Domenico Gilardi. Its main building is decorated with eight Corinthian columns. The house was famous for its musical soirées. It is now the **Museum of the Arts of the People of the Orient**.

Tverskoy Boulevard {F3}: Tverskoy (*Tver*) Boulevard, the next in sequence, begins at the square called **Nikitskiye Gate** and ends at **Pushkin Square** (Metro Pushkinskaya). It is the oldest of the boulevards, contemporary with the mighty oak that, for over 200 years, has witnessed the people and events that have formed this quiet neighbourhood. Pushkin, Lermontov and Tolstoy enjoyed strolling along the boulevard's central walk. In their day, it was one of the city's most select public gardens.

The beginning of the right side of the boulevard is occupied by the buildings of the **Telegraph Agency of the Sovereign States** (TASS). Further along are 19th-century mansions. In **Bogoslovsky Lane**, branching off to the left, stands the 17th- to 18th-century **Church of St John the Divine**, the oldest architectural monument on Tverskoy Boulevard. Not far from where the boulevard crosses **Tverskaya Street** on its right side is the new building of the **Moscow Art Theatre**, built in 1973.

Tverskoy Boulevard comes to an end at Pushkin Square where several streets – Tverskaya, Chekhov and Pushkinskaya – intersect. Under **Pushkin Square** there are three Metro stations – **Gorkovskaya, Pushkinskaya** and **Chekhovskaya** – connected to each other by long underground passages.

Moscow's McDonald's: Tverskoy Boulevard is always swarming with people. This has been especially true since 1989, when the world's biggest **McDonald's restaurant** opened in the premises of the former Lira Café. It is difficult to believe that people spend

Overlooking Pushkin Square.

three to four hours standing in line in order to try a Big Mac at least once in their life. But then, this is no great feat for people who, from their childhood, have been used to standing in line for everything – first with their mammies for nappies and a pram, then for an apartment, a car, clothing, food, a trip abroad, and, finally, for a place in a cemetery. To stand here for several hours is less frustrating, because you can be confident that no one will slam the door in your face and say that the restaurant has run out of beefburgers and that you should come back tomorrow.

From its crossing with **Pushkinskaya Street** to the former **Petrovsky Gate** lies one of the shortest (some 300 metres/990 feet) and widest (about 125 metres/410 feet) part of the boulevard. It was known from the early 19th century as **Strastnoy Boulevard** {**F4**} and was named after the Strastnoy Convent (Convent of the Passion of Jesus) which at one time stood on the site of today's **Rossiya Movie Theatre**.

At the very end of the boulevard on the left side stands a **city hospital**. This imposing building with a portico supported by 12 Ionic columns stretches right up to the Petrovsky Gate. The building was erected in the late 18th century to the design of Matvei Kazakov and later rebuilt by Osip Bove. At one time the house belonged to the Princes Gagarin. In the early 19th century it housed the English Club, made famous by Leo Tolstoy's novel, *War and Peace*. In 1812 the French writer Stendhal stayed in this house. In 1833 it became the New St Catherine's Hospital. The English Club moved to the mansion in Tverskaya Street which now houses the Museum of the Revolution.

At the intersection of Strastnoy Boulevard and Petrovka Street is **Petrovsky Gate Square**, one of the few squares whose appearance has remained practically unchanged since the end of the 19th century.

From the crossing with Petrovka Street down to Trubnaya Square stretches

Wall papers, Arbat-style.

Petrovsky Boulevard {**F5**}, named after the **Vysokopetrovsky Monastery** (Monastery of St Peter on the Hill). The boulevard was laid out in the late 19th century. At its start, on the right side of the boulevard, are former mansions of the aristocratic Tatishchev family.

On the same side, at the end of the boulevard, stands the building of the former Hermitage Hotel and Restaurant. The house was built in the early 19th century and was later rebuilt in the mid-1860s.

The Hermitage, once one of the best-known and most popular restaurants in Moscow, was established by the French chef, Olivier. It was frequented by celebrated writers, composers and various guests of honour who were all given the warmest reception. Both Fyodor Dostoyevsky and Ivan Turgenev were among its patrons. Peter Tchaikovsky held his wedding party in the White Room of the restaurant. After the revolution the building was converted into the **House of the Collective Farmer**. Today it belongs to one of Moscow's theatres.

The boulevard terminates at Trubnaya Square where several streets – Neglinnaya, Tsvetnoy Boulevard, Trubnaya, and Rozhdestvenka – come together. The square was named after the pipe (*Truba* means pipe in Russian) which was laid in the wall, leading from the Neglinka River to the White City.

Today there is no trace whatsoever of this conduit, but the Neglinka still flows under the square and the Tsvetnoy Boulevard. Before the revolution, the most popular pet market in town, *Ptichy Rynok*, was located here and next to it was a slum neighbourhood known as Grachevka.

Branching off to the left from Trubnaya Square is **Tsvetnoy Boulevard** {**F6**}, named after a flower market that used to be here. The round grey building, characteristic of its period, houses the **Mir Movie Theatre**, which is equipped for wide-frame projection and multi-channel stereo sound – one of the first movie theatres of this type in Moscow. The next building, at No. 13, houses the **State Circus**, also known as the Old Circus. It was established in 1919 on the basis of Salamonsky's private circus.

Where market prices are set: Next to the circus is a building that houses still another curious facility, the **Central Market**. It was founded at the very beginning of the 20th century and its new building was erected in the mid-1950s. It is the best-known and most expensive market in Moscow. If they don't have certain goods here, you can be sure you will not find them anywhere else in Moscow. You will see sellers from every part of the country and prices here determine the level of prices at the other Moscow markets.

In general, the market is an integral part of everyday life in all Russian towns. When all that state-run shops have to offer is empty shelves, town life centres round the market. Even during the "stable" period of "stagnation", the market responded sensitively to every slight change in the true value of the national currency, the rouble.

Here you will have the best opportunity to grasp the fluctuating, actual value of the rouble. At present it seems that no administrative corrective measures can hold price increases in check and this poses intractable problems for the average shopper. It is hardly feasible to park an armoured personnel carrier opposite every market counter!

Rozhdestvensky Boulevard {F7}: From Trubnaya Square, Rozhdestvensky Boulevard, which was named after the former **Rozhdestvensky (Nativity) Convent**, rises steeply along the slope of Sretensky Hill. The first street branching off to the right is **Rozhdestvenka Street**. It lies on the route of an old road leading from the Kremlin to the Nativity Convent.

A convent home for the homeless: At its start on the left side, is the ensemble of the Nativity Convent, founded in the late 14th century on the high left bank of the Neglinka River. The ensemble includes the early 16th-century **Cathedral of the Nativity of the Most Holy Virgin**, the **Church of St John Chrysostom** (1677), nuns' cells with a belfry and an over-the-gate church, a typical example of 16th to 17th-century Russian architecture. The convent was lucky not to have been pulled down during the reconstruction of Moscow.

In the mid-1970s it was turned over to the Moscow Institute of Architecture, located nearby, for restoration and conversion to of a museum of Russian art.

All that remained unresolved was the question of money for the restoration. Thus, for more than 20 years, the ancient ensemble has been in enshrouded in scaffolding and abandoned. Although it lacks official support, it is not quite devoid of life. In its dilapidated buildings, students who have no bed in a dormitory, the homeless and assorted members of the so-called "socially dangerous element" have found shelter.

Among the inhabitants of this medieval convent, a rather unusual spiritual atmosphere reigns, to which the average Russian, who has been educated to believe in the unity of views and interests, is unaccustomed. A kind of commune, where practically everything is shared, has developed. Apparently, this pattern of relationships helps people, to a certain extent, to survive in the fearful economic circumstances that have followed the collapse of communism.

It should be noted that this pattern is becoming more widespread in the capital. Quite a few old houses, abandoned by their former tenants, are occupied by squatter communes and the city authorities, having no funds for their restoration, shut their eyes to this phenomenon.

At No. 11 is the **Moscow Institute of Architecture**. The building, from the mid 18th century, was the central part of Count Vorontsov's estate.

Sretensky Boulevard {F8}: Sretensky Boulevard, the shortest in the chain of Moscow boulevards (its length is only 214 metres/700 ft), lies between the **Sretensky Gate** and **Turgenev Square** (Metro Turgenevskaya).

On the right side of the boulevard is a group of houses that are among the most beautiful in Moscow. These are tenement houses built in the early 20th century by the architect Proskurin with funds provided by the Rossiya insurance agency. They are not only distinguished by the original decor of their facades, but are also very comfortable inside. In the years of the Civil War the Russian Telegraph Agency (ROSTA) had its headquarters here.

At the Clear Pond: Sretensky Boulevard ends at Turgenev Square. The square assumed its present appearance quite recently, after the old houses were pulled down and the rubble cleared away. Beyond the distinctive vestibule of the **Chistiye Prudy Metro Station** opens the widest Moscow boulevard, **Chistoprudny Boulevard {F9}**, named after Chistiye Prudy, the rectangular pond situated in the central part of the boulevard. At one time the pond was known

Sretensky Gate at the turn of the century.

as *Poganiye Prudy* (Foul Ponds), for it collected all the sewage from the slaughterhouses and meat shops in the neighbourhood. In 1703 the pond was cleaned out on the orders of Prince Menshikov and since then it has been called *Chistiye Prudy* (Clear Ponds). The pond is in the floodlands of the **Rachka River**, a tributary of the Yauza. Today you will not see the river, however: it has been diverted underground.

In **Telegrafny Lane**, branching off to the right, is the **Church of St Gabriel the Archangel**, also known as the Menshikov Tower, built in the early 18th century in Russian baroque style. Until 1723 the tower was topped with a tall spire, but this was destroyed in a fire caused by lightning. The 79-metre (260-foot) high tower was the second tallest structure in Moscow after the Bell Tower of Ivan the Great.

The building at No. 19 on the opposite side of the boulevard, designed by Roman Klein in the early 20th century, once housed a popular movie theatre, the Coliseum, and earlier still, in the 1930s, the first Moscow Workers' Theater of the Proletkult. Since its reconstruction in the early 1970s, the building has housed the **Sovremennik (Contemporary) Theatre**.

Chistoprudny Boulevard ends at its crossing with **Pokrovka Street** in **Pokrovsky Gate Square**, bearing this name in memory of the Pokrovsky Gate of the White City which once stood at this point.

Pokrovsky Boulevard {F10}: Beyond the square Pokrovsky Boulevard begins. On its left side is a building, erected in the late 18th century, which was formerly the Pokrovsky Barracks. Branching off to the left from this spot is narrow **Durasovsky Lane**, named after the Durasovs, the former owners of the palace at No. 11 which was built in the late 18th century by a disciple of Matvei Kazakov. Since 1932 the building has housed the **Kuibyshev Military Engineering Academy**. On the right side of the boulevard stood a private gymnasium

On Tverskoi Boulevard.

for girls. Today the building houses Soviet Encyclopedia Publishers.

Yauzsky Boulevard {F11}: Pokrovsky Boulevard ends at **Yauzsky Gate Square** and from here it becomes Yauzsky Boulevard, the narrowest of the Moscow boulevards. On its right-hand side is a monumental residential building with a high entrance gate. In front of the arch statues of a worker and a woman collective farmer stand on heavy pedestals. The building was erected in 1936 following the best traditions of the "Stalinist" style.

In the narrow **Petropavlovsky Lane**, branching off to the right, is the **Church of St Peter and Paul**, a monument of early 18th century Moscow baroque. Dominating **Serebryanichesky (Silversmiths) Lane**, named after the jewellers' and carvers' quarter that existed here at one time, is the 17th/18th-century **Trinity Church** with a bell tower, which used to be the principal church in the quarter.

The boulevard becomes ever narrower on its way downhill towards its end at **Yauzsky Gate Square**.

At this spot there are three bridges – the **Bolshoi (Greater) Ustinsky Bridge**, the **Maly (Lesser) Ustinsky Bridge**, and the **Astakhovsky Bridge**. Beyond them lies a historical district of Moscow known as **Zayauzye**. At one time another historical district, **Khitrovka**, the infamous centre of iniquity in Moscow, was situated in the vicinity of the Yauzsky Boulevard and Yauzsky Gate Square. Khitrovka, a district between Yauzsky Boulevard and Solyanka Street, was an area known for its dense doss houses.

Quite a few skilful artisans also lived in this district, the best-known among whom were tailors. It is said that they were able to remedy seemingly hopeless situations, like mending dresses with burned-through holes with such skill that the holes became indiscernible. Such masters, known as "jobbers", charged surprisingly small sums for the "jobs" they did so well.

Life can be difficult for the old.

MONASTERIES AND CONVENTS

The Novodevichy Convent: Grand Duke Vassili III of Moscow made a solemn promise to found a convent on a bend in the Moskva River if he were to recapture the old Russian city of Smolensk from Lithuania. He did, and Smolensk was placed once again under the sceptre of Moscow. Three thousand roubles in silver were contributed from the ducal purse to the construction of the Novodevichy Convent.

Thus one of the most beautiful Moscow monasteries came into being in 1524. Sensibly, it was constructed on the southwestern approaches to Moscow, on the Smolensk Road, to honour the reintegration of the city into the Russian state. Immediately it became a kind of court convent, for its nuns came from the families of Moscow boyars, princes and czars.

From among its oldest structures the **Cathedral of the Smolensk Icon of the Mother of God**, built in the 16th-century on the orders of Vassili III after the pattern of the Dormition Cathedral of the Moscow Kremlin, has survived. Its five domes, cut through with narrow window slits, rest upon a six-pillar base.

The cathedral's iconostasis is the most imposing of all the baroque iconostases in Moscow. Its wooden columns, decorated with an ornament in the form of a climbing grapevine, are made of whole tree trunks. The iconostasis was made on the orders of Czar Boris Godunov. Its oldest icon is the **Smolensk Icon of the Mother of God "Hodegetria"**, dating from the first quarter of the 16th century. Many of the icons were produced by the famous Moscow icon-painter, Simon Ushakov.

The convent has an eventful history. In 1598 Czar Boris Godunov was crowned here. In the late 17th-century Peter the Great imprisoned his power-seeking sister Sophia here: she had been at the head of the *streltsi* mutiny. It was also here that Peter the Great's first wife, the disfavoured Czarina Eudoxia Lopukhina, spent the rest of her life in seclusion. The tombs of both are in the cathedral and its crypt is the burial place of the family of Czar Ivan the Terrible. It was a tradition then to establish cemeteries on monastery grounds.

Preceding pages: the Novodevichy Convent. Left, Kolmenskoye architecture.

The Novodevichy Convent has two cemeteries. The small, older cemetery, with the graves of clergymen, princes, merchants, and heroes of the Patriotic War of 1812, is situated within the convent grounds. The new, larger one lies behind the southern wall of the convent. Since the 19th-century it has been the burial place for notable personalities in Russian history and culture, war heroes and eminent statesmen. Until recently the public was not admitted to this cemetery. The reason for this was prosaic enough: it was the reluctance of the Communist power elite to allow ordinary people to come near the graves of pre-socialist national heroes.

The writers Anton Chekhov, Vladimir Mayakovsky, Nikolai Gogol and Alexander Ostrovsky, the composers Scriabin, Prokofiev and Shostakovich, the painter Serov, the sculptor Mukhina, the actress Yermolova, the directors Stanislavsky and Eisenstein, and also the former Soviet leader Khrushchev and several other Soviet politicians and scientists lie buried here.

The Donskoy Monastery: It was around a field church on the southern approaches to Moscow, which housed the **Don Icon of the Mother of God**, that Boris Godunov gathered his troops before embarking on his final battle with the Crimean Khan, Kazy Girei, who was on yet another raid into the Russian territory. Here, the **Lesser Cathedral** of the future monastery was later built on the orders of Czar Fyodor Ioannovich, the last czar of the Rurik dynasty, to commemorate the ensuing victory.

The foundation stone for the **New Cathedral**, built in Russian baroque style, was laid down a hundred years later, in 1684, in fulfillment of a pledge made by Yekaterina Alexeyevna, Peter the Great's sister. The cathedral is reminiscent of the palaces of ancient Rome. Its walls were painted with frescoes by the Italian artist Claudio. Almost simultaneously with the New Cathedral new walls and towers were erected and the three-tier **Over-the-Gate Church of the Tikhvin Icon of the Mother of God** was built on top of the central entrance. Though the danger of an attack was non-existent, the fortress was built according to all the rules of the science of fortification. The monastery has recently been returned to the Russian Orthodox Church.

The St Daniel Monastery: The St Daniel Monastery was the first place of religious seclusion to be built in Muscovy. It was founded as early as the 13th century by Prince Daniil of Moscow, the youngest son of Alexander Nevsky and known as the founder of the Moscow Principality. Daniil is the only one of the Moscow princes to have been canonised by the Russian Church.

The present walls and towers of the monastery were built in the late 17th century during the period of the monastery's heyday. **Danilovskaya Sloboda** grew and at the **Danilovskoye Cemetery** the **St Daniel Church** was built to the design of Matvei Kazakov. The oldest structure in the monastery grounds is the **Cathedral of the Holy Fathers of the Seven Ecumenical Councils** with the **Church of St Daniel the Stylites** (late 16th to early 17th centuries). The **Trinity Cathedral** was built in the mid-19th century to the design of the architect Osip Bove.

The monastery has a far from ordinary history. It was closed down and then reopened more than once. From the late 19th century onwards it was used as a reformatory for young offenders. After the revolution of 1917 it was closed down and later, in 1928, it was reopened to be used, under the guise of a reception centre for homeless children, as a prison for the children of people repressed by Stalin's regime. In 1985 the monastery

Aerial view of the Novodevichy Convent.

was returned to the Russian Orthodox Church.

When the time came to choose the place to be used as the centre for the celebrations in honour of the Millennium of the Baptism of Rus, Patriarch Pimen chose St Daniel. Having fallen into utter disrepair, the cloister was completely restored in less than five years. Today, the monastery's most significant sacred possession, the reliquary containing the holy relics of the Orthodox Prince St Daniil, is kept in its main cathedral. The monastery has also become the residence of the Patriarch of Moscow and All Russia, whose offices have been transferred here together with the administrative bodies of the Holy Synod.

The New Monastery of the Saviour: The monastery was founded in 1462 in the reign of Ivan III, who unified the Russian lands round Moscow. The czar needed more room in the Kremlin and so he ordered one of the Kremlin convents to be moved outside the city limits (today's Dinamovskaya Street, Proletarskaya Metro Station).

No 15th-century structures have survived in the monastery. Its present stone walls and five towers date back to the 17th century. The main church of the monastery, the **Cathedral of the Transfiguration of the Saviour**, was built in 1645–51. Subsequently, it became the burial place of the boyar family Romano, relatives of the czar. The inner vaults of the cathedral were painted later in the 17th century by masters of the Armoury. Dominating the monastery ensemble is a 70-metre (230-ft) high bell tower. Today the monastery is occupied by the **Restoration Institute** and experimental workshops.

The Krutitskoye Metropolitans' Residence: Towering over the high bank of the Moskva River are the domes and spires of the ensemble of the former Krutitskoye Metropolitans' Residence. The ensemble was given its name after the small hills (*krutitsy*) on which it stands. The Krutitskoye Metropolitans'

The last resting place for famous people: the Novodevichy cemetery.

Residence was founded in the 17th century on the site of a small 15th-century monastery. The 15th-century **Church of the Resurrection**, the mid-17th century five-domed **Cathedral of the Dormition** with a hipped-roof belfry and the Metropolitans' chambers have survived. The **over-the-gate Teremok**, faced in coloured tiles with carved white-stone columns and connected by a gallery to the Dormition Cathedral, is regarded as the gem of the Krutitskoye ensemble.

The Teremok was built in 1688–94 by Osip Startsev as a residence for the Moscow Metropolitans. Today, many of the structures of the ensemble are being restored.

The Simonov Monastery: As with all the other monasteries, this one, which once served as a mighty fortress, was built on the high bank of the river. Established back in 1379, it was named, together with the entire neighbourhood, after its founder, the **Monk Simon**. The monastery was part of the system of defensive fortresses built on the southern approaches to Moscow and, more than once, it saved the city from raids by Tatar-Mongol khans. In 1640 a new stone wall was built round it. In the 1930s, during the reconstruction of Moscow, most of its structures were pulled down and only three towers and the refectory buildings survived. Today the monastery grounds are occupied by the **Cultural Centre of the Likhachev Auto Works** – such was the ease with which the new "proletarian" culture ousted the age-old culture of the Russian people. Everything which was even just a little different from the then leaders' primitive ideas of beauty and cultural values was destroyed without a moment's thought about the future and without any respect for the past.

The Estate in Kuzminki: The ensemble of the estate in Kuzminki emerged in the 18th and 19th centuries. Formerly the land on which it was built belonged to the Simonov Monastery. The estate was erected in the 18th century for the Stroganovs, a family of rich salt merchants. An amazingly large number of celebrated architects and sculptors – including Bazhenov, Gilardi, Kazakov, Voronikhin and Klodt – played a part in its construction.

The age-old forest is crossed by 12 radiating avenues. In the centre of the estate is the main pavilion with a 10-metre (33-ft) arch, the **Egyptian House**, decorated with sculptures of animals and ancient gods, **the stables**, decorated with sculptural pieces by Klodt, and the **Red Palace**. In the early 19th century the estate was bought by the Princes Golitsyn. After the revolution the buildings were occupied by various public and state organisations whose presence caused great damage. In 1986 the ensemble was turned over to the History Museum and restoration work is now under way.

Kolomenskoye: The name of this area goes back to the 13th century. It was the name of the village founded by settlers from the town of Kolomna who had fled

Left, evening light in Kolomenskoye. Right, restoration at the Danilov Monastery.

under the onslaught of the hordes led by Khan Batu. The village soon became a property of the grand duke of Moscow and, subsequently, a royal estate.

Ivan the Terrible ordered an out-of-town palace to be built in Kolomenskoye. The structures that have survived date approximately from the 16th and 17th centuries. The gem of the ensemble is the hipped-roof **Church of the Ascension**, built in 1532. It was the first and most perfect hipped-roof church in Russia. The church served as one of Moscow's watchtowers. Inside the church is a 17th-century iconostasis.

Next to it stands the 16th-century church of **St George the Victorious** with a belfry. The front gate with an over-the- gate chapel has also survived. The clock above the gate was brought here from the Sukhareva Tower, which was destroyed during Stalin's reconstruction of the city. Near the northern gate the 17th-century of the **Kazan Icon of the Mother of God** has survived. Still another church to be seen here, the one that was the model for the magnificent Cathedral of the Protecting Veil in Red Square, is the five-domed **Church of St John the Baptist**, with its octagonal tower-like side chapel.

In 1926 a museum of wooden architecture was set up in the grounds of the ensemble. Peter the Great's house from Arkhangelsk, the entrance tower from St Nicholas Monastery in Karelia, and several other examples of wooden architecture were transferred here. The ensemble is set in a natural park full of ancient trees. Specialists think one of the oaks is a contemporary of Ivan Kalita (14th-century). The park is a favourite recreation haunt for the residents of the capital's southeastern outskirts.

The Estate in Tsaritsyno (between the Lenino and Orekhovo Metro Stations): The history of this locality goes back to the 16th century when Czarina Irina, the wife of Czar Fyodor Ioannovich, had her estate here. It was then that the **Tsaritsyno Ponds** were dug. In the 17th century this area was the property of the

The Danilov Monastery.

boyar family Streshnev, and later of the Princes Golitsyn. In 1712 Peter the Great presented the estate, which was then called *Chornaya Gryaz* (Black Mud), to the Moldavian, Hospodar Kantemir. In 1775 Catherine the Great bought the estate and renamed it Tsaritsyno.

On the orders of the empress, the architect Vasily Bazhenov worked diligently on the estate for eight years. The palace that he built, however, did not meet with Catherine's approval because, according to historians, the symbolism of the decor of the buildings reminded Catherine of the Freemasons, whom she hated. Bazhenov was disgraced and the palace pulled down.

Construction work was resumed in 1786, this time under the direction of another, no less famous architect, Matvei Kazakov, Bazhenov's principal rival. The work was interrupted by the unexpected death of the empress. To this day, the enormous ensemble, the creation of two great Russian architects, has stood unfinished.

The **Lesser Palace**, the **Opera House**, the **Cavaliers' Building**, the **fancy bridges** and the **Grapevine Entrance Gate** have partially survived. In 1988 a Russian Orthodox church, the first church to be built in Soviet times, was constructed near Tsaritsyno to commemorate the Millennium of the Baptism of Rus.

St Andronik Monastery of the Saviour (10 Pryamikov Square, near the Ploshchad Ilyicha Metro Station) is the oldest architectural monument in Moscow after the Kremlin. It was constructed around 1360 on the high left bank of the Yauza River by Aleksy, the first Russian metropolitan, who came to Moscow and ordered a church to be built on the spot where he first sighted the Kremlin. The monastery was named after its first father superior, St Andronik, and the 10th-century icon of the Saviour. The main church, the **Cathedral of the Saviour**, was built in 1420–27 on the site of a wooden church over the common grave of the Russian warriors

Many have turned back to religion.

who fell in the Battle of Kulikovo. The cathedral was painted by Andrei Rublev and by Daniil Chyorny.

It was here that, shortly before his death, Andrei Rublev created his famous icon, *The Old Testament Trinity*, which is now kept in the Tretiakov Gallery, and it is here, within the walls of the monastery, that the earthly remains of the great master lie buried.

In the 16th and 17th centuries, the monastery became a centre of Russian culture where historical chronicles were written and the fundamentals of philosophy were elaborated.

Late in the 17th-century, after Peter the Great married Eudoxia Lopukhina, the Lopukhins, an influential boyar family, became patrons of the monastery. The festive-looking **Church of St Michael the Archangel**, designed in Moscow baroque style, was built in 1694 with funds provided by the Lopukhins. Later, the church became the Lopukhins' family burial place. Behind it stands still another church, the **Church of St Aleksy**, which is dedicated in honour of the founder of the monastery.

On entering the monastery through the main Holy Gate you come to the **Andrei Rublev Museum of Old Russian Culture and Art**, which was opened in 1960. The museum boasts an exquisite and unique collection of painting, artistic embroidery, sculpture, old manuscripts and books.

The Petrovo-Dalneye Estate (Ilyinskoye Highway, near the Tushino Metro Station): Petrovo-Dalneye, also known as **Petrovskoye**, is one of the most beautiful architectural ensembles among Moscow manors. It lies on the bank of the Moskva River. The manor house, built in the early 19th century in neo-Gothic style, belonged to the Golitsyns. The stable buildings and a marvellous park have also survived.

The Ostankino Estate: In the 18th century the family estate of the Counts Sheremetev stood where the Ostankinsky District of the capital stands now. In 1792 Count Nikolai Sheremetev commissioned the Italian architects Camporesi and Blank to build a wooden palace here to house his **serf theatre**. It was the most popular serf theatre in late 18th-century Moscow. Its stage machinery made it possible to transform the auditorium into a ballroom almost instantaneously. The celebrated serf actress Praskovya Kovalyova-Zhemchugova, whom Count Nikolai Sheremetev later married, performed at the theatre.

The numerous rooms of the palace are well-known for their unique decor featuring rich tapestry, tiled stoves, expensive mirrors, fanciful chandeliers and girandoles, and patterned parquet floors, all of which made the palace one of the most beautiful in Moscow.

Today the palace, boasting a fine collection of paintings, engravings and porcelain, functions as a museum of serf art. Near the old pond the **Trinity Church**, built in early baroque style in the 17th century, has survived.

The Izmailovo Estate: This estate, situated on a small artificial island, was from the 14th century, the property of the boyar family Izmailov and, later on, of the Romanovs. It was the favourite out-of-town estate of Czar Alexei Mikhailovich. Surviving structures in its central part include the **Cathedral of the Protecting Veil**, decorated with coloured tiles made from drawings by Polubes. Next to it stands the **Church of the Nativity**. Both were built in the 1670s. The buildings that served as hospitals for veteran soldiers and single officers, participants in the Patriotic War of 1812, were added in 1840. The Eastern and Western Gates and a three-tier bridge tower have also survived.

This entire complex is associated with the name of Peter the Great, who staged his mock manoeuvres here. Today, the **Izmailovo forest-park** has become a favourite spot for artists and various folk craftsmen. Here, as in old Arbat Street, only in greater quantity, everything from paintings to hand-made articles to trophy weapons is on sale.

Rublev's Trinity.

THE GOLDEN RING OF RUSSIA

One wonders where this tourist route got its name. Here, in the central part of the Mid-Russian Plain, there are no gold mines nor has life ever been particularly opulent. Yet the ring of old Russian towns is pure gold in terms of their role in the development of Russian statehood, the advent of enlightenment and the arts; to this day, the towns of the ring are inhabited by a people who have until recently been known as Great Russians. It is here that the origins of the Great-Russian nation and the Russian state are to be found.

Each stopover on this route is a living chronicle, a museum of ancient Russian architecture. The art of building towns has a long history in Russia. Each city (and Russia was once famous for its cities) had its architectural *piece de resistance*, its unique image.

Sergiyev Posad (Zagorsk), **Pereslavl-Zalessky**, **Rostov Velikii**, **Yaroslavl**, **Kostroma**, **Suzdal** and **Vladimir** – each is an architectural pearl and each conjures up myriad names and events in medieval Russian history. They have gone through many difficult periods in their lifetime. Ravaged by Mongol hordes, treacherous neighbours and Polish invaders, they rose from the ashes time after time, as if illustrating the celebrated Russian knack for survival. Their worst years, however, were during the era of "militant atheism" at the outset of Soviet power.

Names have changed: Churches were demolished, monasteries were closed, bells were broken. The towns lost their very names. Ancient **Sergiyev Posad**, which was built around the Trinity-St Sergius Monastery which the Venerable Sergiy of Radonezh founded in the 14th century, was in 1930 renamed **Zagorsk** in honour of a secretary of the Moscow Party Committee. When the town got its ancient name back in 1990, many did not even know that such a name existed. A similar thing happened to the small mid-Russian town of **Bogorodsk** (1781) which, again in 1930, was given the name **Noginsk** in honour of one of Lenin's comrades-in-arms.

Preceding pages: the charms of country life. **Left**, at Sergiyev Posad.

Another terrible blow was delivered to the ancient towns in Middle Russia by industrial development, which started with Lenin's GOELRO plan and continued in the years of Stalin's universal industrialisation. Industrial development at any price reduced this picturesque land, which was once famous for its endless forests, crystal-clear lakes, rivers and churches of white stone, to an ecological disaster area. Now there is a huge concentration of industrial enterprises in the area. **Vladimir Oblast** (District), which is the smallest in the Russian Federation, has the second highest number of enterprises in Russia.

Sergiyev Posad (Zagorsk): Despite all this, the ancient cities survived, having preserved at least part of their matchless monuments and local colour and evoking the air of days long gone by. Gradually, Russia is becoming aware of its historical heritage. People are rediscovering their roots, both in history and in culture. Ancient names are being restored to towns and cities. The church is regaining control of its desecrated cathedrals. Once again, the mellow chimes of church bells call the Russian people to repentance.

An ancient centre of pilgrimage: The voyage around the Golden Ring starts and ends in Moscow. The road you are travelling on has seen endless columns of pilgrims including, at least twice a year, the Grand Prince, and later the czar himself. All hurried to one of the largest monasteries in Russia, the **Trinity-St Sergius Lavra**. The monastery, which is in ill-fated Sergiyev Posad (Zagorsk), is one of the most interesting places in the vicinity of the capital, and lies 71 km (45 miles) to the northeast of Moscow. It was founded by the Venerable **Sergiy of Radonezh** in the 14th century; in 1744 it was awarded the honorary title of "lavra", which in trans-

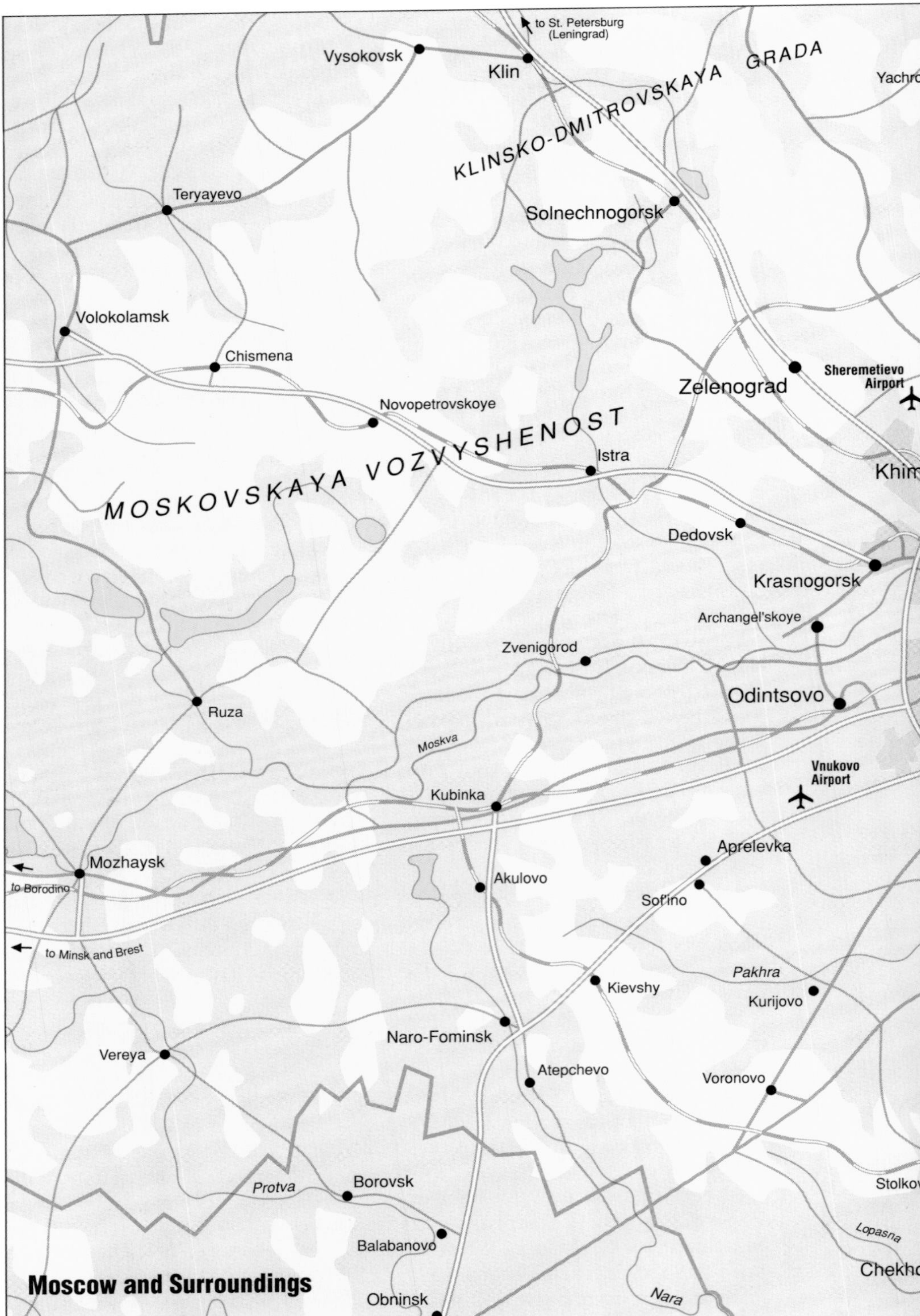
to St. Petersburg
(Leningrad)
Vysokovsk
Klin
KLINSKO-DMITROVSKAYA GRADA
Yachro
Teryayevo
Solnechnogorsk
Volokolamsk
Chismena
Zelenograd
Sheremetievo
Airport
Novopetrovskoye
MOSKOVSKAYA VOZVYSHENOST
Istra
Khim
Dedovsk
Krasnogorsk
Archangel'skoye
Zvenigorod
Odintsovo
Ruza
Moskva
Vnukovo
Airport
Kubinka
Aprelevka
Mozhaysk
to Borodino
Akulovo
Sof'ino
to Minsk and Brest
Pakhra
Kievshy
Kurijovo
Naro-Fominsk
Vereya
Atepchevo
Voronovo
Protva
Borovsk
Stolkov
Lopasna
Balabanovo
Chekho
Nara
Obninsk

Moscow and Surroundings

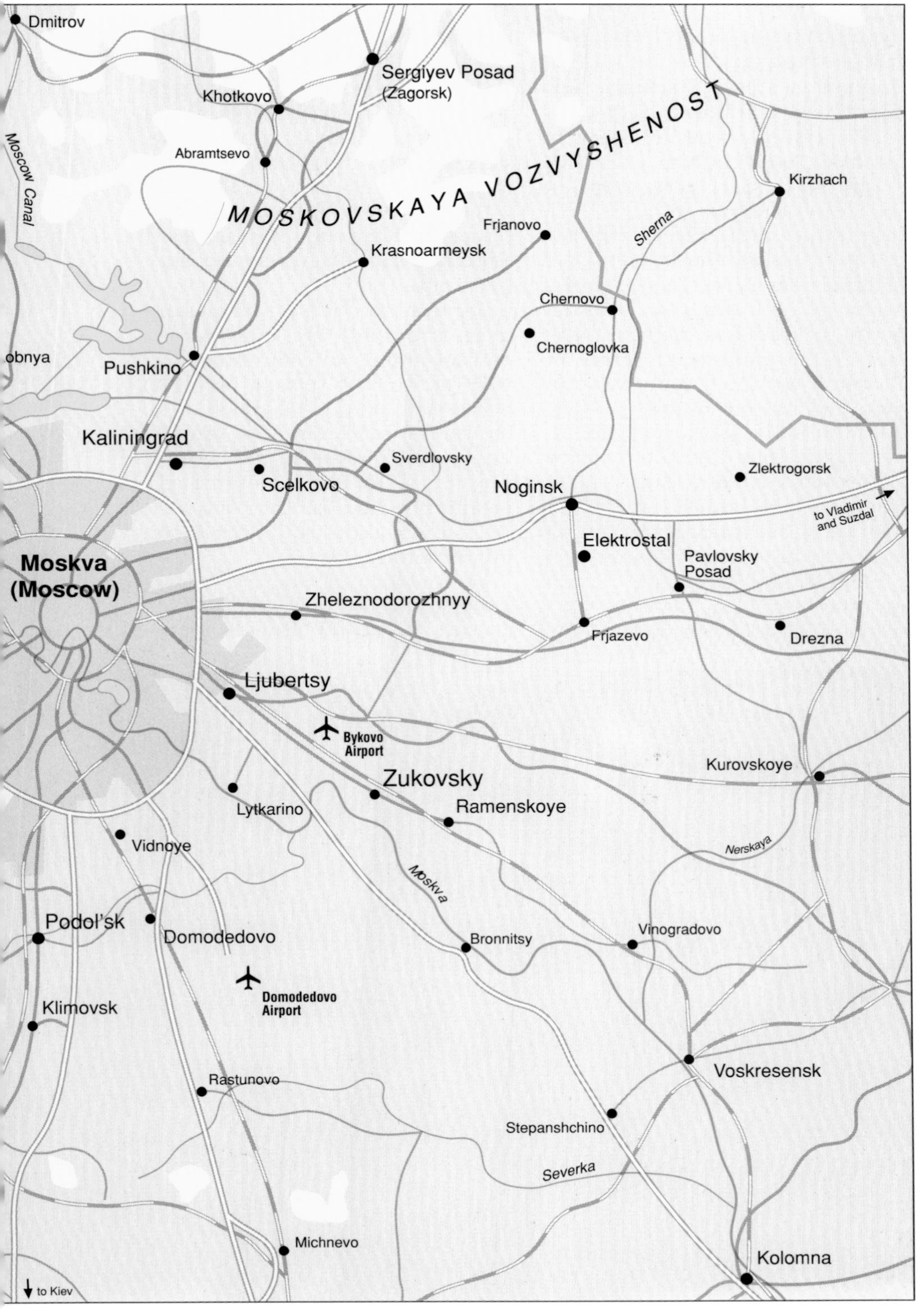

Dmitrov
Sergiyev Posad
(Zagorsk)
Khotkovo
Abramtsevo
Moscow Canal
MOSKOVSKAYA VOZVYSHENOST
Kirzhach
Frjanovo
Sherna
Krasnoarmeysk
Chernovo
Chernoglovka
obnya
Pushkino
Kalinigrad
Sverdlovsky
Scelkovo
Zlektrogorsk
Noginsk
to Vladimir and Suzdal
Elektrostal
Moskva
(Moscow)
Pavlovsky Posad
Zheleznodorozhnyy
Frjazevo
Drezna
Ljubertsy
Bykovo Airport
Kurovskoye
Zukovsky
Lytkarino
Ramenskoye
Vidnoye
Nerskaya
Moskva
Podol'sk
Domodedovo
Vinogradovo
Bronnitsy
Domodedovo Airport
Klimovsk
Voskresensk
Rastunovo
Stepanshchino
Severka
Michnevo
Kolomna
to Kiev

lation from the Greek means "main, most important monastery". There were four such monasteries in Russia: Kievo-Pecherskaya Lavra, Aleksandro-Nevskaya Lavra in St Petersburg, Pochayevo-Uspenskaya (on the Volyn) and the Trinity-St Sergius in Sergiyev Posad.

From monk's cell to national glory: For the Trinity-St Sergius, its many centuries of history started with a tiny monk's cell and a wooden church built by two brothers from Radonezh, Stefan and Varfolomei. The latter took his vows as Sergiy and became the first father superior of the monastery. Sergiy of Radonezh was closely tied to the struggle against the Mongols and the unification of Russia. It was he who gave the blessing to Prince Dmitry of Moscow and his troops on the eve of the battle in Kulikovo Field (1380), which was destined to become a crucial turning point in the history of the Russian state.

Sergiy was canonised after his death and proclaimed "guardian of the Russian land". He was buried in the Trinity Church of the monastery. To this day, pilgrims flock here to bow before his remains, still untouched by decay.

Two centuries later, the monastery resisted the Polish invaders with incredible valour and stubbornness. The citadel stood fast for 16 months. The 15,000-strong army of the Poles were helpless before the fortress, which was protected by some 3,000 monks. This militancy helped to enhance forever the reputation of the monastery.

The monastery also played an enormous cultural role. The Trinity school of "book writing" and colour miniatures dates to the 15th century. Valuable manuscripts formed a collection without parallel at the time. The famous artists Andrei Rublev and Daniil Chyorny, whose names will be mentioned many times in connection with other places along the Golden Ring, lived and worked here.

The architectural ensemble of the Trinity-St Sergius was completed by the late 18th century. Even though it

A Suzdal winter scene.

comprises architectural monuments spanning several centuries, the ensemble itself is surprisingly unified. The oldest building in the monastery is the **Trinity Church** (1422–23), which stands over the grave of Venerable Sergiy of Radonezh.

The church was decorated by Andrei Rublev and Daniil Chyorny. It was for this church that Rublev painted his famous **Trinity**, the icon widely regarded as the masterpiece of medieval icon-painting in Russia (it is now in the Tretiakov Gallery in Moscow). The church served as the focus for further construction. To the east, is the **Dukhovskaya Church** (1476–77), which was built by Pskov masters. For a long time, this church served as a belfry and observation tower. In the centre of the grounds is the five-domed **Uspensky (Assumption) Cathedral** (1559–85), which mainly repeats the forms of the Assumption Cathedral in the Moscow Kremlin. Near the northwest corner of the cathedral is a small, low building – **the crypt of Czar Boris Godunov** and his family. The stone walls of the monastery date from the 16th and 17th centuries. Standing parallel to the south wall is the **Mess** and the **Sergius Church** (1686–92). The highest structure in the "lavra" is the belfry (87 metres/285 ft), which was built in the middle of the 18th century and which was the last architectural component of the ensemble.

Today, Trinity-St Sergius is still a place of pilgrimage for Russia's Christians. Believers come here from all over the land to pay homage to the holy remains of "the guardian of the Russian land". This is the largest monastery today run by the Russian Orthodox Church (some 100 monks).

There is also a **Spiritual Seminary** (founded in 1742) and a **Spiritual Academy** (founded in 1814). About 500 students attend the two establishments; some live in the monastery. In 1969 the town, then called Zagorsk, was opened to tourists. Today, it receives about 700,000 tourists a year. The

The higher echelon of the Russian Orthodox church.

monastery's **museum of history and art** has thousands of items: unique 14th- to 17th-century icons, masterpieces of applied art – wood, bone and stone carvings, embroidery, woven fabrics, folk costumes and jewellery.

Matryoshkas: Besides its reputation as the oldest centre of religion in the country, the town is also famous for its toys. It is here that the first **Russian Matryoshka doll** appeared early this century. You can buy one in any shop. As you leave for the next stop, Pereslavl-Zalessky (138 km/85 miles from Moscow), you'll be able to see the outlines of Sergiyev Posad for a long, long time on the horizon.

Pereslavl-Zalessky was founded in the 12th century by Prince Yury Dolgorukiy, the Long-Armed, the founder of Moscow. The town sprawls on the banks of **Lake Plescheevo**, which resembles a giant crystal dish. One of the most famous people born here was Grand Prince Alexander Nevsky, whose military fame resulted from his victory over Teutonic knights on the ice of Lake Chudskoye in 1242. The town also contributed to Russia's naval fame: it was on the waters of Lake Plescheevo that Peter the Great's "mock flotilla" sailed in 1689–93. The flotilla became the first school of shipbuilding and navigation in Russia. You can familiarise yourself with this page in our history in the **Botik ("Little Boat") Museum Estate**, named after the first vessel built by Russian shipwrights under Peter's guidance.

During the reign of **Grand Prince Ivan Kalita**, **Goritsky Monastery** was founded in Pereslavl-Zalessky (1337–40). The gates of the monastery, which date to the 17th century, are a masterpiece of Russian art. The huge seven-domed **Assumption Cathedral** (1757) is the crown of the ensemble. Surprisingly, the church is painted in the realist tradition. In the centre of the town is the white-stone **Spaso-Preobrazhensky Cathedral** (1152–57), which is the oldest surviving monument in the whole of northeastern Rus. Take a closer look at the **Church of Peter the Metropolitan** (1585). It is an interesting specimen of the tent-roofed churches that were so common in the 16th century.

Rostov Veliki: The next town on your tour is Rostov Velikii (the Great). It lies 203 km/126 miles from Moscow. Rostov has been compared to "a great marvel, a reflection of heaven on the ground". It has also been called "a symphony in stone" and the "eternal city" of Russia. It does indeed rise like some fairy-tale apparition out of the expanse of the Russian landscape beyond the waters of **Lake Nero**. By the end of the first millennium AD it was already a centre of power, with many inhabitants and great wealth. But real fame came when human hands created the 17th-century **Kremlin** (1670–83), which looks as if it had risen from the depths of the lake, so astounding is its glamour and glitter even to the contemporary eye. In the southeastern part of the Kremlin, is the **Church of the Saviour-on-the-Porch** (1675), made particularly interesting by

Happy old man in Pereslavl.

its interior: the walls, iconostasis and the arcade are covered with frescoes, the work of Timofei Yarets of Rostov.

The chimes of Rostov: The White Chamber next to the church resembles the Granovitaya Chamber in the Moscow Kremlin. The **Odigitriiya Church** (1693) stands near the northern wall. The centre of the town is dominated by the 16th-century **Assumption Cathedral** and belfry (1682–87). The latter is a veritable treasury of ancient masterpieces. In all, there are 13 bells with a total weight of 4,500 poods (a pood equals 16 kg or 35 pounds). The largest bell, the **Sysoi**, weighs 2,000 poods.

Rostov has long been famous for its mellow chimes. The velvet voice of the Sysoi alone can be heard 18 versts (one verst is 1.067 km/0.67 miles) away. Later, an individual *Kammerton* was picked for each bell, and music was written for the Rostov chimes. Recordings have been made to the joy of amateur fans and real connoisseurs of the bell-ringer's art.

In addition to its architectural magnificence and bell music, Rostov is also famous for its enamels. The first enamels were manufactured here in the 1780s. The process involves the application of easy-melting paints to a metal base. Today, the works of local artisans are known all over the world.

Yaroslavl (260 km/161 miles from Moscow) is one of the oldest towns in Russia. It was founded in 1010 by Yaroslav the Wise. Situated on a well-fortified cape where the **River Kotorosl** meets the **Volga**, Yaroslavl was an important fortress which controlled the movement of merchant ships along the great Volga. You could get as far as Rostov the Great along the River Kotorosl; from Rostov, an intricate system of minor rivers and patches of dry land, the *voloki*, over which the boats were hauled, took the merchants to Vladimir. Between 1218 and 1463, Yaroslavl was the centre of an independent principality; in 1468 it became part of Muscovy. In the 16th century

Old Russian architecture in Suzdal.

Yaroslavl became the main trading post between Moscow and Archangel (the only port connecting Russia with Western Europe at that time). People said then that Yaroslavl was a corner of Moscow. For a long time, it had the second largest population in Russia.

Yaroslavl is full of historical and cultural monuments and for this reason is called the "Florence of Russia". Stonemasons came to the city in the 13th century, but the oldest buildings still standing date from the 16th century (the **Spaso-Preobrazhensky Monastery**). The golden age came in the 17th century, when a new architectural school was born here. Most churches were built with the money donated by rich merchants. These include the masterpiece of the Yaroslavl school, the **Church of Iliah the Prophet** (1647–50), and the **Church of Ioann Zlatoust** in Korovniki. Both have a smart, dressy look which obviously appealed to the clients. In the 18th century, Yaroslavl gave Moscow the art of theatre. **The Yaroslavl Drama Theatre** bears the name of its founder, Fyodor Volkov.

Kostroma, another town founded by Yury the Long-Armed (1152), is 76 km (47 miles) from Yaroslavl. The town presents a unique opportunity to visit the past century and is a favourite site for the motion-picture industry. The old town was built of wood and no two houses are alike.

One of the ancient structures surviving is the **Ipatievsky Monastery**, founded in the late 13th century. The mighty walls of the monastery protected a unique collection of books, which boasted the *Ipatievskaya Chronicle*, a genuine encyclopaedia of knowledge about ancient Rus.

Ivanovo, "a women's town": The next stop on your way is Ivanovo, called "the town of brides" because 80 percent of the population are women. It is, needless to say, a wonderful place for bachelors.

Suzdal: From Ivanovo, your way lies to Suzdal, the pearl of the Golden Ring

A winter evening in Suzdal.

(89 km/55 miles from Ivanovo, 198 km/128 miles from Moscow). It is one of the oldest towns in the country. It is first mentioned in the chronicles in 1024. In the 12th century under Yury the Long-Armed, the town was the capital of the Rostov-Suzdal Principate. Later, when the capital moved to Vladimir, it remained an important part of the Vladimir-Suzdal Principate, which played a role of immense historical importance. Suzdal suffered many destructive assaults, yet rose from the ashes every time. Wood and stone houses were built by the dozen, and several monasteries were founded.

In the 19th century, when a railroad was built between Moscow and Nizhni Novgorod, it passed Suzdal by. The town avoided becoming an industrial centre and kept its original image intact. There are over 100 architectural monuments of the 13th to 19th centuries crowded into a small area (only 9 square km/3.5 square miles). Suzdal is a unique museum town: in 1983, it received the Golden Apple Prize, which is awarded by an international jury for the preservation of local colour and for the excellence of its tourist facilities.

It is impossible to describe all of Suzdal's highlights, but here are several particularly interesting monuments: the **Bogoroditse-Rozhdestvensky Church** (1222–25) with its matchless Golden Gate; the unforgettable ensembles of the **Spaso-Yevfimiev Monastery** (founded in 1352) with the grave of Dmitry Pozharsky, the leader of the popular army during the Polish invasion, and the belfry whose bells ring every hour; the **Rizopolozhensky Monastery** (founded in 1204) and the **Pokrovsky Monastery** (1364), to which Vasily III, Ivan the Terrible and Peter the Great exiled their wives.

Then there is also the interesting **Vasilievsky Monastery**. Be sure to see the museum of wooden architecture and peasant life, which started with the 18th-century **Nikolskaya and Preobrazhenskaya churches**. Other wooden

Onion towers on a Suzdal church.

structures were later moved here from all over the district.

To take it all in, to breathe the atmosphere of days long gone, climb the steep bank of the **Kamenka River** in front of Pokrovsky Monastery. You'll never forget the panorama of Suzdal. Naturally, your impressions will last longer fortified with a glass or two of *medovukha* – the alcoholic beverage made with honey – that is made only in Suzdal.

Thirty km (19 miles) further is the ancient town of **Vladimir**. Founded in 1104 by **Prince Vladimir Monomakh**, it played an enormous role in the development of the Russian state and national culture. For a long time, Vladimir was capital city to the Vladimir-Suzdal Principate. The town survived two terrible invasions: that of the Mongols in 1238 and that of the Poles early in the 17th century. There are world-famous buildings of white stone here, dating from the 12th century.

The **Uspensky Cathedral**, which was consecrated as the main cathedral in Rus in 1158, saw the rise of Rus and its destruction by the hordes of Batu Khan. There, on the bank of the Kliazma, some of Russia's greatest statesmen, including Alexander Nevsky and Dmitry Donskoy, were crowned and there the regiments of Vladimir received the church's blessing on their way to fight the Teutonic Knights and the Mongols. The walls of the cathedral were decorated by the best artists of ancient Russia. The oldest frescoes date to the 12th and the 13th centuries. In 1408, Andrei Rublev and Daniil Chyorny worked here, covering over 300 sq. metres (3,200 sq. ft) of wall space with brilliant frescoes.

The Golden Gate of Vladimir is another unique monument which was erected in 1164, the heyday of the Vladimir-Suzdal Principate. **Dmitrievsky Church**, made of sawn white stone, is of world architectural fame. It was built in 1194–97 by Vladimir Prince Vsevolod in honor of his newborn son, Dmitry. Another famous place in Vladimir is the **Assumption Cathedral** in the **Kniaginin Convent** (1201), where the wives and daughters of princes were buried. These masterpieces of Russian art set the tone for the architectural picture of Vladimir, whose historical centre contains some 700 lesser architectural creations of the 19th to early 20th centuries.

The best place to end our tour of the Golden Ring is **Bogoliubovo**, 9 km (6 miles) away from Vladimir, near the residence of **Prince Andrei Bogoliubsky** who reigned in 1158–65. Here we find what was probably the greatest creation of Vladimir architects – the **Pokrov-on-Nerl Church** (1164). The cathedral must have seemed a supernatural wonder to anyone in the 12th century, standing as it did over the perilous waters of the flooded river. To this day the sight of the single-domed cathedral, stretching towards the sky and reflected in the waters of the Nerl, evokes tears of admiration in anyone who sees it for the first time.

Left, procession at Sergiyev Posad. Right, church tower at Vladimir.

TRAVEL TIPS

GETTING THERE

BY AIR

Thirty-four international airlines connect Moscow with the rest of the world. Flights take about 9 hours from New York, 4 hours from London or Paris, 3 hours from Frankfurt, 2 hours from Stockholm, 6 hours from Delhi and 8 hours from Peking.

ARRIVING BY PLANE

Sheremetyevo II is Moscow's International Airport and the main entry point to Russia. The airport has great difficulty in handling the growing number of passengers who want to visit as well as the growing number of citizens travelling abroad for business or pleasure – and those who want to emigrate.

Tourists can expect waiting times of an hour or more to pass through passport and customs control. Aeroflot and Lufthansa have recently started a joint venture to modernise the airport and changes can be expected. The newly built Novotel is now open for transit passengers. In the main arrival hall there is an exchange office and an Intourist counter.

At Domodyedovo and at Vnukovo airports there are special Intourist lounges that should provide a little more comfort during waiting hours. If you continue on a domestic flight arrive at the check-in counter at least 40 minutes before the scheduled departure time, otherwise your seat might be taken.

Sheremetyevo information, tel: 155 0922.
Tourist information Sheremetyevo II, tel: 578 9101, 578 5614.
Tourist information Sheremetyevo I, tel: 578 5614.
Domodyedovo information, tel: 323 8652.
Vnukovo information, tel: 436 2967.

The Central Air Terminal allows you to check in your luggage for Aeroflot internal flights and take an Aeroflot bus to the different domestic airports.

FROM THE AIRPORT TO TOWN

The only convenient way to reach Moscow from the airport is by taxi, by reserved hotel car for business travellers, by Intourist bus or by reserved bus for tourists arriving in groups. The official taxi fare, based on a rate of 6 roubles per kilometre plus 6 roubles service charge, should be between 30 and 50 roubles, but don't count on it: the least the taxi drivers will ask is 70 roubles. Foreign travellers will most probably have to pay in hard currency in order to find any form of transport.

AIRLINE OFFICES

Aeroflot, St Petersburgsky Highway 37, tel: 155 0922.
Air France, Korovy Val 7, tel: 237 2325, 237 3344, 237 6777. Monday–Friday 9am–1pm and 2–6pm.
Air India, Korovy Val 7, ground floor, tel: 237 7494, 236 4440. Monday–Friday 9.30am–1pm and 2–5.15pm, Saturday 10am–3pm.
Alitalia, Pushechnaya Street 7, tel: 923 9840. Monday–Friday 9am–1.30pm and 2–5pm, Saturday 9am–12.30pm.
Austrian Airlines, Krasnopresnenskaya nab. 12, Floor 18, tel: 253 8268, 253 1670, 253 1671. Monday–Friday 9am–5.30pm, Saturday 9am–1pm.
British Airways, Krasnopresnenskaya nab. 12, Floor 19, tel: 253 2492. Monday–Friday 9am–5.30pm.
Finnair, Kamergersky Per. 6, tel: 292 8788, 292 3337. Monday–Friday 9am–5pm.
Japan Airlines, Kuznetsky Most Street 3, tel: 921 6448, 921 6648. Monday–Friday 9am–6pm.
KLM, Royal Dutch Airlines, Krasnopresnenskaya nab. 12, Floor 13, tel: 253 2150/51. Monday–Friday 9am–5pm.
Lufthansa, German Airlines, Olimpiysky Prospekt 18/1, Hotel Penta, tel: 975 2501. Monday–Friday 9am–5.30pm.
Delta, Krasnopresnenskaya nab. 12, Floor 11, tel: 253 2658/59. Monday–Friday 9am–5.30pm.
LOT, Polish Airlines, KJorovy Val 7, Office 5, tel: 238 0003, 238 0313. Monday–Friday 9am–6pm, Saturday 9am–5pm.
Sabena, Belgian World Airlines, Hotel Belgrade-II, Floor 7, tel: 248 1214, 2302241, Monday–Friday 9am–1pm and 2–6pm.
SAS, Scandinavian Airlines, Kuznetsky Most Street 3, tel: 925 4747. Monday–Friday 9am–6pm.
Swissair, Krasnopresnenskaya nab. 12, Floor 20, tel: 253 8988, 253 1859. Monday–Friday 8.30am–noon; 1–5.30pm.

These airlines and others also have offices in Sheremetyevo II Airport. Airline tickets can be booked direct with the airline and at the Mezhdunarodnaya, Rossiya and Cosmos Hotels. For further information about international airlines tel: 156 8019.

AEROFLOT

Most visitors who fly to Moscow arrive by Aeroflot, the world's largest airline, which transports more than 110 million passengers a year. Aeroflot is buying the European Airbus which will go into service on certain international routes.

Check-in at Aeroflot counters starts 1œ hours and ends half an hour before departure. Tourists have to pay for their tickets in hard currency.
Aeroflot Information Sheremetyevo II: tel: 575 8816

AEROFLOT OFFICE IN MOSCOW
Moscow: 37, St Petersburg Highway. Tel: 155 0922

AEROFLOT'S INTERNATIONAL OFFICES
Amsterdam: Singel 540. Tel: 245 715
Athens: Ksenofontis 14. Tel: 322 1022
Bangkok: 7 Silom Rd. Tel: 233 6965
Berlin: Unter den Linden 51/53. Tel: 229 1592
Bombay: 7th Brabourn Stadium, Vir Nariman Rd. Tel: 221 682
Brussels: Rue des Colonies 54. Tel: 218 6046
Bucharest: 35 Boulevard Nicolae Balcescu. Tel: 167 431
Budapest: 4 Waci. Tel: 185 892
Copenhagen: 1–3 Vester Farimasgade. Tel: 126 338
Delhi: 18 Barakhamba Rd. Tel: 40 426
Frankfurt: Theaterplatz 2. Tel, 230 771
Helsinki: Mannerheimintie 5. Tel, 659 655
Lisbon: Av. Antonio Augusto de Aguiar 2H–3E. Tel: 561 296
London: 69–72 Piccadilly. Tel: 492 1756
Madrid: 25 Calle Princesa. Tel: 241 9934
Milan: 19 Via Vittor Pisani. Tel: 669 985
Munich: Ludwigstr. 6. Tel: 288 261
Paris: 33 Avenue des Champs Elysees. Tel: 225 4381
Peking: 2–2–42, Jianguomenwai. Tel: 523 581
Prague: 15 Vaclavske Namesti. Tel: 260 862
Rome: 27 Via Leonida Bissolati. Tel: 475 7704
Singapore: 55 Market St. Tel: 336 1757
Sofia: 2 Russky Blvd. Tel: 879 080
Tokyo: 3–19, 1-chome Yaesu, chuo-ku. Tel: 272 8351
Vienna: 10 Parkring. Tel: 521 501
Warsaw: 29 Allee Jerozolimskie. Tel: 281 710
Zurich: 9 Usteristr. Tel: 211 4633

BY SEA

Although Moscow is connected to five seas, by river and canal, and is therefore an international seaport, tourists arriving in Russia by sea arrive at the larger coastal ports and continue from there by train or plane to Moscow.

Several Russian ports accept international passenger liners. St Petersburg on the Baltic Sea is connected with London, Helsinki, Gothenburg, Stockholm and Oslo. Odessa on the Black Sea is on the itinerary of liners from Marseilles, Istanbul, Naples, Barcelona, Malta, Piraeus, Varna, Dubrovnik, Alexandria and Constanta.

Ismail is a Dunai river port connected with Passau, Vienna and Budapest. Nakhodka on the Japan Sea coast is connected with Yokohama, Hongkong, Singapore and Sidney. Additional information about sea routes, schedules and bookings can be obtained from Intourist or Morflot offices.

RIVER TRANSPORT

There are two River Terminals in Moscow, connecting the city with cities as far away as Astrakhan, Perm, Rostov on Don, Ufa, Kazan, Kuibyshev, Saratov, Volgograd, Kasimov and Ryazan:

The Northern River Terminal: 51 St Petersburg Highway, tel: 457 4050.
The Southern River Terminal: Yuri Andropov Prospekt, tel: 118 7955.

BY RAIL

Within the European part of Russia railways are the most important means of passenger transportation. Railways connect the largest CIS cities (Moscow, St Petersburg, Kiev, Minsk) with Western European capitals. Travellers who can spare the time can travel in comfortable first-class sleeping-cars, the pride of the Russian Railways. From Western Europe the train takes about three days to Moscow, with a change of gauge when reaching the railway system.

The most popular among rail routes between the west and Russia is the Helsinki-St Petersburg route (departure 1pm, arrival 9pm) and Helsinki-Moscow (departure 5pm and arrival 9.30 next morning).

There are also transcontinental rail routes, such as those from Moscow to Vladivostok and from Moscow to Peking. They demand an adventurous spirit and a willingness to spend a week on the train contemplating the endless Siberian and Transsiberian (Baikal) landscapes. Food for the trip should be taken along since the food available at the station buffets is often not to the weary traveller's liking.

Of the nine railway stations in Moscow, the most important are:
The **Byelorussian Station**, on Byelorussian Square; from here trains leave for Western Europe, Poland, Byelorussia and Lithuania. **Riga Station** on Prospekt Mira, for the Baltic states.
Kiev Station, Kiev Station Square, for the Ukraine.
St Petersburg Station, on Komsomolskaya Square, for St Petersburg and Helsinki.
Yaroslavl Station, also on Komsomolskaya Square, for the Transsiberian Express.

The high-speed Aurora Express from Moscow to St Petersburg takes five hours for the 650-km (410-mile) journey and is a good alternative to flying.

BY CAR

If you intend to visit Moscow by car you should first get in contact with Intourist. They have worked out a number of routes through the European part of Russia which can easily be negotiated.

Entry points to the CIS are: Brusnichnoe and Torfyanovka on the Finnish border; Brest and Shegini on the Polish; Chop when coming from Czechoslovakia and Hungary, and Porubnoe and Leusheny when coming from Romania. You can also ship your car directly to St Petersburg.

Below are the routes to Moscow which you can use driving your own car. You will, however, have to stick to your schedule, staying overnight only at pre-booked hotels or camp sites. You are not permitted to deviate from your planned route. Intourist will make the necessary arrangements.

From Finland: Torfyanovka–Vyborg–St Petersburg–Moscow. From Western Europe: Brest–Minsk–Smolensk–Moscow and Chop–Uzhgorod–Lvov–Kiev–Orel–Moscow.

If you intend to continue within European Russia you can drive to the Caucasus and the Black Sea, ferrying the car across to Yalta or Odessa and crossing the Ukraine to Czechoslovakia or Poland. Details of this and other routes across the Baltic States, Byelorussia or Moldavia can be found in the book: *Motorists' Guide to the Soviet Union* (Pergamon Publishers Oxford, 1987). It gives details about petrol stations, repair shops, overnight stops and emergency procedures.

Since crossing the border into Turkey is now possible, you can also exit or enter via Anatolia. Whether this route remains open, however, depends on the changing political conditions in the Caucasus.

Sovinterautoservice are the specialists for car travel in the CIS. They tackle every problem a foreigner is likely to experience on Russian roads. Write or phone for detailed information: **Institutski Pereulok** 2–1, Moscow, tel: (007095) 101 496.

During the past few years marked changes have taken place in the quality of services along roads but you should still be cautious of their state. Diverting from the highways (which is still not permitted if the diversion is not on your officially approved itinerary) might get you into some unexpected adventures.

BY BUS

There are no scheduled international bus lines to Moscow, but private tour bus operators run coach tours from the UK, Germany and Finland.

YOUTH TRAVEL

Russia does not have an extensive system of youth hostels. Only big cities have youth hotels belonging to **Sputnik**, the Bureau of International Youth Tourism, once a part of Comsomol but now acting independently. During the summer months, when demand exceeds hostel capacity, Sputnik falls back on unused, inexpensive university dormitories. Contact: **Sputnik** International Youth Travel Bureau, 4 Chaplygina Street.

TRAVEL ESSENTIALS

VISAS & PASSPORTS

You will need a valid passport, an official application form, confirmation of your hotel reservations (for both business travellers and tourists) and three passport photographs, to get your visa from a Russian embassy or consulate. If you apply individually, rather than through a travel agency, you should allow ample time, as it might take up to a month or so to check your papers.

According to the new regulations, this term can be shortened to 48 hours if an applicant is a business traveller or if he has a written invitation (telex and fax are also accepted) from a Russian host. However, it might take the Russian counterpart some time to have the invitation stamped by the local authority.

The visa is not stamped into the passport but onto a separate sheet of paper, consisting of the three sections. The first part is removed when a person enters the country, and the last is taken out when leaving.

There are several types of visa. Transit visas (for not more than 48 hours), tourist, ordinary and multiple entry visas (for two or more visits).

If you go to Russia at the invitation of relatives or friends, you can get a visa for a private journey which presupposes that no hotel reservation is needed. Individual tourists should have their trip organised through Intourist, Mir, Sputnik or their Russian hosts. They need an itinerary listing, in detail, times, places and overnight reservations.

You should carry your passport at all times. Without it you might be prohibited entry to your hotel, your embassy and many other places. Intourist hotels will give you a special hotel card that serves as a permit to enter your hotel and to use its restaurant and currency exchange office.

HEALTH REGULATIONS

Visitors from the USA, Canada, European countries and Japan do not require a health certificate. Visitors from regions infected with yellow fever, especially from Africa and South America, require an international certificate of vaccination against yellow fever. A cholera and tetanus vaccination certificate may also be required. Visitors from certain AIDS-infected regions who are planning to stay in Russia for an extended period can be subjected to an AIDS test.

CUSTOMS

When entering Russia you will have to fill in the customs declaration which must be kept as carefully as a passport during the whole period of your stay on Russian territory. It must be returned to the customs office, along with another declaration, which you fill in on leaving the country.

Customs regulations have been revised several times in the past few years. Customs authorities want to find a compromise between conforming to international customs regulations in the epoch of openness and of preventing the export of large batches of goods bought cheaply in shops for resale in other countries.

The latest edition of the customs regulations prohibits the import and export of weapons and ammunition (excluding approved fowling-pieces and hunting-tackle), and of drugs and devices for their use. It is prohibited to export antiquities and art objects except for those which the visitor imported to the country and declared on entry.

It is permitted to import free and without limitation: 1) gold and the other valuable metals except for gold coins, whose import is prohibited; 2) materials which are of historic, scientific and cultural value; 3) articles approved by the licensees of V/O Vneshposyltorg; 4) foreign currency and foreign currency documents; 5) personal property except for computers and other technical devices (*see below*).

Limited duty-free import:

1) Gifts with a total value less then 500 roubles (about US$770). If you want to bring gifts, which are highly appreciated among Russian people, it is best to choose ballpoint pens, elegant business notebooks, calculators, electronic watches and other inexpensive items. You are recommended not to have more than 10 of the same item if you want to escape time-consuming questions from customs officers.

2) Cars and motorcycles approved according to International Traffic Convention, no more than one unit per family, with the obligation to export the vehicle.

3) Spare parts to the vehicles insured by the Russian international insurance company Ingosstrakh and approved by the documents of Ingosstrakh or Intourist (for the other spare parts duty must be paid).

4) Medicines not registered in the list of the Russian Ministry of Health Protection must be approved by the Russian medical institutions.

5) Personal computers, photocopying apparatus, video-recorders, TVSat systems with the obligation to export (if the obligation is broken and the article is sold in Russia, duty must be paid).

6) Alcohol (limited to persons over 21): spirits 1.5 litres, wines 2 litres.

7) Tobacco (limited to persons over 16): 200 cigarettes or 200 gm of tobacco per person.

Duty-free export:

1) Articles imported by the visitor.

2) Articles bought in the Russian hard currency shops or in rouble-shops for legally exchanged Russian currency (with some limitations – *see below*).

3) Food stuffs with a total value of less than 5 roubles.

4) Alcohol (over 21): spirits 1.5 litres, wines 2 litres per person.

5) Tobacco (over 16): 100 cigarettes or 100 gm tobacco.

It is prohibited to export the following articles bought in rouble-shops: electric cable, instruments, building materials, fur, cloth, carpets, leather clothes, linen, knitted fabric, socks, stockings, umbrellas, plates, dishes, medicines, perfumes, sewing machines, refrigerators, bicycles, cameras, vacuum cleaners, washing machines, children's clothes, boots, all articles produced abroad, valuable metals and jewels.

Some of the customs officers are quite severe in their observation of these regulations. Therefore when you enter or leave the country you must expect a careful examination of all your luggage and you will be asked if any personal items are intended for sale (if the answer is yes, you will have to pay duty).

MONEY MATTERS

Roubles can neither be imported to nor exported from Russia. The newly-established exchange rate for tourists has lessened the importance of the black market. There is no limit to the import of hard currency – which, however, has to be declared on entry. The amount exported should not exceed the amount declared when entering the country. Officially documented but unspent roubles can be reconverted at the hotel exchange counter or at the Sheremetyevo II airport bank. If you intend to do this, you should reckon on spending at least half-an-hour in long queues at the bank counters.

All Intourist hotels have an official exchange counter where you can buy roubles with hard currency cash, traveller's cheques and credit cards. You will need to produce your customs declaration form, where all your money transactions have to be recorded. You will need this form when leaving the country and you should, as with your passport, make sure not to lose it. Most major hotels have bars, restaurants and shops were you can only pay with hard currency, but these transactions do not need to be recorded on your exchange certificate. Leaving the country, customs will check that you have officially exchanged money for the goods you bought and export from the country and that you are not exporting more hard currency than you imported.

CREDIT CARDS

Most tourist related businesses accept major credit cards. American Express runs two cash dispensers in Moscow where card holders can either receive roubles or US dollar traveller's cheques. They are located at 21A Sadovokudrinskaya Street and at the

Sovincentr, Mezhdunarodnaya Hotel, 12 Krasnoprenskaya Embankment.

Intourist hotels, restaurants and co-operative cafés that accept credit cards usually have a notice to this effect at the entrance. In addition to American Express, Diners Club, Visa, Eurocard and Mastercard are also accepted.

LOCAL CURRENCY

Banknotes in Russia are available in the following denominations: 1, 3, 5, 10, 25, 50, 100, 200, 500, 1000 roubles. In 1991, 50 and 100 rouble notes were removed from circulation. One rouble is divided into 100 kopecks and these are issued in denominations of 1, 2, 3, 5, 10, 15, 20 and 50. Platinum and golden coins of 150 and 200 roubles are more the object of numismatists than a matter of real circulation. You should not neglect small coins of 3 and 2 kopecks.

Foreign visitors are sometimes still exposed to propositions from black market dealers, the most active of whom are taxi drivers and restaurant waiters. The black market rate is published regularly by the newly established non-governmental newspaper *Commersant*, which is also published in English.

Despite the fact that the inscription on rouble banknotes says they are accepted for "all fees within Russia's territory" porters and taxi drivers not infrequently demand hard currency payment. You should remember that currency black market activities are unlawful and culprits can be punished severely.

TIPPING

Though Russia as part of the Soviet Union was a socialist state for 70 years, tipping was and still is an accepted practice. Waiters, porters, taxi drivers, especially in Moscow and St Petersburg, have always appreciated tips. 10 percent is the accepted rule.

However, do not tip guides, interpreters or other Intourist personnel. If you want to show your gratitude they will appreciate a small souvenir or gift.

GETTING ACQUAINTED

COUNTRY PROFILE

Area: 6,563,000 sq. miles (17 million sq. km).
Population 1992: 148,000,000. Population density: 33 per sq. mile (86 per sq. km).
GNP per capita: US$8,734, (cf. USA, US$18,951).

Demographics:
Population growth: 0.1 percent
Population 2000: 315,175,000

Age distribution:	Under 15	26.0 percent
	15–65	64.9 percent
	Over 65	9.1 percent

Health:
Life expectancy (M): 65 years
Life expectancy (F): 74 years
Infant mortality: 25.4 per 1,000
Politics:
Type of State: Federation of Republics
Government Leader: President Yeltsin, Boris Nikolayevich.
Languages: Russian, Ukrainian, Turkish languages, Caucasian languages.
Ethnic Groups: Russians 52 percent, Ukrainians 16 percent, others 32 percent.
Religions: Orthodox 18 percent, Muslims 9 percent, Jews 3 percent, atheists (including other confessions like Buddhists, Hare Krishnas etc.) 70 percent.
Natural Resources: crude oil, natural gas, coal, timber, manganese, gold, lead, zinc, nickel, potash, phosphates, mercury.
Agriculture: wheat, rye, oats, potatoes, sugar beets, linseed, sunflower seeds, cotton, flax, cattle, pigs, sheep.
Major Industries: mining, metallurgy, fuels, building materials, chemicals, machinery, aerospace.

CLIMATE

Summer is the best time to visit Moscow, though July and August are the rainiest months. The winter months can be very cold but do have a certain charm that you can enjoy if you bring appropriate clothing.

Average Temperatures:

	°C	°F
January	–18 to –27	0 to –17
February	–13 to –29	9 to –20
March	–3 to –17	27 to 1
April	11 to –3	52 to 27
May	28 to 10	82 to 50
June	35 to 18	95 to 64
July	22 to 9	72 to 48
August	36 to 19	97 to 66
September	24 to 10	75 to 50
October	12 to 1	54 to 34
November	–4 to –11	25 to 12
December	–15 to –22	5 to –8

WHAT TO WEAR

Today Moscow has become an open-minded city, visited by many who are casually dressed. The old guide-book phrase: "When going to Russia, follow a modest and classic style of clothes" is somewhat outdated. You may dress as you would at home.

Waterproof shoes are a necessity in winter, since the traditional Russian frost is not as frosty anymore and is often interrupted by periods of thaw. For business meetings formal dress is obligatory. The dress code is as rigorously enforced and compliance with it is an important matter of status.

TIME ZONE

Moscow time is GMT plus 3 hours. Moscow time is used nearly everywhere west of the Urals.

DURATION OF A HOTEL DAY

The hotel day is from noon to noon. Exceptions are seldom made, especially in hotels that cater for large groups. If you need to stay beyond noon, ask the receptionist or house-lady responsible for your floor; maybe the next group is not due until the evening.

ELECTRICITY

Electrical current in Moscow's tourist hotels is normally 220 V AC. Sockets require a continental type plug. It is best to have a set of adaptors with you. The same is true for batteries. If your appliances depend on a supply of batteries, bring plenty with you, since they might not be available in Moscow, not even for hard currency.

OFFICIAL HOLIDAYS

1 January: New Year Holiday.
7 January: Christmas
23 February: Armed Forces Day.
8 March: International Women's Day.
Easter
1 and 2 May: Day of Spring.
9 May: Victory Day.
7 October: Constitution Day.
7 and 8 November: Bank Holiday.
On 23 February, 1 May, 9 May, 7 November and 12 April (Cosmonaut's Day) there are evening fireworks.

RELIGIOUS HOLIDAYS

The greatest religious holidays of the year include Christmas (celebrated by the Orthodox Church on 7 January) and Easter (a movable holiday celebrated in March–April). Religious holidays are acknowledged officially, and they are recognised by the many people who participate in religious services.

RELIGIOUS SERVICES

Russian Orthodox
The Church of Assumption, Novodevichy Convent, 1 Novodevichy Proyezd.
The Church of the Transfiguration, 17 Krasnobogatyrskaya Street.
The Church of the Resurrection, 15/2 Nezhdanova Street.
The Church of All Saints, 73 St Petersburg Highway.

Anglican
St Andrew's Church, 9 Stankevich St, tel: 231 8511.

American Protestant
5 Olof Palme St, tel: 143 3562.

Catholic
Chapel of Our Lady of Hope, 7/4 Kutuzovsky Prospekt, tel: 243 9621.

German Evangelical
German Embassy, twice a month, tel: 238 1324.

Baptist
Seven Day Adventist and Baptist Churches, 3 Maly Zukovsky Per.

Muslim
7 Vypolzov Lane.

Jewish
Coral Synagogue, 8 Arkhipov Street.

Old Believers' Society
29 Rogozhsky Posyolok.

Society of Evangelic Christian Baptists
3 Maly Vuzovsky Lane.

Greek Orthodox
15a Telegrafny Lane.

RELIGIOUS ORGANISATIONS

Many former churches and religious meeting places are now being reconverted to religious use. Information can be obtained at the following addresses.

Moscow Patriarchate, 5 Chisty Lane, tel: 201 2340.
Armenian Religious Community, 10 Sergei Makeyav Street, tel: 255 5019.
Religious Board of the Buddhists of Russia, 49 Ostozhenka Street, tel: 245 0939.
Jewish Religious Community, 5, 2nd Vysheslavtsev Lane, tel: 289 2325.
Muslim Religious Society, 7 Vypolzov Lane, tel: 281 3866.

COMMUNICATIONS

MEDIA

Kiosks located in Intourist hotels carry the main western daily newspapers (usually one day late).

The most important magazine for any foreigner in Moscow is *Moscow Magazine* (8a Suvorovsky Bldg, House of Journalism, Room 309, tel: 203 3644). This English-language, monthly listings magazine gives all the information you need about upcoming events. *Moscow News*, (16 Tverskaya Street, tel: 292 1432) is an informative and liberal English-language weekly published in Moscow. The first independent Russian-American weekly newspaper is *We*, a joint venture between Izvestia and the Hearst Corporation. *Travel to Russia* is a bimonthly illustrated magazine, published in Russian, German, English and French that carries tips and information about fast-changing travel conditions in the country.

Rossiyskaya Gazeta and *Izvestia* are the leading national government newspapers. If you read Russian, there is also *Komsomolskaya Pravda*, *Trud*, *Nezavisimaya Gazeta* and *Literaturnaya Gazeta*.

RADIO

Radio Moscow broadcasts news in English every hour on the hour. There is also a French station, Radio Nostalgie, that broadcasts 11am–3pm. There are also four new radio stations in Moscow.

TELEVISION

Channel One and Two are national channels, Channel Three is a Moscow Channel and Channel Four is an educational station.

POSTAL SERVICES

The opening times of post offices vary, but most of them are open 8am–7pm or 8pm during the week and from 9am–6pm on Saturdays. They are closed on Sundays. Some post offices, however, work only one shift a day, 9am–3pm or 2–8pm. The mail service in Russia is constantly understaffed.

Not all post offices accept international mail larger than a standard letter. Postal delivery is quite slow, and it may take some two or three weeks for a letter from Moscow to reach Western Europe, sometimes even a month or more to reach the US. A standard letter up to 20 gm to any country costs 5 roubles and a postcard 3 roubles 50 kopecks.

International Post Office: 37a Varshavloye Highway, tel: 114 4645.

Poste restante address: Intourist Communication Dept, Hotel Intourist, 3 Tverskaya Street, Moscow 103009, Poste Restante.

Stamp Collectors: Philatelic items are sold at Dom Knigi, 26 Vozdvishenka Street.

CABLES & TELEGRAMS

Cables to addresses within Russia can be sent from any post office or by phone. International cables can be sent from most post offices, but not by phone.

Central Telegraph office: 7 Tverskaya Street.

TELEPHONE

From a pay phone a local call costs 15 kopecks per 3 minutes. The coin must be inserted before dialling. If you hear a beep-beep tone during the conversation it's time to insert another coin.

For long-distance calls within Russia there are specially marked phones which accept 15 kopeck coins. First dial 8, than the area code and the number. International calls must be booked in advance or at the hotel service bureau.

Area codes within the CIS:

Alma-Ata	327
Ashkhabad	363
Baku	892
Bukhara	365
Donetsk	062
Dushanbe	377
Yerevan	296
Kiev	044
Kishinev	042
Kharkov	057
St Petersburg	812
Lvov	032
Minsk	017
Novosibirsk	383
Odessa	048
Riga	013
Rostov On Don	863
Samarkand	366
Tallinn	014
Tashkent	371
Tbilisi	883
Vilnius	012
Yalta	060

If you are calling Russia from abroad be prepared to try often; the lines are not too good and are always very busy. The country code for Russia is 007.

Some international direct dial numbers:
Moscow 007095; St Petersburg 007812; Minsk 0070172; Tallinn 0070142.

The hotel service bureau will also book international calls: the charge per minute to the US is 6 roubles, to Western European countries 4 roubles. Express calls are charged double.

You can also call direct from the Savoy, Intourist and Mezhdunarodnaya Hotels by using their Comstar telephone booths. This is the fastest but most expensive way to phone. Payments must be made in hard currency. At the Central Telephone and Telegraph Office you'll have to wait for a few hours even though there are direct lines to the US and Canada.

Some useful numbers:
Time: 08
Information (Spravochnaya): 09
Moscow city information: 05
Telegrams by phone: tel: 927 2002
Booking of international calls: tel: 8194
Aeroflot Information: tel: 155 0922
Taxi: tel: 927 0000
Metro Information: tel: 222 2085
Tram and Trolleybus Information: tel: 923 8753
Tourist Information: in English, German and French between 9am and 9pm, tel: 203 6962.

TELEX & FAX

All official institutions and major business representatives have telex numbers and the majority of them now also have telefax. Moscow has a public fax and telex service, where you can send a fax or telex or register a number by which you can be reached if you plan to stay in the city for a long period. Any incoming message will be forwarded either by phone or by local mail. It is the only way to beat the slow mail service. The telex access codes for Russia are 871 from the US and 64 from Europe.

At the **Central Telegraph Office**, 7 Tverskaya Street, you can fax for roubles or you can use **Alphagraphic's**, 50 Tverskaya Street, tel: 251 1215. They charge US$10 per page plus a US$8 phone charge.

COURIER SERVICES

Federal Express, tel: 253 1641; **TNT**, tel: 156 5771; **DHL**, tel: 201 2585 and **UPS**, tel: 430 7069 offer worldwide courier service from Moscow. They charge US$60 per envelope (250 gm/9 ounces).

EMERGENCIES

EMERGENCY NUMBERS

All cities have unified emergency telephone numbers. These numbers can be dialled free of charge from public telephones. Officials responding to these calls will speak little English, so a minimal knowledge of Russian may be needed to make yourself understood.

Fire Guards (Pozharnaya okhrana): 01
Police (Militsia): 02
Ambulance (Skoraya pomoshch): 03
Gas Emergency (Sluzhba gaza): 04

MEDICAL SERVICES

Medical services for tourists are free of charge except for drugs from chemist's shops and in-patient treatment. Doctors at the big hotels speak foreign languages.

The following hospitals charge hard currency for their services: the **Intourist** polyclinic at 2 Gruzinsky Proyezd, tel: 254 4396; the French-operated **International Health Care** polyclinic at 3 Gruzinsky Pereulok, Korpus 2, tel: 253 0703; **UPDK's** polyclinic at 4 Dobrininsky Pereulok, tel: 237 3904 for house or hotel calls; and **Botkin Hospital**, at 5, 2nd Botkinsky Proyzed, tel: 255 0015.

DENTISTS

Medical Interline, Intourist Hotel, Room 2030, 5 Tverskaya Street, tel: 203 9553, is a Western-Russian joint venture; it has Western equipment and is very efficient. A Russian-German joint venture at 26 Durova Street, tel: 928 575, can also be contacted for fast and efficient help.

EYE CLINIC

Svyatoslav Fyodorov's clinic, tel: 484 8120, is world famous for curing shortsightedness with laser microsurgery.

WATER

Moscow's tap water is drinkable but people with stomach problems are advised to drink bottled mineral water. Ice in bars and coffee shops is made from tap water.

24-HOUR PHARMACIES

The **international pharmacy** at 59a Beskud-nekovsky Blvd, tel: 905 4227, sells Western medicine. You can also obtain Western drugs at the Mezhdunarodnaya Hotel supermarket and at the **Unipharm**, 13 Skaterny Pereulok, tel: 202 5071. The pharmacy at 59a Beskudnokovsky Blvd, tel: 905 4227, works with a German company, and can get medicine from abroad.

Local pharmacies: 19-21 Nikolskaya St; 6 Tishinskaya Square, tel: 254 4610; 30 Komsomolsky Prospekt, tel: 246 3030; 32 Myasnitskaya St, tel: 228 3063.

GETTING AROUND

PUBLIC TRANSPORT

Trolley buses and buses run 6am–1am. Trams and the Metro run 5.30am–1am. They all cost 50 kopecks per ride. The salary and price hike in 1993 will, most probably, result in increased fares for the Metro, buses and trolleybuses, even though low transport prices are seen as a major achievement.

Taxis and collectivos are available 24 hours and cost 20 kopecks and 15 kopecks per kilometre respectively. There are 437 bus lines, 82 trolleybus lines and 37 tram lines in Moscow.

Make sure, when using buses, that you carry enough change with you, since you buy the tickets on a self-service basis inside the buses. You can also buy booklets in advance at newspaper kiosks.

METRO

The most convenient local transport is Moscow's Metro. Construction started before the war and was completed in 1955. All Intourist hotels and most others have Metro stations marked "**M**" nearby. The Metro has 148 stations and is 244 km (151 miles) long. It carries an average of 8.7 million people daily. To use it you simply insert a 50-kopeck token coin into the entrance turnstile. Change machines are located in each station.

TAXIS

Though Moscow taxi fares are regulated your chances as a foreigner of being driven for 6 roubles per kilometre are nil. You will probably have to pay in convertible currency.

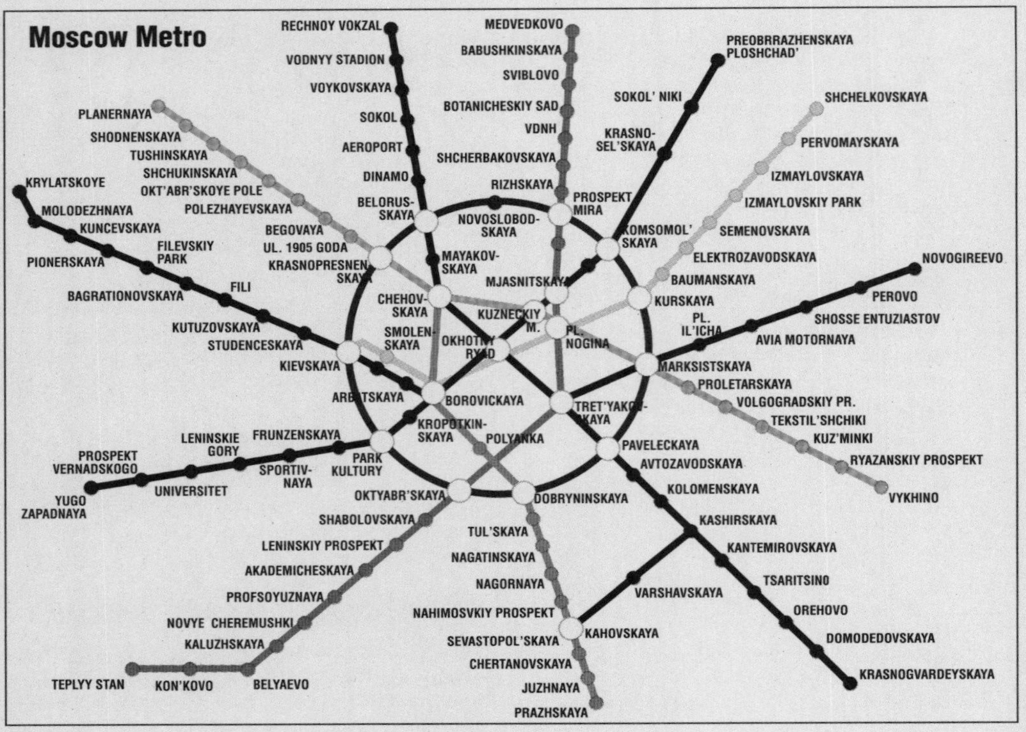

To order a taxi phone the regional number:
For all regions: tel: 927 0000 or 457 9005.
For Leninsky Prospekt and the Southwest: tel: 137 0040.

CAR RENTAL

Both Intourist and joint-venture hotels offer cars with or without a driver. Intourservice, tel: 203 0096, offers chauffeured limousines and so does Mosrent, tel: 248 0251. Other car rentals are Avis, tel: 578 5646 and Hertz, tel: 578 7532.

PETROL STATIONS

Finding a petrol station in Moscow is not easy as they are often hidden away on side roads. Russian cars use 73 or 93 octane fuel while Western cars use 95 octane. Without a voucher the 95 octane fuel is not available. Vouchers valid for 10 litres (2 gallons) of fuel can be obtained from your embassy. The following stations sell diesel fuel and 95 octane petrol:
97–10, Marshal Grecho Prospekt, tel: 144 4787.
105–23, Kashirskoye Highway, tel: 541 8703.

SERVICE STATIONS

Spare parts are available from Stockmann's, 4–8 Zasepsky Val, tel: 233 2602 or at the GlavUPDK workshop, Kiev Street 8, tel: 240 2092. Repairs to foreign-made cars are undertaken at 6, 2nd Selskokhozyaistvenny Proyezd (Mercedes, Volkswagen, BMW and SAAB), tel: 181 0631, or at Kuntsevsky Technical Centre, 16 Gorbunov Street, tel: 448 4268; the Technical Assistance Service, 91 Varshavskoye Highway, tel: 119 8000; the Road Service Station, 165 Mozhaiskoye Highway, tel: 446 1740 or the Road Service Station at km 22, Varshavskoye Highway, tel: 119 8108.

ACCIDENTS

Report to the GAI, 1 Sadovaya Samotechnaya Street, tel: 925 5533.

INSURANCE

To drive in Moscow it is advisable to take out additional Russian car insurance. Claims are paid only in the currency in which you pay for the policy. Contact: Ingosstrakh, 12 Pyatnitskaya Street, tel: 231 1677.

RULES OF THE ROAD

Russia is a signatory to the International Traffic Convention. Rules of the road and road signs correspond in general to international standards. The basic rules, however, are worth mentioning.
1) In Russia traffic drives on the right.
2) It is prohibited to drive a car after consuming even the smallest amount of alcohol. If the driver shows a positive alcohol test, the consequences may be very serious. It is also prohibited to drive a car under the effect of drugs or strong medicines.
3) The driver must have an international driving licence and documents verifying his right to drive the car. These papers must be in Russian and are issued by Intourist.
4) Vehicles, except those rented from Intourist, must carry their national registration code. All must have a national licence plate.
5) The use of the horn is prohibited within city limits except in emergencies.
6) The use of seat belts for the driver and front-seat passenger is compulsory.
7) The speed limit in populated areas (marked by blue-coloured signs indicating "town") is 60 kph (37 mph); on most arterial roads the limit is 90 kph (56 mph). On highways different limits apply and are shown on road signs.
8) You can insure your car in Russia through **Ingosstrakh**, the national insurance company.
9) Foreigners still need a permit to drive more than 40 km (25 miles) beyond city limits and must follow the routes established by Intourist. Permits may be checked by traffic police when exiting the city.

RIVER TRAMS

Between May and September you can travel up and down the Moskva River. There are two routes: one is from the International Trade Centre to the Novospassky Bridge, stopping every few minutes; the other leaves from the Kiev Station Pier and goes to the Rowing Basin in Kutsevo. They leave three times a day: 11.30am, 2.30pm and 5.30pm. The one-way fare is 60 kopecks. If you like travelling on the river, you can also take the Raketa hydrofoil from Gorky Park to the Novospassky Bridge.

TOUR OPERATORS

INTOURIST

Intourist cooperates with more than 700 foreign firms who are agents for Intourist in their respective countries. It offers services in more than 200 cities in all the 15 Ex-USSR republics and runs numerous hotels, motels, camp sites and restaurants. On Intourist's itinerary are more than 600 different tours within Russia. They include local sightseeing trips, thematic tours for history, art and nature lovers, as well as arrangements for recreation and medical treatment, sporting and hunting tours. A visit to the nearest Intourist office will give you a good overall impression of their diversified offerings.

In addition to tours Intourist also runs a car rental service, with or without driver, in the larger cities. Lada, Chaika and Volga cars as well as Ikarus coaches and LAZ, PAZ and RAF buses for 9 to 42 people can be hired.

Intourist also manages 110 hotels, motels and

camp sites for 55,000 guests. A variety of hotels are now being built and reconstructed in cooperation with foreign partners.

Over 5,000 guide-interpreters, speaking more than 30 languages work for Intourist. Since cooperatives are now permitted, Intourist is no longer a monopoly tourist agency; a few small and independent firms have sprung up during the past few years which now serve special interest groups. The most prominent of these, having its headquarters in St Petersburg, is Mir International Centre. But since the itinerary of a journey to Russia has to be confirmed in advance, and submitted together with the visa application, Intourist, with its worldwide net of sales agents, is still carrying out most of the business.

Besides group tours, Intourist also arranges individual journeys to Russia. These trips must be planned on a day-by-day basis in advance, and Intourist arranges transport, accommodation and food. Judging by accommodation prices in those Intourist hotels that are available to foreigners, such individual travel is, however, not inexpensive.

Moscow Intourist Office: tel: 203 6962. June–September Intourist runs a **Cultural Centre** to acquaint visitors with the Russian way of life: 32, Prospekt Obukhovskoi Oborony.

INTOURIST OFFICES

Amsterdam, Honthorststrasse 42, tel: 798 964.
Athens, Stadiou 3, Syntagma Sq., tel: 323 3776.
Berlin, Friedrichstrasse 153A, tel: 229 1948; 15 Kurfürstendamm 63, tel: 880 070.
Budapest, Felszabadulas ter. 1, tel: 180 098.
Brussels, Galerie Ravenstein 2, tel: 513 8234.
Copenhagen, Vester Farimagsgade 6, tel: 112 527.
Delhi, Plot 6/7, Block 50-E, Njaja Marg Chanakiapuri, tel: 609 145.
Frankfurt, Stephanstrasse 1, tel: 285 776.
Helsinki, Etela Esplanaadi 14, tel: 631 875.
London, 292 Regent St, tel: 071-631 1252.
Montreal, 1801 McGill College Ave, Suite 630, tel: 849 6394.
New York, 630 Fifth Ave, Suite 868, tel: 212-757 3884.
Paris, 7 Boulevard de Capucines, tel: 474-24740.
Prague, Stepanska 47, tel: 267 162.
Rome, Piazza Buenos Aires 6/7, tel: 863 892.
Sydney, Underwood House, 37–49 Pitt St, tel: 277 652.
Tokyo, Roppongi Heights, 1–16, 4-chome Roppongi, Minato-ku, tel: 584 6617.
Vienna, Schwedenplatz 3–4, tel: 639 547.
Zurich, Usteristrasse 9, tel: 211 3335.

WHERE TO STAY

HOTELS

When you check into a Russian hotel you will be given a guest card which you should carry at all times. Each hotel has a service bureau which will assist you in all small matters, from medical help to calling a taxi or obtaining theatre tickets, restaurant reservations or arranging international telephone calls. Each floor has a floor clerk who will call a porter, get your laundry done, etc. He or she will also keep your room key while you are away.

Most visitors to Moscow arrive with a confirmed hotel reservation. If, however, you need accommodation during the peak tourist season, when every hotel is booked out, try the **Alexander Blok** (houseboat), tel: 255 9278 or the **Hotel Vizit**, tel: 202 2848; they are both American-Russian joint ventures and you will have a good chance of finding accommodation for hard currency. You can also, try the **New Times** co-operative, tel: 274 4694 or 274 1089. New Times can also issue you with an official invitation, which you need if you want to visit Russia and avoid the high charges of Intourist.
Hotels are listed in the following categories (in brackets after the hotel name): de Luxe, A, B, 1, 2.

Aerostar Hotel (de Luxe), Leningadsky Prospekt 37. Tel: 155 5030.
Belgrad Hotel (B), 5 Smolenskaya Street. Tel: 248 1643.
Budapest Hotel (B), 2/18 Petrovskiye Linii Street. Tel: 924 8820.
Varshava Hotel (B) , 2/1 Leninsky Prospekt. Tel: 238 1970.
Druzhba Hotel (B), 53 Vernadsky Prospekt. Tel: 434 2782.
Izmailovo Hotel (A), 69a Izmailovskoye Highway. Tel: 166 0109.
Intourist National Hotel (A), 3/4 Tverskaya Street, tel: 203 4008; 14/1 Okhotny Ryad, tel: 203 6539.
Cosmos Hotel (de Luxe), 150 Prospekt Mira. Tel: 217 0785.
Leningrad Hotel (B), 21/40 Kalanchevskaya Street. Tel: 975 3008.
Mezhdunarodnaya Hotel (de Luxe), 12 Krasnopresnenskaya Embankment. Tel: 253 2382.
Metropol Hotel (de Luxe), Teatralny Proyezd 1/4. Tel: 927 6000.
Minsk Hotel (B), 22 Tverskaya Street. Tel: 299 1215.

Mozhaiskaya Hotel (B), 165 Mozhaiskoye Highway. Tel: 447 3434.
Molodezhnaya Hotel (1), 25 Dmitrovskoye Highway. Tel: 210 4565.
Moskva Hotel (A), 7 Okhodny Ryad. Tel: 292 1000.
Novotel Hotel Airport (de Luxe), Sheremetievo-2. Tel: 578 9407.
Olympic Penta (de Luxe), 18/1 Olimpiysky Prospekt. Tel: 971 6101.
Orlyonok Hotel (A), 15 Kosygin Street. Tel: 939 8844.
Ostankino Hotel (1), 29 Botanicheskaya Street. Tel: 219 2880.
Pullman Iris Hotel (de Luxe), Korovinskoye Chaussee 10. Tel: 488 8000.
Rossiya Hotel (A), 6 Varvarka Street. Tel: 298 5530.
Salyut Hotel (B), 158 Lenin Prospekt. Tel: 438 6565.
Savoy Hotel (de Luxe), 3 Rozhdestvenka. Tel: 929 8500.
Sevastopol Hotel (B), 1a Bolshaya Yushinskaya Street. Tel: 318 2263.
Sovetskaya Hotel (A), 32/2 Lenin Prospekt. Tel: 250 7255.
Solnechny Hotel (A), includes a camping site, 21st km on Varshavskoye Highway. Tel: 382 1465.
Soyuz Hotel (B), 12 Levoberezhnaya Street. Tel: 457 9004.
Sport Hotel (A), 90/2 Lenin Prospekt. Tel: 131 1191.
Sputnik Hotel (1), 38 Lenin Prospekt. Tel: 938 7106.
Ukraina Hotel (B), 2/1 Kutuzovsky Prospekt. Tel: 243 3030.
Tsentralnaya Hotel (2), 10 Tverskaya Street. Tel: 229 8589.
Yunost Hotel (B), 34 Frunzensky Val. Tel: 242 1980.

PRIVATE

If you stay for a prolonged period you can rent an apartment through an official co-op rental agency:
Yupiter: Butyrsky Val, tel: 250 2300
Astoria: 7 Tverskaya, tel: 229 2300
Rubin: Prospekt Mira, tel: 283 1659
Avangard: Byelomorskaya Street, tel: 455 9210

Many Muscovites like to rent out their apartment for hard currency. This can be a very personal and inexpensive way of staying in the city and avoiding the high Intourist hotel charges.

FOOD DIGEST

WHERE TO EAT

The best food is served in restaurants at Intourist hotels and in the new co-ops, which have already gained popularity among diplomats and business representatives. Most restaurants accept credit cards (American Express, Visa, Carte Blanche, Mastercard).

Moscow is, naturally, the place for Russian cuisine. There are also European (mostly French), American, Indian, Chinese and Japanese restaurants. There are also fast-food buffets where you can grab a quick bite to eat at your Intourist hotel.

Breakfast costs equivalent of 15–30 roubles, and a three- to four-course lunch with tea or coffee will cost between 60 and 100 roubles. Breakfast is usually 7.30am–9am and lunch noon–3pm. There isn't much variety in hotel restaurants. The best remedy is to eat in different restaurants serving national cuisine, such as Russian, Georgian, Ukrainian or Uzbek food. Be sure to book a table in advance (the night before or on the morning of the day) either yourself or through the service bureau of your hotel.

To get into a restaurant, you will find that your passport is as indispensable as it is when you cross the border. If there is a line of people in front of the restaurant you have chosen, or if the doorman tells you the place is full, don't be shy – show your passport. Most restaurants have menus in English and, in 90 percent of the places we recommend, the waiters speak passable English.

Dinner for two in a state-run restaurant costs 20 to 25 roubles (without drinks), and around 50 roubles in a co-op. In hard-currency restaurants, the bill will be between US$20 and $100.

To visit a roubles restaurant, even in one of Intourist's hotels, prepare for slow, lacklustre service. The situation can be remedied – to an extent – with a moderate up-front tip in foreign currency.

Things get really difficult after 10.30pm, when practically all Moscow restaurants close. Fortunately, the power of foreign currency works round-the-clock, and you'll probably find what you want in one of the city's few all-night, hard-currency bars located in several Intourist hotels.

The best source for information on new co-operative restaurants and the whole wining and dining scene in Moscow is the *Moscow Magazine*, a well-researched bi-monthly.

The following restaurants accept payment in

roubles unless marked **(HC)** at the end (which also indicates the very good restaurants). Before going to one of these restaurants, make a reservation by telephone, identify yourself as a foreigner and show your credit card or passport at the door. If you cannot get a table in a rouble restaurant, indicate that you are willing to pay in hard currency. It does open doors.

If you just want to have a drink in a bar or a pub, you are best off in one of the big hotels. The **Cosmos**, the **Moskva**, the **Savoy** and the **Mezhdunarodnaya** are the best places for Western drinks. You will also find many red-lipped ladies around who shouldn't be there according to the law – but the doorman and the Mafia also make their money at these hard currency places.

Aist, 1/8 Malaya Bronnaya, tel: 291 6692. Food from across the CIS.
Alexander Blok, 12 Krasnopresnenskaya Naberezhnaya, tel: 255 9278. Restaurant on a ship with American food (HC).
Aragvi, 6 Tverskaya Street, tel: 229 3762. Great Georgian food.
Arbat, 29 Vozdvishenka, tel: 291 1445. Russian and Western food.
Arlecchino, 15 Druzhinnikovskaya Street, tel: 205 7088. Italian cuisine (HC).
Atrium, 44 Lenin Prospekt, tel: 137 3008. Russian nouvelle cuisine (HC).
Baku, 24 Tverskaya Street, tel: 299 8506. Azerbaijani and Lebanese food.
Café Margarita, 28 Malaya Bronnaya, tel: 299 6539. Homemade pastries and a view of the Patriarch's Pond.
Café Viru, 50 Ostozhenka Street, tel: 246 6107. Estonian café, salads and sandwiches.
Delhi, 23b Krasnaya Presnya Street, tel: 252 1766. Top-class Indian food and Indian entertainment (HC).
Dom na Tverskoy, 12 Gotvalda Street, tel: 251 8419. Russian food in a club/café atmosphere.
Druzhba, 12 Krasnopresnenskaya Naberezhnaya, tel: 255 2970. International food (HC).
El Rincon Espanol, Hotel Moskva, 7 Okhotny Ryad, tel: 292 2893. Bar-restaurant with Spanish food (HC).
Farkhad, 4 Bolshaya Marfinskaya, tel: 218 4136. Authentic Azerbaijani food, but bring your own alcohol. Arabian atmosphere.
Fyodorov's, 36 Prechistenka Street, tel: 201 7500. Moscow's first co-operative restaurant; good food and atmosphere (HC).
Guria, 7 Komsomolsk Prospekt, tel: 246 0378. Georgian co-operative café.
Kolkhida, 6 Sadovo-Samotechnaya, Stroyenie 2, tel: 299 6757. Good Georgian food. Quiet ambiance.
Kuilong, 7 Litovski Boulevard, tel: 425 1111. Vietnamese food.
Lasagne, 40 Pyatnitskaya Street, tel: 231 1085.Italian food (HC).
Livan, 24 Tverskaya Street, tel: 299 8506. Excellent Azerbaijani food (HC).
Manila, 81 Vavilova Street, tel: 132 0055. Filipino food. Hard currency restaurant but also has a rouble room for Russian food.
McDonald's, Pushkin Square, tel: 200 1655. Moscow's most successful fast-food outlet. You may have to stand in line for anything up to an hour.
Mei-Hua, 2/1 Rusakovskaya Street, Stroyenie 1, tel: 264 9574. Chinese restaurant.
Moosh, 2/4 Oktyabrskaya Street, tel: 284 3670. Armenian food, Armenian music.
Olimp, near Luzhniki Stadium, tel: 201 0148. Floor show, food for roubles, alcohol for hard currency.
Peking, 1/2 B. Sadovaya Street, tel: 209 1865. Chinese food.
Pizza Hut, 17 Kutuzovsky Prospekt and 12 Tverskaya Street, tel: 243 1727.
Praga, 2 Arbat, tel: 290 6171. Russian and Czech food, live entertainment.
Razgulyai, 11 Spartakovskaya Street, tel: 267 7613. Russian food, music at weekends (HC).
Red Lion Pub, Hotel Mezhdunarodnaya, 2nd floor. English pub serving British meals and beers (HC).
Ruslan, Vorontsovskaya 32/36, tel: 272 0632. Russian food, gypsy music at weekends.
Russkaya Izba, Ilyinskoye village close to Archangelskoye, tel: 561 4244. Traditional Russian food.
Sakura, Mezhdunarodnaya Hotel, 12 Krasnopresnenskaya Nab., tel: 253 2894. Japanese food and beer. Excellent but expensive (HC).
Savoy, 3 Rozhdestvenka, tel: 928 0450. The classic Western restaurant in Moscow. (HC).
Sayat-Nova, 17 Yasnogorskaya Street, Korpus 1, tel: 426 8511. Georgian specialities. Bring your own alcohol.
Skazka, 1 Tovarishchevsky Pereulok, tel: 271 0998. Expensive traditional Russian food, gypsy music.
Slavyansky Bazaar, 13 Nikolsky Street, tel: 921 1872. Dancing, live entertainment, Russian food.
Sorok Cheterie, 44 St Petersburg Highway, tel: 159 9951. Jazz and Russian food.
Stanislavskogo 2, 2 Stanislavskogo Street, tel: 291 8689. French and Russian food.
Stoleshniki Café, 8 Stoleshniki Pereulok, tel: 229 2050. Tavern with both atmosphere and great food.
Strastnoy 7, 7 Strastnoy Boulevard, tel: 299 0498. Luxurious dining (HC).
Taganka Bar, 15 V. Radishevskaya, tel: 272 4351. Live entertainment, great atmosphere, good food.
Taganka Café, 76 Chkalova Street, tel: 272 7320. Good Russian food, theatre hangout.
Tren-Mos, 21 Komsomolsky Prospekt, tel: 245 1216. Western food, French chef.
Tsentralni, 10 Tverskaya Street, tel: 229 0241. Elegant old dining room, Russian food.
U Margarita , 9 Ryleyeva Street, tel: 291 6063. Violin music, great food, imported beer.
U Nikitskikh Vorot, Gertsena Street at Nikitsky Boulevard, tel: 290 4825. Bistro with live entertainment, good food.
U Pirosmani, 4 Novodevichy Proyzed, tel: 247 1926. Georgian food and wine, violin and piano

music. Food for roubles, alcohol for hard currency.
U Yuzefa, 11/17 Dubininskaya, tel: 238 4646. The only Jewish restaurant in Russia.
Uzbekistan, 29 Neglinnaya Street, tel: 924 6053. Central Asian food.
Viennese Café, Hotel Mezhdunarodnaya, 1st floor, tel: 253 2894. A newly installed western breakfast place (HC).
Villa Peredelkino, Peredelkino, tel: 435 1478. Italian-Russian restaurant (HC).
Vycherny Siluett, 88 Taganskaya Square, tel: 272 2280. Great food, classic atmosphere (HC).
Victoria a.k.a. Hard Rock Café, Zelyoni Theatre, Gorky Park, tel: 237 0709. Owned by a rock-star, Russian-Armenian food.
The Writer's Union, 52 Povarskaya Street, tel: 291 2169. Great Russian food (HC).
Yakimanka, 2/10 Bolshaya Polyanka, Stroyenie 1, tel: 238 8888. Uzbek restaurant.
Zaidi i Poprobui, 124 Prospekt Mira, Korpus 1, tel: 286 750. Russian cuisine. Alcohol for hard currency.
Zaria Vostok, 4 Dvardsati Shesti Bakinskikh Komissarov, Korpus 2, Street, tel: 433 2201. Korean restaurant.
Zolotoi Drako, 64 Plushchikha Street, tel: 248 3602. Szechuan food.

WHAT TO EAT

Most of the different ethnic groups populating the former USSR claim to have their national cuisine, and some of them do genuinely have them. Within this diversity experts say Georgian, Ukrainian, Russian and Central Asian cuisines are the best.

To interpret what is offered on co-op menus, here is a list of typical dishes from all the regions:
Achma Georgian cheese noodles
Baklazhannaya Ikra eggplant dish
Baranina mutton
Basturma Caucasian cured meat
Bliny leavened pancakes
Chicken Tabaka flattened grilled chicken
Golubysky stuffed cabbage
Govyadina beef
Griby mushrooms
Hachapury Georgian pastry
Ikra caviar
Khyleb bread
Kuritsa chicken
Lavash Georgian bread
Lososina salmon
Maslo butter
Morozhenoe ice cream
Osetrina sturgeon
Pelmeni Siberian ravioli
Plov pilav
Shchi cabbage soup
Telyatina veal
Vareniki Ukraininan ravioli
Zakuski hors d'oeuvres
Zharkoe beef stew

GEORGIAN

With *perestroika* Georgian food came to the rest of the former USSR. Many co-operatives in the Caucasus opened restaurants serving Georgian food. This cuisine is famous for its *shashlyk*, *tsyplyata tabaka* (chicken fried under pressure), *basturma* (specially fried meat), *suluguni* (salted cheese) and *satsyvi* (chicken). It can be served with *lavash* (special kind of bread) or with *khachapuri* (roll stuffed with cheese).

UKRAINIAN

Ukrainians are traditionally regarded as poeple who eat a lot – but also well. Among their favourite dishes are *borshch* (beetroot soup with cabbage, meat, mushrooms and other ingredients), *galushky* (small boiled dumplings) and *varenyky zvyshneyu* (curd dumplings with red cherries served with sugar and sour cream). Chicken Kiev (or Kiev Cutlet), known and served throughout the world, is prepared with different spices and garlic. Loved by everyone in the Ukraine is *salo* (salted raw lard spiced with garlic). It is served with black bread. *Kovbasa* (different kinds of smoked sausage) is also popular.

RUSSIAN

Famous dishes include beef Stroganov and Beluga caviar. Russian cuisine includes less refined but no less popular dishes like *bliny* (pancakes served with butter and sour cream, caviar, meat, jam etc.), *shchi* (sour cabbage soup with meat; gourmets prefer this with mustard), *pelmeni* (boiled dumplings with meat) and *kasha* (gruel or porridge of different grains).

DRINKING NOTES

Everyone knows what they drink in Russia: vodka (only available with a ration card since 1991) and tea from the samovar. But there are numerous other drinks within the different national cuisines, though it is not so easy to get these specialities in Moscow.

The Ukrainian traditional alcoholic drink is *gorilka* which resembles vodka. More popular, and more refined, is *gorilka z pertsem,* i.e. *gorilka* with a small red pepper. The traditional non-alcoholic drink is *uzvar* (made of stewed fruit).

Georgians drink dry and semi-dry wines: *Tsinandali*, *Mukuzani*, *Kinzmarauli*, *Alazan Valley*, and *Tvishi* (reported to have been Stalin's favourite). Non-alcoholic drinks include the best Georgian mineral water such as *Borzhomi* and *vody Lagidze* (mineral water with various syrup mixtures).

In summer Russians prefer to drink *kvas*, a refreshing drink prepared from bread fermented with water and yeast.

Central Asians drink *geok chaj* (green tea), the best treatment for a thirst in the hot climate.

The State Tretyakov Gallery

The Tretyakov Gallery was founded by the wealthy merchant and patron of the arts P.M. Tretyakov and bequeathed to the city of Moscow in 1892. It is now housed in a Neo-Russian building that was constructed between 1900 and 1905. A new gallery has recently been completed nearby that exhibits a collection of postrevolutionary art.

The original Tretyakov displays the world's finest collection of Russian icons.

Kiev School: Of the few remaining icons from that period the best known dates back to Demetrius of Thessalonika (12th century).

Byzantine School: The most famous exhibit is the "Virgin of Vladimir". It came to Kiev around 1135, was then moved to Vladimir and in 1390 to the Cathedral of the Assumption in the Moscow Kremlin where it remained until 1930 when it was finally brought to the Tretyakov.

Vladimir-Suzdal School: This school is represented by the "Virgin Great Panagia" (12th century).

Pskov School: "The prophet Elijah in the Wilderness" (13th century)

Novgorod School: "The last Judgement" and "Laying in the Tomb" (both 15th century).

Moscow School: "Boris and Gleb on horseback" (14th century), and by Theophanes the Greek, among others, the famous "Virgin of the Don" (14th century) which Dimitry Donskoy, the founder of Moscow, supposedly had with him during the decisive battle of Kulikovo.

Works by Russia's most famous painters such as Kramskoy, Repin, Polenov and Vasnetsov and many others are exhibited in the other halls, collected in the following order:

Art of the 18th century and first half of the 19th century.

Art of the second half of the 19th century.

Art of the early 20th century.

Ground Floor

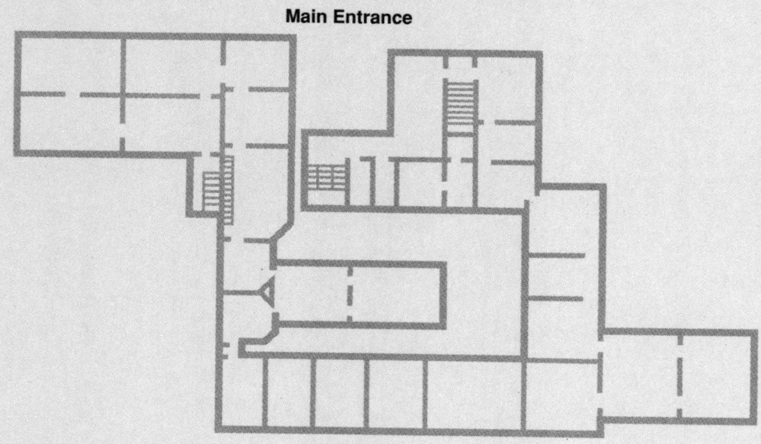

First Floor

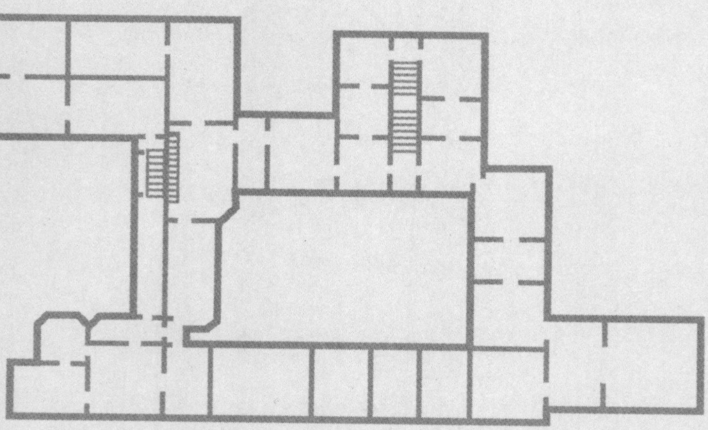

Ground Floor

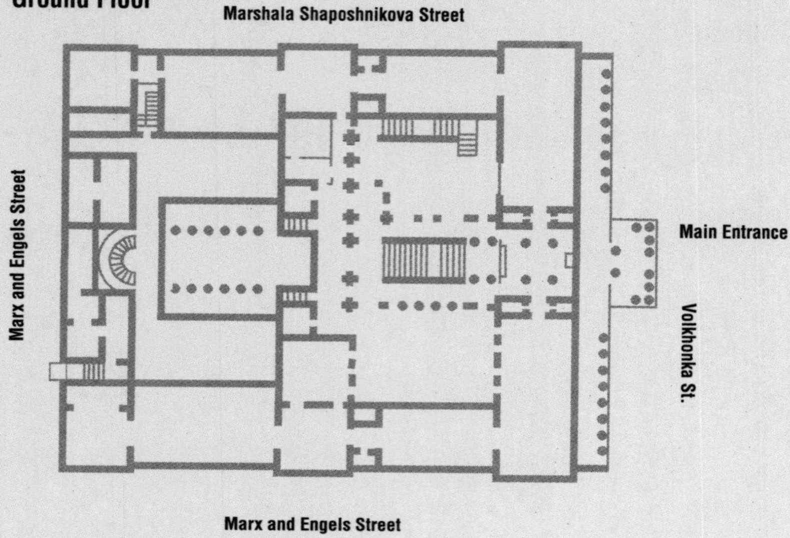

First Floor

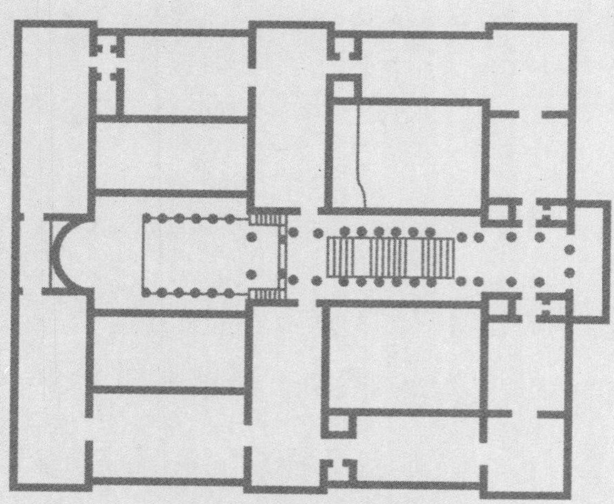

Pushkin Museum of Fine Arts

The Pushkin Museum of Fine Arts was founded by the father of the famous Russian poetess Marina Tsvetaeva (1882-1941). Professor Tsvetaeva wanted it to be a museum and an educational institution for professional artists.
Today the museum has its own rich collection and exhibits paintings from the best known art depositories of the world.

Ground Floor

Ancient art, Egypt and Middle East

Icons, Byzantian schools of the 14th century

Italian art, paintings, 8th to 16th centuries

Collection of Dutch, Flemish and German paintings, 15th to 17th centuries

Ancient Greek and Roman art: many copies

Temporary expositions

First Floor

Ancient Greek art of the 5th century B.C.

Collection of European paintings of the 19th and the beginning of the 20th century

Ancient Greek art, 4th to 1st century B.C.

Copies of some of the world's best known sculpture

Medieval art

Italian Renaissance

Among the masterpieces of the Pushkin Museum are works by Rembrandt, Boticelli, Lucas Cranach the Elder, Goya and Michelangelo.
The second half of the 19th century and the beginning of the 20th century are represented with works by Auguste Renoir, Edgar Degas, Vincent van Gogh, Paul Gauguin, Fernand Leger, Auguste Rodin, Henri Matisse and Pablo Picasso.

CULTURE PLUS

Theatre, ballet and classical music in Moscow certainly deserve at least one of your evenings. The **Bolshoi**, which sells tickets for about US$50 when on tour in the United States (and which is practically off-limits to hapless Moscow theatre buffs), will practically guarantee you a seat for as little as US$10 (booking is no trouble at all).

If, however, you are down on your luck and the Bolshoi is out of town, you can still enjoy a classical opera or ballet at the **Stanislavsky** and **Nemirovich-Danchenko Music Theatre** in Pushkinskaya Street, just a 10-minute walk from the Bolshoi. Many Muscovites, by the way, prefer that theatre's production of Tchaikovsky's *Swan Lake* to the Bolshoi's. The **Theatre of Operetta** is in the same street. It occupies the building that used to belong to the Bolshoi. We only hope that lovers of *The Queen of Chardash* and *The Merry Widow* won't mind that the lyrics are in Russian.

Theatres are popping up in today's Moscow like mushrooms after rain; strange as it may seem, soft-core "erotic" productions are very "in", even though few of them live up to the standards of genuine art. Experts assuage public fears, saying it's just a passing fad, and that only the most professionally competent troupes will survive; meanwhile, theatre managers have the commercial side of things to worry about – the state has no extra money for culture these days.

Another music theatre worthy of your attention, particularly if you have brought your children along on this trip is the **Children's Music Theatre**, which occupies a luxurious edifice decorated by virtuoso *Palekh* artists (across Vernadskogo Prospekt from Moscow University). It is run by Director Natalia Suts, who created the theatre and has stood at its helm ever since she managed, heaven knows how, to convince the proper authorities that musical education for children of all ages was worth paying for. The repertoire of her theatre boasts such hits as *Mowgli*, *Bluebird* and *Chio-Chio-San*.

To hear leading Russian performers (**Sviatoslav Richter**, the **Borodin Quartet**, the **Virtuosos of Moscow**) play organ music, symphonies, piano concertos and chamber music, visit the **Conservatoire's Concert Hall** (the one where International Tchaikovsky Contests are held every four years). The concert hall in Mayakovsky Square also bears Tchaikovsky's name. It is the usual arena for the **Russian Army Song and Dance Ensemble** and for Igor Moiseev's **Folk Dance Ensemble**.

True lovers of painting will probably find it impossible to resist the temptation to become acquainted with art salons and exhibits. Moscow has a great many places besides the **Tretiakov Gallery** (a remarkable collection of classical Russian painting and icons) and the **Pushkin Museum of Fine Arts** (famous for its European painting of the 18th to 20th centuries). For early Russian painting, go to the **Andrei Rublev Museum** in the **Andronnikov Monastery**. Don't forget Moscow's other monasteries – the **Novodevichy Convent** and the **Danilovsky** and **Donskoy Monasteries**.

"Official" Russian painting is on display at the **State Gallery of Art** on Krymskaya Embankment (opposite Gorky Park). It always has interesting exhibitions of Russian modernism and the avant garde. To see some of the "unknown painting" of the 20th century, be sure to visit the **Gallery of the Russian Culture Fund**.

The exhibition hall in Malaya Gruzinskaya Street is where to see the non-conformists, soc-art and abstractionists, many of whom are already known in the West. It is legal for you to buy and take back home any work from official exhibit-sales (the **Mars Gallery** in the western part of Moscow, holds them on a regular basis), or from the hard-currency auctions organised by the Russian Culture Fund. You will find that most Russian artists will gladly sell you works for foreign currency right from their studios; taking such paintings out of Russia, however, may involve a zillion customs formalities.

Another place to buy works of art is **Izmailovo**. Then there is the Russian Montmartre in old **Arbat Street**. You can have your portrait done here in 10 minutes. The usual price is 30 to 50 roubles, although you will probably be asked to pay in dollars. There are also assorted souvenirs *à la Russe* like fake antiques and icons. Arbat Street is also famous for its cafés, street concerts by semi-pro bands and not-yet-famous singers and, of course, poets, whose satirical verse would, until quite recently, have guaranteed them banishment beyond the city limits.

The **Public Museum of the Repressed** (in the House on the Quay, the grey hulk on the Moskva River just beyond Bolshoi Kamenny Bridge if you walk from the Kremlin) will help you find out about the destinies of artists, poets, writers, statesmen, etc., who were turned into non-persons by Stalin's regime in the 1930s. Apartments in this house were given to artists with connections in the Kremlin or even, at one stage or another in their career, with Big Brother himself (Stalin considered himself a patron of the arts). Those times – and the numerous tragedies they brought – are described in Yury Trifonov's *House on the Quay*. Trifonov, who wrote his best books in the 1960s and the 1970s, was barely tolerated by Russian officialdom.

MUSEUMS

ART

State Tretiakov Art Gallery, All-Union Museum Association, 10 Krymsky Val, tel: 231 1362. Open: daily 10am–8pm except Monday. (*See floor plan page 278–279.*)

Tretiakov Art Gallery, 10 Lavrushinsky Lane. Has been closed since 1986; no one knows yet when it will be reopened.

Pushkin Museum of Fine Arts, 12 Volkhonka Street, tel: 203 9578. Open: Tuesday–Saturday 10am–8pm, Sunday 10am–6pm. (*See floor plan page 281–282.*)

Andrei Rublev Museum of Early Russian Art, 10 Pryamikov Square, tel: 278 1429. Open: May–September, Monday and Tuesday 1–8pm, Thursday–Sunday 11am–6pm; October–April, daily except Wednesday 11am–6pm.

Church of the Intercession in Fili, 6 Novozavodskaya Street, tel: 148 4552. Open: May–September, Friday–Sunday 11am–6pm, Monday & Thursday 11am–8pm.

Museum of the Arts of the People of the East, 16 Vorontsovo Polye Street, and 12a Suvorovsky Blvd, tel: 297 4800. Open: Tuesday–Sunday 11am–7pm.

All-Russia Museum of Decorative, Applied and Folk Art, 3, Delegatskaya Street, tel: 923 1741. Open: Monday, Wednesday, Saturday, and Sunday 10am–6pm, Tuesday and Thursday 12.30–8pm.

Folk Art Museum, 7 Stanislavsky Street, tel: 2902114. Open: Tuesday and Thursday 10am–8pm, Wednesday, Friday, Saturday and Sunday 10am–5pm.

ART-ARCHITECTURE

Alexei Shchusev Architecture Museum, 2 Vozdvishenka Street, tel: 291 1978. Open: daily except Monday and Friday 11am–7pm.

Donskoy Monastery, 1 Donskaya Square, tel: 232 0766. Open: May–September, daily except Monday and Friday, 10am – 6pm; October–April 10am–5pm.

Ostankino Palace Museum of Serf Art, 5 Ostankinskaya Street, tel: 283 1165. Open: May–September, daily 10am–5pm; October–April 10am–3pm.

Kol0menskoye Museum Reserve, 39 Yuri Andropov Pr., tel: 115 2767. Open: September–April 11am–5pm; May–August Wednesday and Thursday, 1–8pm, Friday–Sunday 11am–5pm.

Museum of Ceramics and the 18th Century Kuslovo Estate-Museum, 2 Yunost Street, tel: 370 0160. Open: April–September, daily except Monday and Tuesday 11am–7pm; October–March, 10am–4pm.

MEMORIAL

Glinka Museum, 4 Fadeyev Street, tel: 972 3237. Open: Tuesday and Thursday 10am–8pm, Wednesday, Friday, Saturday and Sunday 10am–6pm.

Alexander Herzen Museum, 28 Petrovka Street, tel: 241 5859. Open: Tuesday, Thursday, Saturday and Sunday 11am–5.30pm, Wednesday & Friday 2–8.30pm.

Dostoyevsky Apartment-Museum, 2 Dostoyevsky Street, tel: 281 1085. Open: Thursday, Saturday and Sunday 11am–6pm, Wednesday and Friday 2–9pm.

Mikhail Lermontov House-Museum, 2 Malaya Molchanovka Street, tel: 291 5298. Open: Wednesday and Friday 2–9pm, Thursday, Saturday, Sunday 11am–6pm.

Anton Chekhov House-Museum, 6 Sadovaya-Kudrinskaya Street, tel: 291 6154. Open: Tuesday, Thursday, Saturday and Sunday 11am–6pm, Wednesday and Friday 2–8.30pm.

Alexander Pushkin Museum, 12/2 Prechistenka Street, tel: 202 8531. Open: Tuesday–Friday noon–8pm, Saturday and Sunday 10am–6pm.

Pushkin Apartment-Museum, 53 Arbat Street, tel: 241 3010. Open: Wednesday–Friday noon–6pm, Saturday and Sunday 11am–5pm.

Leo Tolstoy Museum, 11 Prechistenka Street, tel: 202 9338. Open: Tuesday, Thursday, Saturday and Sunday 11am–7pm, Wednesday and Friday noon–8pm.

Leo Tolstoy's Estate-Museum, 21 Leo Tolstoy Street, tel: 246 9444. Open: Wednesday and Friday 11am–6pm, Saturday 10am–4pm.

Maxim Gorky Museum, 25a Vorovsky Street, tel: 290 5130. Open: Wednesday and Friday noon–8pm, Thursday, Saturday and Sunday 10am–5.45pm.

Maxim Gorky Apartment-Museum, 6/2 Povarskaya Street, tel: 290 0535. Open: Wednesday and Friday noon–8pm, Thursday, Saturday and Sunday 10am–5.45pm.

Vladimir Mayakovsky Museum, 3/6 Proyezd Serova, tel: 921 9560. Open: Tuesday, Friday, Saturday and Sunday 10am–6pm, Monday & Thursday noon–8pm.

TECHNICAL

Politechnical Museum, 3/4 Novaya Square, tel: 923 0756. Open: Wednesday, Friday, Saturday, and Sunday 10am–6pm, Tuesday & Thursday noon–8pm.

The Planetarium, 5 Sadovaya Kudrinskaya Street, tel: 254 0153. Open: Monday, Wednesday, Thursday, Friday and Saturday 1–6pm, Sunday 10am–6pm.

Museum of Sport, Lenin Stadium, Luzhniki, East Wing, tel: 201 0747. Open: daily noon– 6pm except Monday.

Zoological Museum, 6 Herzen Street, tel: 203 8923. Open: Tuesday, Saturday and Sunday 10am–5pm, Wednesday and Friday noon–8pm.

The Zoo, 1 Bolshaya Gruzinskaya Street, tel: 2555375. Open: May–August 9am–8pm; September 9am–6pm; October–March 9am–5pm; April 9am–6pm.

Memorial Museum of Cosmonauts, Alley of Cosmonauts, Prospekt Mira, tel: 283 7914. Open: Tuesday–Thursday noon–8pm, Friday–Sunday 11am–5.30pm.

HISTORICAL

Novodevichy Convent, 1 Novodevichy Proyezd, tel: 246 5862. Open: daily 10am–5.30pm, except Monday.
Church of the Trinity in Nikitniki, 3 Nikitnikov Lane, tel: 298 5018. Open: Monday, Friday, Saturday, Sunday 10am–6pm, Wednesday and Thursday noon–8pm.
St Basil's Cathedral Museum, Red Square, tel: 298 3304. Open: daily 9.30am–5.30pm, except Tuesday.
Cathedral of the Annunciation, Kremlin, tel: 202 0341. Open: daily except Thursday, May–September 10am–6pm; October–April 10am–5pm.
Archangel Cathedral, Kremlin, tel: 202 0347. Open: daily except Thursday, May–September 10am–6pm; October–April 10am–5pm.
Church of the Deposition of the Robe, Kremlin, tel: 221 7124. Open: daily except Thursday, May–September 10am–6pm; October–April 10am–5pm.
16th and 17th Century Chambers in Zaryadye, 10 Varvarka Street, tel: 298 3235. Open: Monday, Tuesday, Thursday, Saturday and Sunday noon–8pm, Wednesday and Friday 10am–6pm.
The Armoury, Kremlin. Open: daily except Thursday, May–September 10am–6pm, October–April 10am–5pm.

REVOLUTION

Central Museum of the Revolution, 21 Tverskaya Street, tel: 299 5227. Open: Monday and Friday 11am–7pm., Tuesday, Thursday Saturday and Sunday 10am–6pm.
Central Lenin Museum, 2 Revolution Square, tel: 202 9304. Open: October–April Tuesday, Wednesday and Thursday 11am–7.30pm, Friday–Sunday 10am–6.30pm; May–September Tuesday, Thursday and Friday 10am–6pm, Wednesday 11am–7.30pm.
Krasnaya Presnya Historical Revolutionary Museum, 4 B. Predtechensky Lane, tel: 252 3035. Open: Wednesday and Friday–Sunday 10am–6pm, Tuesday and Thursday 11am–7pm.
Marx and Engels Museum, 5 Marx and Engels Street, tel: 203 0104. Open: Tuesday, Wednesday and Friday noon–7pm, Thursday, Saturday and Sunday 11am–6pm.
State History Museum, 1–2 Red Square, tel: 221 4311. Open: Monday, Thursday, Friday, Saturday and Sunday 10am–6pm, Wednesday 11am–7pm.

MILITARY-HISTORICAL

Museum of the Armed Forces, 2 Sovetskoy Armii Street, tel: 281 4877. Open: Tuesday, Friday, Saturday and Sunday noon–5pm, Wednesday and Thursday noon–7pm.
Frunze Aviation and Cosmonautics House, 4 Krasnoarmeyskaya Street, tel: 212 5461. Open: daily 10am–6pm.
Museum of the Defence of Moscow, 3 Michurinsky Prospect, tel: 427 6056. Open: Tuesday, Thursday, Saturday and Sunday noon–6pm, Wednesday and Friday noon–8pm.
Battle of Borodino Panorama Museum, 38 Kutuzovsky Prospekt, tel: 148 1927. Open: daily except Friday, 10.30am–4pm.

THEATRES

Bolshoi Opera and Ballet Theatre, 1 Teatralnaya Square, tel: 292 9986.
Theatre of Friendship of the People, 22 Tverskoy Boulevard, tel: 203 6222.
Moscow Academic Art Theatre, 3 Kamergersky Per., tel: 229 2546.
Art Theatre Branch, 3 Moskvin Street, tel: 229 9631.
Maly Theatre, 1/6 Teatralnaya Sq., tel: 925 9868.
Maly Theatre Branch, 69 Bolshaya Ordynka Street, tel: 237 2891.
Yevgeny Vakhtangov Drama Theatre, 26 Arbat Street, tel: 241 0744.
Mossoviet Theatre, 16 Bolshaya Sadovaya Street, tel: 299 2035.
Stanislavsky and Nemirovich-Danchenko Musical Theatre, 17 Pushkinskaya Street, tel: 229 4250.
Central Children's Theatre, 2 Teatralnaya Square, tel: 292 0069.
Children's Musical Theatre, 5 Vernadsky Prospekt, tel: 930 5177.
Gogol Drama Theatre, 8a Kazakov Street, tel: 261 5528.
Malaya Bronnaya Drama Theatre, 4 Malaya Bronnaya Street, tel: 290 0482.
New Drama Theatre, 2 Prokhodchikov Street, tel: 182 0347.
Pushkin Drama Theatre, 23 Tverskoy Blvd, tel: 203 4221.
Stanislavsky Drama Theatre, 23 Tverskaya Street, tel: 299 7621.
Chamber Yiddish Musical Theatre, 12 Taganskaya Square, tel: 272 4924.
Chamber Musical Theatre, 71 St Petersburg Highway, tel: 198 7204.
Central Puppet Theatre, 3 Sadovaya-Samotechnaya Street, tel: 299 6313.
Moscow Puppet Theatre, 26 Spartakovskaya Street, tel: 261 2197.
Lenin Komsomol Theatre, 6 Chekhov Street, tel: 2999668.
Mayakovsky Theatre, 19 Herzen Street, tel: 290 4232.
Mime Theatre, 39/4 Izmailovsky Blvd, tel: 163 8130.
Miniature Theatre, 3 Karetny Ryad Street, tel: 209 2076.
Satiricon Theatre, 2 Triumphalnaya Square, tel: 299 9042.
Romen Gypsy Theatre, 32 St Petersburg Highway, tel: 250 7334.
Circus, 7 Vernadsky Prospekt, tel: 930 2815.
Old Circus, 13 Tsvetnoy Boulevard, tel: 200 6889.

CONCERT HALLS

Conservatoire Grand Hall, 13 Herzen Street, tel: 229 7412.
Central Concert Hall (Hotel Rossiya), 1 Moskvoretskaya Embankment, tel: 203 9093.

Tchaikovsky Concert Hall, 20 Tverskaya Street, tel: 299 3487.
Gnesins Institute Concert Hall, 1 Paliashvili Street, tel: 290 6737.
Olympic Village Concert Hall, 1 Michurinsky Prospekt, tel: 437 5660.
Cathedral of the Sign, 8a Varvarka Street, tel: 298 3462.
Church of Vlasii, 20 Rylyev Street, tel: 241 2380.

PERMANENT EXHIBITIONS

Russian Exhibition of Economic Achievements, tel: 181 9162 (Metro VDNKh). Open: 10am–9pm.
Permanent Moscow Town Planning Exhibition, 6, 2nd Brestskaya Street, tel: 258 7724. Open: daily 11am–5pm.
Central Exhibition Hall, 1, 50th Anniversary of the October Revolution Square, tel: 202 9304. Open: 11am–8pm.
Exhibition Hall of the Union of Artists, 20 Kutsnetsky Most, tel: 228 1844. Open: Tuesday–Friday 1–8pm, Saturday and Sunday noon–7.30pm.
Central House of the Union of Artists, 10–14 Krymsky Val, tel: 238 9634. Open: 11am–8pm.

GALLERIES

To export any work of art you must have it appraised by the Export Commission, 29 Chekhova Street. Antiques (everything that was made before 1945 is regarded as antique), cannot be taken out of the country, nor can any book printed before 1977.

Gallery Art Moderne, 39 Bolshaya Ordynka, tel: 233 1551.
The First Gallery, 7 Strastnoy Blvd, tel: 299 0498.
Today Gallery, 35 Arbat, tel: 248 4976.
Malaya Gruzinskaya 28, 28 Malaya Gruzinskaya, tel: 253 5627.
Sadovniki Gallery, 35 Akademika Millionshchikova, tel: 112 1161.
Bulvar Yana Raisina 19/1, 19/1 Boulevard Yana Raisina, tel: 493 1467.
Mars Art Gallery, 32 Malaya Filyovskaya, tel: 144 9368.
Oktyabrskaya Ulitsa 28, 28 Oktyabrskaya Street, tel: 289 2491.
Gruelman Gallery, 2/6, 15 Dimitrova Street, tel: 238 8492.
Petrovskiye Linii Centre, 1/20 Petrovskiye Linii Street, tel: 200 5441.

CINEMAS

Cinemas open at 9 or 10am and show films all day long. Tickets cost between 30 and 70 kopecks and are half-price until 4pm.

Varshava, 1 Ganetski Square, tel: 156 4227.
Gorizont, 21/10 Komsomolsky Prospekt, tel: 245 3143.
Zaryadye, 1 Moskvoretskaya Embankment, tel: 298 5686.
Illusion (showing old films), 1/15 Kotelnycheskaya Embankment, tel: 227 4339
Cosmos, 109 Prospekt Mira, tel: 283 5636.
Circular Cinerama, Exhibition of Economic Achievements, tel: 181 9525.
MIR, 11 Tsvetnoy Boulevard, tel: 200 1695.
Moskva, 2 Mayakovsky Square, tel: 251 7222.
Rossiya, 2 Pushkinskaya Square, tel: 229 2111.
Khudozhestvenny, 14 Arbat Square, tel: 291 9624.

PARKS

Muscovites love both their circuses – the New and the Old. The **Old Circus**, built way back in the 1880s, has recently reopened its doors after several years of reconstruction. There are also seasonal circuses, which usually unfold their tents in **Gorky Park**, now of worldwide fame thanks to Martin Cruz Smith's bestselling novel.

The park is one of the city's largest (240 hectares/593 acres) and easily the most popular in town. It offers a unique blend of history, architecture, and nature. Stretching for 7 km/4 miles along the right bank of the Moskva River from Krymsky Bridge to the mouth of the Setun River, it is divided into three parts – the main part, Neskuchny Garden, and Sparrow Hills (called Lenin Hills since 1935).

The main part is an amusement area with open-air concert stages, discos, numerous rides and computer-game parlours patronised by young and old alike, ponds with row-boats for rent, and open-air sports facilities. You can also ride a horse (under the watchful eye of instructors from the Yamskoi Dvor Co-op), play ping-pong or tennis and jog through the park if you feel like it. Come winter, the ponds become a huge, 8-hectare/20-acre skating-rink. You can rent a pair of skates or skis.

Things get a little difficult when you grow hungry, even though the park has abundant restaurants and cafeterias. Don't count on lunch; the most you can hope for is an ice cream (much sweeter and with many more calories than Western Europeans are used to) and a Pepsi.

Neskuchny Garden and the **Sparrow Hills** were popular in the 19th century. The place, with its 150-year-old oaks, limes and maples, with its ruins of what in the 18th and the 19th centuries were estates of Moscow's aristocrats and with its ensemble of the former **Andreevsky Monastery**, has traditionally attracted the older generation, people who appreciate peace and quiet.

On Sparrow Hills (next to the university), there are several observation platforms which offer **spectacular views of Moscow** on a good day; these are much-favoured by tourists and newlyweds, who come here have their photos taken. When the weather is hot, the embankment turns into a crowded beach. In winter, the place crawls with skiing fanatics; there are trampoline jumpers and "walruses", as aficionados of winter-time bathing are known in Russia.

Those who are addicted to fishing will notice the lonely figures with rods in their hands. Putting it mildly, you must be a queer fish to fish these waters. It takes an untold quantity of patience to sit there for hours, waiting for a bite that never comes. As usual, there is more satisfaction to be had if you are willing to pay for it: there is a co-operative fisherman's paradise in mid-town Moscow, on the **Patriarch's Ponds** which Mikhail Bulgakov immortalised in his novel *Master and Margarita*.

Gorky Park: 9 Krymsky Val (Metro Park Kultury).
Izmailovo Park: 17 Narodny Prospekt (Metro Izmailovo Park).
Sokolniki Park: 1 Sokolnichesky Val (Metro Sokolniki).
Botanical Garden: 4 Botanicheskaya Street (Metro Shcherbakovskaya).
Druzhba Forest Park: 90 St Petersburg Highway (Metro Rechnoy Vokzal).
Serebryany Bor Forest Park: Serebryany Bor, 1st Line (Metro Polezhayevskaya).
Hermitage Garden: 3 Karetny Ryad Street (Metro Novoslobodskaya).

NIGHTLIFE

CASINOS

Back in the hopeless gloom of the Middle Ages, the Catholic Church made a revealing discovery: there was money to be made from man's age-old affinity for forbidden fruit. The Vatican's crafty strategists devised a kind of insurance against the nine circles of hell. This remarkable document, popular among medieval con artists, rapists and other assorted blackguards, was known as "the indulgence".

Fortunately, the ingenious brainchild of the Holy See did not get lost on history's dusty shelves: it gained a new lease on life in today's Moscow – with a vengeance. The state of things is perfectly described by the currently popular saying, "You must not, but if you want it badly enough, do it!" But – there is a catch, of course – only for dollars or any other freely convertible currency.

While running a casino is punishable by up to 5 years' imprisonment or exile (and gamblers caught on the premises risk heavy fines), the law applies only to rouble casinos. Lovers of excitement on green cloth easily acquire their "indulgences" for greenbacks; payment in bills bearing the stern faces of US presidents or members of the British Royal Family buys you the perfectly legal right to play roulette or blackjack in either of Moscow's two casinos (located in the **Leningrad Hotel** and the **Savoy**).

SHOPPING

OPENING TIMES

The absence of a variety of goods, and sometimes the absence of even the most basic goods, very often determines the opening hours of shops. If there are no goods, the shop need not open!

Many small shops have a one-hour lunchbreak sometime between noon and 5pm. Larger shops are continuously open Monday–Saturday from between 7am and 9am to 8pm and 9pm; certain food stores might be open until 10pm. Book shops and other specialty shops open around 10am and are open until 7pm or 8pm, with a break usually between 2pm and 3pm. On Sundays all shops, with the exception of some food stores, are closed.

SHOPS FOR FOREIGNERS

For foreigners who are horrified by the empty counters of Russian shops there are special shops where they can buy nearly everything for hard currency. There are now different chains of such shops in hotels, international airports and at certain points in the big cities called *Beriozka* (birch tree), *Sadko*, etc. The sales personnel in these shops usually speak English, French or German.

They offer Russian goods of comparatively good quality. Furs, glass, ceramics, vodka, Crimean and Georgian wines, and many goods that are hard to come by in other city shops are all available if paid for in hard currency.

Beriozka prices are based on the so called *invalyutny rouble* and correspond with prices in Western European countries, though they are lower for Russian goods than for imported ones. These hard currency shops accept cash, traveller's cheques as well as most credit cards.

SOUVENIRS

Inexpensive souvenirs, toys and other knick-knacks abound on Russian streets. Many co-operatives prefer not to rack their brain with such difficult matters as food production or the maintenance of computers. Instead they produce souvenirs, hair combs and belts. They also imitate the labels of designer jeans, so Russian youngsters can sew them on their pants to get an imitation feeling of freedom.

All this cheap stuff is sold on the streets, in the vicinity of department stores and in co-operative markets. Souvenirs abound also in special shops for foreigners, art-salons and curio shops. But beware of the problems waiting you at customs. The danger of confiscation is quite real. For all goods bought in a Russian shop, including those you bought at "Beriozka", you must keep the receipt to show that you paid in hard currency and that the goods are not antique. According to the new Russian regulations antiquities and art works may not be exported.

Russkiy Uzory (embroideries etc.), 16, Petrovka Street, tel: 923 1883. Open: 11am–7pm.
Russky Souvenir (applied art), 9 Kutuzovsky Prospekt, tel: 243 6986. Open: 11am–8pm. Closed Sunday.
Russky Lyon (linen goods), 29 Komsomolsky Prospekt, tel: 242 5925.
Podarky, 29 Vozdvishenka Street (formerly Kalinin Prospekt), tel: 203 1111 and 37. **Tverskaya**, (former Gorky) Street, tel: 200 5704. Open: 9am–8pm.
Art Salon, (prints, pictures, etc.), 6 Ukrainsky Boulevard, tel: 243 6357 and 24 Kutyzovsky Prospekt, tel: 249 1242. Open: 10am–7pm.

JEWELLERY

Beriozka, 12 Tverskaya (formerly Gorky) Street, tel: 229 9785. Open: 9am–8pm.
Malakhitovaya Shkatulka, 24 Vozdvishenka (formerly Kalinin Prospekt), tel: 291 2272.
Yantar (amber), 14 Gruzinsky Val Street, tel: 151 4430. Open: 10am–7pm.
Agat (agate), 16–18 Bolshaya Kolkhoznaya Square, tel: 207 2200. Open: 10am–7pm.
Almaz (diamonds), 14 Stoleshnikov Lane, tel: 925 6666. Open: 10am–7pm.
Samotsvety (semi-precious stones), 35 Arbat Street, tel: 241 0765. Open: 9am–8pm.
Rubin (rubies), 78 St Petersburg Prospekt, tel: 151 4306. Open: 11am–8pm.
Izumrud (emeralds), 23 Lomonosov Prospekt, tel: 130 3272. Open: 10am–7pm.

DEPARTMENT STORES

GUM, 3 Red Square, tel: 921 5763. Open: 8am–9pm.
TsUM, 2 Petrovka Street, tel: 292 1157. Open: 8am–9pm.
Moskva Department Store, 54 Lenin Prospekt, tel: 137 0018. Open: 8am–9pm.
Detsky Mir, 2 Okhotny Ryad, tel: 927 2007. Open: 8am–9pm.

BOOKSHOPS

The following bookshops should all have foreign-language books available.
Druzhba, 15 Tverskaya, tel: 229 5383. Open daily: 11am–8pm.
House of Books, 26 Vozdvishenka, tel: 290 4507. Open: 10am–8pm.
Progress, 17 Zubovsky Boulevard, tel: 246 9976,. Open: 10.30am–7.30pm.
Planeta, 8–10 Vesnin Street, tel: 241 9587. Open: 10am–7pm.

BERIOZKA (HARD CURRENCY SHOPS)

Hotel Mezhdunarodnaya, 12 Krasnopresnenskaya Embankment, tel: 253 8152. Open: 9am–8pm.
Hotel Rossiya, East Building, 1 Moskvoretskaya Embankment, tel: 298 5715. Open: 9am–8pm.
Hotel Salyut, 158 Leninsky Prospekt, tel: 438 1671. Open: 9am–8pm.
Tsentralny Dom Turista, 146 Lenin Prospekt, tel: 438 3444, open: 9am–8pm; 25a Luzhnetsky Proyezd, tel: 246 5270, open: 9am–8pm; 9 Kutuzovsky Prospekt, tel: 243 3730. Open: 9am–6pm; 16 Bolshaya Dorogomilovskaya Street, tel: 243 1016, Open: 10am–7pm.

SPECIAL ITEMS

Rifle Jeans Store, 10 Kuznetsky Most. Open: Monday–Saturday 10am–7pm. Hard currency only.
Fur Kommissioni, 30 Pushkinskaya Street. Open: Monday–Saturday 9am–8pm.
Poster Shop, 4 Arbat Open: Monday–Saturday 10am–7pm.
Record Store Sovetskaya Muzika, 4 Sadovaya-Triumfalnya. Open: Monday–Saturday 9am–7pm.

FARMERS' MARKETS

Baumansky, 47/1 Baumanskaya Street, tel: 265 2335.
Bolshoi Kolkhorzny, 8 Kukhrikov Lane, tel: 240 0355.
Danilovsky, 78 Mytnaya Street, tel: 239 2641.
St Petersburgsky, 11 Chasovayas Street, tel: 151 7871.
Tsentralny, 15 Tsvetnoy Boulevard, tel: 200 0166.
Yaroslavsky, 122 Prospekt Mira, tel: 283 3228.

FOOD

Stockmann's, 4–8 Zatsepsky Val Street, tel: 233 2606. This is where the Western community in Moscow gets its food. Western goods, groceries, newspapers, toiletries come by truck from Finland every Wednesday and Friday. Open: daily 10am–8pm. Credit cards only.
Sadko, 16 Bolshaya Dorogomilovskaya Street, tel: 243 6601. Foreign food items, long queues, hard currency only. Open: Monday–Saturday 10am–8pm, Sunday 10am–6pm.

Food Beriozka, Hotel Mezhdunarodnaya, 12 Krasnopresnenskaya Naberezhnaya. A well-stocked supermarket selling local and imported goods, including meat, fruit, beer and delicatessen. Hard currency only. Open: Monday–Saturday 10am–6pm.
Morozko shops, 2 L. Tolstogo Street. Frozen food. Open: Monday–Saturday 9am–7pm.

SPORTS

Even though Moscow hosts up to 2,000 assorted athletic competitions each year, sports lovers – both foreign and local – have never had it easy here. Most sports facilities are departmental property and therefore off-limits to all but the most persistent. It is easier to take a swim in the pool of your Intourist hotel, or play tennis on the open courts in Luzhniki, or, if you prefer, to jog in one of the parks. Meanwhile, the "gold rush" for foreign currency that is presently sweeping across the country has penetrated the minds of sports-establishment managers, who compete among themselves for foreign clientele from embassies and trade missions. There is hope that, as hard-currency services grow in volume, sports establishments will soon start catering to individual clients.

SPECTATOR

All the big Moscow sports complexes such as the Dynamo Stadium, St Petersburg Highway 36 (tel: 212 7092) and the Palace of Sports, Luzhniki (tel: 201 0955), host sports events ranging across the board, from soccer to athletics. Check with the hotel service counter. Information about upcoming international events can be obtained through Goskomsport, tel: 201 1535.

PARTICIPANT

Bowling: Both the Cosmos Hotel and the Mezhdunarodnaya Hotel have bowling alleys, but you must pay in hard currency.

Cross-country skiing: At the larger Moscow parks: Izmailovsky, Sokolniki and Gorky. For downhill skiing and sledging you should go to the Lenin Hills.

Fitness: The Chaika Sports Complex on Kropotinskaya Naberezhnaya 3–5, next to the Moskva Swimming Pool has a gym, a sauna and tennis courts. All the Moscow parks are ideal for jogging: Gorky Park, Krimsky Val 9; Izmailovsky Park, Narodny Prospekt 17; Sokolniki Park, Sokolnichesky Val 1 and the Botanical Gardens, Botanicheskata Street 4. The Cosmos, the Rossiya and the Mezhdunarodnaya Hotels also have swimming pools, saunas and gyms for their guests.

Golf: Tambo Golf Club, Dovzhenko Street 1, tel: 147 6254.

Horse-back riding: There is a riding co-operative, Yamskoe Dvor, at Lenin Prospekt 35, tel: 135 8255. Otherwise there is the Hippodrome at Bogovaya Street 22, the Urozhay Riding Centre at Sokolriki Park and the Equestrian Sports Centre in the Bitsevo Forest Park.

Skating: In Gorky Park, at the Young Pioneer Stadium (Metro Dynamo), on the ice-rink on the St Petersburg Highway and on the Patriarch's Pond. If you need skates you might find them in the Detsky Mir store.

Squash: The Indian and the American Embassies have courts open to members and their friends. For a US$30 fee you can become a member at the Indian squash club.

Swimming: The Cosmos Hotel has an indoor swimming pool open to non-residents. At the Basin Moskva Swimming Pool you can enjoy swimming outdoors in a heated pool even when the temperatures around you are far below freezing point.

Weight lifting: Beneath the Lenin Stadium (Metro Sportivnaya).

PHOTOGRAPHY

The diplomat services agency (UPDK) and some newly appearing co-operative laboratories develop Agfa, Kodak and Fuji films (E6 process). Ask at your hotel service counter or at the photography shops at 61 Liteiny Prospekt or 92 Nevsky Prospekt. Some Beriozka shops sell Ektachrome and Fuji films. Kodachrome, which needs a special development process, is not officially available.

You should not take photographs of military installations or from aircraft – nor, to be on the safe side, should you take them from a train. The interpretation of what constitutes a military installation rests with the officials. You will have to be cautious and, if possible, ask your guide or interpreter before you take a picture of a bridge, a railway station, an airport or anything else that might be assessed as a special security object.

Local film types, such as ORWO NC-21 or DC 4 colour film, can only be developed in Russia. If you want film developed in Moscow try the film shop on Arbat, 43. The processing services offered at official shops, such as the Zenith at the Metro Sokolniki, usually take too long for short-stay visitors.

LANGUAGE

Russian is one of the 130 languages used by the peoples of the former USSR. It is the mother tongue of some 150 million Russians and the state language of the Russian Federation (RSFSR). Most Russian citizens understand Russian.

From a linguistic point of view, Russian belongs to the Slavonic branch of the Indo-European family of languages; English, German, French, Spanish and Hindi are its relatives.

Historically Russian is a comparatively young language. The appearance of the language in its present shape, based on the spoken language of the Eastern Slavs and the Church-Slavonic written language, is attributed to the 11th to 14th centuries.

Modern Russian has absorbed numerous foreign words and they form a considerable group within the Russian vocabulary. Very few tourists will be puzzled by Russian words like telefon, televizor, teatr, otel, restoran, kafe, taxi, metro or aeroport.

The thing that usually intimidates people on their first encounter with Russian is the alphabet. In fact it is easy to come to terms with after a little practice, and the effort is worthwhile if you want to make out the names of streets and shop signs.

The Russian (or Cyrillic) alphabet was created by two brothers, philosophers and public figures, Constantine (St Cyril) and Methodius; both were born in Solun (now Thessaloniki in Greece). Their purpose was to facilitate the spread of Greek liturgical books in Slavonic speaking countries. Today the Cyrillic alphabet, with different modifications, is used in the Ukrainian, Byelorussian, Bulgarian, Serbian and in some other languages.

It is important that you reproduce the accent (marked here with the sign ’ before each stressed vowel) correctly to be understood well.

ALPHABET

printed letter		sounds, as in	Russian name of letter
А	**а**	a, archaeology	a
Б	**б**	b, buddy	be
В	**в**	v, vow	v
Г	**г**	g, glad	ge
Д	**д**	d, dot (the tip of the tongue close to the teeth, not the alveoli)	de
Е	**е**	e, get	ye
Ё	**ё**	yo, yoke	yo
Ж	**ж**	zh, composure	zhe
З	**з**	z, zest	ze
И	**и**	i, ink	i
Й	**й**	j, yes	jot
К	**к**	k, kind	ka
Л	**л**	l, life (but a bit harder)	el’
М	**м**	m, memory	em
Н	**н**	n, nut	en
О	**о**	o, optimum	o
П	**п**	p, party	pe
Р	**р**	r (rumbling, as in Italian, the tip of the tongue is vibrating)	er
С	**с**	s, sound	es
Т	**т**	t, title (the tip of the tongue close to the teeth, not the alveoli)	te
У	**у**	u, nook	u
Ф	**ф**	f, flower	ef
Х	**х**	kh, hawk	ha
Ц	**ц**	ts (pronounced conjointly)	tse
Ч	**ч**	ch, charter	che
Ш	**ш**	sh, shy	sha
Щ	**щ**	shch (pronounced shcha conjointly)	
ъ		(the hard sign)	
Ы	**ы**	y (pronounced with the same position of a tongue as when pronouncing G,K)	y
		(the soft sign)	
Э	**э**	e, ensign	e
Ю	**ю**	yu, you	yu
Я	**я**	ya, yard	ya

NUMBERS

1	adín	**один**
2	dva	**два**
3	tri	**три**
4	chityri	**четыре**
5	pyat’	**пять**
6	shes’t’	**шесть**
7	sem	**семь**

8	vósim	**восемь**
9	d'évit'	**девять**
10	d'ésit'	**десять**
11	adínatsat'	**одиннадцать**
12	dvinátsat'	**двенадцать**
13	trinátsat'	**тринадцать**
14	chityrnatsat'	**четырнадцать**
15	pitnátsat'	**пятнадцать**
16	shysnátsat'	**шестнадцать**
17	simnátsat'	**семнадцать**
18	vasimnátsat'	**восемнадцать**
19	divitnátsat'	**девятнадцать**
20	dvátsat'	**двадцать**
21	dvatsat' adin	**двадцать один**
30	trítsat'	**тридцать**
40	sórak	**сорок**
50	pidisyat	**пятьдесят**
60	shyz'disyat	**шестьдесят**
70	s'émdisyat	**семьдесят**
80	vósimdisyat	**восемьдесят**
90	divinósta	**девяносто**
100	sto	**сто**
101	sto adin	**сто один**
200	dv'és'ti	**двести**
300	trísta	**триста**
400	chityrista	**четыреста**
500	pitsót	**пятьсот**
600	shyssót	**шестьсот**
700	simsót	**семьсот**
800	vasimsót	**восемьсот**
900	divitsót	**девятьсот**
1,000	tysicha	**тысяча**
2,000	dve tysichi	**две тысяч и**
10,000	d'ésit' tysich	**десять тысяч**
100,000	sto tysich	**сто тысяч**
1,000,000	milión	**миллион**
1,000,000,000	miliárd	**миллиард**

PRONOUNS

I/We
ya/my
я/мы

You
ty (singular, informal)
vy (plural, or formal singular)
ты /вы

He/She/They
on/aná/aní
он/она/они

My/Mine
moj (object masculine)
mayá (object feminine)
mayó (neutral or without marking the gender)
maí (plural)
мой/моя/моё/мои

Our/Ours
nash/násha/náshe/náshy (resp.)
наш/наша/наше/наши

Your/Yours
tvoj etc. (see My)
vash etc. (see Our)
твой/ваш

His/Her, Hers/Their, Theirs
jivó/jiyó/ikh
его/её/их

Who?
khto?
Кто?

What?
shto?
Что?

FORMS OF GREETINGS

Forms of Address: Modern Russian has no established and universally used forms of address. The old revolutionary form *tavárishch* (comrade), still used amongst some party members, lacks popularity with the rest of the population.

One way is to say: *Izviníte, skazhíte pozhálsta...* (Excuse me, tell me, please...) or *Izvinite, mózhna sprasít'...* (Excuse me, can I ask you...).

If you want to look original and to show your penetration into the depths of history of courteous forms, you can appeal to the man *súdar'* (sir), and to the woman *sudárynya* (madam). Many people want to restore these pre-revolutionary forms of address in modern Russian society. If you know the name of the father of the person you talk to, the best and the most neutral way is to use these both when addressing him (her): "Mikhál Sirgéich" to Mr Gorbachev and "Raísa Maxímavna" to his spouse.

In business circles you can use forms *gaspadín* to a man and *gaspazhá* to a woman. The English forms of address "Mister" or "Sir" are also acceptable.

You can hear common parlance forms *Maladói chilavék!* (Young man!) and *Dévushka!* (Girl!) to a person of any age and also *Zhénshchina!* (Woman!) to women on the bus, in a shop or at the market. These forms should be avoided in conversation.

Hello!
zdrástvuti (neutral, often accompanied by shaking hands, but it is not necessary)
Здравствуйте!

alo! (by telephone only)
Алло!

zdrástvuj (to one person, informal)
Здравствуй!

priv'ét! (informal)
Привет!

Good afternoon/Good evening
dóbry den'/dobry véchir
Добрый день/Добрый вечер

Good morning/Good night
dobrae útra/dobraj nóchi (= Sleep well)
Доброе утро/Доброй ночи

Good bye
dasvidán'ye (neutral)
До свиданья

chao! (informal)
Чао!

paká! (informal, literally means "until")
Пока!

Good luck to you!
shchislíva!
Счастливо!

What is your name?
kak vas (tibya) zavút?/kak váshe ímya ótchistva? (the second is formal)
Как вас (тебя) зовут?/Как ваще имя и отчество?

My name is... /I am...
minya zavut... /ya...
Меня зовут... /Я...

It's a pleasure
óchin' priyatna
Очень приятно

Good/Excellent
kharashó/privaskhódna
хорощо/отлично

Do you speak English?
vy gavaríti pa anglíski?
Вы говорите по-английски?
I don't understand/I didn't understand
ya ni panimáyu/ya ni pónyal
Я не понимаю/Я не понял

Repeat, please
pavtaríti pazhálsta
Повторите, пожалуйста

What do you call this?
kak vy éta nazyváiti?
Как вы это называете?

How do you say...?
kak vy gavaríti...?
Как вы говорите...?

Please/Thank you (very much)
pazhálsta/(bal'shóe) spasíba
Пожалуйста/(Больщое) спасибо

Excuse me
izviníti
Извините

GETTING AROUND

Where is the...?
gd'e (nakhóditsa)...?
Где находится...?

beach
plyazh
...пляж

bathroom
vánnaya
...ванная

bus station
aftóbusnaya stántsyja/aftavakzál
...автобусная станция/автовокзал

bus stop
astanófka aftóbusa
...остановка автобуса

airport
airapórt
...аэропорт

railway station
vakzál/stántsyja (in small towns)
...вокзал/станция

post office
póchta
...почта

police station
...milítsyja
...милиция

ticket office
bil'étnaya kássa
...билетная касса

marketplace
rynak/bazár
...рынок/базар

embassy/consulate
pasól'stva/kónsul'stva
...посольство/консульство

Where is there a...?
gd'e z'd'es'...?
Где здесь...?

currency exchange
abm'én val'úty
...обмен валюты

pharmacy
apt'éka
...аптека

(good) hotel
(kharóshyj) atél'/(kharoshaya) gastínitsa
...(хороший) отель (хорошая) гостиница

restaurant
ristarán
...ресторан

bar
bar
...бар

taxi stand
stayanka taxí
...стоянка такси

subway station
mitró
...метро

service station
aftazaprávachnaya stantsyja/aftasárvis
...автозаправочная станция

newsstand
gaz'étnyj kiósk
...газетный киоск

public telephone
tilifón
...телефон

hard currency shop
val'útnyj magazín
...валютный магазин

supermarket
univirsám
...универсам

department store
univirmák
...универмаг

hairdresser
parikmákhirskaya
...парикмахерская

jeweller
yuvilírnyj magazin
...ювелирный магазин

hospital
bal'nítsa
...больница

Do you have...?
u vas jes't'...?
У вас ес...

I (don't) want...
ya (ni) khachyu...
Я (не) хо чу...

I want to buy...
ya khachyu kupít'...
Я хочу купить...

Where can I buy...?
gd'e ya magú kupít'...?
Где я могу купить...?

cigarettes
sigaréty
...сигареты

wine
vinó
...вино

film
fotoplyonku
...фотоплёнку

a ticket for...
bilét na...
...билет на...

this
éta
...это

postcards/envelopes
atkrytki/kanv'érty
...открытки/конверты

a pen/a pencil
rúchku/karandásh
...ручку/карандаш

soap/shampoo
myla/shampún'
...мыло/шампунь

aspirin
aspirín
...аспирин

I need...
mn'e núzhna...
Мне нужно...

I need a doctor/a mechanic
mn'e núzhyn dóktar/aftamikhánik
Мне нужен доктор/автомеханик

I need help
mn'e nuzhná pómashch'
Мне нужна помощь

Car/Plane/Train/Ship
mashyna/samal'yot/póist/karábl'
маъшина/самолёт/поезд/корабль

A ticket to...
bil'ét do...
билет до...

How can I get to...
kak ya magu dabrátsa do...
Как я могу добраться до...

Please, take me to...
pazhalsta atvizíti minya...
Пожалуйста, отвезите меня...

What is this place called?
kak nazyváitsa eta m'ésta?
Как называется это место?

Where are we?
gd'e my?
Где мы?

Stop here
astanavíti z'd'es'
Остановите здесь

Please wait
padazhdíti pazhalsta
Подождите, пожалуйста

When does the train [plane] leave?
kagdá atpravl'yaitsa poist [samalyot]?
Когда отправляется поезд (самолёт)?

I want to check my luggage
ya khachyu prav'érit' bagázh
Я хочу проверить багаж

Where does this bus go?
kudá id'yot état aftóbus?
Куда идёт этот автобус?

SHOPPING

How much does it cost?
skól'ka eta stóit?
Сколько это стоит?

That's very expensive
eta óchin' dóraga
Это о чень дорого

A lot, many/A little, few
mnóga/mála
много/мало

It (doesn't) fits me
eta mn'e (ni) padkhódit
Это мне (не) подходит

AT THE HOTEL

I have a reservation
u minya zakázana m'esta
У меня заказана комната

I want to make a reservation
ya khachyu zakazát' m'esta
Я хочу заказать место

A single (double) room
adnam'éstnuyu (dvukhmestnuyu) kómnatu
одноместную (двухместную) комнату

I want to see the room
ya khachyu pasmatrét' nómer
Я хо чу посмотреть номер

Key/Suitcase/Bag
klyuch/chimadán/súmka
ключ /чемодан/сумка

AT THE RESTAURANT

Waiter/Menu
afitsyánt/minyu
официант/меню

I want to order...
ya khachyu zakazat'...
Я хочу заказать

breakfast/lunch/supper
záftrak/ab'ét/úzhyn
завтрак/обед/ужин

the house specialty
fírminnaya blyuda
фирменное блюдо

mineral water/juice
minirál'naya vadá/sok
минерал'ьная вода/сок

coffee/tea/beer
kófe/chai/píva
кофе/ чай/пиво

What do you have to drink (alcoholic)?
shto u vas jes't' vypit'?
Что у вас есть выпить?

Ice/Fruit/Dessert
marózhynaya/frúkty/disért
можо еное/фрукты/десссерт

Salt/Pepper/Sugar
sol'/périts/sákhar
соль/перец/сахар

Beef/Pork/Chicken/Fish/Shrimp
gavyadina/svinína/kúritsa/ryba/kriv'étki
говядина/свинина/курица/рыба/креветки

Vegetables/Rice/Potatoes
óvashchi/ris/kartófil'
овощи/рис/картофель

Bread/Butter/Eggs
khleb/másla/yajtsa
хлеб/масло/яйца

Soup/Salad/Sandwich/Pizza
sup/salát/butyrbrót/pitsa
суп/салат/бутерброд/пицца

a plate/a glass/a cup/a napkin
tar'élka/stakán/cháshka/salf'étka
тарелка/стакан/чашка/салфетка

The bill, please
shchyot pazhalsta
Счёт, пожалуйста

Well done/Not so good
fkúsna/ták sibe
вкусно/так себе

I want my change, please
zdáchu pazhalsta
Сдачу, пожалуйста

MONEY

I want to exchange currency (money)
ya khachyu abmin'át' val'yutu (d'én'gi)
Я хочу обменять валюту (деньги)

Do you accept credit cards?
vy prinimáiti kridítnyi kártachki?
Вы принимаете кредитные карточки ?

Can you cash a traveller's cheque?
vy mózhyti razminyat' darózhnyj chek?
Вы можете разменять дорожный чек?

What is the exchange rate?
kakój kurs?
Какой курс?

TIME

What time is it?
katóryj chas?
Который час?

Just a moment, please
adnú minútachku
Одну минуточку

How long does it take?
skól'ka vrémini eta zanimáit?
Сколько времени это занимает?

Hour/day/week/month
chas/den'/nid'élya/m'ésits
час/день/неделя/месяц

At what time?
f kakóe vrémya?
В какое время?

At 1:00/at 8am/at 6pm
f chas/ v vósim utrá/f shés't' chisóf v'échira
в час/в восемь утра/в шесть часов вечера

This (last, next) week
eta (próshlaya, sl'édujshchiya) nid'elya
эта (прошлая, следующая) неделя

Yesterday/Today/Tomorrow
fchirá/sivód'nya/záftra
вчера/сегодня/завтра

Sunday
vaskris'én'je
воскресенье

Monday
panid'él'nik
понедельник

Tuesday
ftórnik
вторник

Wednesday
sridá
среда

Thursday
chitv'érk
четверг

Friday
pyatnitsa
пятница

Saturday
subóta
суббота

The weekend
vykhadnyi dni
выходные дни

SIGNS & INSCRIPTIONS

вход/выход/входа нет
fkhot/vykhat/fkhóda n'et
Entrance/Exit/No Entrance

туалет/уборная
tual'ét/ubórnaya
Toilet/Lavatory

Ж (З) / М (М)
dlya zhén'shchin/dlya mushchín
Ladies/Gentlemen

зал ожидания
zal azhidán'ya
Waiting hall

занято/свободно
zánita/svabódna
Occupied/Free

касса
kassa
Booking office/cash desk

медпункт
mitpúnt
Medical Services

справочное бюро
správachnae bzuro
Information

вода для питья
vadá dlya pit'ya
Drinking Water

вокзал
vakzál
Terminal/Railway station

открыто/закрыто
atkryta/zakryta
Open/Closed

запрещается/опасно
zaprishchyaitsa/apásna
Prohibited/Danger

продукты/гастроном
pradúkty/gastranóm
Grocery

булочная/кондитерская
búlachnaya/kan'dítirskaya
Bakery/Confectionery

закусочная/столовая
zakúsachnaya/stalóvaya
Refreshment room/Canteen

самообслуживание
samaapslúzhivan'je
Self-service

баня/прачечная/химчистка
bánya/práchichnaya/khimchístka
Bath-House/Laundry/Chemical Cleaning

книги/культтовары
knígi/kul'taváry
Books/Stationery

мясо/птица
m'ása/ptítsa
Meat/Poultry

обувь
óbuf'
Shoe-Store

овощи/фрукты
óvashchi/frúkty
Green-Grocery/Fruits

универмаг/универсам
univirmák/univirsám
Department Store/Supermarket

ткани/цветы
tkani/tsvity
Fabrics/Flowers

TRANSLATION SERVICES

Translation services are available through Intourist or one of the local co-operative translation bureaus: Inlingua, 2/1 Semyonovskaya Nab., tel: 360 0874 or Interpret, 1st Kadashevsky Per., tel: 231 1020.

FURTHER READING

HISTORY

Catherine the Great, by J.T. Alexander. Oxford University Press, 1989.
Stalin, Man of Contradiction, by K.N. Cameron. Strong Oak Press, 1989.
History of Soviet Russia, by E.H. Carr. Pelican, 3 vols, first published 1953.
A History of the Soviet Union, by G. Hosking. Fontana/Collins, 1990.
The Making of Modern Russia, by L. Kochan and R. Abraham. Penguin, 1983.

POLITICS

Voices of Glasnost, by S. Cohen and K. van den Heuvel. Norton, 1989.
The Other Russia, by Michael Glenny and Norman Stone. Faber & Faber, 1990.
Perestroika, by M.S. Gorbachev. Fontana, 1987.
Towards a Better World, by M.S. Gorbachev. Richardson and Steirman, 1987.
Soviet Union: Politics, Economics and Society, by R.J. Hill. Pinter Publishers, 1989.
Glasnost in Action, by A. Nove. Unwin Hyman, 1989.
Against the Grain, by Boris Yeltsin. Jonathan Cape, 1990.

BIOGRAPHY/MEMOIRS

The Making of Andrei Sakharov, by G. Bailey. Penguin, 1990.
Alone Together, by Elena Bonner. Collins Harvill, 1986.
An English Lady at the Court of Catherine the Great, ed. by A.G. Gross. Crest Publications, 1989.
On the Estate: Memoirs of Russia Before the Revolution, ed. by Olga Davydoff Bax. Thames & Hudson, 1986.
Into the Whirlwind Within a Whirlwind, by Eugenia Ginzburg. Collins Harvill, 1989.
In the Beginning, by Irina Ratushinskaya. Hodder & Stoughton, 1990.
Ten Days that Shook the World, by John Reed. Penguin, first published 1919.
The Gulag Archipelago, by Alexander Solzhenitsyn. Collins Harvill, 1988.
Russia: Despatches from the Guardian Correspondent in Moscow, by Martin Walker. Abacus, 1989.

ART

A History of Russian Painting, by A. Bird. Phaidon, 1987.
Russian Art of the Avant Garde, by J.E. Bowlt. Thames & Hudson, 1988.
New Worlds: Russian Art and Society 1900–37, by D. Elliot. Thames & Hudson, 1986.
The Kremlin and its Treasures, by Rodimzeva, Rachmanov and Raimann. Phaidon, 1989.
Russian Art from Neoclassicism to the Avant Garde, by D.V. Sarabianov. Thames & Hudson, 1990.
Street Art of the Revolution, by V. Tolstoy, I. Bibikova and C. Cooke. Thames & Hudson, 1990.
The Art of Central Asia. Aurora Art Publishers, 1988.
Folk Art in the Soviet Union. Abrams/Aurora, 1990.
The Hermitage. Aurora, 1987.
Masterworks of Russian Painting in Soviet Museums. Aurora, 1989.

TRAVEL, GEOGRAPHY & NATURAL HISTORY

First Russia, Then Tibet, by Robert Byron. Penguin, first published 1905.
Caucasian Journey, by Negley Farson. Penguin, first published 1951.
Sailing to St Petersburg, by R. Foxall. Grafton, 1990.
The Natural History of the USSR, by Algirdas Kynstautas. Century Hutchinson, 1987.
Portrait of the Soviet Union, by Fitzroy Maclean. Weidenfeld and Nicolson, 1988.
Atlas of Russia and the Soviet Union, by R. Millner-Gulland with N. Dejevsky. Phaidon, 1989.
The Big Red Train, by Eric Newby. Picador, 1989.
The USSR: From an Original Idea by Karl Marx. Faber & Faber, 1983.
Journey into Russia, by Laurens van der Post. Penguin, first published 1964.
Among the Russians, by Colin Thubron. Penguin, first published 1983.
Ustinov in Russia, by Peter Ustinov. Michael O Mara Books, 1987.
Motorist's Guide to the Soviet Union, Pergamon Publishers, 1987.
The Nature of the Soviet Union: Landscapes, Flora and Fauna. Mokslas Publishing 1987.
Russia. Bracken Books, 1989.
USSR: The Economist Guide. Hutchinson Business Books, 1990.

LITERATURE

The Brothers Karamazov; The Idiot, by Fyodor Dostoevsky.
Doctor Zhivago, by Boris Pasternak.
Children of the Arbat, by Anatoli Rybakov. Hutchinson, 1988.
And Quiet Flows the Don; The Don Flows Home to the Sea, by Mikhail Sholokov.
War and Peace; Anna Karenina, by Leo Tolstoy.

USEFUL ADDRESSES

RUSSIAN MISSIONS ABROAD

Argentina, 1741 Rodriges Penya, Buenos Aires. Tel: 421 552.
Australia, Griffis, 70 Canberra Ave, Canberra. Tel: 956 6408.
Austria, 45–47 Reisnerstrasse, Vienna. Tel: 721 229.
Belgium, 66 Avenue de Fre, 1180 Bruxelles. Tel: 373 3569, 374 3406.
Canada, 285 Sharlotta Street, Ottawa. Tel: 235 4341.
Denmark, 3–5 Christianiagade, Copenhagen. Tel: 125 585.
Finland, 6 Tehtaankatu, Helsinki. Tel: 661 876.
France, 16 Boulevard Lann 40/50, Paris. Tel: 450 40550.
Germany, *Embassy:* 2 Waldstrasse 42, 5300 Bonn. Tel: 312 086.
Consulate: 76 Am Feenteich, 2000 Hamburg. Tel: 229 5301.
Greece, 28 Nikiforu Litra Street, Paleo Psyhico, Athens. Tel: 672 6130, 672 5235.
India, Shantipath Street, Chanakiapury, Delhi. Tel: 606 026.
Ireland, 186 Orwell Road, Dublin. Tel: 975 748.
Italy, 5 Via Gaeta, Rome. Tel: 494 1681.
Japan, Minato-ku, Adzabu-dai 2-1-1, T-106, Tokyo. Tel: 583 4224.
Netherlands, 2 Andries Bickerweh, The Hague. Tel: 345 1300.
New Zealand, Carory, 57 Messines Road, Wellington. Tel: 766 113.
Norway, 2 Dramensveien 74, Oslo. Tel: 553 278.
Singapore, 51 Nassim Road, Singapore 1025. Tel: 235 1834.
Spain, 6 & 14 Maestro Ripol, Madrid. Tel: 411 0706, 262 2264.
Sweden, 31 Ervelsgatan, Stockholm. Tel: 813 0440.
Switzerland, 37 Brunnadenrein 3006, Bern. Tel: 440 566.
Thailand, 108 Sathorn Nua, Bangkok. Tel: 258 0628.
Turkey, Caryagdy, Soc. 5, Ankara. Tel: 139 2122.
United Kingdom, 5, 13 & 18 Kensington Palace Gardens, London. Tel: (71) 229 3628.
USA, *Embassy:* 1125 16th Street, 20036 Washington DC. Tel: 628 7551, 628 8548, 628 6412.
Consulate: 2790 Green Street, San Francisco. Tel: 922 6644.

CHAMBERS OF COMMERCE

V/O Expocentr, 1a Sokolnichesky Val. Tel: 268 7083.
V/O Sovincentr, 12 Krasnopresnenskaya Embankment. Tel: 256 6303.
International Trade Centre, 1–2 Mezhdunarodnaya. Tel: 923 4323.
British-Russian Chamber of Commerce, 1904 World Trade Centre, 12 Krasnopresnenskaya Embankment. Tel: 253 2554.
American-Russian Trade and Economic Council, 3 Shevchenko Embankment. Tel: 243 5470.
Franco-Russian CoC, 4/17 (3rd floor) Pokrovsky Boulevard. Tel: 297 9092.
Italian-Russian CoC, 7 Vesnin Street. Tel: 241 6517.
Finnish-Russian CoC, 4/17 (2nd floor) Pokrovsky Boulevard. Tel: 294 2032.

TRAVEL AGENTS

UNITED STATES

Four Winds Travel, 175 Fifth Ave, New York, NY 10010.
Lindblad Travel Inc., 1 Sylvan Rd North, Westport, CT 06880.
Russian Travel Bureau Inc, 245 E. 44th St, New York, NY 10017.

UNITED KINGDOM

American Express Co. Inc, 6 Haymarket, London SW1.
Voyages Jules Verne, 10 Glentworth St, London NW1 5 PG.
P & O Holidays, 77 New Oxford St, London WC1.
London Walkabout Club, 20–22 Craven Terrace, Lancaster Gate, London W2.

GERMANY

Hansa Tourist, Hamburger Strasse 132, 2000 Hamburg 76.
Lindex Reisen, Rauchstrasse 5, 8000 München.
Intratours, Eiserne Hand 19, 6000 Frankfurt 1.
GeBeCo-Reisen, Eckernförder Strasse 93, 2300 Kiel.

SINGAPORE

Folke von Knobloch, 126 Telok Ayer, #02-01 Gat House, Singapore.

ART/PHOTO CREDITS

Photography by

Page 58, 175	**APN**
74, 75, 76, 77, 184	**Associated Press/Topham**
46	**Bogdanovich**
141L, 141R, 154/155, 168	**Bodo Bondzio**
33, 51, 165, 237	**Contrast**
63, 66, 159, 196, 200	**Duclos/Contrast**
65, 185, 213, 239, 249	**Hires/Contrast**
9, 61, 68, 70, 116, 179, 190, 197	**Jasmin/Contrast**
56/57, 123	**Poderni/Contrast**
55, 147, 152, 172, 180, 234	**Shone/Contrast**
122	**Shone-Blanche/Contrast**
Cover, 3, 14/15, 16/17, 18/19, 20/21, 60, 64, 67, 72, 94/95, 96, 97, 98, 99, 100, 102, 106/107, 108/109 110/111, 112/113, 114, 115, 130, 131, 132, 134/135, 138, 143, 145, 148/149, 150, 153, 156, 161, 163, 164, 166, 167, 169, 170/171, 176, 177, 178, 186, 188, 191, 192/193, 194, 198, 199, 202, 203, 204/205, 206, 208, 209, 210, 211, 214/215, 218, 219, 220/221, 222, 224, 225, 228, 229, 230, 232, 235, 236, 238, 242/243, 244, 248, 250, 251, 252, 253	**Fritz Dressler**
27, 28, 29, 36R, 37, 40, 41, 42/43, 45, 126/127	**Govorukhin**
30, 34/35, 36L, 38, 44, 50, 52, 82L, 82R, 83L, 83R, 85, 87, 93, 101, 103, 104, 139, 140L, 140R, 142, 160, 164, 183, 227, 241, 254	**Gudenko**
22, 24, 25, 48, 49, 54, 71, 73, 84, 86, 90, 128, 181, 189, 216	**Jürgens**
124/125, 146	**Wilhelm Klein**
62, 69	**Knob**

Maps **Berndtson & Berndtson**

Illustrations **Klaus Geisler**

Visual Consultant **V. Barl**

INDEX

A

B

C

K

L

M

N

O

P

R

S

T

U – W

Y

Z

A
B
C
E
F
G
H
I
J
a
b
d
e
f
g
h
i
j
k
l